P9-ECS-364

Student's Solutions Manual

to accompany

HOWARD ANTON
C A L C U L U S
Third Edition

Prepared by

ALBERT HERR

JOHN WILEY & SONS
New York Chichester Brisbane Toronto Singapore

Copyright © 1988 by John Wiley & Sons, Inc.

All rights reserved.

Reproduction or translation of any part of
this work beyond that permitted by Section
107 or 108 of the 1976 United States Copyright
Act without the permission of the copyright
owner is unlawful. Requests for permission or
further information should be addressed to the
Permissions Department, John Wiley & Sons, Inc.

ISBN 0 471 62741 0

Printed in the United States of America

10 9 8 7 6 5 4 3 2

CONTENTS

Chapter 1	Coordinates, Graphs, Lines	1
Chapter 2	Functions and Limits	37
Chapter 3	Differentiation	63
Chapter 4	Applications of Differentiation	95
Chapter 5	Integration	147
Chapter 6	Applications of the Definite Integral	173
Chapter 7	Logarithm and Exponential Functions	205
Chapter 8	Inverse Trigonometric and Hyperbolic Functions	237
Chapter 9	Techniques of Integration	251
Chapter 10	Improper Integrals; L'Hôpital's Rule	287
Chapter 11	Infinite Series	303
Chapter 12	Topics in Analytic Geometry	345
Chapter 13	Polar Coordinates and Parametric Equations	365
Chapter 14	Three-Dimensional Space; Vectors	383
Chapter 15	Vector-Valued Functions	411
Chapter 16	Partial Derivatives	437
Chapter 17	Multiple Integrals	471
Chapter 18	Topics in Vector Calculus	497
Chapter 19	Second-Order Differential Equations	519
Appendix 1	Trigonometry Review	533
Appendix 2	Supplementary Material	541

CHAPTER 1

EXERCISE SET 1.1

1. (a)

4

(b)

−3

(c)

−1   7

(d)

−3   0   3

(e)

−3   3

(f)

−3   3

3. (a) rational
   (c) integer, rational
   (e) integer, rational
   (g) rational

   (b) integer, rational
   (d) rational
   (f) irrational
   (h) integer, rational

5. (a) always correct (add −3 to both sides of a ≤ b)
   (b) not always correct (correct only if a = b = 0)
   (c) not always correct (correct only if a = b = 0)
   (d) always correct (multiply both sides of a ≤ b by 6)
   (e) not always correct (correct only if a ≥ 0)
   (f) always correct (multiply both sides of a ≤ b by the nonnegative
       quantity a^2)

7. (a)   x = 0.123123123···
       1000x = 123.123123123···
        999x = 123
           $x = \dfrac{123}{999} = \dfrac{41}{333}$

   (c)   x = 38.07818181···

       100x = 3807.81818181···

        99x = 3769.74 = $\dfrac{3769.74}{99}$

           = $\dfrac{376974}{9900} = \dfrac{20943}{550}$

   (b)   x = 12.7777···
        10x = 127.7777···
         9x = 115
          $x = \dfrac{115}{9}$

   (d) 0.4296000··· = 0.4296

       = $\dfrac{4296}{10000}$

       = $\dfrac{537}{1250}$

9. (a) all values because $a = a$ is always valid
   (b) none

11. (a) yes, because $a \leq b$ is true if $a < b$
    (b) no, because $a < b$ is false if $a = b$ is true

13. (a) {x: x is a positive odd integer}
    (b) {x: x is an even integer}
    (c) {x: x is irrational}
    (d) {x: x is an integer and $7 \leq x \leq 10$}

15. (a) empty because there are no real values of x for which $x^2 = -1$
    (b) not empty because -1 and 1 are in the set
    (c) empty because $x > 3$ and $x < 3$ are contradictory
    (d) not empty because 3 is in the set

17. (a) false, there are points inside the triangle that are not inside the circle
    (b) true, all points inside the triangle are also inside the square
    (c) true   (d) false   (e) true
    (f) true, a is inside the circle   (g) true

19. $3x - 2 < 8$
    $$3x < 10$$
    $$x < \frac{10}{3}$$
    $S = (-\infty, 10/3)$

21. $4 + 5x \leq 3x - 7$
    $$2x \leq -11$$
    $$x \leq -\frac{11}{2}$$
    $S = (-\infty, -11/2\,]$

23. $3 \leq 4 - 2x < 7$
    $$-1 \leq -2x < 3$$
    $$\frac{1}{2} \geq x > -\frac{3}{2}$$
    $S = (-3/2, 1/2\,]$

25.
$$\frac{x}{x-3} < 4$$
$$\frac{x}{x-3} - 4 < 0$$
$$\frac{x - 4(x-3)}{x-3} < 0$$
$$\frac{12 - 3x}{x-3} < 0$$
$$\frac{4 - x}{x-3} < 0$$

$$S = (-\infty, 3) \cup (4, +\infty)$$

27.
$$\frac{3x+1}{x-2} < 1$$
$$\frac{3x+1}{x-2} - 1 < 0$$
$$\frac{3x+1 - (x-2)}{x-2} < 0$$
$$\frac{2x+3}{x-2} < 0$$
$$\frac{x + 3/2}{x-2} < 0$$

$$S = (-3/2, 2)$$

29.
$$\frac{4}{2-x} \le 1$$
$$\frac{4 - (2-x)}{2-x} \le 0$$
$$\frac{x+2}{2-x} \le 0$$

$$S = (-\infty, -2] \cup (2, +\infty)$$

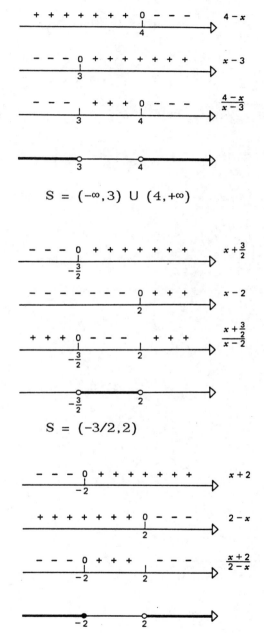

31. $x^2 > 9$
          $x^2 - 9 > 0$
   $(x + 3)(x - 3) > 0$

33. $(x - 4)(x + 2) > 0$

35. $x^2 - 9x + 20 \leq 0$
      $(x - 4)(x - 5) \leq 0$

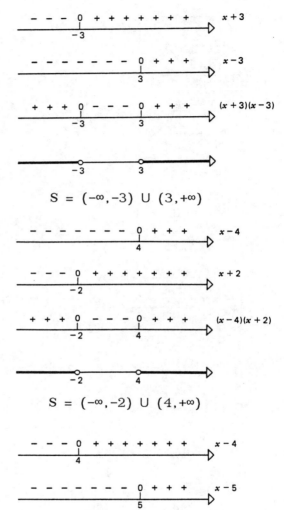

$S = (-\infty, -3) \cup (3, +\infty)$

$S = (-\infty, -2) \cup (4, +\infty)$

$S = [4, 5]$

37.
$$\frac{2}{x} < \frac{3}{x - 4}$$

$$\frac{2}{x} - \frac{3}{x - 4} < 0$$

$$\frac{2(x - 4) - 3x}{x(x - 4)} < 0$$

$$\frac{-x - 8}{x(x - 4)} < 0$$

$$\frac{x + 8}{x(x - 4)} > 0$$

$$S = (-8,0) \cup (4, +\infty)$$

39. By trial-and-error we find that $x = 2$ is a root of the equation $x^3 - x^2 - x - 2 = 0$ so $x - 2$ is a factor of $x^3 - x^2 - x - 2$. By long division we find that $x^2 + x + 1$ is another factor so $x^3 - x^2 - x - 2 = (x - 2)(x^2 + x + 1)$. The linear factors of $x^2 + x + 1$ can be determined by first finding the roots of $x^2 + x + 1 = 0$ by the quadratic formula. These roots are complex numbers so $x^2 + x + 1 \neq 0$ for all real x; thus $x^2 + x + 1$ must be always positive or always negative. Since $x^2 + x + 1$ is positive when $x = 0$, it follows that $x^2 + x + 1 > 0$ for all real x. Hence

$$x^3 - x^2 - x - 2 > 0$$
$$(x - 2)(x^2 + x + 1) > 0$$
$$x - 2 > 0$$
$$x > 2$$

so $S = (2, +\infty)$

41. $\sqrt{x^2 + x - 6}$ is real if $x^2 + x - 6 \geq 0$. Factor to get $(x + 3)(x - 2) \geq 0$ which has as its solution $x \leq -3$ or $x \geq 2$.

43. (a) Assume m and n are rational, then $m = \dfrac{p}{q}$ and $n = \dfrac{r}{s}$ where p, q, r, and
     s are integers

$$\text{so} \quad m + n = \frac{p}{q} + \frac{r}{s} = \frac{ps + rq}{qs}$$

which is rational because ps + rq and qs are integers.

(b) (proof by contradiction) Assume m is rational and n is
     irrational, then $n = \dfrac{p}{q}$ where p and q are integers. Suppose that
     m + n is rational, then $m + n = \dfrac{r}{s}$ where r and s are integers

$$\text{so} \quad n = \frac{r}{s} - m = \frac{r}{s} - \frac{p}{q} = \frac{rq - ps}{sq}$$

But rq − ps and sq are integers, so n is rational which contradicts
the assumption that n is irrational.

45. (by example) Given that $\sqrt{2}$ is irrational, then

$(-1)\sqrt{2} = -\sqrt{2}$ is irrational (Exercise 44, part (b))

but $\sqrt{2} + (-\sqrt{2}) = 0$, which is rational;

$\sqrt{2} + \sqrt{2} = 2\sqrt{2}$, which is irrational (Exercise 44, part (b));

$(\sqrt{2})(\sqrt{2}) = 2$, which is rational;

$1 + \sqrt{2}$ is irrational (Exercise 43, part (b))

and so $\sqrt{2}(1 + \sqrt{2}) = \sqrt{2} + 2$, which is irrational (Exercise 43, part (b)).

47. (I) The average of two rational numbers is rational:

Assume m and n are rational; then $m = \dfrac{p}{q}$ and $n = \dfrac{r}{s}$ where p,
q, r, and s are integers. The average of m and n is

$$\frac{1}{2}(m + n) = \frac{1}{2}(p/q + r/s) = \frac{ps + rq}{2qs}$$

which is rational because ps + rq and 2qs are integers.

(II) The average of two irrational numbers can be rational or
     irrational:

By example,

$\sqrt{2}$, and $-\sqrt{2}$ are irrational and their average

is $\dfrac{1}{2}(\sqrt{2} + (-\sqrt{2})) = \dfrac{1}{2}(0) = 0$, which is rational

$\sqrt{2}$ and $\sqrt{2}$ are irrational and their average

is $\dfrac{1}{2}(\sqrt{2} + \sqrt{2}) = \sqrt{2}$, which is irrational.

49. $8x^3 - 4x^2 - 2x + 1$ can be factored by grouping terms:

$$(8x^3 - 4x^2) - (2x - 1) = 4x^2(2x - 1) - (2x - 1)$$
$$= (2x - 1)(4x^2 - 1)$$
$$= (2x - 1)^2(2x + 1).$$

The problem, then, is to solve $(2x - 1)^2(2x + 1) < 0$. By inspection, $x = \frac{1}{2}$ is not a solution. If $x \neq \frac{1}{2}$, then $(2x - 1)^2 > 0$ and it follows that $2x + 1 < 0$, $2x < -1$, $x < -\frac{1}{2}$, so $S = (-\infty, -1/2)$.

51. If $a < b$, then $ac < bc$ because c is positive; if $c < d$, then $bc < bd$ because b is positive, so $ac < bd$ (part (a), Theorem 1.1.3)

EXERCISE SET 1.2

1. (a) 7   (b) $\sqrt{2}$ (c) k^2 (d) k^2

3. $|x - 3| = |3 - x| = 3 - x$ if $3 - x \geq 0$, which is true if $x \leq 3$.

5. All real values of x because $x^2 + 9 > 0$.

7. $|3x^2 + 2x| = |x(3x + 2)| = |x||3x + 2|$. If $|x||3x + 2| = x|3x + 2|$, then $|x||3x + 2| - x|3x + 2| = 0$, $(|x| - x)|3x + 2| = 0$, so either $|x| - x = 0$ or $|3x + 2| = 0$. If $|x| - x = 0$, then $|x| = x$, which is true for $x \geq 0$. If $|3x + 2| = 0$, then $x = -2/3$. The statement is true for $x \geq 0$ or $x = -2/3$.

9. $\sqrt{(x + 5)^2} = |x + 5| = x + 5$ if $x + 5 \geq 0$, which is true if $x \geq -5$.

13. (a) $|7 - 9| = |-2| = 2$ (b) $|3 - 2| = |1| = 1$
     (c) $|6 - (-8)| = |14| = 14$
     (d) $|-3 - \sqrt{2}| = |-(3 + \sqrt{2})| = 3 + \sqrt{2}$
     (e) $|-4 - (-11)| = |7| = 7$ (f) $|-5 - 0| = |-5| = 5$

15. (a) B is 6 units to the left of A; $b = a - 6 = -3 - 6 = -9$
     (b) B is 9 units to the right of A; $b = a + 9 = -2 + 9 = 7$
     (c) B is 7 units from A; either $b = a + 7 = 5 + 7 = 12$ or
         $b = a - 7 = 5 - 7 = -2$.   Since it is given that $b > 0$, it follows
         that $b = 12$.

17. $|6x - 2| = 7$

Case 1:   Case 2:
$6x - 2 = 7$ $6x - 2 = -7$
$\quad 6x = 9$ $\quad 6x = -5$
$\quad\quad x = 3/2$ $\quad\quad x = -5/6$

19. $|6x - 7| = |3 + 2x|$

Case 1:   Case 2:
$6x - 7 = 3 + 2x$ $6x - 7 = -(3 + 2x)$
$\quad 4x = 10$ $\quad 8x = 4$
$\quad\quad x = 5/2$ $\quad\quad x = 1/2$

21. $|9x| - 11 = x$

Case 1:   Case 2:
$9x - 11 = x$ $-9x - 11 = x$
$\quad 8x = 11$ $\quad -10x = 11$
$\quad\quad x = 11/8$ $\quad\quad x = -11/10$

23. $\left|\dfrac{x + 5}{2 - x}\right| = 6$

Case 1:   Case 2:
$\dfrac{x + 5}{2 - x} = 6$ $\dfrac{x + 5}{2 - x} = -6$
$x + 5 = 12 - 6x$ $x + 5 = -12 + 6x$
$\quad 7x = 7$ $\quad -5x = -17$
$\quad\quad x = 1$ $\quad\quad x = 17/5$

25. $\quad\quad |x + 6| < 3$ 27. $\quad\quad |2x - 3| \leq 6$
     $-3 < x + 6 < 3$ $-6 \leq 2x - 3 \leq 6$
     $-9 < \quad x \quad < -3$ $-3 \leq \quad 2x \quad \leq 9$
     $S = (-9, -3)$ $-3/2 \leq \quad x \quad \leq 9/2$
                                            $S = [-3/2, 9/2]$

29. $|x + 2| > 1$

    <u>Case 1:</u>
    $x + 2 > 1$
    $x > -1$

    <u>Case 2:</u>
    $x + 2 < -1$
    $x < -3$

    $S = (-\infty, -3) \cup (-1, +\infty)$

31. $|5 - 2x| \geq 4$

    <u>Case 1:</u>
    $5 - 2x \geq 4$
    $-2x \geq -1$
    $x \leq 1/2$

    <u>Case 2:</u>
    $5 - 2x \leq -4$
    $-2x \leq -9$
    $x \geq 9/2$

    $S = (-\infty, 1/2] \cup [9/2, +\infty)$

33. $\dfrac{1}{|x - 1|} < 2,\ x \neq 1$

    $|x - 1| > 1/2$

    <u>Case 1:</u>
    $x - 1 > 1/2$
    $x > 3/2$

    <u>Case 2:</u>
    $x - 1 < -1/2$
    $x < 1/2$

    $S = (-\infty, 1/2) \cup (3/2, +\infty)$

35. $\dfrac{3}{|2x - 1|} \geq 4,\ x \neq 1/2$

    $\dfrac{|2x - 1|}{3} \leq \dfrac{1}{4}$

    $|2x - 1| \leq 3/4$

    $-3/4 \leq 2x - 1 \leq 3/4$

    $1/4 \leq\ \ 2x\ \ \leq 7/4$

    $1/8 \leq\ \ \ x\ \ \ \leq 7/8$

    $S = [1/8, 1/2) \cup (1/2, 7/8]$

37. $|x + 3| < |x - 8|$

    $(x + 3)^2 < (x - 8)^2$

    $x^2 + 6x + 9 < x^2 - 16x + 64$

    $22x < 55$

    $x < 5/2$

    $S = (-\infty,\ 5/2)$

39.
$$|4x| \geq |7 - 6x|$$
$$(4x)^2 \geq (7 - 6x)^2$$
$$16x^2 \geq 49 - 84x + 36x^2$$
$$-20x^2 + 84x - 49 \geq 0$$
$$20x^2 - 84x + 49 \leq 0$$
$$(2x - 7)(10x - 7) \leq 0$$
$$(x - 7/2)(x - 7/10) \leq 0$$

$$S = [7/10, 7/2]$$

41.
$$\left| \frac{x - \frac{1}{2}}{x + \frac{1}{2}} \right| < 1, \quad x \neq -\frac{1}{2}$$

$$\left| x - \frac{1}{2} \right| < \left| x + \frac{1}{2} \right|$$
$$(x - \frac{1}{2})^2 < (x + \frac{1}{2})^2$$
$$x^2 - x + \frac{1}{4} < x^2 + x + \frac{1}{4}$$
$$-2x < 0$$
$$x > 0$$
$$S = (0, +\infty)$$

43.
$$\frac{1}{|x - 4|} < \frac{1}{|x + 7|}, \quad x \neq 4 \text{ or } -7$$

$$|x + 7| < |x - 4|$$
$$x^2 + 14x + 49 < x^2 - 8x + 16$$
$$22x < -33$$
$$x < -3/2; \quad \text{but } x = -7 \text{ is excluded}$$
so $\quad S = (-\infty, -7) \cup (-7, -3/2)$

45. $\sqrt{(x^2 - 5x + 6)^2} = x^2 - 5x + 6 \quad$ if $\quad x^2 - 5x + 6 \geq 0$ or, equivalently, if $(x - 2)(x - 3) \geq 0$. $\quad x \in (-\infty, 2] \cup [3, +\infty)$

47. If $z = |x - 3|$ then $z^2 - 4z = 12$, $z^2 - 4z - 12 = 0$, $(z - 6)(z + 2) = 0$, so $z = 6$ or $z = -2$. If $z = 6$ then $|x - 3| = 6$, so $x = 9$ or $x = -3$. If $z = -2$ then $|x - 3| = -2$ which is impossible. The solutions are -3 and 9.

49. If $|3x - 4| < 5$ then $-5 < 3x - 4 < 5$, $-1 < 3x < 9$, $-1/3 < x < 3$, $-2/3 < x - 1/3 < 8/3$, so $|x - 1/3| < 8/3$.

51. $|a - b| = |a + (-b)|$

$\qquad \le |a| + |-b| \qquad$ (triangle inequality)

$\qquad = |a| + |b|$.

53. From Exercise 52

(i) $\qquad |a| - |b| \le |a - b|$;

but $\qquad |b| - |a| \le |b - a| = |a - b|$, so

(ii) $\qquad |a| - |b| \ge -|a - b|$.

Combining (i) and (ii)

$-|a - b| \le |a| - |b| \le |a - b|$,

so $\qquad ||a| - |b|| \le |a - b|$

55. $\qquad -4 < x < 2$

$\qquad 3 < x + 7 < 9$

$\qquad 1/3 > \dfrac{1}{x + 7} > 1/9$, but $\dfrac{1}{x + 7} = \left|\dfrac{1}{x + 7}\right|$ because $\dfrac{1}{x + 7} > 0$, thus

$\left|\dfrac{1}{x + 7}\right| < \dfrac{1}{3}$ so $M = \dfrac{1}{3}$.

57. $\left|\dfrac{x + 3}{x - 3}\right| = \dfrac{|x + 3|}{|x - 3|} \qquad$ if $\qquad -\dfrac{3}{4} \le x \le \dfrac{1}{4}$, $\qquad$ then

$\qquad -\dfrac{15}{4} \le x - 3 \le -\dfrac{11}{4}$

$\qquad |x - 3| \ge \dfrac{11}{4}$

$\qquad \dfrac{1}{|x - 3|} \le \dfrac{4}{11}$

(i) $\qquad \dfrac{|x + 3|}{|x - 3|} \le \dfrac{4}{11} |x + 3|$,

but

$$\frac{9}{4} \leq x + 3 \leq \frac{13}{4}$$

$$|x + 3| \leq \frac{13}{4}$$

$$\frac{4}{11} |x + 3| \leq \frac{4}{11} \cdot \frac{13}{4} = \frac{13}{11},$$

thus from (i) it follows that $\left|\frac{x + 3}{x - 3}\right| \leq \frac{13}{11}$ so $M = \frac{13}{11}$.

EXERCISE SET 1.3

1.

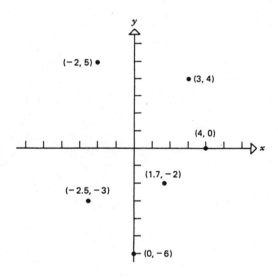

3. (a) x = 2   (b) y = -3

(c) x $\geq$ 0

(d) y = x

(e) y $\geq$ x

(f) |x| $\geq$ 1

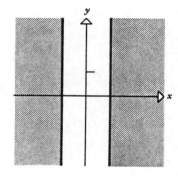

5. Let A(-1,4), B(6,4), and C(-1,9) be the given vertices. Line segment AB is horizontal (A and B have the same ordinate) and line segment AC is vertical (A and C have the same abscissa). Thus AB and AC form a right angle and so AB and AC are sides of the rectangle. The fourth vertex must lie on the vertical line through B and the horizontal line through C, which identifies it as the point (6,9).

7. (a) vertical; $|-3 - \sqrt{2}| = 3 + \sqrt{2}$

    (b) vertical; $|-4 - (-11)| = 7$

    (c) vertical; $|-5 - 0| = 5$

9. (a) d = $\sqrt{(1 - 7)^2 + (9 - 1)^2} = \sqrt{36 + 64} = \sqrt{100} = 10$

    (b) $(\dfrac{7 + 1}{2}, \dfrac{1 + 9}{2}) = (4,5)$

11.   (a) $d = \sqrt{[-7 - (-2)]^2 + [-4 - (-6)]^2} = \sqrt{25 + 4} = \sqrt{29}$

(b) $(\dfrac{-2 + (-7)}{2}, \dfrac{-6 + (-4)}{2}) = (-9/5, -5)$

13.   Let $A(5,-2)$, $B(6,5)$, and $C(2,2)$ be the given vertices and a, b, and c the lengths of the sides opposite these vertices; then

$$a = \sqrt{(2 - 6)^2 + (2 - 5)^2} = \sqrt{25} = 5$$

$$b = \sqrt{(2 - 5)^2 + (2 + 2)^2} = \sqrt{25} = 5$$

Triangle ABC is isosceles because it has two equal sides $(a = b)$.

15. $P_1(0,-2)$, $P_2(-4,8)$, and $P_3(3,1)$ all lie on a circle whose center is $C(-2,3)$ if the points P_1, P_2 and P_3 are equidistant from C.   Denoting the distances between P_1, P_2, P_3 and C by d_1, d_2 and d_3 we find that

$$d_1 = \sqrt{(0 + 2)^2 + (-2 - 3)^2} = \sqrt{29},$$

$$d_2 = \sqrt{(-4 + 2)^2 (8 - 3)^2} = \sqrt{29},$$

$$d_3 = \sqrt{(3 + 2)^2 + (1 - 3)^2} = \sqrt{29}.$$

So P_1, P_2 and P_3 lie on a circle whose center is $C(-2,3)$ because $d_1 = d_2 = d_3$.

17.   (a) $0^2 -2(0) + 4 = 4$, yes   (b) $(-3)^2 - 2(-3) + 7 = 22$, no

(c) $(1/2)^2 - 2(1/2) + 19/4 = 1/4 - 1 + 19/4 = 4$, yes

(d) $(1 + \sqrt{5 - t})^2 - 2(1 + \sqrt{5 - t}) + t$

$= 1 + 2\sqrt{5 - t} + (5 - t) - 2 - 2\sqrt{5 - t} + t = 4$, yes

19.   (a) y-axis, because $(-x)^4 = 2y^3 + y$ gives $x^4 = 2y^3 + y$.

(b) origin, because $(-y) = \dfrac{(-x)}{3 + (-x)^2}$ gives $y = \dfrac{x}{3 + x^2}$.

(c) x-axis, y-axis, and origin because $(-y)^2 = |x| - 5$, $y^2 = |-x| - 5$, and $(-y)^2 = |-x| - 5$ all give $y^2 = |x| - 5$.

21. $y = 6 - x$

23. $y = 4 - x^2$

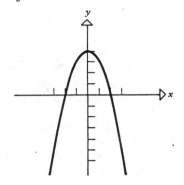

25. $y = \sqrt{x - 4}$

27. $y = |x - 3|$

29. $x^2 y = 2$

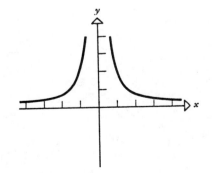

31. $4x^2 + 16y^2 = 16$

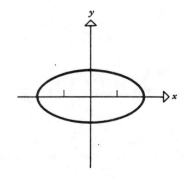

33. $(x - y)(x + y) = 0$

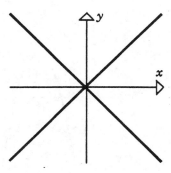

The union of the graphs of
$x - y = 0$ and $x + y = 0$.

35. $u = 3v^2$

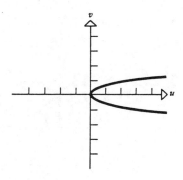

37.   P(x,y) is on the perpendicular bisector of the line segment connecting
(-2,1) and (4,-3) if and only if P is equidistant from (-2,1) and
(4,-3):

$$\sqrt{(x + 2)^2 + (y - 1)^2} = \sqrt{(x - 4)^2 + (y + 3)^2}$$
$$(x + 2)^2 + (y - 1)^2 = (x - 4)^2 + (y + 3)^2$$
$$x^2 + 4x + 4 + y^2 - 2y + 1 = x^2 - 8x + 16 + y^2 + 6y + 9$$
$$12x - 8y - 20 = 0$$
$$3x - 2y - 5 = 0$$

39.   If (2,k) is equidistant from (3,7) and (9,1), then

$$\sqrt{(2 - 3)^2 + (k - 7)^2} = \sqrt{(2 - 9)^2 + (k - 1)^2}$$
$$1 + (k - 7)^2 = 49 + (k - 1)^2$$

$$1 + k^2 - 14k + 49 = 49 + k^2 - 2k + 1$$
$$-12k = 0$$
$$k = 0$$

41. The coordinates of M, the midpoint
of the hypotenuse, are (a/2,b/2). The
distance between M and vertices A

and B is $\frac{1}{2}\sqrt{a^2 + b^2}$. By the distance

formula, the distance between vertex
O and midpoint M is

$$\sqrt{(a/2)^2 + (b/2)^2} = \frac{1}{2}\sqrt{a^2 + b^2}$$

which is equal to the distance
between M and vertices A and B.

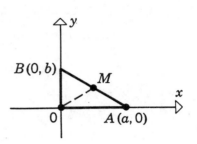

EXERCISE SET 1.4

1. (a) $m = \dfrac{4 - 2}{3 - (-1)} = \dfrac{1}{2}$ (b) $m = \dfrac{1 - 3}{7 - 5} = -1$

(c) $m = \dfrac{\sqrt{2} - \sqrt{2}}{-3 - 4} = 0$ (d) $m = \dfrac{12 - (-6)}{-2 - (-2)} = \dfrac{18}{0}$,

not defined

3. (a) $\tan \dfrac{\pi}{6} = 1/\sqrt{3}$ (b) $\tan 135^\circ = -1$ (c) $\tan 60^\circ = \sqrt{3}$

5. (a) 153° (b) 45° (c) 117° (d) 89°

7.

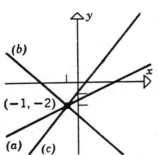

9. Let m' be the slope of line L'.

   (a) $m' = \dfrac{5 - 8}{2 - 1} = -3$, $m' = m$ so L' is parallel to L.

   (b) $m' = \dfrac{4 - 5}{3 - 6} = \dfrac{1}{3}$, $mm' = -1$ so L' is perpendicular to L.

   (c) $m' = \dfrac{1 - 0}{-2 - 1} = -\dfrac{1}{3}$, $m' \neq m$ and $mm' \neq -1$ so L' is neither parallel nor perpendicular to L.

11. Use the points $(1,2)$ and (x,y) to calculate the slope: $\dfrac{y - 2}{x - 1} = 3$

   (a) if $x = 5$, then

   $$\dfrac{y - 2}{5 - 1} = 3$$
   $$y - 2 = 12$$
   $$y = 14$$

   (b) if $y = -2$, then

   $$\dfrac{-2 - 2}{x - 1} = 3$$
   $$x - 1 = -4/3$$
   $$x = -1/3$$

13. Using $(3,k)$ and $(-2,4)$ to calculate the slope, we find $\dfrac{k - 4}{3 - (-2)} = 5$, $k - 4 = 25$, $k = 29$.

15. (a) The line through $(1,1)$ and $(-2,-5)$ has slope $m_1 = \dfrac{-5 - 1}{-2 - 1} = 2$, the line through $(1,1)$ and $(0,-1)$ has slope $m_2 = \dfrac{-1 - 1}{0 - 1} = 2$.
   The given points lie on a line because $m_1 = m_2$.

   (b) The line through $(-2,4)$ and $(0,2)$ has slope $m_1 = \dfrac{2 - 4}{0 + 2} = -1$, the line through $(-2,4)$ and $(1,5)$ has slope $m_2 = \dfrac{5 - 4}{1 + 2} = \dfrac{1}{3}$.
   The given points do not lie on a line because $m_1 \neq m_2$.

17. Let $P(x,0)$ be a point on the x-axis. If AP is perpendicular to BP then $m_{AP}m_{BP} = -1$, so

   $$\dfrac{0 - 2}{x - 1} \cdot \dfrac{0 - 3}{x - 8} = -1$$

   which yields $x^2 - 9x + 14 = 0$, $(x - 7)(x - 2) = 0$, thus $x = 2$ or $x = 7$.

19. Show that opposite sides are parallel by showing that they have the same slope:

   using $(3,-1)$ and $(6,4)$, $m_1 = 5/3$;

   using $(6,4)$ and $(-3,2)$, $m_2 = 2/9$;

using (-3,2) and (-6,-3), m_3 = 5/3;

using (-6,-3) and (3,-1), m_4 = 2/9.

Opposite sides are parallel because m_1 = m_3 and m_2 = m_4.

21. (a) Draw the line through the point of intersection of L_1 and L_2, parallel to the x-axis (see figure). Based on the figure we see that $\phi_2 = \theta + \phi_1$,

so $\theta = \phi_2 - \phi_1$.

(b) if neither L_1 nor L_2 is vertical, then their slopes m_1 and m_2 both exist and are given by

$m_1 = \tan\phi_1$ and $m_2 = \tan\phi_2$

so $\tan\theta = \tan(\phi_2 - \phi_1)$

$$= \frac{\tan\phi_2 - \tan\phi_1}{1 + \tan\phi_2 \tan\phi_1} \quad (\text{trigonometric identity})$$

$$= \frac{m_2 - m_1}{1 + m_1 m_2}$$

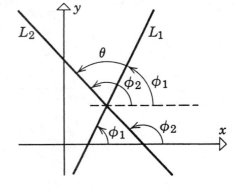

if $m_1 m_2 \neq -1$ (that is, if L_1 and L_2 are not perpendicular).

23. (a) $m_1 = \frac{1}{3}$, $m_2 = \frac{4}{5}$; $\tan\theta = \frac{\frac{4}{5} - \frac{1}{3}}{1 + (1/3)(4/5)} = \frac{7}{19}$

(b) $m_1 = 5$, $m_2 = -0.7$; $\tan\theta = \frac{-0.7 - 5}{1 + (5)(-0.7)} = \frac{5.7}{2.5} = \frac{57}{25}$

(c) $m_1 = -6$, $m_2 = -2$; $\tan\theta = \frac{-2 - (-6)}{1 + (-6)(-2)} = \frac{4}{13}$

25. (a) $20°$ (b) $66°$ (c) $17°$

27. The slope of the line through A and B is $m_1 = -\frac{3}{4}$ and that through A and C is $m_2 = \frac{1}{2}$. Let m denote the slope of the line L that bisects the angle A (see figure). Then

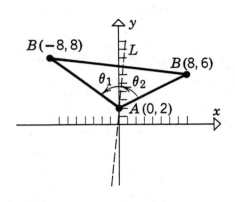

$$\theta_1 = \theta_2$$
$$\tan \theta_1 = \tan \theta_2$$
$$\frac{m_1 - m}{1 + m_1 m} = \frac{m - m_2}{1 + m_2 m}$$
$$\frac{-\frac{3}{4} - m}{1 - \frac{3}{4}m} = \frac{m - \frac{1}{2}}{1 + \frac{1}{2}m}$$
$$\frac{-3 - 4m}{4 - 3m} = \frac{2m - 1}{2 + m}$$
$$(-3 - 4m)(2 + m) = (2m - 1)(4 - 3m)$$
$$-4m^2 - 11m - 6 = -6m^2 + 11m - 4$$
$$2m^2 - 22m - 2 = 0$$
$$m^2 - 11m - 1 = 0$$

and from the quadratic formula $m = \frac{11 \pm \sqrt{121 + 4}}{2} = \frac{11 \pm \sqrt{125}}{2}$, $m \approx 11.09$ or -0.09, of which only the line with a slope of 11.09 is consistent with the figure.

29. Let L_3 be a line that is perpendicular to L_1, then $m_3 = -1/m_1$. But $m_2 = -1/m_1$, so L_2 is parallel to L_3 because they have the same slope. If L_3 is perpendicular to L_1, then so is L_2.

EXERCISE SET 1.5

1. (a)   (b)

    (c)   (d)

3. (a)   (b)   (c)

5. (a) m = 3, b = 2   (b) m = $-\frac{1}{4}$, b = 3

    (c) y = $-\frac{3}{5}$ x + $\frac{8}{5}$ (d) m = 0, b = 1

        so m = $-\frac{3}{5}$, b = $\frac{8}{5}$

    (e) y = $-\frac{b}{a}$ x + b so m = $-\frac{b}{a}$, y-intercept b

7. (a) $m = \tan \phi = \sqrt{3}$, $\phi = 60°$ (b) $m = \tan \phi = -2$, $\phi = 117°$

9. $y = -2x + 4$

11. The slope m of the line must equal the slope of $y = 4x - 2$, thus $m = 4$ so the equation is $y = 4x + 7$.

13. The slope m of the line must be the negative reciprocal of the slope of $y = 5x + 9$, thus $m = -1/5$ and the equation is $y = -\frac{1}{5} x + 6$.

15. $y - 4 = \frac{-7 - 4}{1 - 2} (x - 2) = 11(x - 2)$, $y = 11x - 18$.

17. The slope of the line segment joining (2,8) and (−4,6) is $\frac{6 - 8}{-4 - 2} = \frac{1}{3}$ so the slope of the perpendicular bisector is −3. The midpoint of the line segment is (−1,7) so an equation of the bisector is $y - 7 = -3(x + 1)$; $y = -3x + 4$.

19. $m = \tan \frac{\pi}{6} = \frac{1}{\sqrt{3}}$ so $y = \frac{1}{\sqrt{3}} x - 3$.

21. The line passes through (0,2) and (−4,0), thus $m = \frac{0 - 2}{-4 - 0} = \frac{1}{2}$ so $y = \frac{1}{2} x + 2$.

23. $y = 1$

25. The line is vertical because $\phi = \frac{\pi}{2}$, so $x = 5$.

27. (a) $m_1 = 4$, $m_2 = 4$; parallel because $m_1 = m_2$.

    (b) $m_1 = 2$, $m_2 = -\frac{1}{2}$; perpendicular because $m_1 m_2 = -1$.

    (c) $m_1 = \frac{5}{3}$, $m_2 = \frac{5}{3}$; parallel because $m_1 = m_2$.

    (d) If $A \neq 0$ and $B \neq 0$, then $m_1 = -A/B$, $m_2 = B/A$ and the lines are perpendicular because $m_1 m_2 = -1$. If either A or B (but not both) is zero, then the lines are perpendicular because one is horizontal and the other is vertical.

    (e) $m_1 = 4$, $m_2 = 1/4$; neither.

29. An equation of the line through $(1,4)$ and $(2,1)$ is $y = -3x + 7$. It crosses the y-axis at $y = 7$, and the x-axis at $x = 7/3$, so the area of the triangle is $\frac{1}{2}(7)(7/3) = 49/6$.

31. (a) $\begin{cases} 2x + 3y = 5 \\ y = -1 \end{cases}$,

replacing y by -1 in the first equation gives $2x - 3 = 5$, $2x = 8$, $x = 4$, so the point is $(4,-1)$.

(b) $\begin{cases} 4x + 3y = -2 \\ 5x - 2y = 9 \end{cases}$,

multiply the first equation through by 2 and the second by 3 to get the system

$\begin{cases} 8x + 6y = -4 \\ 15x - 6y = 27 \end{cases}$,

then, by adding these equations, $23x = 23$, $x = 1$, and substitution of this into the first of the original equations gives $4 + 3y = -2$, $3y = -6$, $y = -2$ so the point is $(1,-2)$.

33. Method (see figure): write an equation of the line that goes through the given point and that is perpendicular to the given line; find the point P where this line intersects the given line; find the distance between P and the given point.

The slope of the given line is $4/3$, so the slope of a line perpendicular to it is $-3/4$. The line through $(2,1)$ having a slope of $-3/4$ is $y - 1 = -\frac{3}{4}(x - 2)$ or, after simplification, $3x + 4y = 10$ which when solved simultaneously with $4x - 3y + 10 = 0$ yields $(-2/5, 14/5)$ as the point of intersection. The distance d between $(-2/5, 14/5)$ and $(2,1)$ is

$$d = \sqrt{(2 + 2/5)^2 + (1 - 14/5)^2} = 3.$$

35. If $B = 0$, then the line $Ax + C = 0$ is vertical and $x = -C/A$ for each point on the line. The line through (x_0, y_0) and perpendicular to the

given line is horizontal and intersects the given line at the point $(-C/A, y_0)$. The distance d between $(-C/A, y_0)$ and (x_0, y_0) is

$$d = \sqrt{(x_0 + C/A)^2 + (y_0 - y_0)^2} = \sqrt{\frac{(Ax_0 + C)^2}{A^2}} = \frac{|Ax_0 + C|}{\sqrt{A^2}}$$

which is the value of $\dfrac{|Ax_0 + By_0 + C|}{\sqrt{A^2 + B^2}}$ for $B = 0$.

If $B \neq 0$, then the slope of the given line is $-A/B$ and the line through (x_0, y_0) and perpendicular to the given line is

$$y - y_0 = \frac{B}{A}(x - x_0),$$
$$Ay - Ay_0 = Bx - Bx_0,$$
$$Bx - Ay = Bx_0 - Ay_0.$$

The point of intersection of this line and the given line is obtained by solving the system

$$\begin{cases} Ax + By = -C \\ Bx - Ay = Bx_0 - Ay_0 \end{cases}.$$

Multiply the first equation through by A and the second by B and add the results to get

$$(A^2 + B^2)x = B^2x_0 - ABy_0 - AC$$

$$x = \frac{B^2x_0 - ABy_0 - AC}{A^2 + B^2}$$

Similarly, by multiplying by B and $-A$, we get

$$y = \frac{-ABx_0 + A^2y_0 - BC}{A^2 + B^2}.$$

The square of the distance d between (x, y) and (x_0, y_0) is

$$d^2 = \left[x_0 - \frac{B^2x_0 - ABy_0 - AC}{A^2 + B^2} \right]^2 + \left[y_0 - \frac{-ABx_0 + A^2y_0 - BC}{A^2 + B^2} \right]^2$$

$$= \frac{(A^2x_0 + ABy_0 + AC)^2}{(A^2 + B^2)^2} + \frac{(ABx_0 + B^2y_0 + BC)^2}{(A^2 + B^2)^2}$$

$$= \frac{A^2(Ax_0 + By_0 + C)^2 + B^2(Ax_0 + By_0 + C)^2}{(A^2 + B^2)^2}$$

$$= \frac{(Ax_0 + By_0 + C)^2(A^2 + B^2)}{(A^2 + B^2)^2} = \frac{(Ax_0 + By_0 + C)^2}{A^2 + B^2}$$

so $d = \dfrac{|Ax_0 + By_0 + C|}{\sqrt{A^2 + B^2}}$.

37. $d = \dfrac{|5(8) + 12(4) - 36|}{\sqrt{5^2 + 12^2}} = \dfrac{|52|}{\sqrt{169}} = \dfrac{52}{13} = 4.$

39. Method (see figure): Find an equation
of the perpendicular bisector of the
line segment joining A(3,3) and B(7,-3).
All points on this perpendicular bisector
are equidistant from A and B, thus find
where it intersects the given line.
The midpoint of AB is (5,0), the
slope of AB is -3/2 thus the slope
of the perpendicular bisector is 2/3
so an equation is

$$y - 0 = \frac{2}{3}(x - 5)$$
$$3y = 2x - 10$$
$$2x - 3y - 10 = 0.$$

The solution of the system
$$\begin{cases} 4x - 2y + 3 = 0 \\ 2x - 3y - 10 = 0 \end{cases}$$
gives the point (-29/8,-23/4).

41. (a) If we plot C along the horizontal axis and F along the vertical
axis, then any point on the line relating C and F can be denoted by
(C,F). From the fact that (0,32) and (100,212) are on the line, it
follows that the slope is $\dfrac{212 - 32}{100 - 0} = \dfrac{180}{100} = \dfrac{9}{5}$ and that the

F-intercept is 32. An equation of the line is $F = \dfrac{9}{5}C + 32.$

(b) Solving $F = \frac{9}{5}C + 32$ for C we get $C = \frac{5}{9}F - \frac{160}{9}$, which identifies

the slope as $\frac{5}{9}$.

43. To prove that the graph of $Ax + By + C = 0$ is a straight line consider
two cases, $B = 0$ and $B \neq 0$. If $B \neq 0$, solve for y to get $y = -\frac{A}{B}x - \frac{C}{B}$
which is the slope-intercept form of a line with $m = -A/B$ and $b = -C/B$.
If $B = 0$, then $A \neq 0$ because A and B cannot both be zero. Thus we get
$Ax + C = 0$, or $x = -C/A$ which is an equation of a line parallel to the
y-axis. In either case the graph of $Ax + By + C = 0$ is a straight line.
    Conversely, consider any straight line in the xy-plane. If it is
vertical, then it has an equation of the form $x = a$, or $x - a = 0$ which
is like $Ax + By + C = 0$ with $A = 1$, $B = 0$, and $C = -a$. If the line is
not vertical, then it can be written in slope-intercept form as
$y = mx + b$, or $mx - y + b = 0$ which is like $Ax + By + C = 0$ with $A = m$,
$B = -1$, and $C = b$.

EXERCISE SET 1.6

1. (a) center (0,0), radius 5   (b) center (1,4), radius 4
    (c) center (-1,-3), radius $\sqrt{5}$ (d) center (0,-2), radius 1

3. $(x - 3)^2 + (y - (-2))^2 = 4^2$, $(x - 3)^2 + (y + 2)^2 = 16$

5. $r = 8$ because the circle is tangent to the x-axis, so
   $(x + 4)^2 + (y - 8)^2 = 64$.

7. (0,0) is on the circle, so $r = \sqrt{(-3 - 0)^2 + (-4 - 0)^2} = 5$;
   $(x + 3)^2 + (y + 4)^2 = 25$.

9. The center is the midpoint of the line segment joining (2,0) and (0,2)
   so the center is $(\frac{2 + 0}{2}, \frac{0 + 2}{2}) = (1,1)$. The radius is

   $r = \sqrt{(2 - 1)^2 + (0 - 1)^2} = \sqrt{2}$, so $(x - 1)^2 + (y - 1)^2 = 2$.

11. $(x^2 - 2x) + (y^2 - 4y) = 11$
$(x^2 - 2x + 1) + (y^2 - 4y + 4) = 11 + 1 + 4$
$(x - 1)^2 + (y - 2)^2 = 16$; center $(1,2)$ and radius 4.

13. $2(x^2 + 2x) + 2(y^2 - 2y) = 0$
$2(x^2 + 2x + 1) + 2(y^2 - 2y + 1) = 2 + 2$
$(x + 1)^2 + (y - 1)^2 = 2$; center $(-1,1)$ and radius $\sqrt{2}$.

15. $(x^2 + 2x) + (y^2 + 2y) = -2$
$(x^2 + 2x + 1) + (y^2 + 2y + 1) = -2 + 1 + 1$
$(x + 1)^2 + (y + 1)^2 = 0$; the point $(-1,-1)$.

17. $x^2 + y^2 = \frac{1}{9}$; center $(0,0)$ and radius $\frac{1}{3}$.

19. $x^2 + (y^2 + 10y) = -26$
$x^2 + (y^2 + 10y + 25) = -26 + 25$
$x^2 + (y + 5)^2 = -1$; no graph

21. $16(x^2 + \frac{5}{2}x) + 16(y^2 + y) = 7$
$16(x^2 + \frac{5}{2}x + \frac{25}{16}) + 16(y^2 + y + \frac{1}{4}) = 7 + 25 + 4$
$(x + 5/4)^2 + (y + 1/2)^2 = 9/4$; center $(-5/4,-1/2)$ and radius $3/2$.

23. (a) $y^2 = 16 - x^2$, so $y = \pm\sqrt{16 - x^2}$. The bottom half is $y = -\sqrt{16 - x^2}$.
(b) Complete the square in y to get $(y - 2)^2 = 3 - 2x - x^2$, so
$y - 2 = \pm\sqrt{3 - 2x - x^2}$, or $y = 2 \pm \sqrt{3 - 2x - x^2}$. The top half is
$y = 2 + \sqrt{3 - 2x - x^2}$.

25.   (a)

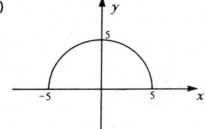

(b) $y = \sqrt{5 + 4x - x^2}$

$= \sqrt{5 - (x^2 - 4x)}$

$= \sqrt{5 + 4 - (x^2 - 4x + 4)}$

$= \sqrt{9 - (x - 2)^2}$

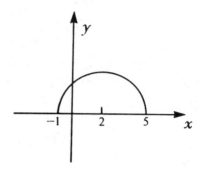

27.   The tangent line is perpendicular to the radius at the point.   The slope of the radius is 4/3, so the slope of the perpendicular is –3/4.   An equation of the tangent line is $y - 4 = -\frac{3}{4}(x - 3)$, or $y = -\frac{3}{4}x + \frac{25}{4}$.

29.   (a) The center of the circle is at (0,0) and its radius is $\sqrt{20} = 2\sqrt{5}$.

The distance between P and the center is $\sqrt{(-1)^2 + (2)^2} = \sqrt{5}$ which is less than $2\sqrt{5}$, so P is inside the circle.

(b) Draw the diameter of the circle that passes through P, then the shorter segment of the diameter is the shortest line that can be drawn from P to the circle, and the longer segment is the longest line that can be drawn from P to the circle (can you prove it?).

Thus, the smallest distance is $2\sqrt{5} - \sqrt{5} = \sqrt{5}$, and the largest is $2\sqrt{5} + \sqrt{5} = 3\sqrt{5}$.

31. $y = x^2 + 2$

33. $y = x^2 + 2x - 3$

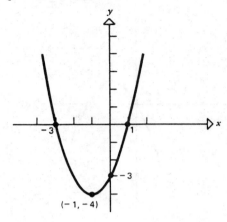

35. $y = -x^2 + 4x + 5$

37. $y = (x - 2)^2$

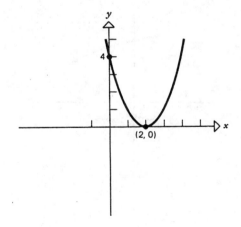

39. $x^2 - 2x + y = 0$

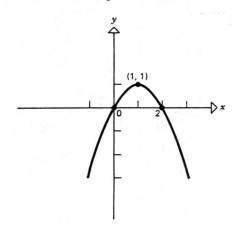

41. $y = 3x^2 - 2x + 1$

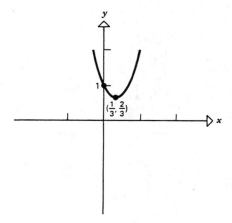

43. $x = -y^2 + 2y + 2$

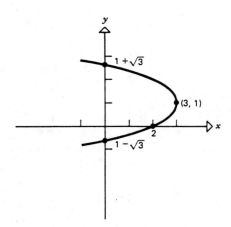

45. (a) $x^2 = 3 - y$, $x = \pm\sqrt{3 - y}$.   The
         right half is $x = \sqrt{3 - y}$.

     (b) Complete the square in x to
         get $(x - 1)^2 = y + 1$,
         $x = 1 \pm \sqrt{y + 1}$.   The left half
         is $x = 1 - \sqrt{y + 1}$.

47. (a)

     (b)

49. (a) $s = 32t - 16t^2$

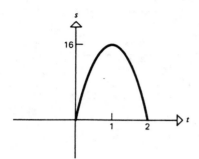

(b) The ball will be at its highest point when $t = 1$ sec; it will rise 16 ft.

51. (a) $[(x - 4)^2 + (y - 1)^2] + [(x - 2)^2 + (y + 5)^2] = 45$

$x^2 - 8x + 16 + y^2 - 2y + 1 + x^2 - 4x + 4 + y^2 + 10y + 25 = 45$

$2x^2 + 2y^2 - 12x + 8y + 1 = 0$, which is a circle.

(b) $2(x^2 - 6x) + 2(y^2 + 4y) = -1$

$2(x^2 - 6x + 9) + 2(y^2 + 4y + 4) = -1 + 18 + 8$

$(x - 3)^2 + (y + 2)^2 = 25/2$; center $(3, -2)$, radius $5/\sqrt{2}$.

SUPPLEMENTARY EXERCISES, CHAPTER 1

1. (a) $(-3, 5]$

(b) $-1 < 0 \le x^2$ for all x, thus $-1 < x^2 \le 9$ is equivalent to $x^2 \le 9$, so $|x| \le 3$. The interval is $[-3, 3]$.

(c) $x^2 \ge \frac{1}{4}$ is equivalent to $|x| \ge \frac{1}{2}$, which in interval notation is $(-\infty, -1/2] \cup [1/2, +\infty)$.

3. (a) $2x^2 - 5x > 3$

$2x^2 - 5x - 3 > 0$

$(2x + 1)(x - 3) > 0$

$(x + 1/2)(x - 3) > 0$

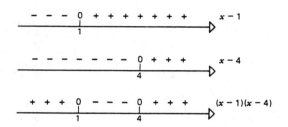

$$S = (-\infty, -1/2) \cup (3, +\infty)$$

(b) $x^2 - 5x + 4 \leq 0$

$(x - 1)(x - 4) \leq 0$

$$S = [1, 4]$$

5.   (a) $\dfrac{|x| - 1}{|x| - 2} \leq 0$

Case 1: $x \geq 0$,

$\dfrac{x - 1}{x - 2} \leq 0$

$$S_1 = [1, 2)$$

Case 2: $x < 0$,

$\dfrac{-x - 1}{-x - 2} \leq 0; \quad \dfrac{x + 1}{x + 2} \leq 0$

$$S_2 = (-2, -1]$$

$$S = S_1 \cup S_2 = (-2, -1] \cup [1, 2)$$

(b)
$$|x - 1| \leq 2|x + 2|$$
$$x^2 - 2x + 1 \leq 4(x^2 + 4x + 4)$$
$$-3x^2 - 18x - 15 \leq 0$$
$$x^2 + 6x + 5 \geq 0$$
$$(x + 1)(x + 5) \geq 0$$

$$S = (-\infty, -5] \cup [-1, +\infty)$$

7.   (a) $a = -2$ and $b = 1$

(b) $a^2 < b^2$, $a^2 - b^2 < 0$, $(a + b)(a - b) < 0$; if $a < b$ then $a - b < 0$ so $(a + b)(a - b) < 0$ if $a + b > 0$

9. $x^2 \leq x^2 + y^2$ because $y^2 \geq 0$, thus $\sqrt{x^2} \leq \sqrt{x^2 + y^2}$ and so $|x| \leq \sqrt{x^2 + y^2}$. Similarly, $|y| \leq \sqrt{x^2 + y^2}$. The right triangle with vertices $(0,0)$, (x,y), and $(x,0)$ has legs of lengths $|x|$ and $|y|$, and a hypotenuse of length $\sqrt{x^2 + y^2}$. The lengths of the legs cannot exceed the length of the hypotenuse.

11. (a)
$$xy = x^2$$
$$xy - x^2 = 0$$
$$x(y - x) = 0,$$
so $x = 0$ or $y - x = 0$

(b) $y(x - 1) = x^2 - 1$
$$y(x - 1) - (x^2 - 1) = 0$$
$$y(x - 1) - (x - 1)(x + 1) = 0$$
$$(x - 1)(y - x - 1) = 0,$$
so $x - 1 = 0$ or $y - x - 1 = 0$

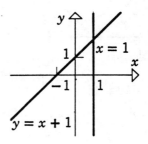

13. (a) Complete the square:
$(x^2 - 2x + 1) + (y^2 - 6y + 9) \geq 6 + 1 + 9$
$(x - 1)^2 + (y - 3)^2 \geq 16$,
all points on or outside the
circle with center (1,3) and
radius 4.

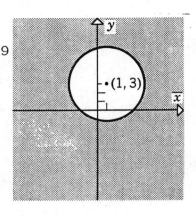

(b) $x + |y - 2| = 1$
If $y \geq 2$, then $x + y - 2 = 1$
$y = -x + 3.$
If $y < 2$, then $x - (y - 2) = 1$
$y = x + 1.$

15. The curves intersect at (x,y) where $y = x^2$ and $y = x + 2$, so $x^2 = x + 2$,
$x^2 - x - 2 = 0$, $(x + 1)(x - 2) = 0$, $x = -1$ or $x = 2$. The points of
intersection are $(-1,1)$ and $(2,4)$.

17. If $x \geq 2$, then $y = x - 2$;   19.
if $x < 2$, then $y = -x + 2$.

21. The radius is $r = \sqrt{(4 - 1)^2 + (-2 - 2)^2} = 5$, so
$(x - 1)^2 + (y - 2)^2 = 25$.

23. Let C(h,k) be the center, then $|h - 6| = 5$ and $|k - 7| = 5$, so $h = 6 \pm 5$ and $k = 7 \pm 5$. There are four circles, $(x - h)^2 + (y - k)^2 = 25$, where $h = 1$ or 11 and $k = 2$ or 12.

25. $(x^2 + 4x + 4) + (y^2 + 2y + 1) = -5 + 4 + 1$
$(x + 2)^2 + (y + 1)^2 = 0$; the point $(-2,-1)$.

27. $(x^2 - 3x + \frac{9}{4}) + (y^2 + 2y + 1) = -4 + \frac{9}{4} + 1$
$(x - \frac{3}{2})^2 + (y + 1)^2 = -\frac{3}{4}$; no graph

29. (a) $m = \frac{-4 - 4}{-3 - 3} = \frac{4}{3}$, so $y - 4 = \frac{4}{3}(x - 3)$, $y = \frac{4}{3}x$;

$d = \sqrt{(-6)^2 + (-8)^2} = 10$; midpoint: $(\frac{3 - 3}{2}, \frac{4 - 4}{2}) = (0,0)$.

(b) $m = \frac{-4 - 4}{3 - 3} = \frac{-8}{0}$ which is not defined, so the vertical line $x = 3$;

$d = |-4 - 4| = 8$; midpoint: $(\frac{3 + 3}{2}, \frac{4 - 4}{2}) = (3,0)$.

(c) $m = \frac{4 - 4}{-3 - 3} = 0$, so $y = 4$; $d = |-3 - 3| = 6$;

midpoint: $(\frac{3 - 3}{2}, \frac{4 + 4}{2}) = (0,4)$.

(d) $m = \frac{3 - 4}{4 - 3} = -1$, so $y - 4 = -(x - 3)$, $y = -x + 7$;

$d = \sqrt{(1)^2 + (-1)^2} = \sqrt{2}$; midpoint: $(\frac{3 + 4}{2}, \frac{4 + 3}{2}) = (7/2, 7/2)$.

31. Equation of circle is $x^2 + y^2 = 25$, equation of line is $y = -\frac{3}{4}x$.

Eliminate y: $x^2 + (-\frac{3}{4}x)^2 = 25$, $x^2 + \frac{9}{16}x^2 = 25$, $\frac{25}{16}x^2 = 25$, $x^2 = 16$,

so $x = \pm 4$. The points of intersection are $(-4,3)$ and $(4,-3)$.

33. (a) vertical line, so $\phi = 90°$.   (b) horizontal line, so $\phi = 0°$.

(c) $y = -x + \frac{1}{2}$ thus $m = -1$, so $\phi = 135°$.

(d) $y = \sqrt{3}x - 4$ thus $m = \sqrt{3}$, so $\phi = 60°$.

35. $m = \frac{-3 + 3}{4 - 2} = 0$, so $y = -3$.

37. For $x + 2y = 3$, $m = -\frac{1}{2}$. A parallel line through the origin is
     $y - 0 = -\frac{1}{2}(x - 0)$, $y = -\frac{1}{2}x$.

39. L: $m = -2$, so $y = -2(x - 1)$;
     L': $m = \frac{1}{2}$ because L' is perpendicular to L, so $y = \frac{1}{2}x - 3$.
     L meets L' when $-2(x - 1) = \frac{1}{2}x - 3$, which gives $x = 2$. But $y = -2$ when $x = 2$, so the point is $(2,-2)$.

41. L: $y - 1 = \frac{2}{5}(x - 3)$, $y = \frac{2}{5}x - \frac{1}{5}$;
     L': $(-\frac{8}{3}, 0)$ and $(0,-4)$ are on the line, so $m = -\frac{3}{2}$ and therefore
     $$y = -\frac{3}{2}\left(x + \frac{8}{3}\right) = -\frac{3}{2}x - 4.$$
     L meets L' when $\frac{2}{5}x - \frac{1}{5} = -\frac{3}{2}x - 4$, which gives $x = -2$. But $y = -1$ when $x = -2$, so the point is $(-2,-1)$.

43. Label the points as A(5,6), B(−4,3), C(−3,−2), and D(6,1). Then $m_{AB} = \frac{1}{3}$, $m_{BC} = -5$, $m_{CD} = -\frac{1}{3}$, and $m_{DA} = -5$, so ABCD is a parallelogram because opposite sides are parallel ($m_{AB} = m_{CD}$, $m_{BC} = m_{DA}$). It is not a rectangle because sides AB and BC do not form a right angle ($m_{AB} \neq -1/m_{BC}$).

CHAPTER 2

EXERCISE SET 2.1

1.   (a) 14   (b) 50   (c) 2
     (d) 11   (e) $3a^2 + 6a + 5$ (f) $27t^2 + 2$

3.   (a) $2(-4) = -8$ (b) 1/4   (c) $2(0) = 0$
     (d) $2(3) = 6$ (e) $2(2.9) = 5.8$ (f) $1/(t^2 + 5)$

5. $(-\infty,3) \cup (3,+\infty)$ 7. $(-\infty,-\sqrt{3}] \cup [\sqrt{3},+\infty)$

9. $\dfrac{x-1}{x+2} \geq 0$ if $x < -2$ or $x \geq 1$; domain: $(-\infty,-2) \cup [1,+\infty)$

11. $(-\infty,+\infty)$ 13. $[5,+\infty) \cap (-\infty,8] = [5,8]$

15. $x^2 - 2x + 5 = 0$ has no real solutions so $x^2 - 2x + 5$ is always positive
     or always negative.   If $x = 0$, then $x^2 - 2x + 5 = 5 > 0$;
     domain: $(-\infty,+\infty)$.

17. $(-\infty,0) \cup (0,+\infty)$ 19. $[0,+\infty)$

21.   all real values of x except those for which $\sin x = 1$;
     domain: all x except $x = \pi/2 + 2k\pi$, $k = 0, \pm 1, \pm 2, \cdots$

23.   domain: $(-\infty,3]$; range: $[0,+\infty)$ 25.   domain: $[-2,2]$; range: $[0,2]$

27.   domain: $[0,+\infty)$; range: $[3,+\infty)$ 29.   domain: $(-\infty,+\infty)$; range: $[3,+\infty)$

31.   domain: $(-\infty,+\infty)$; range: $(-\infty,+\infty)$

33.   domain: $(-\infty,+\infty)$; range: $[-3,3]$

35.   domain: $(-\infty,+\infty)$; range: $[1,3]$

37. If $x < 0$, then $|x| = -x$ so $f(x) = -x + 3x + 1 = 2x + 1$. If $x \geq 0$, then $|x| = x$ so $f(x) = x + 3x + 1 = 4x + 1$;

$$f(x) = \begin{cases} 2x + 1, & x < 0 \\ 4x + 1, & x \geq 0 \end{cases}.$$

39. If $x < 0$, then $|x| = -x$ and $|x - 1| = 1 - x$ so $g(x) = -x + 1 - x = 1 - 2x$. If $0 \leq x < 1$, then $|x| = x$ and $|x - 1| = 1 - x$ so $g(x) = x + 1 - x = 1$. If $x \geq 1$, then $|x| = x$ and $|x - 1| = x - 1$ so $g(x) = x + x - 1 = 2x - 1$;

$$g(x) = \begin{cases} 1 - 2x, & x < 0 \\ 1, & 0 \leq x < 1 \\ 2x - 1, & x \geq 1 \end{cases}.$$

41. $\sqrt{3x - 2} = 6$, $3x - 2 = 36$, $3x = 38$, $x = 38/3$.

43. $x^2 + 5 = 7$, $x^2 = 2$, $x = \pm\sqrt{2}$.

45. $\cos x = 1$, $x = 2k\pi$, $k = 0, \pm 1, \pm 2, \cdots$

47. $\sin \sqrt{x} = 1/2$, $\sqrt{x} = \pi/6 + 2k\pi$ or $5\pi/6 + 2k\pi$ for $k = 0, 1, 2, \cdots$ so $x = (1/6 + 2k)^2 \pi^2$ or $(5/6 + 2k)^2 \pi^2$ for $k = 0, 1, 2, \cdots$

49. $A = \pi r^2$, but $C = 2\pi r$ so $r = \dfrac{C}{2\pi}$; $A = \dfrac{C^2}{4\pi}$.

51. (a) $S = 6x^2$

(b) $V = x^3$ so $x = V^{1/3}$; substitute into (a) to get $s = 6V^{2/3}$.

53. Multiplication of the numerator and denominator of a fraction by the same number is valid only if the number is not zero, so $(1 - 1/x)/(1 + 1/x) = (x - 1)/(x + 1)$ if $x \neq 0$.

55. $\dfrac{(x + 2)(x^2 - 1)}{(x + 2)(x + 1)} = \dfrac{(x + 2)(x + 1)(x - 1)}{(x + 2)(x + 1)} = x - 1$, $x \neq -1$ or -2

57. $\dfrac{x + 1 + \sqrt{x + 1}}{\sqrt{x + 1}} = \dfrac{\sqrt{x + 1}(\sqrt{x + 1} + 1)}{\sqrt{x + 1}} = \sqrt{x + 1} + 1,\ x \neq -1$

59. $\dfrac{x^3 + 2x^2 - 3x}{(x - 1)(x + 3)} = \dfrac{x(x^2 + 2x - 3)}{(x - 1)(x + 3)} = \dfrac{x(x - 1)(x + 3)}{(x - 1)(x + 3)} = x,\ x \neq -3 \text{ or } 1$

61. (a) monomial, polynomial, rational, explicit algebraic
    (b) explicit algebraic   (c) rational, explicit algebraic
    (d) polynomial, rational, explicit algebraic

63. (a) explicit algebraic   (b) rational, explicit algebraic
    (c) monomial, polynomial, rational, explicit algebraic

    (d) explicit algebraic ($|x| = \sqrt{x^2}$)

EXERCISE SET 2.2

1. (a) −4, −3, −2, 2, 3   (b) 0, 4
   (c) $-4 \le x \le -3,\ -2 \le x \le 2,\ x \ge 3$ (d) $x \le -4,\ -3 \le x \le -2,\ 2 \le x \le 3$

3.

5.

7.

9.

11.

13.

15.

17.

19.

21.

23.

25.

27.

29.

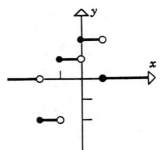

31. $x \leq 2$: $f(x) = 2x + (2 - x) = x + 2$,
     $x > 2$: $f(x) = 2x + (x - 2) = 3x - 2$;

$$f(x) = \begin{cases} x + 2, & x \leq 2 \\ 3x - 2, & x > 2 \end{cases}$$

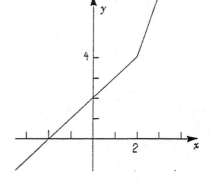

33. $x < 3$: $g(x) = (5 - x) - (3 - x) = 2$,
    $3 \leq x < 5$: $g(x) = (5 - x) - (x - 3) = 8 - 2x$,
        $x \geq 5$: $g(x) = (x - 5) - (x - 3) = -2$;

$$g(x) = \begin{cases} 2, & x < 3 \\ 8 - 2x, & 3 \leq x < 5 \\ -2, & x \geq 5 \end{cases}$$

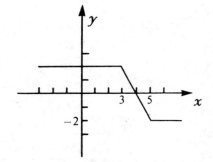

35.   If $0 \leq x \leq 1$, $A = \frac{1}{2}(2x)(x) = x^2$;
    if $x > 1$, $A = 1 + 2(x - 1) = 2x - 1$;

$$A = \begin{cases} x^2, & 0 \leq x \leq 1 \\ 2x - 1, & x > 1 \end{cases}$$

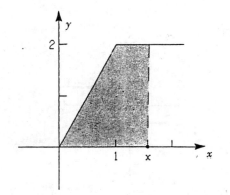

37. $g(x) = \begin{cases} 2, & x < -1 \\ 1 - x, & -1 \leq x < 1 \\ \frac{1}{2}(x - 1), & x \geq 1 \end{cases}$

39.   (a)

(b)

(c)

(d)

41. (a)

(b)

43. (a) both; $y = -2x - 4$, $x = -\frac{1}{2}y - 2$ (b) y is a function of x; $y = \dfrac{1}{x^{2/3}}$

(c) neither

(d) both; $y = \dfrac{1}{2x}$, $x = \dfrac{1}{2y}$

45. (a) $y = 1/x^2$ (b) $x = \dfrac{1 - y}{1 + y}$ (c) $y^2 + 2xy + x^2 = 0$

$$x(1 + y) = 1 - y$$
$$x + xy = 1 - y$$
$$xy + y = 1 - x$$
$$y(x + 1) = 1 - x$$
$$y = \frac{1 - x}{1 + x}$$

$$(y + x)^2 = 0$$
$$y + x = 0$$
$$y = -x$$

47. Treat $y^2 + 4xy + 1 = 0$ as a quadratic equation in y and use the quadratic formula to solve for y:

$$y = \frac{-4x \pm \sqrt{16x^2 - 4}}{2} = -2x \pm \sqrt{4x^2 - 1}.$$

There are two real values of y for each x for which $|x| > 1/2$.

49. (a) both   (b) a function of x
    (c) a function of y   (d) neither

EXERCISE SET 2.3

1. (a) $f(t) = t^2 + 1$

   (b) $f(t + 2) = (t + 2)^2 + 1 = t^2 + 4t + 5$

   (c) $f(x + 2) = (x + 2)^2 + 1 = x^2 + 4x + 5$

   (d) $f(1/x) = (1/x)^2 + 1 = 1/x^2 + 1$

   (e) $f(x + h) = (x + h)^2 + 1 = x^2 + 2h + h^2 + 1$

   (f) $f(-x) = (-x)^2 + 1 = x^2 + 1$

   (g) $f(\sqrt{x}) = (\sqrt{x})^2 + 1 = x + 1, \; x \geq 0$

   (h) $f(3x) = (3x)^2 + 1 = 9x^2 + 1$

3. (a) $x^2 + 2x + 1$ (b) $-x^2 + 2x - 1$

   (c) $2x(x^2 + 1)$ (d) $\dfrac{2x}{x^2 + 1}$

   (e) $2(x^2 + 1)$ (f) $4x^2 + 1$

5. (a) $\sqrt{x + 1} + x - 2$ (b) $\sqrt{x + 1} - x + 2$

   (c) $(x - 2)\sqrt{x + 1}$ (d) $\dfrac{\sqrt{x + 1}}{x - 2}$

   (e) $\sqrt{x - 1}$ (f) $\sqrt{x + 1} - 2$

7. (a) $\sqrt{x - 2} + \sqrt{x - 3}$ (b) $\sqrt{x - 2} - \sqrt{x - 3}$

   (c) $\sqrt{x - 2} \, \sqrt{x - 3}$ (d) $\dfrac{\sqrt{x - 2}}{\sqrt{x - 3}}$

   (e) $\sqrt{\sqrt{x - 3} - 2}$ (f) $\sqrt{\sqrt{x - 2} - 3}$

9. (a) $(f + g)(x) = \begin{cases} 0, & x < 0 \\ 2x, & x \geq 0 \end{cases}$

(b) $(f - g)(x) = \begin{cases} -2x, & x < 0 \\ 0, & x \geq 0 \end{cases}$

(c) $(f \cdot g)(x) = \begin{cases} -x^2, & x < 0 \\ x^2, & x \geq 0 \end{cases}$

(d) $(f/g)(x) = \begin{cases} -1, & x < 0 \\ 1, & x > 0 \end{cases}$

11. (a) $4x - 15$

(b) $4x^2 - 20x + 25$

13. (a) $f(g(x)) = 1/(x^2 + 1)$, which is defined for all x.

(b) $g(x) = x^2 + 2$, for example
(c) g must be defined for all x and g(x) must never equal zero.

15. Show that $(f\circ(g\circ h))(x) = ((f\circ g)\circ h)(x)$:

$(f\circ(g\circ h))(x) = f((g\circ h)(x)) = f(g(h(x)))$,

$((f \circ g) \circ h)(x) = (f \circ g)(h(x)) = f(g(h(x)))$.

Thus $(f \circ (g \circ h))(x) = ((f \circ g) \circ h)(x)$ so $f \circ (g \circ h) = (f \circ g) \circ h$.

17. $g(x) = \sqrt{x}$, $h(x) = x + 2$ 19. $g(x) = x^7$, $h(x) = x - 5$

21. $g(x) = |x|$, $h(x) = x^2 - 3x + 5$

23. $g(x) = x^2$, $h(x) = \sin x$

25. $g(x) = \dfrac{3}{5 + x}$, $h(x) = \cos x$

27. $g(x) = \dfrac{x}{3 + x}$, $h(x) = \tan x$

29. $f(x) = \sqrt{x}$, $g(x) = 3 - x^2$, $h(x) = \sin x$

31. $z = x + 1$ so $x = z - 1$;
      $f(z) = (z - 1)^2 + 3(z - 1) + 5 = z^2 + z + 3$; $f(x) = x^2 + x + 3$.

33. $f(g(x)) = f(2x - 1) = 0$ only if $2x - 1 = -1$ or $2x - 1 = 2$, so $x = 0$ or
      $x = 3/2$.

35. $f(g(x)) = \sqrt{g(x) + 5} = 3|x|$, $g(x) + 5 = 9x^2$, $g(x) = 9x^2 - 5$.

37.   Let $x = y = 0$, then $f(0 + 0) = f(0) = f(0) - f(0)$ so $f(0) = 0$.
      Let $x = 0$, then $f(0 + y) = f(y) = f(0) - f(y)$ so $2f(y) = f(0) = 0$;
      $f(y) = 0$ for all real y and hence $f(x) = 0$ for all real x.

39.   (a) $f(-x) = (-x)^2 = x^2 = f(x)$, even
      (b) $f(-x) = (-x)^3 = -x^3 = -f(x)$, odd
      (c) $f(-x) = |-x| = |x| = f(x)$, even   (d) $f(-x) = -x + 1$, neither
      (e) $f(-x) = \dfrac{(-x)^5 - (-x)}{1 + (-x)^2} = \dfrac{-x^5 + x}{1 + x^2} = -\dfrac{x^5 - x}{1 + x^2} = -f(x)$, odd
      (f) $f(-x) = 2 = f(x)$, even

41.   (a) even   (b) odd   (c) odd   (e) neither

43.   (a) If f and g are even, then
          $(f \cdot g)(-x) = f(-x)g(-x) = f(x)g(x) = (f \cdot g)(x)$ so $f \cdot g$ is even.

(b) If f and g are odd, then
$(f \cdot g)(-x) = f(-x)g(-x) = [-f(x)][-g(x)] = f(x)g(x) = (f \cdot g)(x)$ so
$f \cdot g$ is even.

(c) If f is even and g is odd, then
$(f \cdot g)(-x) = f(-x)g(-x) = f(x)[-g(x)] = -f(x)g(x) = -(f \cdot g)(x)$ so
$f \cdot g$ is odd.

EXERCISE SET 2.4

1. (a) −1   (b) 3   (c) does not exist
   (d) 1   (e) −1   (f) 3

3. (a) 1   (b) 1   (c) 1
   (d) 1   (e) −∞   (f) +∞

5. (a) 0   (b) 0   (c) 0
   (d) 3   (e) +∞   (f) +∞

7. (a) −∞   (b) +∞   (c) does not exist
   (d) not defined (e) 2   (f) 0

9. (a) −∞   (b) −∞   (c) −∞
   (d) 1   (e) 2   (f) 2

11. (a) 0   (b) 0   (c) 0
    (d) 0   (e) does not exist (f) does not exist

13. all values except −4

EXERCISE SET 2.5

1. 7   3. π 5. 36

7. $\sqrt{109}$ 9. 14   11. 0

13. $\lim\limits_{x \to 4} \dfrac{x^2 - 16}{x - 4} = \lim\limits_{x \to 4} \dfrac{(x + 4)(x - 4)}{x - 4} = \lim\limits_{x \to 4} (x + 4) = 8$

15. $\displaystyle\lim_{x\to1^+} \frac{x^4 - 1}{x - 1} = \lim_{x\to1^+} \frac{(x - 1)(x^3 + x^2 + x + 1)}{x - 1}$

$\qquad\qquad = \displaystyle\lim_{x\to1^+} (x^3 + x^2 + x + 1) = 4$

17. $\displaystyle\lim_{x\to-1} \frac{x^2 + 6x + 5}{x^2 - 3x - 4} = \lim_{x\to-1} \frac{(x + 1)(x + 5)}{(x + 1)(x - 4)} = \lim_{x\to-1} \frac{x + 5}{x - 4} = -\frac{4}{5}$

19. $\displaystyle\lim_{x\to+\infty} \frac{3x + 1}{2x - 5} = \lim_{x\to+\infty} \frac{3 + 1/x}{2 - 5/x} = \frac{3}{2}$ 21. 0

23. $\displaystyle\lim_{x\to-\infty} \frac{x - 2}{x^2 + 2x + 1} = \lim_{x\to-\infty} \frac{1/x - 2/x^2}{1 + 2/x + 1/x^2} = 0$

25. $\displaystyle\lim_{x\to-\infty} \sqrt{\frac{5x^2 - 2}{x + 3}} = \lim_{x\to-\infty} \frac{-\sqrt{5 - 2/x^2}}{1 + 3/x} = -\sqrt{5}$

27. $\displaystyle\lim_{y\to-\infty} \frac{2 - y}{\sqrt{7 + 6y^2}} = \lim_{y\to-\infty} \frac{2/y - 1}{-\sqrt{7/y^2 + 6}} = 1/\sqrt{6}$

29. $\displaystyle\lim_{x\to-\infty} \frac{\sqrt{3x^4 + x}}{x^2 - 8} = \lim_{x\to-\infty} \frac{\sqrt{3 + 1/x^3}}{1 - 8/x^2} = \sqrt{3}$

31. $+\infty$ 33. does not exist   35. $-\infty$

37. $+\infty$ 39. does not exist

41. $\displaystyle\lim_{x\to4^-} \frac{3 - x}{x^2 - 2x - 8} = \lim_{x\to4^-} \frac{3 - x}{(x - 4)(x + 2)} = +\infty$

43. $\displaystyle\lim_{x\to+\infty} \frac{7 - 6x^5}{x + 3} = \lim_{x\to+\infty} \frac{7/x - 6x^4}{1 + 3/x} = -\infty$

45. $\displaystyle\lim_{t\to+\infty} \frac{6 - t^3}{7t^3 + 3} = \lim_{t\to+\infty} \frac{6/t^3 - 1}{7 + 3/t^3} = -1/7$

47. $\lim\limits_{x \to 0^-} \dfrac{x}{|x|} = \lim\limits_{x \to 0^-} \dfrac{x}{(-x)} = \lim\limits_{x \to 0^-} (-1) = -1$

49. $\lim\limits_{x \to 9} \dfrac{x - 9}{\sqrt{x} - 3} = \lim\limits_{x \to 9} \dfrac{(\sqrt{x} - 3)(\sqrt{x} + 3)}{\sqrt{x} - 3} = \lim\limits_{x \to 9} (\sqrt{x} + 3) = 6$

51. (a) $\lim\limits_{x \to 3^-} f(x) = \lim\limits_{x \to 3^-} (x - 1) = 2$ (b) $\lim\limits_{x \to 3^+} f(x) = \lim\limits_{x \to 3^+} (3x - 7) = 2$

    (c) 2

53. $\lim\limits_{x \to 3} h(x) = \lim\limits_{x \to 3} (x^2 - 2x + 1) = 4$

55. (a) The limit of a difference is the difference of limits if the latter limits exist, in this problem these limits do not exist.

    (b) $\lim\limits_{x \to 0^+} \left(\dfrac{1}{x} - \dfrac{1}{x^2}\right) = \lim\limits_{x \to 0^+} \dfrac{x - 1}{x^2} = -\infty$

57. $\lim\limits_{x \to 0} \dfrac{\sqrt{x + 3} - \sqrt{3}}{x} = \lim\limits_{x \to 0} \dfrac{(x + 3) - 3}{x(\sqrt{x + 3} + \sqrt{3})} = \lim\limits_{x \to 0} \dfrac{1}{\sqrt{x + 3} + \sqrt{3}} = \dfrac{1}{2\sqrt{3}}$

59. $\lim\limits_{x \to +\infty} (\sqrt{x^2 + ax} - x) = \lim\limits_{x \to +\infty} \dfrac{(x^2 + ax) - x^2}{\sqrt{x^2 + ax} + x} = \lim\limits_{x \to +\infty} \dfrac{ax}{\sqrt{x^2 + ax} + x}$

    $= \lim\limits_{x \to +\infty} \dfrac{a}{\sqrt{1 + a/x} + 1} = a/2$

61. $\lim\limits_{x \to +\infty} (\sqrt{x^2 + 3} - x) = \lim\limits_{x \to +\infty} \dfrac{(x^2 + 3) - x^2}{\sqrt{x^2 + 3} + x} = \lim\limits_{x \to +\infty} \dfrac{3}{\sqrt{x^2 + 3} + x} = 0$

63. $\lim\limits_{x \to 0^+} \sin(1/x)$ does not exist because $1/x \to +\infty$ and so $\sin(1/x)$ oscillates

    between 1 and -1.

65. $\lim\limits_{x \to +\infty} \sin x$ does not exist because $\sin x$ oscillates between 1 and -1.

67. $\lim\limits_{x \to +\infty} \cos(1/x) = 1$ because $1/x \to 0^+$ and so $\cos(1/x)$ approaches 1.

69. $r(x) = p(x)/q(x)$ if $r(x)$ is a rational function, where $p(x)$ and $q(x)$
     are polynomials. We know that $\lim\limits_{x \to a} p(x) = p(a)$ and $\lim\limits_{x \to a} q(x) = q(a)$ so if

$g(a) \neq 0$ then $\lim\limits_{x \to a} r(x) = \lim\limits_{x \to a} \dfrac{p(x)}{q(x)} = \dfrac{\lim\limits_{x \to a} p(x)}{\lim\limits_{x \to a} q(x)} = \dfrac{p(a)}{q(a)} = r(a)$.

If $q(a) = 0$, then $r(a)$ is not defined and thus cannot equal $\lim\limits_{x \to a} r(x)$, so

$\lim\limits_{x \to a} r(x) = r(a)$ only if $r(a)$ is defined.

EXERCISE SET 2.6

1. $\quad |2x - 8| < 0.1$
    if $2|x - 4| < 0.1$
    or if $|x - 4| < 0.05$,
    so $\delta = 0.05$

3. $|(7x + 5) - (-2)| < 0.01$
    if $\quad |7x + 7| < 0.01$
    or if $\quad 7|x + 1| < 0.01$
    or if $\quad |x + 1| < \dfrac{1}{700}$

    so $\delta = \dfrac{1}{700}$

5. Suppose $x \neq 2$, then $\left|\dfrac{x^2 - 4}{x - 2} - 4\right| = |(x + 2) - 4| = |x - 2|$. Thus

    $\left|\dfrac{x^2 - 4}{x - 2} - 4\right| < 0.05$ if $0 < |x - 2| < 0.05$, so $\delta = 0.05$.

7. $|x^2 - 16| = |(x + 4)(x - 4)| = |x + 4||x - 4|$. If we restrict δ so
    that $\delta \leq 1$, then
    $$|x - 4| < 1$$
    $$3 < x < 5$$
    $$7 < x + 4 < 9$$
    $$|x + 4| < 9$$
    $$|x + 4||x - 4| \leq 9|x - 4|.$$
    Thus $|x^2 - 16| < 0.001$ if $9|x - 4| < 0.001$, or if $|x - 4| < \dfrac{1}{9000}$,

    so $\delta = \dfrac{1}{9000}$.

9. $\left|\dfrac{1}{x} - \dfrac{1}{5}\right| = \left|\dfrac{5 - x}{5x}\right| = \dfrac{|x - 5|}{|5x|}$. If we restrict δ so that $\delta \le 1$, then

$$|x - 5| < 1$$
$$4 < x < 6$$
$$20 < 5x < 30$$
$$\frac{1}{20} > \frac{1}{5x} > \frac{1}{30}$$
$$\frac{1}{|5x|} < \frac{1}{20}$$
$$\frac{|x - 5|}{5|x|} \le \frac{|x - 5|}{20}.$$

Thus $\left|\dfrac{1}{x} - \dfrac{1}{5}\right| < 0.05$ if $\dfrac{|x - 5|}{20} < 0.05$, or if $|x - 5| < 1$, so $\delta = 1$.

11. $|3x - 15| < \epsilon$ if $3|x - 5| < \epsilon$, or if $|x - 5| < \dfrac{\epsilon}{3}$, so $\delta = \dfrac{\epsilon}{3}$.

13. Suppose $x \ne 0$, then $\left|\dfrac{x^2 + x}{x} - 1\right| = |(x + 1) - 1| = |x|$. Thus

$\left|\dfrac{x^2 + 1}{x} - 1\right| < \epsilon$ if $0 < |x| < \epsilon$, so $\delta = \epsilon$.

15. $|2x^2 - 2| = 2|x + 1||x - 1|$. If we restrict δ so that $\delta \le 1$, then

$$|x - 1| < 1$$
$$0 < x < 2$$
$$1 < x + 1 < 3$$
$$|x + 1| < 3$$
$$2|x + 1||x - 1| \le 6|x - 1|.$$

Thus $|2x^2 - 2| < \epsilon$ if $6|x - 1| < \epsilon$, or if $|x - 1| < \dfrac{\epsilon}{6}$, so
$\delta = \min(\epsilon/6, 1)$.

17. $\left|\dfrac{1}{x} - 3\right| = \left|\dfrac{3}{x}\right|\left|\dfrac{1}{3} - x\right| = \left|\dfrac{3}{x}\right|\left|x - \dfrac{1}{3}\right|$. If we restrict δ so that $\delta \le \dfrac{1}{4}$,
then

$$\left|x - \frac{1}{3}\right| < \frac{1}{4}$$
$$\frac{1}{12} < x < \frac{7}{12}$$

$$12 > \frac{1}{x} > \frac{12}{7}$$

$$36 > \frac{3}{x} > \frac{36}{7}$$

$$\left|\frac{3}{x}\right| < 36$$

$$\left|\frac{3}{x}\right|\left|x - \frac{1}{3}\right| \leq 36\left|x - \frac{1}{3}\right|.$$

Thus $\left|\frac{1}{x} - 3\right| < \epsilon$ if $36\left|x - \frac{1}{3}\right| < \epsilon$, or if $\left|x - \frac{1}{3}\right| < \frac{\epsilon}{36}$, so $\delta = \min(\epsilon/36, 1/4)$.

19. $|\sqrt{x} - 2| = \left|\frac{\sqrt{x} - 2}{1} \frac{\sqrt{x} + 2}{\sqrt{x} + 2}\right| = \frac{|x - 4|}{\sqrt{x} + 2}$. If we restrict δ so that $\delta \leq 4$, then

$$|x - 4| < 4$$

$$0 < x < 8$$

$$0 < \sqrt{x} < \sqrt{8}$$

$$2 < \sqrt{x} + 2 < \sqrt{8} + 2$$

$$\frac{1}{2} > \frac{1}{\sqrt{x} + 2} > \frac{1}{\sqrt{8} + 2}$$

$$\frac{1}{\sqrt{x} + 2} < \frac{1}{2}$$

$$\frac{|x - 4|}{\sqrt{x} + 2} \leq \frac{1}{2}|x - 4|$$

Thus $|\sqrt{x} - 2| < \epsilon$ if $\frac{1}{2}|x - 4| < \epsilon$, or if $|x - 4| < 2\epsilon$, so $\delta = \min(2\epsilon, 4)$.

21. If $x \neq 1$, then $|f(x) - 3| = |(x + 2) - 3| = |x - 1|$.
Thus $|f(x) - 3| < \epsilon$ of $0 < |x - 1| < \epsilon$ so $\delta = \epsilon$.

23. Assume there is a number L such that $\lim_{x \to 0} f(x) = L$. Then there exists a number $\delta > 0$ such that $|f(x) - L| < \frac{1}{8}$ whenever $0 < |x - 0| < \delta$; in particular, $x = \frac{\delta}{2}$ and $x = -\frac{\delta}{2}$ are two such values of x. But $f(\delta/2) = \frac{1}{8}$

and $f(-\delta/2) = -\frac{1}{8}$ so $\left|\frac{1}{8} - L\right| < \frac{1}{8}$ and $\left|-\frac{1}{8} - L\right| < \frac{1}{8}$ or, equivalently,

$0 < L < \frac{1}{4}$ and $-\frac{1}{4} < L < 0$, which is impossible.

25. Assume there is a number L such that $\lim_{x \to 1} \frac{1}{x-1} = L$. Then there

    exists a number $\delta > 0$ such that $\left|\frac{1}{x-1} - L\right| < 1$ whenever

    $0 < |x - 1| < \delta$; in particular, $x = 1 + \frac{\delta}{\delta+1}$ and $x = 1 - \frac{\delta}{\delta+1}$ are two

    such values of x. So $\left|\frac{\delta+1}{\delta} - L\right| < 1$ and $\left|-\frac{\delta+1}{\delta} - L\right| < 1$ or

    equivalently, $\frac{1}{\delta} < L < \frac{1}{\delta} + 2$ and $-2 - \frac{1}{\delta} < L < -\frac{1}{\delta}$, which is impossible.

27. $|x^2 - 9| = |x + 3||x - 3|$. If $\delta \le 2$, then

$$|x - 3| < 2$$
$$1 < x < 5$$
$$4 < x + 3 < 8$$
$$|x + 3| < 8$$
$$|x + 3||x - 3| \le 8|x - 3|.$$

Thus $|x^2 - 9| < \epsilon$ if $8|x - 3| < \epsilon$, or if $|x - 3| < \frac{\epsilon}{8}$, so $\delta = \min(\epsilon/8, 2)$

EXERCISE SET 2.7

1. continuous on (d), (e), (f); discontinuous at x = 2 on (a), (b), (c)

3. continuous on (b), (d), (f); discontinuous at x = 1 and 3 on (a), at
   x = 1 on (c), at x = 3 on (e)

5. none   7. none   9. x = ±4

11. x = ±3   13. none   15. none

17. (a) f is continuous everywhere for any k, except perhaps at x = 1;
     $$\lim_{x \to 1^-} f(x) = \lim_{x \to 1^-} (7x - 2) = 5, \quad \lim_{x \to 1^+} f(x) = \lim_{x \to 1^+} kx^2 = k, \text{ and}$$
     f(1) = 5 thus $\lim_{x \to 1} f(x) = f(1)$ if k = 5, so f is continuous
     everywhere if k = 5.

     (b) $\lim_{x \to 2^-} f(x) = \lim_{x \to 2^-} kx^2 = 4k, \quad \lim_{x \to 2^+} f(x) = \lim_{x \to 2^+} (2x + k) = 4 + k,$
     and f(2) = 4k, so $\lim_{x \to 2} f(x) = f(2)$ if 4k = 4 + k, k = 4/3.

19. (a) If c > 0, then $\lim_{x \to c} f(x) = \lim_{x \to c} \sqrt{x} = \sqrt{c} = f(c)$; also

     $$\lim_{x \to 0^+} f(x) = \lim_{x \to 0^+} \sqrt{x} = 0 = f(0) \text{ so } f(x) \text{ is continuous on } [0, +\infty).$$

     (b) $\sqrt{g(x)}$ is continuous by Theorem 2.7.6 because it is the composition
     of $\sqrt{x}$ with g(x) where $\sqrt{x}$ is continuous on $[0, +\infty)$ and g(x) is
     continuous and nonnegative.

21. x = 0, ±1, ±2, ...

23. If f and g are continuous at c, then $\lim_{x \to c} f(x) = f(c)$ and
     $\lim_{x \to c} g(x) = g(c)$, so
     (a) $\lim_{x \to c} (f + g)(x) = \lim_{x \to c} [f(x) + g(x)]$
     $$= \lim_{x \to c} f(x) + \lim_{x \to c} g(x)$$
     $$= f(c) + g(c) = (f + g)(c)$$
     therefore f + g is continuous at c.
     (b) Similar to part (a) with + replaced by −.
     (c) $\lim_{x \to c} (f \cdot g)(x) = \lim_{x \to c} [f(x) g(x)]$
     $$= \left[\lim_{x \to c} f(x)\right]\left[\lim_{x \to c} g(x)\right]$$
     $$= f(c) g(c) = (f \cdot g)(c)$$
     therefore f·g is continuous at c.

25. (a) Let $f(x) = \begin{cases} 0, & x < 2 \\ 1, & x \geq 2 \end{cases}$ and $g(x) = \begin{cases} 1, & x < 2 \\ 0, & x \geq 2 \end{cases}$; f and g are
     discontinuous at x = 2, but f + g is continuous at x = 2.

If $f(x) = \begin{cases} 0, & x < 2 \\ 1, & x \geq 2 \end{cases}$ and $g(x) = \begin{cases} 1, & x < 2 \\ 2, & x \geq 2 \end{cases}$; then f, g, and f + g are discontinuous at x = 2.

(b) Replace f + g by f·g everywhere in part (a).

27. If f(a) and f(b) have opposite signs then 0 is between f(a) and f(b). From Theorem 2.7.9 there is at least one number x in [a,b] such that f(x) = 0. But f(a) ≠ 0 and f(b) ≠ 0 by assumption, so there is at least one solution of f(x) = 0 in the interval (a,b).

29. $f(x) = x^3 + x^2 - 2x - 1$ is continuous on [−1,1], f(−1) = 1 and f(1) = −1 have opposite signs so Theorem 2.7.10 applies.

31. From the graph there are two real solutions, one in (−2,−1.5) and the other in (1,1.5). For the one in (−2,−1.5) we find

x	−2.0	−1.9	−1.8	−1.7	−1.6
y	−9	−6.13	−3.70	−1.65	0.05

so one solution is in (−1.7,−1.6); use −1.65 to approximate it. For the one in (1,1.5) we find

x	1.0	1.1	1.2	1.3	1.4
y	3	2.44	1.73	0.84	−0.24

so this solution is in (1.3,1.4); use 1.35 to approximate it.

33. If $f(x) = x^2 - 5$, then f(2) = −1 and f(3) = 4 have opposite signs so $\sqrt{5}$ is in (2,3).

(a) f(2.2) = −0.16 and f(2.3) = 0.29 so $\sqrt{5}$ is in (2.2, 2.3); use 2.25 to approximate it with an error of at most 0.05.

(b) f(2.23) = −0.03 and f(2.24) = 0.02 so $\sqrt{5}$ is in (2.23,2.24); use 2.235 to approximate it with an error of at most 0.005.

35. $f(x) = \begin{cases} x + 1, & 0 \leq x \leq 1 \\ x - 3, & 1 < x \leq 2 \end{cases}$, for example.

37. If $p(x) = a_n x^n + a_{n-1} x^{n-1} + \cdots + a_1 x + a_0$ where $a_n \neq 0$ and n is odd, then either $\lim\limits_{x \to +\infty} p(x) = +\infty$ and $\lim\limits_{x \to -\infty} p(x) = -\infty$, or $\lim\limits_{x \to +\infty} p(x) = -\infty$ and $\lim\limits_{x \to -\infty} p(x) = +\infty$, depending on whether $a_n > 0$ or $a_n < 0$, respectively. In either case, there are numbers a and b with $p(x)$ continuous on $[a,b]$ where $p(a)$ and $p(b)$ have opposite signs so that $p(x) = 0$ has at least one real solution in (a,b).

EXERCISE SET 2.8

1. none

3. $x = n\pi$; $n = 0, \pm1, \pm2, \cdots$

5. $x = n\pi$; $n = 0, \pm1, \pm2, \cdots$

7. none

9. discontinuous if $\sin x = 1/2$, so $x = \pi/6 + 2n\pi$ or $x = 5\pi/6 + 2n\pi$; $n = 0, \pm1, \pm2, \cdots$

11. $\sin(g(x))$ is the composition of $\sin x$ with $g(x)$; $\sin x$ is continuous, so by theorem 2.7.6 $\sin(g(x))$ is continuous at every point where $g(x)$ is continuous.

13. $\lim\limits_{x \to +\infty} \sin\left(\dfrac{\pi x}{2 - 3x}\right) = \sin\left(\lim\limits_{x \to +\infty} \dfrac{\pi x}{2 - 3x}\right) = \sin(-\pi/3) = -\sqrt{3}/2$

15. $\lim\limits_{\theta \to 0} \dfrac{\sin 3\theta}{\theta} = \lim\limits_{\theta \to 0} 3 \dfrac{\sin 3\theta}{3\theta} = 3 \lim\limits_{\theta \to 0} \dfrac{\sin 3\theta}{3\theta} = 3(1) = 3.$

17. $\lim\limits_{x \to 0} \dfrac{\tan 7x}{\sin 3x} = \lim\limits_{x \to 0} \dfrac{\dfrac{\sin 7x}{\cos 7x}}{\sin 3x} = \lim\limits_{x \to 0} \dfrac{1}{\cos 7x} \dfrac{\sin 7x}{\sin 3x}$

$= \lim\limits_{x \to 0} \dfrac{1}{\cos 7x} \dfrac{7 \dfrac{\sin 7x}{7x}}{3 \dfrac{\sin 3x}{3x}}$

$= \dfrac{7}{3} \lim\limits_{x \to 0} \dfrac{1}{\cos 7x} \dfrac{\lim\limits_{x \to 0} \dfrac{\sin 7x}{7x}}{\lim\limits_{x \to 0} \dfrac{\sin 3x}{3x}} = \dfrac{7}{3} (1) \dfrac{(1)}{(1)} = \dfrac{7}{3}.$

19. $\lim\limits_{h\to 0} \dfrac{h}{\tan h} = \lim\limits_{h\to 0} \dfrac{h}{\dfrac{\sin h}{\cos h}} = \lim\limits_{h\to 0} \dfrac{h\cos h}{\sin h} = \lim\limits_{h\to 0} \dfrac{\cos h}{\dfrac{\sin h}{h}}$

$= \dfrac{\lim\limits_{h\to 0} \cos h}{\lim\limits_{h\to 0} \dfrac{\sin h}{h}} = \dfrac{1}{1} = 1.$

21. $\lim\limits_{\theta\to 0} \dfrac{\theta^2}{1 - \cos\theta} = \lim\limits_{\theta\to 0} \dfrac{\theta^2}{1 - \cos\theta}\dfrac{1 + \cos\theta}{1 + \cos\theta} = \lim\limits_{\theta\to 0} \dfrac{\theta^2(1 + \cos\theta)}{1 - \cos^2\theta}$

$= \lim\limits_{\theta\to 0} \dfrac{\theta^2(1 + \cos\theta)}{\sin^2\theta} = \lim\limits_{\theta\to 0} \dfrac{1 + \cos\theta}{\dfrac{\sin^2\theta}{\theta^2}}$

$= \dfrac{\lim\limits_{\theta\to 0}(1 + \cos\theta)}{\lim\limits_{\theta\to 0}\left(\dfrac{\sin\theta}{\theta}\right)^2} = \dfrac{1 + 1}{1^2} = 2.$

23. $\lim\limits_{\theta\to 0} \dfrac{\theta}{\cos\theta} = \dfrac{\lim\limits_{\theta\to 0}\theta}{\lim\limits_{\theta\to 0}\cos\theta} = \dfrac{0}{1} = 0.$

25. Use the identity $1 - \cos\theta = 2\sin^2\dfrac{\theta}{2}$:

$\lim\limits_{h\to 0} \dfrac{1 - \cos 5h}{\cos 7h - 1} = \lim\limits_{h\to 0} \dfrac{2\sin^2(5h/2)}{-2\sin^2(7h/2)} = -\lim\limits_{h\to 0} \dfrac{(5/2)^2\dfrac{\sin^2(5h/2)}{(5h/2)^2}}{(7/2)^2\dfrac{\sin^2(7h/2)}{(7h/2)^2}}$

$= -\dfrac{25}{49}\dfrac{\lim\limits_{h\to 0}\left[\dfrac{\sin(5h/2)}{5h/2}\right]^2}{\lim\limits_{h\to 0}\left[\dfrac{\sin(7h/2)}{7h/2}\right]^2} = -\dfrac{25}{49}\cdot\dfrac{1^2}{1^2} = -\dfrac{25}{49}.$

27. $\displaystyle\lim_{x\to 0^-} f(x) = \lim_{x\to 0^-} \frac{\tan kx}{x} = \lim_{x\to 0^-} \frac{k}{\cos kx} \frac{\sin kx}{kx} = k;$

$\displaystyle\lim_{x\to 0^+} f(x) = \lim_{x\to 0^+} (3x + 2k^2) = 2k^2;$ $f(0) = 2k^2;$ so $\displaystyle\lim_{x\to 0} f(x) = f(0)$ if

$2k^2 = k,\ 2k^2 - k = 0,\ k(2k - 1) = 0;\ k = 1/2.$

29. (a) If $t = \dfrac{1}{x}$, then $x = \dfrac{1}{t}$ and $x \to +\infty$ as $t \to 0^+$

so $\displaystyle\lim_{x\to +\infty} x\sin\frac{1}{x} = \lim_{t\to 0^+} \frac{\sin t}{t} = 1.$

(b) If $t = \dfrac{1}{x}$, then $x = \dfrac{1}{t}$ and $x \to -\infty$ as $t \to 0^-$

so $\displaystyle\lim_{x\to -\infty} (1 - \cos\frac{1}{x}) = \lim_{t\to 0^-} \frac{1 - \cos t}{t} = 0.$

(c) If $t = \pi - x$, then $x = \pi - t$ and $x \to \pi$ as $t \to 0$

so $\displaystyle\lim_{x\to \pi} \frac{\pi - x}{\sin x} = \lim_{t\to 0} \frac{t}{\sin(\pi - t)} = \lim_{t\to 0} \frac{t}{\sin t} = 1.$

31. Let $t = x - 1$, then $x = t + 1$ and

$\displaystyle\lim_{x\to 1} \frac{\sin(\pi x)}{x - 1} = \lim_{t\to 0} \frac{\sin(\pi t + \pi)}{t} = \lim_{t\to 0} \frac{-\sin \pi t}{t} = -\pi \lim_{t\to 0} \frac{\sin \pi t}{\pi t} = -\pi$

33. $\displaystyle\lim_{x\to 0}(1 - x^2) = 1$ and $\displaystyle\lim_{x\to 0} \cos x = 1$ so by the pinching theorem

$\displaystyle\lim_{x\to 0} f(x) = 1.$

35. If $x > 0$, then $xL \le xf(x) \le xM$; if $x < 0$, then $xL \ge xf(x) \ge xM$.
So $\displaystyle\lim_{x\to 0} xf(x) = 0$ by the pinching theorem because $\displaystyle\lim_{x\to 0^+} xL = \lim_{x\to 0^+} xM = 0$

and $\displaystyle\lim_{x\to 0^-} xL = \lim_{x\to 0^-} xM = 0.$

37. If $h < 0$, then $-h > 0$ so from (7)

$\cos(-h) < \dfrac{\sin(-h)}{-h} < 1$

$$\cos h \quad < \frac{-\sin h}{-h} \quad < 1$$

$$\cos h \quad < \frac{\sin h}{h} \quad < 1.$$

39. $f(x) = x - \cos x$ is continuous on $[0, \pi/2]$; $f(0) = -1$ and $f(\pi/2) = \pi/2$ have opposite signs so Theorem 2.7.10 applies.

SUPPLEMENTARY EXERCISES CHAPTER 2

1. $\sqrt{4 - x^2}$ is real if and only if $4 - x^2 \geq 0$, thus $4 \geq x^2$, so the domain is $|x| \leq 2$; $f(-\sqrt{2}) = \sqrt{2}$, $f(0) = 2$, $f(\sqrt{3}) = 1$.

3. $f(x) = \dfrac{(x - 1)}{(x + 2)(x - 1)}$, domain: all x except -2 and 1; $f(0) = 1/2$, $f(1)$ is not defined, $f(2) = 1/4$.

5. domain: all x; $f(0) = -1$, $f(2) = 3$, $f(4) = \sqrt{3}$.

7. (a) $f(x^2) - (f(x))^2 = \dfrac{3 - x^2}{x^2} - \left(\dfrac{3 - x}{x}\right)^2 = \dfrac{3 - x^2}{x^2} - \dfrac{9 - 6x + x^2}{x^2}$

$$= \dfrac{-2x^2 + 6x - 6}{x^2}$$

(b) $f(x + 3) - [f(x) + f(3)] = \dfrac{3 - (x + 3)}{x + 3} - \left[\dfrac{3 - x}{x} + \dfrac{3 - 3}{3}\right]$

$$= - \dfrac{9}{x(x + 3)}$$

(c) $f(1/x) - 1/f(x) = \dfrac{3 - 1/x}{1/x} - \dfrac{x}{3 - x} = 3x - 1 - \dfrac{x}{3 - x} = \dfrac{3x^2 - 9x + 3}{x - 3}$

(d) $f(f(x)) = f(\dfrac{3 - x}{x}) = \dfrac{3 - \dfrac{3 - x}{x}}{\dfrac{3 - x}{x}} = \dfrac{4x - 3}{3 - x}$

9.

domain: all x

range: $y = -\pi$

11.

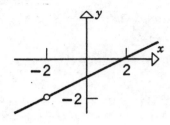

$$f(x) = \frac{x^2 - 4}{2x + 4} = \frac{1}{2}(x - 2),$$

$x \neq -2$
domain: all x except -2
range: all y except -2

13.

domain: $x \geq -1/3$

range: $y \leq 0$

15.

$$f(x) = \frac{2x - 4}{x^2 - 4} = \frac{2}{x + 2}, \; x \neq 2$$

domain: all x except -2, 2
range: all y except 0, 1/2

17.   Some possible answers are:

(a) $h(x) = x^3$, $g(x) = x^2 + 3$; $h(x) = x^6$, $g(x) = x + 3$

(b) $h(x) = x^2 + 1$, $g(x) = \sqrt{x}$; $h(x) = x^2$, $g(x) = \sqrt{x + 1}$

(c) $h(x) = 3x + 2$, $g(x) = \sin x$; $h(x) = 3x$, $g(x) = \sin(x + 2)$

19. (a) −1
     (b) does not exist
     (c) 1
     (d) 0
     (e) −∞ (does not exist)
     (f) 0
     (g) 0
     (h) −∞ (does not exist)

21. $f(x) = \sqrt{2 - x}$ is defined for $x \leq 2$ and $\lim\limits_{x \to a} f(x) = \sqrt{2 - a}$ if $a < 2$, so
     $\lim\limits_{x \to a} f(x) = 2,1,0$ for $a = -2,1,2^-$. Because $f(x)$ is not defined for
     $x > 2$, $\lim\limits_{x \to 2^+} f(x)$ and $\lim\limits_{x \to +\infty} f(x)$ do not exist. Finally, $\lim\limits_{x \to -\infty} f(x) = +\infty$,
     so this limit does not exist.

23. $f(x) = \dfrac{x^2 - 25}{x - 5} = x + 5,\ x \neq 5$, so
     $\lim\limits_{x \to a} f(x) = \lim\limits_{x \to a} (x + 5) = a + 5 = 5,10,0,10,0$ for $a = 0,5^+,-5^-,5,-5$.
     Also, $\lim\limits_{x \to -\infty} f(x) = -\infty$ and $\lim\limits_{x \to +\infty} f(x) = +\infty$, so neither of these limits
     exist.

25. $\lim\limits_{x \to 0} \dfrac{\tan ax}{\sin bx} = \lim\limits_{x \to 0} \dfrac{\sin ax}{\sin bx} \dfrac{1}{\cos ax} = \lim\limits_{x \to 0} \dfrac{a[(\sin ax)/(ax)]}{b[(\sin bx)/(bx)]} \dfrac{1}{\cos ax} = \dfrac{a}{b}.$

27. $\lim\limits_{\theta \to 0} \dfrac{\sin 2\theta}{\theta^2} = \lim\limits_{\theta \to 0} \dfrac{\sin 2\theta}{2\theta} \dfrac{2}{\theta}$, but $\dfrac{\sin 2\theta}{2\theta} \to 1$ as $\theta \to 0$ and $\left|\dfrac{2}{\theta}\right| \to +\infty$ as $\theta \to 0$
     so the limit does not exist.

29. $\lim\limits_{x \to 0^+} \dfrac{\sin x}{\sqrt{x}} = \lim\limits_{x \to 0^+} \sqrt{x}\left(\dfrac{\sin x}{x}\right) = (0)(1) = 0$

31. $\lim\limits_{x \to 0} \dfrac{3x - \sin(kx)}{x} = \lim\limits_{x \to 0} \left[3 - k\,\dfrac{\sin(kx)}{kx}\right] = 3 - k$

CHAPTER 3

EXERCISE SET 3.1

1. (a) $m_{sec} = \dfrac{f(4) - f(3)}{4 - 3} = \dfrac{\frac{1}{2}(4)^2 - \frac{1}{2}(3)^2}{1} = \dfrac{7}{2}$

    (b) $m_{tan} = \lim_{h \to 0} \dfrac{f(3 + h) - f(3)}{h}$ (c)

    $= \lim_{h \to 0} \dfrac{\frac{1}{2}(3 + h)^2 - \frac{9}{2}}{h}$

    $= \lim_{h \to 0} \dfrac{3h + \frac{1}{2}h^2}{h}$

    $= \lim_{h \to 0}(3 + \frac{1}{2}h) = 3;$

    tangent line: $y - 9/2 = 3(x - 3),$
    $y = 3x - 9/2.$

3. (a) $m_{sec} = \dfrac{f(3) - f(2)}{3 - 2} = \dfrac{(1/3) - (1/2)}{1} = -\dfrac{1}{6}$

    (b) $m_{tan} = \lim_{h \to 0} \dfrac{f(2 + h) - f(2)}{h}$ (c)

    $= \lim_{h \to 0} \dfrac{\frac{1}{2 + h} - \frac{1}{2}}{h}$

    $= \lim_{h \to 0} \dfrac{2 - (2 + h)}{2h(2 + h)}$

    $= \lim_{h \to 0} -\dfrac{1}{2(2 + h)} = -\dfrac{1}{4};$

    tangent line: $y - \dfrac{1}{2} = -\dfrac{1}{4}(x - 2),$

    $y = -\dfrac{1}{4}x + 1.$

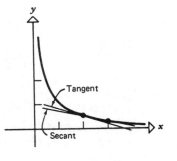

5. (a) $m_{tan} = \lim\limits_{h \to 0} \dfrac{f(x_0 + h) - f(x_0)}{h} = \lim\limits_{h \to 0} \dfrac{(x_0 + h)^3 - x_0^3}{h}$

$= \lim\limits_{h \to 0} \dfrac{(x_0^3 + 3x_0^2 h + 3x_0 h^2 + h^3) - x_0^3}{h}$

$= \lim\limits_{h \to 0}(3x_0^2 + 3x_0 h + h^2) = 3x_0^2.$

(b) $m_{tan} = 3(5)^2 = 75$ at $(5, 5^3) = (5,125)$, so $y - 125 = 75(x - 5)$, or $y = 75x - 250.$

(c) $y - x_0^3 = 3x_0^2(x - x_0)$, $y = 3x_0^2 x - 2x_0^3.$

7. (a) $m_{tan} = \lim\limits_{h \to 0} \dfrac{f(x_0 + h) - f(x_0)}{h}$

$= \lim\limits_{h \to 0} \dfrac{[(x_0 + h)^2 + (x_0 + h)] - (x_0^2 + x_0)}{h}$

$= \lim\limits_{h \to 0} \dfrac{2x_0 h + h^2 + h}{h} = \lim\limits_{h \to 0}(2x_0 + h + 1) = 2x_0 + 1.$

(b) $m_{tan} = 2(2) + 1 = 5$ at $(2,6)$ so $y - 6 = 5(x - 2)$, or $y = 5x - 4.$

(c) $y - (x_0^2 + x_0) = (2x_0 + 1)(x - x_0)$, $y = (2x_0 + 1)x - x_0^2$

9. From the figure:

(a) The particle is moving faster at time t_0 because the slope of the tangent to the curve at t_0 is greater than that at t_2.

(b) The initial velocity is 0 because the slope of a horizontal line is 0.

(c) The particle is speeding up because the slope increases as t increases from t_0 to t_1.

(d) The particle is slowing down because the slope decreases as t increases from t_1 to t_2.

11. It is a straight line with slope equal to the velocity.

13. (a) $5(40)^3 = 320,000$ ft.

(b) average velocity $= 320,000/40 = 8,000$ ft/sec.

(c) $5t^3 = 135$ when the rocket has gone 135 ft, so $t^3 = 27$, $t = 3$ sec;
average velocity $= 135/3 = 45$ ft/sec.

(d) $f'(40) = \lim_{h \to 0} \dfrac{5(40 + h)^3 - 5(40)^3}{h} = \lim_{h \to 0} \dfrac{5(4800h + 120h^2 + h^3)}{h}$

$\quad = \lim_{h \to 0} 5(4800 + 120h + h^2) = 24,000$ ft/sec.

15. (a) average velocity $= \dfrac{6(4)^4 - 6(2)^4}{4 - 2} = 720$ ft/min.

(b) $f'(2) = \lim_{h \to 0} \dfrac{6(2 + h)^4 - 6(2)^4}{h} = \lim_{h \to 0} \dfrac{6(32h + 24h^2 + 8h^3 + h^4)}{h}$

$\quad = \lim_{h \to 0} 6(32 + 24h + 8h^2 + h^3) = 192$ ft/min.

17. (a) $x_0 = 1$, $x_1 = 4$; $y_0 = 2(1)^2 - 1 = 1$, $y_1 = 2(4)^2 - 1 = 31$,

$\dfrac{y_1 - y_0}{x_1 - x_0} = \dfrac{31 - 1}{4 - 1} = 10.$

(b) $y'(1) = \lim_{h \to 0} \dfrac{[2(1 + h)^2 - 1] - [2(1)^2 - 1]}{h} = \lim_{h \to 0} (4 + 2h) = 4.$

19. (a) $r_0 = 1$, $r_1 = 2$; $A_0 = \pi(1)^2 = \pi$, $A_1 = \pi(2)^2 = 4\pi$,

$\dfrac{A_1 - A_0}{r_1 - r_0} = \dfrac{4\pi - \pi}{2 - 1} = 3\pi.$

(b) $A'(2) = \lim_{h \to 0} \dfrac{\pi(2 + h)^2 - \pi(2)^2}{h} = \lim_{h \to 0} \pi(4 + h) = 4\pi.$

21. $m_{sec} = \dfrac{f(x_1) - f(x_0)}{x_1 - x_0} = \dfrac{x_1^2 - x_0^2}{x_1 - x_0} = \dfrac{(x_1 + x_0)(x_1 - x_0)}{x_1 - x_0} = x_1 + x_0, \; x_1 \neq x_0$

m_{sec} approaches $2x_0$ as x_1 approaches x_0, thus $m_{tan} = 2x_0$ so

$|m_{tan} - m_{sec}| = |2x_0 - (x_1 + x_0)| = |x_0 - x_1| = |x_1 - x_0|.$

EXERCISE SET 3.2

1. $f'(x) = \lim\limits_{h \to 0} \dfrac{3(x + h)^2 - 3x^2}{h} = \lim\limits_{h \to 0} \dfrac{3(x^2 + 2xh + h^2) - 3x^2}{h}$

$= \lim\limits_{h \to 0} \dfrac{6xh + 3h^2}{h} = \lim\limits_{h \to 0} (6x + 3h) = 6x.$

3. $f'(x) = \lim\limits_{h \to 0} \dfrac{(x + h)^3 - x^3}{h} = \lim\limits_{h \to 0} \dfrac{x^3 + 3x^2 h + 3xh^2 + h^3 - x^3}{h}$

$= \lim\limits_{h \to 0} \dfrac{3x^2 h + 3xh^2 + h^3}{h} = \lim\limits_{h \to 0}(3x^2 + 3xh + h^2) = 3x^2.$

5. $f'(x) = \lim\limits_{h \to 0} \dfrac{\sqrt{x + h + 1} - \sqrt{x + 1}}{h}$

$= \lim\limits_{h \to 0} \dfrac{\sqrt{x + h + 1} - \sqrt{x + 1}}{h} \cdot \dfrac{\sqrt{x + h + 1} + \sqrt{x + 1}}{\sqrt{x + h + 1} + \sqrt{x + 1}}$

$= \lim\limits_{h \to 0} \dfrac{(x + h + 1) - (x + 1)}{h(\sqrt{x + h + 1} + \sqrt{x + 1})} = \lim\limits_{h \to 0} \dfrac{h}{h(\sqrt{x + h + 1} + \sqrt{x + 1})}$

$= \lim\limits_{h \to 0} \dfrac{1}{\sqrt{x + h + 1} + \sqrt{x + 1}} = \dfrac{1}{2\sqrt{x + 1}}.$

7. $f'(x) = \lim\limits_{h \to 0} \dfrac{\dfrac{1}{x + h} - \dfrac{1}{x}}{h} = \lim\limits_{h \to 0} \dfrac{\dfrac{x - (x + h)}{x(x + h)}}{h}$

$= \lim\limits_{h \to 0} \dfrac{-h}{hx(x + h)} = \lim\limits_{h \to 0} - \dfrac{1}{x(x + h)} = - \dfrac{1}{x^2}.$

9. $f'(x) = \lim_{h \to 0} \dfrac{[a(x + h)^2 + b] - [ax^2 + b]}{h}$

$= \lim_{h \to 0} \dfrac{ax^2 + 2axh + ah^2 + b - ax^2 - b}{h}$

$= \lim_{h \to 0} \dfrac{2axh + ah^2}{h} = \lim_{h \to 0} (2ax + ah) = 2ax.$

11. $f'(x) = \lim_{h \to 0} \dfrac{\dfrac{1}{\sqrt{x + h}} - \dfrac{1}{\sqrt{x}}}{h} = \lim_{h \to 0} \dfrac{\sqrt{x} - \sqrt{x + h}}{h\sqrt{x}\sqrt{x + h}}$

$= \lim_{h \to 0} \dfrac{x - (x + h)}{h\sqrt{x}\sqrt{x + h}(\sqrt{x} + \sqrt{x + h})} = \lim_{h \to 0} \dfrac{-1}{\sqrt{x}\sqrt{x + h}(\sqrt{x} + \sqrt{x + h})}$

$= -\dfrac{1}{2x^{3/2}}.$

13. $f'(3) = 6(3) = 18;\ f(3) = 3(3)^2 = 27$ so $y - 27 = 18(x - 3)$, $y = 18x - 27.$

15. $f'(0) = 3(0)^2 = 0;\ f(0) = 0^3 = 0$ so $y - 0 = (0)(x - 0),\ y = 0.$

17. $f'(8) = \dfrac{1}{2\sqrt{8 + 1}} = \dfrac{1}{6};\ f(8) = \sqrt{8 + 1} = 3$ so $y - 3 = \dfrac{1}{6}(x - 8)$,

$y = \dfrac{1}{6}x + \dfrac{5}{3}.$

19. (a) $\dfrac{dy}{dx} = \lim_{h \to 0} \dfrac{[4(x + h)^2 + 2] - [4x^2 + 2]}{h}$

$= \lim_{h \to 0} \dfrac{4x^2 + 8xh + 4h^2 + 2 - 4x^2 - 2}{h} = \lim_{h \to 0} (8x + 4h) = 8x.$

(b) $\dfrac{dy}{dx}\bigg|_{x=1} = 8(1) = 8.$

21. $f'(t) = \lim\limits_{h \to 0} \dfrac{f(t + h) - f(t)}{h}$

$= \lim\limits_{h \to 0} \dfrac{[4(t + h)^2 + (t + h)] - [4t^2 + t]}{h}$

$= \lim\limits_{h \to 0} \dfrac{4t^2 + 8th + 4h^2 + t + h - 4t^2 - t}{h}$

$= \lim\limits_{h \to 0} \dfrac{8th + 4h^2 + h}{h} = \lim\limits_{h \to 0} (8t + 4h + 1) = 8t + 1.$

23. $\dfrac{dA}{d\lambda} = \lim\limits_{h \to 0} \dfrac{[3(\lambda + h)^2 - (\lambda + h)] - [3\lambda^2 - \lambda]}{h}$

$= \lim\limits_{h \to 0} \dfrac{3\lambda^2 + 6\lambda h + 3h^2 - \lambda - h - 3\lambda^2 + \lambda}{h}$

$= \lim\limits_{h \to 0} \dfrac{6\lambda h + 3h^2 - h}{h} = \lim\limits_{h \to 0} (6\lambda + 3h - 1) = 6\lambda - 1.$

25. (a) D   (b) F   (c) B
     (d) C   (e) A   (f) E

27.   29.

31.

33.

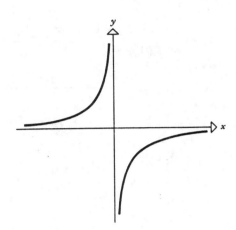

35. $\lim_{x \to 0} f(x) = \lim_{x \to 0} \sqrt[3]{x} = 0 = f(0),$

so f is continuous at x = 0.

$$\lim_{h \to 0} \frac{f(0 + h) - f(0)}{h} = \lim_{h \to 0} \frac{\sqrt[3]{h} - 0}{h}$$

$$= \lim_{h \to 0} \frac{1}{h^{2/3}} = +\infty,$$

so f'(0) does not exist.

37. $\lim_{x \to 1^-} f(x) = \lim_{x \to 1^+} f(x) = f(1),$ so

f is continuous at x = 1.

$$\lim_{h \to 0^-} \frac{f(1 + h) - f(1)}{h} = \lim_{h \to 0^-} \frac{[(1 + h)^2 + 1] - 2}{h}$$

$$= \lim_{h \to 0^-} (2 + h) = 2;$$

$$\lim_{h \to 0^+} \frac{f(1 + h) - f(1)}{h} = \lim_{h \to 0^+} \frac{2(1 + h) - 2}{h}$$

$$= \lim_{h \to 0^+} 2 = 2,$$

so f'(1) = 2.

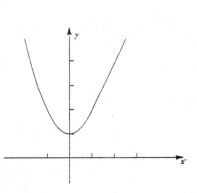

39. f is continuous at x = 1 because it is differentiable there, thus
$\lim_{h \to 0} f(1 + h) = f(1)$ and so $f(1) = 0$ because $\lim_{h \to 0} \dfrac{f(1 + h)}{h}$ exists.

$f'(1) = \lim_{h \to 0} \dfrac{f(1 + h) - f(1)}{h} = \lim_{h \to 0} \dfrac{f(1 + h)}{h} = 5.$

EXERCISE SET 3.3

1. $28x^6$

3. $24x^7 + 2$

5. 0

7. $-\dfrac{1}{3}(7x^6 + 2)$

9. $3ax^2 + 2bx + c$

11. $24x^{-9} + 1/\sqrt{x}$

13. $y = x^{-3} + x^{-7}$ so $\dfrac{dy}{dx} = -3x^{-4} - 7x^{-8}$

15. $\dfrac{dy}{dx} = (3x^2 + 6) \dfrac{d}{dx}(2x - \tfrac{1}{4}) + (2x - \tfrac{1}{4}) \dfrac{d}{dx}(3x^2 + 6)$

$= (3x^2 + 6)(2) + (2x - \tfrac{1}{4})(6x) = 18x^2 - \dfrac{3}{2}x + 12$

17. $\dfrac{dy}{dx} = (x^3 + 7x^2 - 8) \dfrac{d}{dx}(2x^{-3} + x^{-4}) + (2x^{-3} + x^{-4}) \dfrac{d}{dx}(x^3 + 7x^2 - 8)$

$= (x^3 + 7x^2 - 8)(-6x^{-4} - 4x^{-5}) + (2x^{-3} + x^{-4})(3x^2 + 14x)$

$= -15x^{-2} - 14x^{-3} + 48x^{-4} + 32x^{-5}$

19. $12x(3x^2 + 1)$

21. $\dfrac{dy}{dx} = -\dfrac{1}{(5x - 3)^2} \dfrac{d}{dx}(5x - 3)$

$= -\dfrac{5}{(5x - 3)^2}$

23. $\dfrac{dy}{dx} = \dfrac{(2x + 1) \dfrac{d}{dx}(3x) - (3x) \dfrac{d}{dx}(2x + 1)}{(2x + 1)^2}$

$= \dfrac{(2x + 1)(3) - (3x)(2)}{(2x + 1)^2} = \dfrac{3}{(2x + 1)^2}$

25. $\dfrac{dy}{dx} = \dfrac{(x+3)\frac{d}{dx}(2x-1) - (2x-1)\frac{d}{dx}(x+3)}{(x+3)^2}$

$= \dfrac{(x+3)(2) - (2x-1)(1)}{(x+3)^2} = \dfrac{7}{(x+3)^2}$

27. $\dfrac{dy}{dx} = (\dfrac{3x+2}{x}) \dfrac{d}{dx}(x^{-5}+1) + (x^{-5}+1)\dfrac{d}{dx}(\dfrac{3x+2}{x})$

$= (\dfrac{3x+2}{x})(-5x^{-6}) + (x^{-5}+1)\left[\dfrac{x(3)-(3x+2)(1)}{x^2}\right]$

$= (\dfrac{3x+2}{x})(-5x^{-6}) + (x^{-5}+1)(-\dfrac{2}{x^2})$

29. (a) $g'(x) = \sqrt{x}f'(x) + \dfrac{1}{2\sqrt{x}}f(x)$, $g'(4) = (2)(-5) + \dfrac{1}{4}(3) = -37/4$

(b) $g'(x) = \dfrac{xf'(x) - f(x)}{x^2}$, $g'(4) = \dfrac{(4)(-5)-3}{16} = -23/16$

31. $32t$ 　　　　　　　　　　33. $3\pi r^2$

35. $\dfrac{ds}{dt} = \dfrac{(t^3+7)\frac{d}{dt}(t) - t\frac{d}{dt}(t^3+7)}{(t^3+7)^2} = \dfrac{(t^3+7)(1) - t(3t^2)}{(t^3+7)^2} = \dfrac{7-2t^3}{(t^3+7)^2}$

37. $F = GmMr^{-2}$, $\dfrac{dF}{dr} = -2GmMr^{-3} = -\dfrac{2GmM}{r^3}$

39. (a) $dy/dx = 21x^2 - 10x + 1$, $d^2y/dx^2 = 42x - 10$

(b) $dy/dx = 24x - 2$, $d^2y/dx^2 = 24$

(c) $dy/dx = -1/x^2$, $d^2y/dx^2 = 2/x^3$

(d) $y = 35x^5 - 16x^3 - 3x$, $dy/dx = 175x^4 - 48x^2 - 3$,
$d^2y/dx^2 = 700x^3 - 96x$

41. (a) $y' = -5x^{-6} + 5x^4$, $y'' = 30x^{-7} + 20x^3$, $y''' = -210x^{-8} + 60x^2$

(b) $y = x^{-1}$, $y' = -x^{-2}$, $y'' = 2x^{-3}$, $y''' = -6x^{-4}$

(c) $y' = 3ax^2 + b$, $y'' = 6ax$, $y''' = 6a$

43.   (a) $f'(x) = 6x$, $f''(x) = 6$, $f'''(x) = 0$, $f'''(2) = 0$

(b) $\dfrac{dy}{dx} = 30x^4 - 8x$, $\dfrac{d^2y}{dx^2} = 120x^3 - 8$, $\dfrac{d^2y}{dx^2}\bigg|_{x=1} = 112$

(c) $\dfrac{d}{dx}[x^{-3}] = -3x^{-4}$, $\dfrac{d^2}{dx^2}[x^{-3}] = 12x^{-5}$, $\dfrac{d^3}{dx^3}[x^{-3}] = -60x^{-6}$,

$\dfrac{d^4}{dx^4}[x^{-3}] = 360x^{-7}$, $\dfrac{d^4}{dx^4}[x^{-3}]\bigg|_{x=1} = 360$

45. $y' = 3x^2 + 3$, $y'' = 6x$, and $y''' = 6$ so

$y''' + xy'' - 2y' = 6 + x(6x) - 2(3x^2 + 3) = 6 + 6x^2 - 6x^2 - 6 = 0$.

47.   The graph has a horizontal tangent at points where $\dfrac{dy}{dx} = 0$, but

$\dfrac{dy}{dx} = x^2 - 3x + 2 = (x - 1)(x - 2) = 0$ if $x = 1, 2$.   The corresponding

values of y are 5/6 and 2/3 so the tangent line is horizontal at (1,5/6) and (2,2/3).

49. $y - 2 = 5(x + 3)$, $y = 5x + 17$.

51. $m_{tan} = \dfrac{dy}{dx} = 2ax + b$ so $2a + b = 8$ when $x = 1$.   Also $5 = a + b$ because
(1,5) is on the curve.   Solve the pair of equations $2a + b = 8$ and
$a + b = 5$ to get $a = 3$ and $b = 2$.

53.   The points $(-1,1)$ and $(2,4)$ are on the secant line so its slope is
$(4 - 1)/(2 + 1) = 1$.   The slope of the tangent line to $y = x^2$ is $y' = 2x$
so $2x = 1$, $x = 1/2$.

55. $y' = -2x$, so at any point (x_0, y_0) on $y = 1 - x^2$ the tangent line is

$y - y_0 = -2x_0(x - x_0)$, or $y = -2x_0 x + x_0^2 + 1$.   The point (2,0) is to be

on the line, so $0 = -4x_0 + x_0^2 + 1$, $x_0^2 - 4x_0 + 1 = 0$.   Use the

quadratic formula to get $x_0 = \dfrac{4 \pm \sqrt{16 - 4}}{2} = 2 \pm \sqrt{3}$.

57. $y' = 3ax^2 + b$; the tangent line at $x = x_0$ is

$y - y_0 = (3ax_0^2 + b)(x - x_0)$ where $y_0 = ax_0^3 + bx_0$. Solve with

$y = ax^3 + bx$ to get

$$(ax^3 + bx) - (ax_0^3 + bx_0) = (3ax_0^2 + b)(x - x_0)$$

$$ax^3 + bx - ax_0^3 - bx_0 = 3ax_0^2x - 3ax_0^3 + bx - bx_0$$

$$x^3 - 3x_0^2x + 2x_0^3 = 0$$

$$(x - x_0)(x^2 + 2x_0 - 2x_0^2) = 0$$

$$(x - x_0)^2(x + 2x_0) = 0.$$

so $x = -2x_0$.

59. $y' = -\dfrac{1}{x^2}$; the tangent line at $x = x_0$ is $y - y_0 = -\dfrac{1}{x_0^2}(x - x_0)$, or

$y = -\dfrac{x}{x_0^2} + \dfrac{2}{x_0}$. The tangent line crosses the x-axis at $2x_0$, the y-axis

at $2/x_0$, so that the area of the triangle is $\dfrac{1}{2}(2/x_0)(2x_0) = 2$.

61. $(f - g)'(x) = (f + (-1)g)'(x)$
$\qquad\qquad\quad = f'(x) + ((-1)g)'(x) \qquad$ by Theorem 3.3.4
$\qquad\qquad\quad = f'(x) + (-1)g'(x) \qquad$ by Theorem 3.3.3
$\qquad\qquad\quad = f'(x) - g'(x)$
$\qquad\qquad\quad = (f' - g')(x)$ so $(f - g)' = f' - g'$

63. (a) $2(1 + x^{-1})(x^{-3} + 7) + (2x + 1)(-x^{-2})(x^{-3} + 7)$
$\qquad\quad + (2x + 1)(1 + x^{-1})(-3x^{-4})$

$\quad$ (b) $-5x^{-6}(x^2 + 2x)(4 - 3x)(2x^9 + 1) + x^{-5}(2x + 2)(4 - 3x)(2x^9 + 1)$
$\qquad\quad + x^{-5}(x^2 + 2x)(-3)(2x^9 + 1) + x^{-5}(x^2 + 2x)(4 - 3x)(18x^8)$

(c) $(x^7 + 2x - 3)^3 = (x^7 + 2x - 3)(x^7 + 2x - 3)(x^7 + 2x - 3)$ so

$\frac{d}{dx}(x^7 + 2x - 3)^3 = (7x^6 + 2)(x^7 + 2x - 3)(x^7 + 2x - 3)$

$+ (x^7 + 2x - 3)(7x^6 + 2)(x^7 + 2x - 3)$

$+ (x^7 + 2x - 3)(x^7 + 2x - 3)(7x^6 + 2)$

$= 3(7x^6 + 2)(x^7 + 2x - 3)^2$

(d) $(x^2 + 1)^{50} = (x^2 + 1)(x^2 + 1)\cdots(x^2 + 1)$, where $(x^2 + 1)$ occurs 50 times so

$\frac{d}{dx}(x^2 + 1)^{50} = [(2x)(x^2 + 1)\cdots(x^2 + 1)] + [(x^2 + 1)(2x)\cdots$

$(x^2 + 1)] + \cdots + [(x^2 + 1)(x^2 + 1)]\cdots(2x)$

$= 2x(x^2 + 1)^{49} + 2x(x^2 + 1)^{49} + \cdots + 2x(x^2 + 1)^{49}$

$= 100x(x^2 + 1)^{49}$ because $2x(x^2 + 1)^{49}$

occurs 50 times.

65. $2(2x^3 - 5x^2 + 7x - 2)(6x^2 - 10x + 7)$

67.   f is continuous at 1 because $\lim\limits_{x \to 1^-} f(x) = \lim\limits_{x \to 1^+} f(x) = f(1)$,

also $\lim\limits_{x \to 1^-} f'(x) = \lim\limits_{x \to 1^-} 2x = 2$ and $\lim\limits_{x \to 1^+} f'(x) = \lim\limits_{x \to 1^+} \frac{1}{2\sqrt{x}} = \frac{1}{2}$ so f is not differentiable at 1.

69.   (a) $f(x) = 3x - 2$ if $x \geq 2/3$, $f(x) = -3x + 2$ if $x < 2/3$ so f is differentiable everywhere except perhaps at 2/3. f is continuous at 2/3, also $\lim\limits_{x \to 2/3^-} f'(x) = \lim\limits_{x \to 2/3^-} (-3) = -3$ and

$\lim\limits_{x \to 2/3^+} f'(x) = \lim\limits_{x \to 2/3^+} (3) = 3$ so f is not differentiable at $x = 2/3$.

(b) $f(x) = x^2 - 4$ if $|x| \geq 2$, $f(x) = -x^2 + 4$ if $|x| < 2$ so f is differentiable everywhere except perhaps at ± 2. f is continuous at -2 and 2, also $\lim\limits_{x \to 2^-} f'(x) = \lim\limits_{x \to 2^-} (-2x) = -4$ and

$\lim\limits_{x \to 2^+} f'(x) = \lim\limits_{x \to 2^+} (2x) = 4$ so f is not differentiable at $x = 2$.

Similarly, f is not differentiable at $x = -2$.

71. (a) $f'(x) = nx^{n-1}$, $f''(x) = n(n-1)x^{n-2}$,
$f'''(x) = n(n-1)(n-2)x^{n-3}, \ldots,$
$f^{(n)}(x) = n(n-1)(n-2)\cdots 1$.

(b) From part (a), $f^{(k)}(x) = k(k-1)(k-2)\cdots 1$ so $f^{(k+1)}(x) = 0$ thus $f^{(n)}(x) = 0$ if $n > k$.

(c) From parts (a) and (b), $f^{(n)}(x) = a_n n(n-1)(n-2)\cdots 1$.

73. (a) $\dfrac{d^2}{dx^2}[cf(x)] = \dfrac{d}{dx}\left[\dfrac{d}{dx}[cf(x)]\right] = \dfrac{d}{dx}\left[c\,\dfrac{d}{dx}[f(x)]\right]$

$$= c\,\dfrac{d}{dx}\left[\dfrac{d}{dx}[f(x)]\right] = c\,\dfrac{d^2}{dx^2}[f(x)]$$

$$\dfrac{d^2}{dx^2}[f(x)+g(x)] = \dfrac{d}{dx}\left[\dfrac{d}{dx}[f(x)+g(x)]\right]$$

$$= \dfrac{d}{dx}\left[\dfrac{d}{dx}[f(x)] + \dfrac{d}{dx}[g(x)]\right]$$

$$= \dfrac{d^2}{dx^2}[f(x)] + \dfrac{d^2}{dx^2}[g(x)]$$

(b) yes, by repeated application of the procedure illustrated in part (a).

75. (a) The argument is based on the assumption that $\lim_{h\to 0} f(x+h) = f(\lim_{h\to 0}(x+h))$ which is not necessarily true.

(b) If $h \neq 0$, then $1+h \neq 1$ thus $f(1+h) = 1+h$ so $\lim_{h\to 0} f(1+h) = \lim_{h\to 0}(1+h) = 1$. But $f(1) = 3$ so $\lim_{h\to 0} f(1+h) \neq f(1)$.

77. (a) If a function is differentiable at a point then it is continuous at that point, thus f' is continuous on (a,b) and consequently so is f.

(b) f and all its derivatives up to $f^{(n-1)}(x)$ are continuous on (a,b).

EXERCISE SET 3.4

1. $-2\sin x - 3\cos x$

3. $f'(x) = \dfrac{x(\cos x) - \sin x(1)}{x^2} = \dfrac{x\cos x - \sin x}{x^2}$

5. $f'(x) = x^3(\cos x) + (\sin x)(3x^2) - 5(-\sin x) = x^3\cos x + (3x^2 + 5)\sin x$

7. $\sec x \tan x - \sqrt{2}\sec^2 x$

9. $f'(x) = \sec x\,(\sec^2 x) + (\tan x)(\sec x \tan x) = \sec^3 x + \sec x \tan^2 x$

11. $1 + 4\csc x \cot x - 2\csc^2 x$

13. $f'(x) = \dfrac{(1 + \csc x)(-\csc^2 x) - \cot x(0 - \csc x \cot x)}{(1 + \csc x)^2}$

    $= \dfrac{\csc x(-\csc x - \csc^2 x + \cot^2 x)}{(1 + \csc x)^2}$

    but $1 + \cot^2 x = \csc^2 x$ (identity) thus $\cot^2 x - \csc^2 x = -1$

    so $f'(x) = \dfrac{\csc x(-\csc x - 1)}{(1 + \csc x)^2} = -\,\dfrac{\csc x}{1 + \csc x}$

15. $f(x) = \sin^2 x + \cos^2 x = 1$ (identity)   so $f'(x) = 0$

17. $f(x) = \dfrac{\tan x}{1 + x \tan x}$ (because $\sin x \sec x = (\sin x)(\frac{1}{\cos x}) = \tan x$),

    $f'(x) = \dfrac{(1 + x \tan x)(\sec^2 x) - \tan x[x(\sec^2 x) + (\tan x)(1)]}{(1 + x \tan x)^2}$

    $= \dfrac{\sec^2 x - \tan^2 x}{(1 + x \tan x)^2} = \dfrac{1}{(1 + x \tan x)^2}$ (because $\sec^2 x - \tan^2 x = 1$)

19. $dy/dx = -x\sin x + \cos x,$
    $d^2y/dx^2 = -x\cos x - \sin x - \sin x = -x\cos x - 2\sin x$

21. $dy/dx = x(\cos x) + (\sin x)(1) - 3(-\sin x) = x \cos x + 4 \sin x$,

$d^2y/dx^2 = x(-\sin x) + (\cos x)(1) + 4 \cos x = -x \sin x + 5 \cos x$

23. $dy/dx = (\sin x)(-\sin x) + (\cos x)(\cos x) = \cos^2 x - \sin^2 x$,

$d^2y/dx^2 = (\cos x)(-\sin x) + (\cos x)(-\sin x) - [(\sin x)(\cos x)$

$+ (\sin x)(\cos x)]$

$= -4 \sin x \cos x$

25. (a) $f'(x) = - \sin x$; $f'(x) = 0$ when $\sin x = 0$ so $x = n\pi$,
     $n = 0, \pm 1, \pm 2, \cdots$.

(b) $f'(x) = - \csc^2 x$; $f'(x) = 0$ when $\csc x = 0$, but $\csc x = 0$ has no
     solutions.

(c) $f'(x) = - \csc x \cot x$; $f'(x) = 0$ when $\cot x = 0$ so $x = \dfrac{\pi}{2} + n\pi$,
     $n = 0, \pm 1, \pm 2, \cdots$.

27. Let $f(x) = \tan x$, then $f'(x) = \sec^2 x$.

(a) $f(0) = 0$ and $f'(0) = 1$ so $y - 0 = (1)(x - 0)$, $y = x$.

(b) $f(\dfrac{\pi}{4}) = 1$ and $f'(\dfrac{\pi}{4}) = 2$ so $y - 1 = 2(x - \dfrac{\pi}{4})$, $y = 2x - \dfrac{\pi}{2} + 1$.

(c) $f(- \dfrac{\pi}{4}) = - 1$ and $f'(- \dfrac{\pi}{4}) = 2$ so $y + 1 = 2(x + \dfrac{\pi}{4})$, $y = 2x + \dfrac{\pi}{2} - 1$.

29. In each part, f is differentiable throughout its domain because f' can
     be determined there by use of the various derivative formulas that have
     been presented in the text; f is not differentiable elsewhere.
     (a) all x   (b) all x
     (c) $x \neq \pi/2 + n\pi$, $n = 0, \pm 1, \pm 2, \cdots$ (d) $x \neq n\pi$, $n = 0, \pm 1, \pm 2, \cdots$
     (e) $x \neq \pi/2 + n\pi$, $n = 0, \pm 1, \pm 2, \cdots$ (f) $x \neq n\pi$, $n = 0, \pm 1, \pm 2, \cdots$
     (g) $x \neq \pi + 2n\pi$, $n = 0, \pm 1, \pm 2, \cdots$ (h) $x \neq n\pi/2$, $n = 0, \pm 1, \pm 2, \cdots$
     (i) all x

31. $f'(x) = -\sin x$, $f''(x) = -\cos x$, $f'''(x) = \sin x$, and $f^{(4)}(x) = \cos x$ with
     higher order derivatives repeating this pattern, so $f^{(n)}(x) = \sin x$ for
     $n = 3, 7, 11, \cdots$

33. $\displaystyle \lim_{x \to 0} \frac{\tan(x + y) - \tan y}{x} = \lim_{h \to 0} \frac{\tan(y + h) - \tan y}{h} = \frac{d}{dy}(\tan y) = \sec^2 y$

35. Let t be the radian measure, then $h = \frac{180}{\pi} t$ and $\cos h = \cos t$,
    $\sin h = \sin t$.

    (a) $\lim\limits_{h \to 0} \dfrac{\cos h - 1}{h} = \lim\limits_{t \to 0} \dfrac{\cos t - 1}{180 t/\pi} = \dfrac{\pi}{180} \lim\limits_{t \to 0} \dfrac{\cos t - 1}{t} = 0.$

    (b) $\lim\limits_{h \to 0} \dfrac{\sin h}{h} = \lim\limits_{t \to 0} \dfrac{\sin t}{180 t/\pi} = \dfrac{\pi}{180} \lim\limits_{t \to 0} \dfrac{\sin t}{t} = \dfrac{\pi}{180}.$

    (c) $\dfrac{d}{dx} [\sin x] = \sin x \lim\limits_{h \to 0} \dfrac{\cos h - 1}{h} + \cos x \lim\limits_{h \to 0} \dfrac{\sin h}{h}$

    $\qquad = \sin x(0) + \cos x(\pi/180) = \dfrac{\pi}{180} \cos x.$

EXERCISE SET 3.5

1. $f'(x) = 37(x^3 + 2x)^{36} \dfrac{d}{dx} (x^3 + 2x) = 37(x^3 + 2x)^{36}(3x^2 + 2)$

3. $f'(x) = -2(x^3 - \dfrac{7}{x})^{-3} \dfrac{d}{dx} (x^3 - \dfrac{7}{x}) = -2(x^3 - \dfrac{7}{x})^{-3}(3x^2 + \dfrac{7}{x^2})$

5. $f(x) = 4(3x^2 - 2x + 1)^{-3},$
    $f'(x) = -12(3x^2 - 2x + 1)^{-4}\dfrac{d}{dx}(3x^2 - 2x + 1)$

    $\qquad = -12(3x^2 - 2x + 1)^{-4}(6x - 2) = \dfrac{24(1 - 3x)}{(3x^2 - 2x + 1)^4}$

7. $f'(x) = \dfrac{1}{2\sqrt{4 + 3\sqrt{x}}} \dfrac{d}{dx} (4 + 3\sqrt{x}) = \dfrac{3}{4\sqrt{x}\sqrt{4 + 3\sqrt{x}}}$

9. $f'(x) = \cos(x^3) \dfrac{d}{dx} (x^3) = 3x^2\cos(x^3)$

11. $f'(x) = \sec^2(4x^2) \dfrac{d}{dx} (4x^2) = 8x\sec^2(4x^2)$

13. $f'(x) = 20 \cos^4 x \dfrac{d}{dx} (\cos x) = 20 \cos^4 x(-\sin x) = -20 \cos^4 x \sin x$

15. $f'(x) = \cos(1/x^2) \frac{d}{dx} (1/x^2) = -\frac{2}{x^3} \cos(1/x^2)$

17. $f'(x) = 4 \sec(x^7) \frac{d}{dx} [\sec(x^7)] = 4 \sec(x^7)\sec(x^7)\tan(x^7) \frac{d}{dx} (x^7)$

$= 28x^6 \sec^2(x^7)\tan(x^7)$

19. $f'(x) = \dfrac{1}{2\sqrt{\cos(5x)}} \dfrac{d}{dx} [\cos(5x)] = -\dfrac{5\,\sin(5x)}{2\sqrt{\cos(5x)}}$

21. $f'(x) = -3[x + \csc(x^3 + 3)]^{-4} \frac{d}{dx} [x + \csc(x^3 + 3)]$

$= -3[x + \csc(x^3 + 3)]^{-4}[1 - \csc(x^3 + 3)\cot(x^3 + 3) \frac{d}{dx} (x^3 + 3)]$

$= -3[x + \csc(x^3 + 3)]^{-4}[1 - 3x^2\csc(x^3 + 3)\cot(x^3 + 3)]$

23. $f'(x) = x^2 \cdot \dfrac{-2x}{2\sqrt{5 - x^2}} + 2x\sqrt{5 - x^2} = \dfrac{x(10 - 3x^2)}{\sqrt{5 - x^2}}$

25. $f'(x) = x^3(2 \sin 5x) \frac{d}{dx} (\sin 5x) + 3x^2 \sin^2 5x$

$= 10x^3 \sin 5x \cos 5x + 3x^2 \sin^2 5x$

27. $f'(x) = x^5 \sec(\frac{1}{x})\tan(\frac{1}{x}) \frac{d}{dx} (\frac{1}{x}) + \sec(\frac{1}{x})(5x^4)$

$= x^5 \sec(\frac{1}{x})\tan(\frac{1}{x})(-\frac{1}{x^2}) + 5x^4 \sec(\frac{1}{x})$

$= -x^3 \sec(\frac{1}{x})\tan(\frac{1}{x}) + 5x^4 \sec(\frac{1}{x})$

29. $f'(x) = -\sin(\cos x) \frac{d}{dx} (\cos x) = -\sin(\cos x)(-\sin x)$

$= \sin(\cos x)\sin x$

31. $f'(x) = 3\cos^2(\sin 2x) \frac{d}{dx} [\cos(\sin 2x)]$

$= 3\cos^2(\sin 2x)[-\sin(\sin 2x)] \frac{d}{dx} (\sin 2x)$

$= -6\cos^2(\sin 2x)\sin(\sin 2x)\cos 2x$

33. $f'(x) = (5x + 8)^{13} 12(x^3 + 7x)^{11} \dfrac{d}{dx} (x^3 + 7x)$

$\qquad + (x^3 + 7x)^{12} 13(5x + 8)^{12} \dfrac{d}{dx} (5x + 8)$

$\qquad = 12(5x + 8)^{13}(x^3 + 7x)^{11}(3x^2 + 7) + 65(x^3 + 7x)^{12}(5x + 8)^{12}$

35. $f'(x) = 3\left[\dfrac{x - 5}{2x + 1}\right]^2 \dfrac{d}{dx} \left[\dfrac{x - 5}{2x + 1}\right] = 3\left[\dfrac{x - 5}{2x + 1}\right]^2 \cdot \dfrac{11}{(2x + 1)^2} = \dfrac{33(x - 5)^2}{(2x + 1)^4}$

37. $f'(x) = \dfrac{(2x + 1)^{-3}(-8)(4x^2 - 1)^{-9}(8x) - (4x^2 - 1)^{-8}(-3)(2x + 1)^{-4}(2)}{(2x + 1)^{-6}}$

$\qquad = \dfrac{-64x(2x + 1)^{-3}(4x^2 - 1)^{-9} + 6\,(4x^2 - 1)^{-8}(2x + 1)^{-4}}{(2x + 1)^{-6}}$

39. $f'(x) = 5[x \sin 2x \tan^4 (x^7)]^4 \dfrac{d}{dx} [x \sin 2x \tan^4 (x^7)]$

$\qquad = 5[x \sin 2x \tan^4 (x^7)]^4 [x \cos 2x \dfrac{d}{dx} (2x) + \sin 2x$

$\qquad\qquad + 4 \tan^3 (x^7) \dfrac{d}{dx} \tan(x^7)]$

$\qquad = 5[x \sin 2x \tan^4 (x^7)]^4 [2x \cos 2x + \sin 2x + 28x^6 \tan^3(x^7)\sec^2(x^7)]$

41. $\dfrac{dy}{dx} = x(-\sin(5x)) \dfrac{d}{dx} (5x) + \cos(5x) - 2 \sin x \dfrac{d}{dx} (\sin x)$

$\qquad = -5x \sin(5x) + \cos(5x) - 2 \sin x \cos x$

$\qquad = -5x \sin(5x) + \cos(5x) - \sin(2x),$

$\dfrac{d^2y}{dx^2} = -5x \cos(5x) \dfrac{d}{dx} (5x) - 5 \sin(5x) - \sin(5x) \dfrac{d}{dx} (5x) - \cos(2x) \dfrac{d}{dx} (2x)$

$\qquad = -25x \cos(5x) - 10 \sin(5x) - 2 \cos(2x)$

43. $\dfrac{dy}{dx} = -3x \sin 3x + \cos 3x;$ if $x = \pi$ then $y = -\pi$ and $\dfrac{dy}{dx} = -1$

so $y + \pi = - (x - \pi),$ $y = -x.$

45. $\dfrac{dy}{dx} = -3 \sec^3(\pi/2 - x) \tan(\pi/2 - x);$ if $x = -\pi/2$ then $y = -1$

and $\dfrac{dy}{dx} = 0$ so $y + 1 = (0)(x + \pi/2),$ $y = -1.$

47. (I) Use $\dfrac{dy}{dt} = \dfrac{dy}{dx}\dfrac{dx}{dt}$ to get $\dfrac{dy}{dx} = \dfrac{dy/dt}{dx/dt} = \dfrac{2t}{5}$, but

$x = 5t + 2$ gives $t = (x - 2)/5$ so $\dfrac{dy}{dx} = \dfrac{2}{25}(x - 2)$.

(II) Use $t = (x - 5)/5$ to get $y = (x - 2)^2/25$, so $\dfrac{dy}{dx} = \dfrac{2}{25}(x - 2)$.

49. (I) $\dfrac{dy}{du} = \dfrac{dy}{dx}\dfrac{dx}{du}$ so $\dfrac{dy}{dx} = \dfrac{dy/du}{dx/du} = \dfrac{9\sin^2 u \cos u}{-u^{-2}} = -9u^2\sin^2 u \cos u$,

but $u = 1/x$ so $dy/dx = -9\sin^2(1/x)\cos(1/x)/x^2$.

(II) $y = 3\sin^3(1/x)$ because $u = 1/x$, so dy/dx is the same as in part (I).

51. $y = \cot^3(\pi - \theta) = -\cot^3\theta$ so $dy/dx = 3\cot^2\theta\,\csc^2\theta$.

53. $\dfrac{d}{d\omega}[a\cos^2\pi\omega + b\sin^2\pi\omega] = -2\pi a\cos\pi\omega\sin\pi\omega + 2\pi b\sin\pi\omega\cos\pi\omega$

$\qquad\qquad = \pi(b - a)(2\sin\pi\omega\cos\pi\omega) = \pi(b - a)\sin 2\pi\omega$

55. (a) If $x > 0$ then $|x| = x$ so $d(|x|)/dx = 1$, if $x < 0$ then $|x| = -x$ so $d(|x|)/dx = -1$.

(b) With $u = \sin x$, $\dfrac{d}{dx}(|\sin x|) = \dfrac{d}{dx}(|u|) = \dfrac{d}{du}(|u|)\dfrac{du}{dx}$

$= \dfrac{d}{du}(|u|)\cos x = \begin{cases} \cos x, & u > 0 \\ -\cos x, & u < 0 \end{cases} = \begin{cases} \cos x, & \sin x > 0 \\ -\cos x, & \sin x < 0 \end{cases}$

$= \begin{cases} \cos x, & 0 < x < \pi \\ -\cos x, & -\pi < x < 0. \end{cases}$

57. (a) For $x \neq 0$, $f'(x) = x(\cos\tfrac{1}{x})(-\tfrac{1}{x^2}) + \sin\tfrac{1}{x} = -\tfrac{1}{x}\cos\tfrac{1}{x} + \sin\tfrac{1}{x}$

(b) $\lim\limits_{x\to 0} x\sin\tfrac{1}{x} = 0 = f(0)$

(c) $\lim\limits_{h\to 0}\dfrac{f(0 + h) - f(0)}{h} = \lim\limits_{h\to 0}\dfrac{h\sin\tfrac{1}{h}}{h} = \lim\limits_{h\to 0}\sin\tfrac{1}{h}$, which does not exist

59. $(f \circ g)'(x) = f'(g(x))g'(x)$ so $(f \circ g)'(0) = f'(g(0))g'(0)$
$$= f'(0)(3) = (2)(3) = 6.$$

61. $F'(x) = f'(g(x))g'(x) = f'(\sqrt{3x - 1})\ \dfrac{3}{2\sqrt{3x - 1}}$

$$= \frac{\sqrt{3x - 1}}{(3x - 1) + 1}\ \frac{3}{2\sqrt{3x - 1}} = \frac{1}{2x}$$

63. $\dfrac{d}{dx}[f(3x)] = f'(3x)\dfrac{d}{dx}(3x) = 3f'(3x) = 6x$, so $f'(3x) = 2x$. Let $z = 3x$
to get $f'(z) = \dfrac{2}{3} z$; $\dfrac{d}{dx}[f(x)] = f'(x) = \dfrac{2}{3} x$.

65. $\dfrac{d}{dx}[f(g(h(x)))] = \dfrac{d}{dx}[f(g(u))],\quad u = h(x)$

$$= \frac{d}{du}[f(g(u))]\frac{du}{dx} = f'(g(u))g'(u)\frac{du}{dx}$$
$$= f'(g(h(x)))g'(h(x))h'(x)$$

EXERCISE SET 3.6

1. $y = (2x - 5)^{1/3}$; $dy/dx = \dfrac{2}{3}(2x - 5)^{-2/3}$

3. $dy/dx = \dfrac{3}{2}\left[\dfrac{x - 1}{x + 2}\right]^{1/2}\dfrac{d}{dx}\left[\dfrac{x - 1}{x + 2}\right] = \dfrac{9}{2(x + 2)^2}\left[\dfrac{x - 1}{x + 2}\right]^{1/2}$

5. $dy/dx = x^3(-\dfrac{2}{3})(5x^2 + 1)^{-5/3}(10x) + 3x^2(5x^2 + 1)^{-2/3}$

$$= \frac{1}{3} x^2(5x^2 + 1)^{-5/3}(25x^2 + 9)$$

7. $dy/dx = \dfrac{5}{2}[\sin(3/x)]^{3/2}[\cos(3/x)](-3/x^2)$

$$= -\ \frac{15[\sin(3/x)]^{3/2}\cos(3/x)}{2x^2}$$

9. $dy/dx = \sec^2[(2x - 1)^{-1/3}](-\frac{1}{3})(2x - 1)^{-4/3}(2)$

$= -\frac{2}{3}(2x - 1)^{-4/3}\sec^2[(2x - 1)^{-1/3}]$

11. $2x + 2y\frac{dy}{dx} = 0$ so $\frac{dy}{dx} = -\frac{x}{y}$

13. $x^2\frac{dy}{dx} + 2xy + 3x(3y^2)\frac{dy}{dx} + 3y^3 - 1 = 0$

$(x^2 + 9xy^2)\frac{dy}{dx} = 1 - 2xy - 3y^3$ so $\frac{dy}{dx} = \frac{1 - 2xy - 3y^3}{x^2 + 9xy^2}$

15. $-\frac{1}{y^2}\frac{dy}{dx} - \frac{1}{x^2} = 0$ so $\frac{dy}{dx} = -\frac{y^2}{x^2}$

17. $\frac{1}{2\sqrt{x}} + \frac{1}{2\sqrt{y}}\frac{dy}{dx} = 0$ so $\frac{dy}{dx} = -\frac{\sqrt{y}}{\sqrt{x}}$

19. $35(x^2 + 3y^2)^{34}(2x + 6y\frac{dy}{dx}) = 1$

$70x(x^2 + 3y^2)^{34} + 210y(x^2 + 3y^2)^{34}\frac{dy}{dx} = 1$ so $\frac{dy}{dx} = \frac{1 - 70x(x^2 + 3y^2)^{34}}{210y(x^2 + 3y^2)^{34}}$

21. $3x\frac{dy}{dx} + 3y = \frac{3}{2}(x^3 + y^2)^{1/2}(3x^2 + 2y\frac{dy}{dx})$

$[3x - 3y(x^3 + y^2)^{1/2}]\frac{dy}{dx} = \frac{9}{2}x^2(x^3 + y^2)^{1/2} - 3y$

so $\frac{dy}{dx} = \frac{(3/2)x^2(x^3 + y^2)^{1/2} - y}{x - y(x^3 + y^2)^{1/2}}$

23. $\cos(x^2y^2)[x^2(2y)\frac{dy}{dx} + 2xy^2] = 1$, $\frac{dy}{dx} = \frac{1 - 2xy^2\cos(x^2y^2)}{2x^2y\cos(x^2y^2)}$

25. $3\tan^2(xy^2 + y)\sec^2(xy^2 + y)(2xy\frac{dy}{dx} + y^2 + \frac{dy}{dx}) = 1$

so $\frac{dy}{dx} = \frac{1 - 3y^2\tan^2(xy^2 + y)\sec^2(xy^2 + y)}{3(2xy + 1)\tan^2(xy^2 + y)\sec^2(xy^2 + y)}$

27. $\frac{1}{2}[1 + \sin^3(xy^2)]^{-1/2}[3\sin^2(xy^2)][\cos(xy^2)](2xy\frac{dy}{dx} + y^2) = \frac{dy}{dx}$,

   mulitiply through by $2\sqrt{1 + \sin^3(xy^2)}$ and solve for $\frac{dy}{dx}$

   to get $\frac{dy}{dx} = \dfrac{3y^2\sin^2(xy^2)\cos(xy^2)}{2\sqrt{1 + \sin^3(xy^2)} - 6xy\sin^2(xy^2)\cos(xy^2)}$

29. $\frac{dy}{dx} = -\dfrac{3x^2y + y^3}{x^3 + 3y^2x}$; $\left.\frac{dy}{dx}\right|_{(1,2)} = -\frac{14}{13}$

31. $\frac{dy}{dx} = \dfrac{2y^{1/3}}{2x^{1/3} + 3x^{1/3}y^{1/3}}$; $\left.\frac{dy}{dx}\right|_{(1,-1)} = 2$

33. If $xy = 8$, then $y = 8/x$ so $dy/dx = -8/x^2$. By implicit differentiation
   we get $dy/dx = -y/x$. In both cases, $\left.dy/dx\right|_{(2,4)} = -2$.

35. If $x^2 + y^2 = 1$, then $y = -\sqrt{1 - x^2}$ goes through the point $(1/\sqrt{2}, -1/\sqrt{2})$
   so $dy/dx = x/\sqrt{1 - x^2}$. By implicit differentiation we get $dy/dx = -x/y$.
   In both cases, $\left.dy/dx\right|_{(1/\sqrt{2},-1/\sqrt{2})} = 1$.

37. Apply the quadratic formula to $y^2 - 3xy + 2x^2 - 4 = 0$ to get
   $y = (3x \pm \sqrt{9x^2 - 4(2x^2 - 4)})/2 = (3x \pm \sqrt{x^2 + 16})/2$, of which only
   $y = (3x - \sqrt{x^2 + 16})/2$ contains the point $(3,2)$ so
   $dy/dx = [3 - x(x^2 + 16)^{-1/2}]/2$. By implicit differentiation we get
   $dy/dx = (3y - 4x)/(2y - 3x)$. In both cases, $\left.dy/dx\right|_{(3,2)} = 6/5$.

39. $\dfrac{dy}{dx} = -\dfrac{x^2}{y^2}$,

$$\dfrac{d^2y}{dx^2} = -\dfrac{y^2(2x) - x^2(2y\,dy/dx)}{y^4} = -\dfrac{2xy^2 - 2x^2y(-x^2/y^2)}{y^4} = -\dfrac{2x(y^3 + x^3)}{y^5},$$

but $x^3 + y^3 = 1$ so $\dfrac{d^2y}{dx^2} = -\dfrac{2x}{y^5}$.

41. $\dfrac{dy}{dx} = \dfrac{y}{y - x}$,

$$\dfrac{d^2y}{dx^2} = \dfrac{(y - x)(dy/dx) - y(dy/dx - 1)}{(y - x)^2} = \dfrac{(y - x)(\frac{y}{y - x}) - y\,(\frac{y}{y - x} - 1)}{(y - x)^2}$$

$$= \dfrac{y^2 - 2xy}{(y - x)^3} \text{ but } y^2 - 2xy = -3 \text{ , so } \dfrac{d^2y}{dx^2} = -\dfrac{3}{(y - x)^3}$$

43. $\dfrac{dy}{dx} = \dfrac{\cos y}{1 + x\sin y}$,

$$\dfrac{d^2y}{dx^2} = \dfrac{(1 + x\sin y)(-\sin y)(dy/dx) - (\cos y)[(x\cos y)(dy/dx) + \sin y]}{(1 + x\sin y)^2}$$

$$= -\dfrac{2\sin y\cos y + (x\cos y)(2\sin^2 y + \cos^2 y)}{(1 + x\sin y)^3},$$

but $x\cos y = y$, $2\sin y\cos y = \sin 2y$, and $\sin^2 y + \cos^2 y = 1$ so

$$\dfrac{d^2y}{dx^2} = -\dfrac{\sin 2y + y(\sin^2 y + 1)}{(1 + x\sin y)^3} .$$

45. $4a^3 \dfrac{da}{dt} - 4t^3 = 6(a^2 + 2at\dfrac{da}{dt})$, solve for $\dfrac{da}{dt}$ to get $\dfrac{da}{dt} = \dfrac{2t^3 + 3a^2}{2a^3 - 6at}$

47. $2a^2\omega \dfrac{d\omega}{d\lambda} + 2b^2\lambda = 0$ so $\dfrac{d\omega}{d\lambda} = -\dfrac{b^2\lambda}{a^2\omega}$

49. By the chain rule, $\frac{dy}{dx} = \frac{dy}{dt}\frac{dt}{dx}$. Use implicit differentiation on

$2y^3t + t^3y = 1$ to get $\frac{dy}{dt} = -\frac{2y^3 + 3t^2y}{6ty^2 + t^3}$, but $\frac{dt}{dx} = \frac{1}{\cos t}$

so $\frac{dy}{dx} = -\frac{2y^3 + 3t^2y}{(6ty^2 + t^3)\cos t}$.

51. Let $P(x_0, y_0)$ be a point where a line through the orgin is tangent to the

curve $x^2 - 4x + y^2 + 3 = 0$. Implicit differentiation applied to the
equation of the curve gives $dy/dx = (2 - x)/y$. At P the slope of the
curve must equal the slope of the line so $(2 - x_0)/y_0 = y_0/x_0$, or

$y_0^2 = 2x_0 - x_0^2$. But $x_0^2 - 4x_0 + y_0^2 + 3 = 0$ because (x_0, y_0) is on the

curve, and elimination of y_0^2 in the latter two equations gives

$x_0^2 - 4x_0 + (2x_0 - x_0^2) + 3 = 0$, $x_0 = 3/2$ which when substituted into

$y_0^2 = 2x_0 - x_0^2$ yields $y_0^2 = 3/4$, so $y_0 = \pm \sqrt{3}/2$. The slopes of the lines

are $(\pm\sqrt{3}/2)/(3/2) = \pm\sqrt{3}/3$ and their equations are $y = (\sqrt{3}/3)x$ and
$y = -(\sqrt{3}/3)x$.

53. (a) The equation does not define y as an implicit function of x at the
         points where the curve crosses the x-axis. If $y = 0$ then
         $8x^4 = 100x^2$, $8x^2(x^2 - 25/2) = 0$ so $x = 0, \pm 5/\sqrt{2}$. The points are
         $(0,0)$, $(-5/\sqrt{2},0)$, and $(5/\sqrt{2},0)$.

EXERCISE SET 3.7

1.   (a) $\Delta y = (x + \Delta x)^2 - x^2$
        $= (2 + 1)^2 - 2^2$
        $= 9 - 4 = 5$
     (b) $dy = 2x\,dx$
        $= 2(2)(1) = 4$

(c)

3.   (a) $\Delta y = \dfrac{1}{x + \Delta x} - \dfrac{1}{x}$
        $= \dfrac{1}{1 + 0.5} - \dfrac{1}{1}$
        $= -\dfrac{1}{3}$
     (b) $dy = -\dfrac{1}{x^2}\,dx$
        $= -\dfrac{1}{1^2}(0.5) = -0.5$

(c)

5. $dy = 3x^2 dx$;
     $\Delta y = (x + \Delta x)^3 - x^3 = x^3 3x^2\Delta x + 3x(\Delta x)^2 + (\Delta x)^3 - x^3$
        $= 3x^2\Delta x + 3x(\Delta x)^2 + (\Delta x)^3$

7. $dy = (2x - 2)dx$;
     $\Delta y = [(x + \Delta x)^2 - 2(x + \Delta x) + 1] - [x^2 - 2x + 1]$
        $= x^2 + 2x\,\Delta x + (\Delta x)^2 - 2x - 2\Delta x + 1 - x^2 + 2x - 1$
        $= 2x\,\Delta x + (\Delta x)^2 - 2\Delta x$

9. $dy = (12x^2 - 14x + 2)dx$

11. $dy = x\,d(\cos x) + \cos x\,dx = x(-\sin x)dx + \cos x\,dx$
        $= (-x\sin x + \cos x)dx$

13. $\lim\limits_{\Delta x \to 0} \dfrac{(x + \Delta x)^2 - x^2}{\Delta x} = \dfrac{d}{dx}(x^2) = 2x$

15. $\lim\limits_{\Delta x \to 0} \dfrac{\sin(\pi + \Delta x) - \sin\pi}{\Delta x} = \dfrac{d}{dx}(\sin x)\Big|_{x=\pi} = \cos\pi = -1$

17. $f(x) = x^4$, $f'(x) = 4x^3$, $x_0 = 3$, $\Delta x = 0.02$;

$(3.02)^4 \approx 3^4 + (108)(0.02) = 81 + 2.16 = 83.16.$

19. $f(x) = \sqrt{x}$, $f'(x) = \dfrac{1}{2\sqrt{x}}$, $x_0 = 64$, $\Delta x = 1$;

$\sqrt{65} \approx \sqrt{64} + \dfrac{1}{16}(1) = 8 + \dfrac{1}{16} = 8.0625.$

21. $f(x) = \sqrt{x}$, $f'(x) = \dfrac{1}{2\sqrt{x}}$, $x_0 = 81$, $\Delta x = -0.1$;

$\sqrt{80.9} \approx \sqrt{81} + \dfrac{1}{18}(-0.1) \approx 8.9944.$

23. $f(x) = \sqrt[3]{x}$, $f'(x) = \dfrac{1}{3}x^{-2/3}$, $x_0 = 8$, $\Delta x = 0.06$;

$\sqrt[3]{8.06} \approx \sqrt[3]{8} + \dfrac{1}{12}(0.06) = 2.005.$

25. $f(x) = \cos x$, $f'(x) = -\sin x$, $x_0 = \pi/6$, $\Delta x = \pi/180$;

$\cos 31^\circ \approx \cos 30^\circ + \left(-\dfrac{1}{2}\right)\left(\dfrac{\pi}{180}\right) = \dfrac{\sqrt{3}}{2} - \dfrac{\pi}{360} \approx 0.8573.$

27. $f(x) = \sin x$, $f'(x) = \cos x$, $x_0 = \pi/4$, $\Delta x = -\pi/180$;

$\sin 44^\circ \approx \sin 45^\circ + \dfrac{1}{\sqrt{2}}\left(-\dfrac{\pi}{180}\right) = \dfrac{1}{\sqrt{2}} - \dfrac{\pi}{180\sqrt{2}} \approx 0.6947.$

29. $dy = \dfrac{3}{2\sqrt{3x - 2}}\,dx$, $x = 2$, $dx = 0.03$; $\Delta y \approx dy = \dfrac{3}{4}(0.03) = 0.0225.$

31. $dy = \dfrac{1 - x^2}{(x^2 + 1)^2}\,dx$, $x = 2$, $dx = -0.04$; $\Delta y \approx dy = \left(-\dfrac{3}{25}\right)(-0.04) = 0.0048.$

33. (a) $A = x^2$ where x is the length of a side;

$dA = 2x\,dx = 2(10)(\pm 0.1) = \pm 2$ ft^2.

(b) relative error in $x \approx \dfrac{dx}{x} = \dfrac{\pm 0.1}{10} = \pm 0.01$ so percentage error in

$x \approx \pm 1\%$; relative error in $A \approx \dfrac{dA}{A} = \dfrac{2x\,dx}{x^2} = 2\,\dfrac{dx}{x} = 2(\pm 0.01) = \pm 0.02$

so percentage error in $A \approx \pm 2\%$.

35. (a) $x = 10 \sin \theta$, $y = 10 \cos \theta$
(see figure),

$dx = 10(\cos \theta)d\theta$

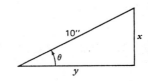

$= 10(\cos \dfrac{\pi}{6})(\pm \dfrac{\pi}{180})$

$= 10(\dfrac{\sqrt{3}}{2})(\pm \dfrac{\pi}{180}) \approx \pm 0.151"$,

$dy = -10(\sin \theta)d\theta = -10\,(\sin \dfrac{\pi}{6})(\pm \dfrac{\pi}{180}) = -10(\dfrac{1}{2})(\pm \dfrac{\pi}{180}) \approx \pm 0.087"$.

(b) relative error in $x \approx \dfrac{dx}{x} = \dfrac{10(\cos \theta)d\theta}{10 \sin \theta} = (\cot \theta)d\theta$

$= (\cot \dfrac{\pi}{6})(\pm \dfrac{\pi}{180}) = \sqrt{3}(\pm \dfrac{\pi}{180}) \approx \pm 0.030$ so percentage error in

$x \approx \pm 3.0\%$; relative error in $y \approx \dfrac{dy}{y} = \dfrac{-10(\sin \theta)d\theta}{10 \cos \theta} = -(\tan \theta)d\theta$

$= -(\tan \dfrac{\pi}{6})(\pm \dfrac{\pi}{180}) = -\dfrac{1}{\sqrt{3}}(\pm \dfrac{\pi}{180}) \approx \pm 0.010$ so percentage error in

$y \approx \pm 1.0\%$.

37. $\dfrac{dR}{R} = \dfrac{(-2k/r^3)dr}{(k/r^2)} = -2\,\dfrac{dr}{r}$, but $\dfrac{dr}{r} \approx \pm 0.05$ so $\dfrac{dR}{R} \approx -2(\pm 0.05) = \pm 0.10$;

percentage error in $R \approx \pm 10\%$.

39. $V = x^3$ where x is the length of a side; $\dfrac{dV}{V} = \dfrac{3x^2 dx}{x^3} = 3\,\dfrac{dx}{x}$, but

$\dfrac{dx}{x} \approx \pm 0.02$ so $\dfrac{dV}{V} \approx 3(\pm 0.02) = \pm 0.06$; percentage error in $V \approx \pm 6\%$.

41. $A = \dfrac{1}{4}\pi D^2$ where D is the diameter of the circle; $\dfrac{dA}{A} = \dfrac{(\pi D/2)dD}{\pi D^2/4} = 2\,\dfrac{dD}{D}$,

but $\dfrac{dA}{A} \approx \pm 0.01$ so $2\,\dfrac{dD}{D} \approx \pm 0.01$, $\dfrac{dD}{D} \approx \pm 0.005$; maximum permissible

percentage error in $D \approx \pm 0.5\%$.

43. V = volume of cylindrical rod = $\pi r^2 h = \pi r^2(15) = 15\pi r^2$; approximate ΔV
by dV if $r = 2.5$ and $dr = \Delta r = 0.001$. $dV = 30\pi r\, dr = 30\pi(2.5)(0.001)$
≈ 0.236 cm^3.

45. $f(x) = \dfrac{1}{1 + x}$, $f'(x) = - \dfrac{1}{(1 + x)^2}$, $x_0 = 0$, $\Delta x = x$;
$f(x_0 + \Delta x) \approx f(x_0) + f'(x_0)\Delta x$ so $f(x) \approx 1 - x$.
  $x = 0.1$: $1/(1 + x) = 0.9091$, $1 - x = 0.9$
  $x = -0.02$: $1/(1 + x) = 1.0204$, $1 - x = 1.02$

47. $y = x^k$, $dy = kx^{k-1}dx$; $\dfrac{dy}{y} = \dfrac{kx^{k-1}dx}{x^k} = k\,\dfrac{dx}{x}$

SUPPLEMENTARY EXERCISES CHAPTER 3

1. $f'(x) = \lim_{h\to0} \dfrac{k(x + h) - kx}{h} = \lim_{h\to0} k = k$

3. $f'(x) = \lim_{h\to0} \dfrac{\sqrt{9 - 4(x + h)} - \sqrt{9 - 4x}}{h} = \lim_{h\to0} \dfrac{[9 - 4(x + h)] - [9 - 4x]}{h(\sqrt{9 - 4(x + h)} + \sqrt{9 - 4x}\,)}$

$= \lim_{h\to0} \dfrac{- 4}{\sqrt{9 - 4(x + h)} + \sqrt{9 - 4x}} = - \dfrac{2}{\sqrt{9 - 4x}}$

5. $\dfrac{d}{dx}\left(|x|^3\right)\Big|_{x=0} = \lim_{h\to0} \dfrac{|0 + h|^3 - |0|^3}{h} = \lim_{h\to0} \dfrac{|h|^3}{h} = \lim_{h\to0} h|h| = 0$

7. $y - (-1) = 5(x - 3)$, $y = 5x - 16$.

9. (a) $2f(x)f'(x) - 3g'(x^2)(2x)\Big|_{x=1} = 12$

  (b) $f(x)g'(x) + f'(x)g(x)\Big|_{x=1} = -7$

  (c) $\dfrac{g(x)f'(x) - f(x)g'(x)}{g^2(x)}\Big|_{x=-2} = 9$

(d) $\dfrac{f(x)g'(x) - g(x)f'(x)}{f^2(x)}\Bigg|_{x=-2} = -\dfrac{9}{4}$

(e) $f'(g(x))g'(x)\Big|_{x=1} = f'(g(1))g'(1) = f'(-2)(-1) = 5$

(f) $f'(g(x))g'(x)\Big|_{x=-2} = f'(g(-2))g'(-2) = f'(1)(7) = 21$

(g) $g'(f(x))f'(x)\Big|_{x=-2} = g'(f(-2))f'(-2) = g'(-2)(-5) = -35$

(h) $g'(g(x))g'(x)\Big|_{x=-2} = g'(g(-2))g'(-2) = g'(1)(7) = -7$

(i) $f'(g(4 - 6x)g'(4 - 6x)(-6)\Big|_{x=1} = f'(g(-2))g'(-2)(-6)$
$$= f'(1)(7)(-6) = -126$$

(j) $3g^2(x)g'(x)\Big|_{x=1} = 3(-2)^2(-1) = -12$

(k) $\dfrac{1}{2}[f(x)]^{-1/2}f'(x)\Big|_{x=1} = \dfrac{1}{2}(1)^{-1/2}(3) = \dfrac{3}{2}$

(l) $f'(-x/2)(-1/2)\Big|_{x=-2} = -\dfrac{3}{2}$

11. $f'(x) = \dfrac{(x^2 + 2x)4(x - 3)^3 - (x - 3)^4(2x + 2)}{(x^2 + 2x)^2}$

$= \dfrac{(x - 3)^3[4(x^2 + 2x) - (x - 3)(2x + 2)]}{(x^2 + 2x)^2}$

$= \dfrac{(x - 3)^3(2x^2 + 12x + 6)}{(x^2 + 2x)^2} = \dfrac{2(x - 3)^3(x^2 + 6x + 3)}{(x^2 + 2x)^2}$

so $f'(x) = 0$ if $x - 3 = 0$ or if $x^2 + 6x + 3 = 0$; the solution of
$x - 3 = 0$ is $x = 3$, and the solution of $x^2 + 6x + 3 = 0$ is $x = -3 \pm \sqrt{6}$.

13. $f'(x) = 3\left[\dfrac{3x + 1}{x^2}\right]^2 \dfrac{x^2(3) - (3x + 1)(2x)}{x^4} = -\dfrac{3(3x + 2)(3x + 1)^2}{x^7}$
so $f'(x) = 0$ if $x = -2/3, -1/3$.

15. $f'(x) = x^{1/2}(1/3)(x^2 + x + 1)^{-2/3}(2x + 1) + (x^2 + x + 1)^{1/3}(1/2)x^{-1/2}$

$= \frac{1}{6} x^{-1/2}(x^2 + x + 1)^{-2/3}[2x(2x + 1) + 3(x^2 + x + 1)]$

$= \frac{7x^2 + 5x + 3}{6x^{1/2}(x^2 + x + 1)^{2/3}}$

but $7x^2 + 5x + 3 = 0$ has no real solutions so there are no values of x for which $f'(x) = 0$.

17. $\frac{d}{dx} (\sqrt{2} x^{-2} - \frac{2}{5} x^{-1}) = -2\sqrt{2} x^{-3} + \frac{2}{5} x^{-2}$

19. $z = (2 \sin r \cos r)^2 = \sin^2 2r$ so $\frac{dz}{dr} = 2(\sin 2r)(\cos 2r)(2) = 2 \sin 4r$

and $\frac{dz}{dr}\Big|_{r=\pi/6} = 2 \sin(2\pi/3) = \sqrt{3}$

21. $u = \left[\frac{x - 1}{x}\right]^2 = (1 - x^{-1})^2$ so $\frac{du}{dx} = 2(1 - x^{-1})(x^{-2}) = 2(x - 1)/x^3$.

23. $\frac{d}{dx} (\sec^2 x - \tan^2 x) = \frac{d}{dx} (1) = 0$

25. $F(x) = \frac{2x + 4x^3}{1 + 2x^2} = \frac{2x(1 + 2x^2)}{1 + 2x^2} = 2x$ so $F'(x) = 2$.

27. $m_{tan} = f'(x) = 1 + x^{-2}$, and the slope of $2x - y = 5$ is 2 so we want $1 + x^{-2} = 2$ which gives $x^2 = 1$, $x = \pm 1$.

29. $m_{tan} = f'(x) = 2(x + 2)$ so at $(x_0, f(x_0))$ the tangent line is

$y - f(x_0) = 2(x_0 + 2)(x - x_0)$, or $y - (x_0 + 2)^2 = 2(x_0 + 2)(x - x_0)$.
But if the line passes through the orgin then $x = 0$, $y = 0$ must satisfy the latter equation thus $-(x_0 + 2)^2 = -2x_0(x_0 + 2)$ which leads to $(x_0 + 2)(x_0 - 2) = 0$ so $x_0 = -2, 2$.

31. $m_{\tan} = f'(x) = 3 - \sec^2 x$, and the slope of $y - x = 2$ is 1 so we want

$3 - \sec^2 x = 1$ which gives $\sec^2 x = 2$, $\sec x = \pm\sqrt{2}$, $x = \pi/4 + k\pi/2$ where $k = 0, \pm 1, \pm 2, \cdots$.

33. $\Delta x = 0 - (-\pi/4) = \pi/4$, $\Delta y = y\big|_{x=0} - y\big|_{x=-\pi/4} = 0 - (-1) = 1$,

$dy = \sec^2(-\pi/4)(\pi/4) = \pi/2$.

35. (a) Consider $y = f(x) = \sqrt[3]{x}$ with $x = -8$ and $dx = -0.25 = -1/4$, then
$f(-8.25) \approx f(-8) + dy$,
$\sqrt[3]{-8.25} \approx \sqrt[3]{-8} + \frac{1}{3}(-8)^{-2/3}(-1/4) = -2 - 1/48 = -97/48$.

(b) Consider $y = f(x) = \cot x$ (x in radians) with $x = 45° = \pi/4$ radians
and $dx = 1° = \pi/180$ radians, then $f(\pi/4 + 180/\pi) \approx f(\pi/4) + dy$,
$\cot 46° \approx \cot 45° + (-\csc^2 45°)(\pi/180) = 1 - \pi/90$.

37. $h = 12 \sin\theta$ thus $dh = 12 \cos\theta \, d\theta$ so, with $\theta = 60° = \pi/3$ radians and
$d\theta = -1° = -\pi/180$ radians, $dh = 12\cos(\pi/3)(-\pi/180) = -\pi/30$ ft.

39. (a) $dW/dt\big|_{t=5} = 200(t - 15)\big|_{t=5} = -2000$ so water is running out at the
rate of 2000 gal/min.

(b) average rate of change of $W = (W\big|_{t=5} - W\big|_{t=0})/5$
$= (10{,}000 - 22{,}500)/5 = -2500$
so water flows out at an average rate of 2500 gal/min during the
first 5 minutes.

41. $2xy \dfrac{dy}{dx} + y^2 = \cos(x + 2y)(1 + 2\dfrac{dy}{dx})$ so $\dfrac{dy}{dx} = \dfrac{\cos(x + 2y) - y^2}{2xy - 2\cos(x + 2y)}$.
$\dfrac{dy}{dx}\Big|_{(0,0)} = -\dfrac{1}{2}$, the tangent line is $y = -x/2$.

43. $x = y = 1$ satisfies both equations so they intersect at the point $(1,1)$.
For $2x^2 + 3y^2 = 5$, $\dfrac{dy}{dx} = -\dfrac{2x}{3y}$ so $\dfrac{dy}{dx}\Big|_{(1,1)} = -\dfrac{2}{3}$.

For $y^2 = x^3$, $\frac{dy}{dx} = \frac{3x^2}{2y}$ so $\left.\frac{dy}{dx}\right|_{(1,1)} = \frac{3}{2}$.

The tangent lines are perpendicular at $(1,1)$ because the slope of one curve is the negative reciprocal of the slope of the other curve.

45. $y = \cos x - 3\sin x$, $y' = -\sin x - 3\cos x$,
$y'' = -\cos x + 3\sin x$, $y''' = \sin x + 3\cos x$
so $y''' + y'' + y' + y = (-3 - 1 + 3 + 1)\sin x + (1 - 3 - 1 + 3)\cos x = 0$.

CHAPTER 4

EXERCISE SET 4.1

1. (a) $A = x^2$, so $\frac{dA}{dt} = 2x \frac{dx}{dt}$

    (b) Find $\frac{dA}{dt}\Big|_{x=3}$ given that $\frac{dx}{dt}\Big|_{x=3} = 2$.

    From part (a), $\frac{dA}{dt}\Big|_{x=3} = 2(3)(2) = 12$ ft^2/min.

3. (a) $V = \pi r^2 h$, so $\frac{dV}{dt} = \pi(r^2 \frac{dh}{dt} + 2rh \frac{dr}{dt})$.

    (b) Find $\frac{dV}{dt}\Big|_{\substack{h=6, \\ r=10}}$ given that $\frac{dh}{dt}\Big|_{\substack{h=6, \\ r=10}} = 1$ and $\frac{dr}{dt}\Big|_{\substack{h=6, \\ r=10}} = -1$.

    From part (a), $\frac{dV}{dt}\Big|_{\substack{h=6, \\ r=10}} = \pi[10^2(1) + 2(10)(6)(-1)] = -20\pi$ in^3/sec;

    the volume is decreasing.

5. (a) $\tan\theta = \frac{y}{x}$, so $\sec^2\theta \frac{d\theta}{dt} = \frac{x\frac{dy}{dt} - y\frac{dx}{dt}}{x^2}$, $\frac{d\theta}{dt} = \frac{\cos^2\theta}{x^2}(x\frac{dy}{dt} - y\frac{dx}{dt})$.

    (b) Find $\frac{d\theta}{dt}\Big|_{\substack{x=2, \\ y=2}}$ given that $\frac{dx}{dt}\Big|_{\substack{x=2, \\ y=2}} = 1$ and $\frac{dy}{dt}\Big|_{\substack{x=2, \\ y=2}} = -\frac{1}{4}$.

    When $x = 2$ and $y = 2$, $\tan\theta = 2/2 = 1$

    so $\theta = \frac{\pi}{4}$ and $\cos\theta = \cos\frac{\pi}{4} = \frac{1}{\sqrt{2}}$. Thus from part (a),

    $\frac{d\theta}{dt}\Big|_{\substack{x=2, \\ y=2}} = \frac{(1/\sqrt{2})^2}{2^2}[2(-\frac{1}{4}) - 2(1)] = -\frac{5}{16}$ radians/sec;

    θ is decreasing.

7. Let A be the area swept out, and θ the angle through which the minute hand has rotated. Find $\frac{dA}{dt}$ given that $\frac{d\theta}{dt} = \frac{\pi}{30}$ radians/min. The area is $A = \frac{1}{2} r^2 \theta = 8\theta$, so $\frac{dA}{dt} = 8 \frac{d\theta}{dt} = \frac{4\pi}{15}$ in^2/min.

9. Find $\frac{dr}{dt}\Big|_{A=9}$ given that $\frac{dA}{dt} = 6$. From $A = \pi r^2$ we get $\frac{dA}{dt} = 2\pi r \frac{dr}{dt}$ so $\frac{dr}{dt} = \frac{1}{2\pi r} \frac{dA}{dt}$. If $A = 9$ then $\pi r^2 = 9$, $r = 3/\sqrt{\pi}$ so $\frac{dr}{dt}\Big|_{A=9} = \frac{1}{2\pi(3/\sqrt{\pi})} (6) = 1/\sqrt{\pi}$ mph.

11. Find $\frac{dV}{dt}\Big|_{r=9}$ given that $\frac{dr}{dt} = -15$. From $V = \frac{4}{3} \pi r^3$ we get $\frac{dV}{dt} = 4\pi r^2 \frac{dr}{dt}$ so $\frac{dV}{dt}\Big|_{r=9} = 4\pi(9)^2(-15) = -4860\pi$. Air must be removed at the rate of 4860π cm^3/min.

13. Find $\frac{dx}{dt}\Big|_{y=5}$ given that $\frac{dy}{dt} = -2$.
    From $x^2 + y^2 = 13^2$ we get
    $2x \frac{dx}{dt} + 2y \frac{dy}{dt} = 0$ so $\frac{dx}{dt} = -\frac{y}{x} \frac{dy}{dt}$.
    Use $x^2 + y^2 = 169$ to find that $x = 12$
    when $y = 5$ so
    $$\frac{dx}{dt}\Big|_{y=5} = -\frac{5}{12} (-2) = \frac{5}{6} \text{ ft/sec}.$$

15. Let x be the length of each edge, S the surface area, and V the volume. Find $\frac{dS}{dt}\Big|_{x=5}$ given that $\frac{dV}{dt}\Big|_{x=5} = 2$. $S = 6x^2$ and $V = x^3$, so $x = V^{1/3}$, $S = 6V^{2/3}$, $\frac{dS}{dt} = 4V^{-1/3} \frac{dV}{dt} = \frac{4}{x} \frac{dV}{dt} = \frac{4}{5} (2) = 8/5$ in^2/min.

17. With ϕ and x as shown in the figure, find $\dfrac{d\phi}{dt}\bigg|_{x=3000}$ given that

$\dfrac{dx}{dt}\bigg|_{x=3000} = 500.$ But $\tan \phi = \dfrac{x}{3000}$

thus $\sec^2\phi \; \dfrac{d\phi}{dt} = \dfrac{1}{3000} \dfrac{dx}{dt}$

$\dfrac{d\phi}{dt} = \dfrac{\cos^2\phi}{3000} \dfrac{dx}{dt},$

$\phi = \pi/4$ when $x = 3000$ so $\cos \phi = 1/\sqrt{2}$ and $\dfrac{d\phi}{dt}\bigg|_{x=3000} = \dfrac{1/2}{3000} (500) = 1/12$

radian/sec.

19. Find $\dfrac{dh}{dt}\bigg|_{h=16}$ given that $\dfrac{dV}{dt} = 20.$

The volume of water in the tank at a depth h is $V = \dfrac{1}{3} \pi r^2 h.$ Use similar triangles (see figure) to get $\dfrac{r}{h} = \dfrac{10}{24}$

so $r = \dfrac{5}{12} h$ thus

$V = \dfrac{1}{3} \pi (\dfrac{5}{12} h)^2 h = \dfrac{25}{432} \pi h^3,$

$\dfrac{dV}{dt} = \dfrac{25}{144} \pi h^2 \dfrac{dh}{dt}; \; \dfrac{dh}{dt} = \dfrac{144}{25\pi h^2} \dfrac{dV}{dt},$

$\dfrac{dh}{dt}\bigg|_{h=16} = \dfrac{144}{25\pi(16)^2} (20) = \dfrac{9}{20\pi}$ ft/min.

21. Find $\dfrac{dV}{dt}\bigg|_{h=10}$ given that $\dfrac{dh}{dt} = 5.$

$V = \dfrac{1}{3} \pi r^2 h,$ but $r = \dfrac{1}{2} h$ so

$V = \dfrac{1}{3} \pi (\dfrac{h}{2})^2 h = \dfrac{1}{12} \pi h^3, \; \dfrac{dV}{dt} = \dfrac{1}{4} \pi h^2 \dfrac{dh}{dt},$

$\dfrac{dV}{dt}\bigg|_{h=10} = \dfrac{1}{4} \pi (10)^2 (5) = 125\pi$ ft^3/min.

23. With s and h as shown in the figure, we want to find $\frac{dh}{dt}$ given that

$\frac{ds}{dt} = 500$. From the figure,

$h = s \sin 30° = \frac{1}{2} s$

so $\frac{dh}{dt} = \frac{1}{2} \frac{ds}{dt} = \frac{1}{2} (500) = 250$ mph.

25. Find $\frac{dy}{dt}$ given that $\frac{dx}{dt}\Big|_{y=125} = -12$.

From $x^2 + 10^2 = y^2$ we get

$2x \frac{dx}{dt} = 2y \frac{dy}{dt}$ so $\frac{dy}{dt} = \frac{x}{y} \frac{dx}{dt}$.

Use $x^2 + 100 = y^2$ to find that

$x = \sqrt{15,525} = 15\sqrt{69}$ when $y = 125$ so $\frac{dy}{dt} = \frac{15\sqrt{69}}{125} (-12) = -\frac{36\sqrt{69}}{25}$.   The

rope must be pulled at the rate of $\frac{36\sqrt{69}}{25}$ ft/min.

27. Find $\frac{dx}{dt}\Big|_{\theta=\pi/4}$ given that $\frac{d\theta}{dt} = \frac{2\pi}{10} = \frac{\pi}{5}$

radians/sec. $x = 4 \tan \theta$ (see figure)

so $\frac{dx}{dt} = 4 \sec^2\theta \frac{d\theta}{dt}$,

$\frac{dx}{dt}\Big|_{\theta=\pi/4} = 4(\sec^2 \frac{\pi}{4})(\frac{\pi}{5}) = 8\pi/5$ kilometers/sec.

29. We wish to find $\left.\dfrac{dz}{dt}\right|_{\substack{x=2,\\y=4}}$ given that

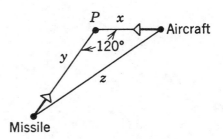

$\dfrac{dx}{dt} = -600$ and $\left.\dfrac{dy}{dt}\right|_{\substack{x=2,\\y=4}} = -1200$ (see

figure). From the law of cosines,

$$z^2 = x^2 + y^2 - 2xy \cos 120°$$
$$= x^2 + y^2 - 2xy\left(-\tfrac{1}{2}\right)$$
$$= x^2 + y^2 + xy,$$

so

$$2z\,\frac{dz}{dt} = 2x\,\frac{dx}{dt} + 2y\,\frac{dy}{dt} + x\,\frac{dy}{dt} + y\,\frac{dx}{dt}$$
$$\frac{dz}{dt} = \frac{1}{2z}\left[(2x + y)\,\frac{dx}{dt} + (2y + x)\,\frac{dy}{dt}\right].$$

When $x = 2$ and $y = 4$, $z^2 = 2^2 + 4^2 + (2)(4) = 28$, so $z = \sqrt{28} = 2\sqrt{7}$, thus

$$\left.\frac{dz}{dt}\right|_{\substack{x=2,\\y=4}} = \frac{1}{2(2\sqrt{7})}\left[(2(2) + 4)(-600) + (2(4) + 2)(-1200)\right]$$

$$= -\frac{4200}{\sqrt{7}} = -600\sqrt{7}\text{ mph};$$

the distance between missile and aircraft is decreasing at the rate of $600\sqrt{7}$ mph.

31. (a) We want $\left.\dfrac{dy}{dt}\right|_{\substack{x=1,\\y=2}}$ given that $\left.\dfrac{dx}{dt}\right|_{\substack{x=1,\\y=2}} = 6$. For convenience, first

rewrite the equation as $xy^3 = \dfrac{8}{5} + \dfrac{8}{5}\,y^2$ then

$$3xy^2\,\frac{dy}{dt} + y^3\,\frac{dx}{dt} = \frac{16}{5}\,y\,\frac{dy}{dt}, \quad \frac{dy}{dt} = \frac{y^3}{\frac{16}{5}\,y - 3xy^2}\,\frac{dx}{dt}$$

so $\left.\dfrac{dy}{dt}\right|_{\substack{x=1,\\y=2}} = \dfrac{2^3}{\frac{16}{5}(2) - 3(1)2^2}\,(6) = -60/7$ units/sec.

(b) falling, because $\dfrac{dy}{dt} < 0$.

33. The coordinates of P are $(x, 2x)$, so the distance between P and the point $(3, 0)$ is $D = \sqrt{(x-3)^2 + (2x-0)^2} = \sqrt{5x^2 - 6x + 9}$. Find $\left.\dfrac{dD}{dt}\right|_{x=3}$ given that $\left.\dfrac{dx}{dt}\right|_{x=3} = -2$.

$\dfrac{dD}{dt} = \dfrac{5x - 3}{\sqrt{5x^2 - 6x + 9}} \dfrac{dx}{dt}$, so $\left.\dfrac{dD}{dt}\right|_{x=3} = \dfrac{12}{\sqrt{36}}(-2) = -4$ units/sec.

35. Find $\left.\dfrac{dS}{dt}\right|_{s=10}$ given that $\left.\dfrac{ds}{dt}\right|_{s=10} = -2$. From $\dfrac{1}{s} + \dfrac{1}{S} = \dfrac{1}{6}$ we get

$-\dfrac{1}{s^2}\dfrac{ds}{dt} - \dfrac{1}{S^2}\dfrac{dS}{dt} = 0$, so $\dfrac{dS}{dt} = -\dfrac{S^2}{s^2}\dfrac{ds}{dt}$. If $s = 10$, then $\dfrac{1}{10} + \dfrac{1}{S} = \dfrac{1}{6}$ which

gives $S = 15$. $\left.\dfrac{dS}{dt}\right|_{s=10} = -\dfrac{225}{100}(-2) = 4.5$ cm/sec. The image is moving away from the lens.

37. Let r be the radius, V the volume, and A the surface area of a sphere. Show that $\dfrac{dr}{dt}$ is a constant given that $\dfrac{dV}{dt} = -kA$, where k is a positive constant. Because $V = \dfrac{4}{3}\pi r^3$,

$$\frac{dV}{dt} = 4\pi r^2 \frac{dr}{dt} \tag{1}$$

But it is given that $\dfrac{dV}{dt} = -kA$ or, because $A = 4\pi r^2$, $\dfrac{dV}{dt} = -4r^2 k$ which when substituted into equation (1) gives $-4\pi r^2 k = 4\pi r^2 \dfrac{dr}{dt}$, $\dfrac{dr}{dt} = -k$.

39. Extend sides of cup to complete the cone and let V_0 be the volume of the portion added, then (see figure)

$V = \frac{1}{3} \pi r^2 h - V_0$ where

$\frac{r}{h} = \frac{4}{12} = \frac{1}{3}$ so $r = \frac{1}{3} h$ and

$V = \frac{1}{3} \pi (\frac{h}{3})^2 h - V_0 = \frac{1}{27} \pi h^3 - V_0$,

$\frac{dV}{dt} = \frac{1}{9} \pi h^2 \frac{dh}{dt}$, $\frac{dh}{dt} = \frac{9}{\pi h^2} \frac{dV}{dt}$,

$\frac{dh}{dt}\Big|_{h=9} = \frac{9}{\pi(9)^2} (2) = \frac{2}{9\pi}$ cm/sec.

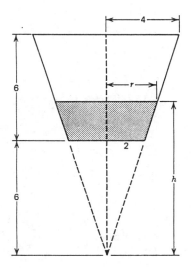

EXERCISE SET 4.2

1. (a) (d,f)
    (c) (a,b), (c,e)

    (b) (a,d), (f,g)
    (d) (b,c), (e,g)

3. $f'(x) = 2x - 5$
    $f''(x) = 2$

    (a) $(5/2,+\infty)$
    (b) $(-\infty,5/2)$
    (c) $(-\infty,+\infty)$
    (d) none
    (e) none

5. $f'(x) = 3(x + 2)^2$
    $f''(x) = 6(x + 2)$

    (a) $(-\infty,-2)$, $(-2,+\infty)$
    (b) none
    (c) $(-2,+\infty)$
    (d) $(-\infty,-2)$
    (e) -2

7. $f'(x) = 9(x^2 - 4/9)$
    $f''(x) = 18x$

    (a) $(-\infty,-2/3)$, $(2/3,+\infty)$
    (b) $(-2/3,2/3)$
    (c) $(0,+\infty)$
    (d) $(-\infty,0)$
    (e) 0

9. $f'(x) = 12x^2(x - 1)$
    $f''(x) = 36x(x - 2/3)$

    (a) $(1,+\infty)$
    (b) $(-\infty,0)$, $(0,1)$
    (c) $(-\infty,0)$, $(2/3,+\infty)$
    (d) $(0,2/3)$
    (e) 0, $2/3$

11. $f'(x) = -\sin x$

   $f''(x) = -\cos x$

   (a) $(\pi, 2\pi)$
   (b) $(0, \pi)$
   (c) $(\pi/2, 3\pi/2)$
   (d) $(0, \pi/2)$, $(3\pi/2, 2\pi)$
   (e) $\pi/2$, $3\pi/2$

13. $f'(x) = \sec^2 x$

   $f''(x) = 2\sec^2 x \tan x$

   (a) $(-\pi/2, \pi/2)$
   (b) none
   (c) $(0, \pi/2)$
   (d) $(-\pi/2, 0)$
   (e) 0

15. $f'(x) = \dfrac{1}{3}(x + 2)^{-2/3}$

   $f''(x) = -\dfrac{2}{9}(x + 2)^{-5/3}$

   (a) $(-\infty, -2)$, $(-2, +\infty)$
   (b) none
   (c) $(-\infty, -2)$
   (d) $(-2, +\infty)$
   (e) -2

17. $f'(x) = \dfrac{4(x + 1)}{3x^{2/3}}$

   $f''(x) = \dfrac{4(x - 2)}{9x^{5/3}}$

   (a) $(-1, 0)$, $(0, +\infty)$
   (b) $(-\infty, -1)$
   (c) $(-\infty, 0)$, $(2, +\infty)$
   (d) $(0, 2)$
   (e) 0, 2

19. (a)   (b)   (c)

21. (a)   (b)

23. $f'(x) = 4(x - a)^3$, $f''(x) = 12(x - a)^2$; no inflection points.

25. $f(x_1) - f(x_2) = x_1{}^2 - x_2{}^2 = (x_1 + x_2)(x_1 - x_2) < 0$ if $x_1 < x_2$ for x_1, x_2 in $[0, +\infty)$, so $f(x_1) < f(x_2)$ and thus increasing.

27. $f(x_1) - f(x_2) = \sqrt{x_1} - \sqrt{x_2} = \dfrac{x_1 - x_2}{\sqrt{x_1} + \sqrt{x_2}} < 0$ if $x_1 < x_2$ for x_1, x_2 in $[0, +\infty)$, so $f(x_1) < f(x_2)$ and thus increasing.

29. If $x_1 < x_2$ where x_1 and x_2 are in I, then $f(x_1) < f(x_2)$ and $g(x_1) < g(x_2)$, so $f(x_1) + g(x_1) < f(x_2) + g(x_2)$, or equivalently, $(f + g)(x_1) < (f + g)(x_2)$. Thus $f + g$ is increasing on I.

31. For example, $f(x) = x$ and $g(x) = 2x$ on $(-\infty, +\infty)$.

33. $f''(x) = 6ax + 2b = 6a(x + \dfrac{b}{3a})$, $f''(x) = 0$ when $x = -\dfrac{b}{3a}$. f changes its direction of concavity at $x = -\dfrac{b}{3a}$ so $-\dfrac{b}{3a}$ is an inflection point.

EXERCISE SET 4.3

1. $f'(x) = 2x - 5$, $f'(x) = 0$ when $x = 5/2$ (stationary point).

3. $f'(x) = 3x^2 + 6x - 9 = 3(x + 3)(x - 1)$, $f'(x) = 0$ when $x = -3$, 1 (stationary points).

5. $f'(x) = 4x(x^2 - 3)$, $f'(x) = 0$ when $x = 0$, $\pm\sqrt{3}$ (stationary points).

7. $f'(x) = (2 - x^2)/(x^2 + 2)^2$, $f'(x) = 0$ when $x = \pm\sqrt{2}$ (stationary points).

9. $f'(x) = \dfrac{2}{3} x^{-1/3} = 2/(3x^{1/3})$, $f'(x)$ does not exist when $x = 0$.

11. $f'(x) = -3 \sin 3x$, $f'(x) = 0$ when $\sin 3x = 0$, $3x = n\pi$, $n = 0$, ± 1, ± 2, $\cdots$ $x = n\pi/3$, $n = 0$, ± 1, ± 2, $\cdots$ (stationary points).

13. $f'(x) = 4 \sin 2x \cos 2x = 2 \sin 4x$,
     $f'(x) = 0$ when $4x = n\pi$, $x = n\pi/4$, $n = 1,2,3,\cdots,7$ (stationary points)

15. $f'(x) = \dfrac{4(x + 1)}{3x^{2/3}}$, $f'(x) = 0$ when $x = -1$ (stationary point), $f'(x)$ does
     not exist when $x = 0$.

17. $f'(x) = -2(x + 2)$; critical point $x = -2$

     (a) $f'(x)$:
     $$\underset{-2}{\underline{+ + + 0 - - -}}$$
     (b) $f''(x) = -2$; $f''(-2) < 0$

     $f(-2) = 5$; relative max of 5 at $x = -2$

19. $f'(x) = 2 \sin x \cos x = \sin 2x$; critical points $x = \pi/2$, π, $3\pi/2$

     (a) $f'(x)$:
     $$\underset{\pi/2 \quad\;\; \pi \quad\;\; 3\pi/2}{\underline{+ + + 0 - - - 0 + + + 0 - - -}}$$
     (b) $f''(x) = 2 \cos 2x$; $f''(\pi/2) < 0$, $f''(\pi) > 0$, $f''(3\pi/2) < 0$
        $f(\pi/2) = f(3\pi/2) = 1$, $f(\pi) = 0$;
        relative min of 0 at $x = \pi$, relative max of 1 at $x = \pi/2$, $3\pi/2$

21. $f'(x) = 3x^2 + 5$; no relative extrema because there are no critical
     points.

23. $f'(x) = (x - 1)(3x - 1)$; critical points $x = 1$, $1/3$
     $f''(x) = 6x - 4$; $f''(1) > 0$, $f''(1/3) < 0$
     relative min of 0 at $x = 1$, relative max of 4/27 at $x = 1/3$

25. $f'(x) = 4x(1 - x^2)$; critical points $x = 0$, 1, -1
     $f''(x) = 4 - 12x^2$; $f''(0) > 0$, $f''(1) < 0$, $f''(-1) < 0$
     relative min of 0 at $x = 0$, relative max of 1 at $x = 1$, -1

27. $f'(x) = \dfrac{4}{5} x^{-1/5}$; critical point $x = 0$

     relative min of 0 at $x = 0$ (first derivative test)

29. $f'(x) = 2x/(x^2 + 1)^2$; critical point $x = 0$
     relative min of 0 at $x = 0$

31. $f'(x) = 2x$ if $|x| > 2$, $f'(x) = -2x$ if $|x| < 2$,
     $f'(x)$ does not exist when $x = \pm 2$; critical points $x = 0$, 2, -2
     relative min of 0 at $x = 2$, -2, relative max of 4 at $x = 0$

33. $f'(x) = -\sin 2x$; critical points $x = 0$, $\pm\pi/2$, $\pm\pi$, $\pm 3\pi/2$, $\cdots$
    relative min of 0 at $x = \pm\pi/2$, $\pm 3\pi/2$, $\cdots$,
    relative max of 1 at $x = 0$, $\pm\pi$, $\pm 2\pi$, $\cdots$

35. $f'(x) = 2x \sec^2(x^2 + 1)$; critical point $x = 0$
    relative min of $\tan 1$ at $x = 0$

37. $f'(x) = 2\cos 2x$ if $\sin 2x > 0$, $f'(x) = -2\cos 2x$ if $\sin 2x < 0$,
    $f'(x)$ does not exist when $x = \pi/2$, π, $3\pi/2$;
    critical points $x = \pi/4$, $3\pi/4$, $5\pi/4$, $7\pi/4$, $\pi/2$, π, $3\pi/2$
    relative min of 0 at $x = \pi/2$, π, $3\pi/2$
    relative max of 1 at $x = \pi/4$, $3\pi/4$, $5\pi/4$, $7\pi/4$

39. Let $f(x) = x^2 + \dfrac{k}{x}$, then $f'(x) = 2x - \dfrac{k}{x^2} = \dfrac{2x^3 - k}{x^2}$. f has a relative

    extremum when $2x^3 - k = 0$, so $k = 2x^3 = 2(3)^3 = 54$.

41. $f(x) = -x^4$ has a relative maximum at $x = 0$,

    $f(x) = x^4$ has a relative minimum at $x = 0$,

    $f(x) = x^3$ has neither at $x = 0$;
    $f'(0) = 0$ for all three functions.

43. (a)

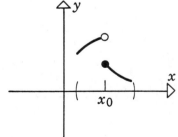

$f(x_0)$ is not an extreme value

(b)

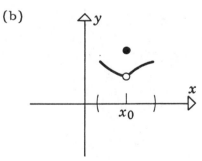

$f(x_0)$ is a relative maximum

(c)

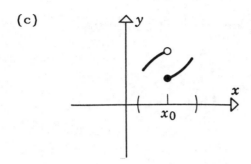

$f(x_0)$ is a relative minimum

EXERCISE SET 4.4

1. $y = x^2 - 2x - 3$
    $y' = 2(x - 1)$
    $y'' = 2$

3. $y = x^3 - 3x + 1$
    $y' = 3(x^2 - 1)$
    $y'' = 6x$

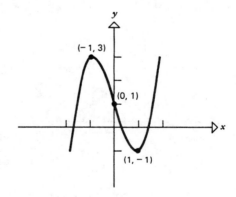

5. $y = x^3 + 3x^2 + 5$

    $y' = 3x(x + 2)$

    $y'' = 6(x + 1)$

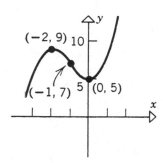

7. $y = 2x^3 - 3x^2 + 12x + 9$

    $y' = 6(x^2 - x + 2)$

    $y'' = 12(x - 1/2)$

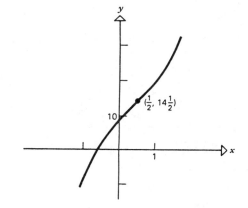

9. $y = (x - 1)^4$

    $y' = 4(x - 1)^3$

    $y'' = 12(x - 1)^2$

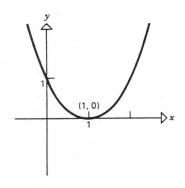

11. $y = x^4 + 2x^3 - 1$

    $y' = 4x^2(x + 3/2)$

    $y'' = 12x(x + 1)$

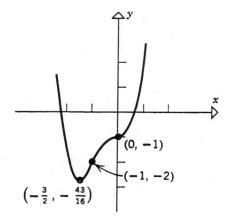

13. $y = x^4 - 3x^3 + 3x^2 + 1$
      $y' = x(4x^2 - 9x + 6)$
      $y'' = 12(x - 1/2)(x - 1)$

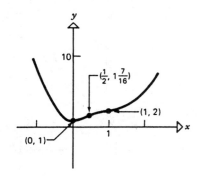

15. $y = x^3(3x^2 - 5)$
      $y' = 15x^2(x^2 - 1)$
      $y'' = 30x(2x^2 - 1)$

17. $y = x(x - 1)^3$
      $y' = (4x - 1)(x - 1)^2$
      $y'' = 6(2x - 1)(x - 1)$

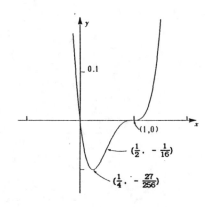

19. $y = 2x/(x - 3)$
      $y' = -6(x - 3)^2$
      $y'' = 12/(x - 3)^3$

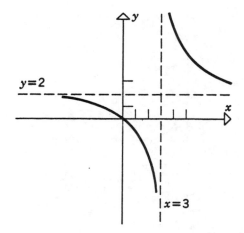

21. $y = \dfrac{x^2}{x^2 - 1}$

$y' = -\dfrac{2x}{(x^2 - 1)^2}$

$y'' = \dfrac{2(3x^2 + 1)}{(x^2 - 1)^3}$

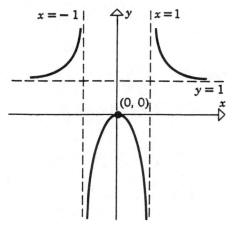

23. $y = \dfrac{x}{x^2 + 1}$

$y' = \dfrac{1 - x^2}{(x^2 + 1)^2}$

$y'' = \dfrac{2x(x^2 - 3)}{(x^2 + 1)^3}$

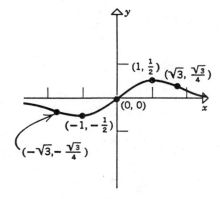

25. $y = (x - 1)/(x - 2)$

$y' = -1/(x - 2)^2$

$y'' = 2/(x - 2)^3$

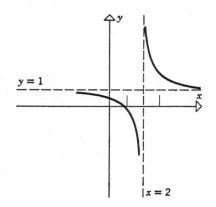

27. $y = x^2 - \dfrac{1}{x} = \dfrac{x^3 - 1}{x}$

$y' = \dfrac{2x^3 + 1}{x^2}$,

$y' = 0$ when $x = -\sqrt[3]{\dfrac{1}{2}}$

≈ -0.8

$y'' = \dfrac{2(x^3 - 1)}{x^3}$

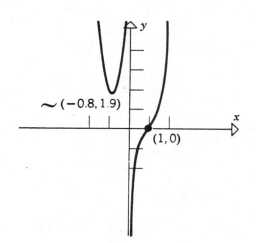

$\sim (-0.8, 1.9)$

$(1, 0)$

29. $y = \dfrac{1 - x}{x^2}$

$y' = \dfrac{x - 2}{x^3}$

$y'' = \dfrac{2(3 - x)}{x^4}$

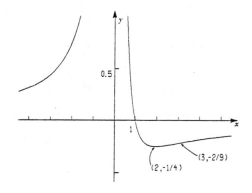

0.5

1

$(3, -2/9)$

$(2, -1/4)$

31. $y = \dfrac{x - 1}{x^2 - 4}$

$y' = -\dfrac{x^2 - 2x + 4}{(x^2 - 4)^2}$

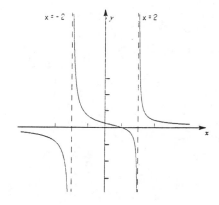

$x = -2$ $x = 2$

33. $y = \dfrac{(x - 1)^2}{x^2}$

 $y' = \dfrac{2(x - 1)}{x^3}$

 $y'' = \dfrac{2(3 - 2x)}{x^4}$

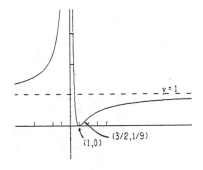

35. $y = 3 - \dfrac{4}{x} - \dfrac{4}{x^2}$

 $y' = \dfrac{4(x + 2)}{x^3}$

 $y'' = -\dfrac{8(x + 3)}{x^4}$

37. $y = \dfrac{x^3 - 1}{x^3 + 1}$

$y' = \dfrac{6x^2}{(x^3 + 1)^2}$

$y'' = \dfrac{12x(1 - 2x^3)}{(x^3 + 1)^3}$

39. $y = \dfrac{x^2 - 2}{x} = x - \dfrac{2}{x}$

so $y = x$ is an oblique asymptote

$y' = \dfrac{x^2 + 2}{x^2}$

$y'' = -\dfrac{4}{x^3}$

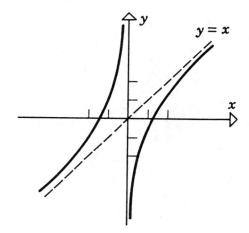

41. $y = \dfrac{(x - 2)^3}{x^2} = x - 6 + \dfrac{12x - 8}{x^2}$

so $y = x - 6$ is an oblique asymptote

$y' = \dfrac{(x - 2)^2(x + 4)}{x^3}$

$y'' = \dfrac{24(x - 2)}{x^4}$

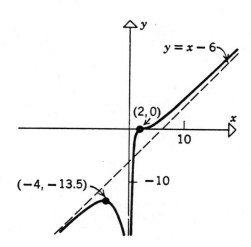

43. $y = x + 1 - \dfrac{1}{x} - \dfrac{1}{x^2} = \dfrac{(x - 1)(x + 1)^2}{x^2}$

$y = x + 1$ is an oblique asymptote

$y' = \dfrac{(x + 1)(x^2 - x + 2)}{x^3}$

$y'' = -\dfrac{2(x + 3)}{x^4}$

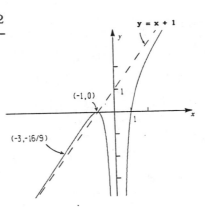

EXERCISE SET 4.5

1. $y = (x - 2)^{1/3}$

$y' = \dfrac{1}{3}(x - 2)^{-2/3}$

$y'' = -\dfrac{2}{9}(x - 2)^{-5/3}$

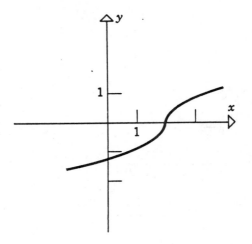

3. $y = x^{1/5}$

$y' = \dfrac{1}{5}x^{-4/5}$

$y'' = -\dfrac{4}{25}x^{-9/5}$

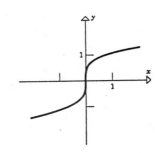

5. $y = x^{4/3}$

$y' = \frac{4}{3} x^{1/3}$

$y'' = \frac{4}{9} x^{-2/3}$

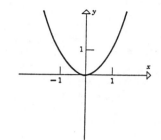

7. $y = 1 - x^{2/3}$

$y' = -\frac{2}{3} x^{-1/3}$

$y'' = \frac{2}{9} x^{-4/3}$

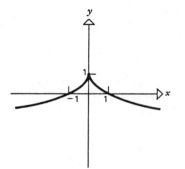

9. $y = \sqrt{x^2 - 1}$

$y' = \frac{x}{\sqrt{x^2 - 1}}$

$y'' = -\frac{1}{(x^2 - 1)^{3/2}}$

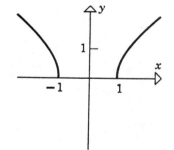

11. $y = 2x + 3x^{2/3}$

$y' = 2 + 2x^{-1/3}$, $y' = 0$
     when $x = -1$

$y'' = -\frac{2}{3} x^{-4/3}$

13. $y = x(3 - x)^{1/2}$

$y' = \dfrac{3(2 - x)}{2\sqrt{3 - x}}$

$y'' = \dfrac{3(x - 4)}{4(3 - x)^{3/2}}$

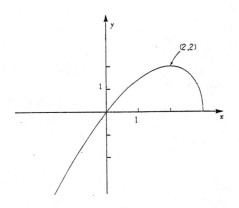

15. $y = \dfrac{8(\sqrt{x} - 1)}{x}$

$y' = \dfrac{4(2 - \sqrt{x})}{x^2}$

$y'' = \dfrac{2(3\sqrt{x} - 8)}{x^3}$

17. $y = \dfrac{\sqrt{x}}{x - 3}$

$y' = -\dfrac{x + 3}{2\sqrt{x}(x - 3)^2}$

$y'' = \dfrac{3(x^2 + 6x - 3)}{4x^{3/2}(x - 3)^3}$

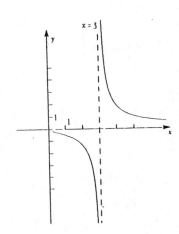

19. y = x − cos x
     y' = 1 + sin x, y' = 0
          when x = −π/2 + 2nπ
     y" = cos x, y" = 0
          when x = π/2 + nπ

21. y = sin x + cos x
     y' = cos x − sin x, y' = 0
          when x = π/4 + nπ
     y" = −sin x − cos x, y" = 0
          when x = 3π/4 + nπ

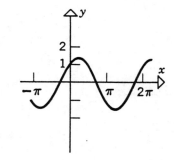

23. y = sin²x, 0 ≤ x ≤ 2π
     y' = 2 sin x cos x = sin 2x
     y" = 2 cos 2x

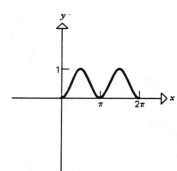

25. (a) retain all points for which
          $f(x) \geq 0$; take the mirror
          image through the x-axis of
          all points for which $f(x) < 0$.

     (b)

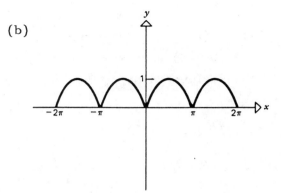

EXERCISE SET 4.6

1. $f'(x) = 8x - 4$, $f'(x) = 0$ when $x = 1/2$; $f(0) = 1$, $f(1/2) = 0$, $f(1) = 1$
    so the maximum value is 1 at $x = 0, 1$ and the minimum value is 0 at
    $x = 1/2$.

3. $f'(x) = 3(x - 1)^2$, $f'(x) = 0$ when $x = 1$; $f(0) = -1$, $f(1) = 0$, $f(4) = 27$
    so the maximum value is 27 at $x = 4$ and the minimum value is -1 at
    $x = 0$.

5. $f'(x) = 3/(4x^2 + 1)^{3/2}$, thus there are no critical points;
    $f(-1) = -3/\sqrt{5}$, $f(1) = 3/\sqrt{5}$ so the maximum value is $3/\sqrt{5}$ at $x = 1$ and the
    minimum value is $-3/\sqrt{5}$ at $x = -1$.

7. $f'(x) = \dfrac{5(8 - x)}{3x^{1/3}}$, $f'(x) = 0$ when $x = 8$ and $f'(x)$ does not exist when
    $x = 0$; $f(-1) = 21$, $f(0) = 0$, $f(8) = 48$, $f(20) = 0$ so the maximum value
    is 48 at $x = 8$ and the minimum value is 0 at $x = 0, 20$.

9. $f'(x) = 1 - \sec^2 x$, $f'(x) = 0$ for x in $(-\pi/4, \pi/4)$ when $x = 0$;
    $f(-\pi/4) = 1 - \pi/4$, $f(0) = 0$, $f(\pi/4) = \pi/4 - 1$ so the maximum value is
    $1 - \pi/4$ at $x = -\pi/4$ and the minimum value is $\pi/4 - 1$ at $x = \pi/4$.

11. $f'(x) = 2\sec x \tan x - \sec^2 x = (2\sin x - 1)/\cos^2 x$, $f'(x) = 0$ for x in
$(0, \pi/4)$ when $x = \pi/6$; $f(0) = 2$, $f(\pi/6) = \sqrt{3}$, $f(\pi/4) = 2\sqrt{2} - 1$ so the
maximum value is 2 at $x = 0$ and the minimum value is $\sqrt{3}$ at $x = \pi/6$.

13. $f(x) = 1 + |9 - x^2| = \begin{cases} 10 - x^2, & |x| \leq 3 \\ -8 + x^2, & |x| > 3 \end{cases}$, $f'(x) = \begin{cases} -2x, & |x| < 3 \\ 2x, & |x| > 3 \end{cases}$ thus
$f'(x) = 0$ when $x = 0$, $f'(x)$ does not exist for x in $(-5,1)$ when $x = -3$
because $\lim\limits_{x \to -3^-} f'(x) \neq \lim\limits_{x \to 3^+} f'(x)$ (see Theorem preceding Exercise 67,
Section 3.3); $f(-5) = 17$, $f(-3) = 1$, $f(0) = 10$, $f(1) = 9$ so the maximum
value is 17 at $x = -5$ and the minimum value is 1 at $x = -3$.

15. $f'(x) = 2x - 3$; critical point $x = 3/2$. Minimum value $f(3/2) = -13/4$,
no maximum.

17. $f'(x) = 12x^2(1 - x)$; critical points $x = 0, 1$. Maximum value $f(1) = 1$,
no minimum because $\lim\limits_{x \to +\infty} f(x) = -\infty$.

19. $(x^2 - 1)^2$ can never be less than zero because it is the square of
$x^2 - 1$; the minimum value is 0 for $x = \pm 1$, no maximum because
$\lim\limits_{x \to +\infty} f(x) = +\infty$.

21. No maximum or minimum because $\lim\limits_{x \to +\infty} f(x) = +\infty$ and $\lim\limits_{x \to -\infty} f(x) = -\infty$.

23. $f'(x) = -1/x^2$; no maximum or minimum because there are no critical
points in $(0, +\infty)$.

25. $f'(x) = x(x + 2)/(x + 1)^2$; critical point $x = -2$ in $(-5, -1)$. Maximum
value $f(-2) = -4$, no minimum.

27. $f(x)$ is periodic with period π, so first consider the interval $[0, \pi]$.
$f'(x) = 4\cos 2x + 4\cos 4x$, $f'(x) = 0$ when $\cos 2x + \cos 4x = 0$, but
$\cos 4x = 2\cos^2 2x - 1$ (trig identity) so

$$2 \cos^2 2x + \cos 2x - 1 = 0$$
$$(2 \cos 2x - 1)(\cos 2x + 1) = 0$$
$$\cos 2x = 1/2 \quad \text{or} \quad \cos 2x = -1.$$

From $\cos 2x = 1/2$, $2x = \pi/3$ or $5\pi/3$ so $x = \pi/6$ or $5\pi/6$.
From $\cos 2x = -1$, $2x = \pi$ so $x = \pi/2$.

$f(0) = 0$, $f(\pi/6) = 3\sqrt{3}/2$, $f(\pi/2) = 0$, $f(5\pi/6) = -3\sqrt{3}/2$, $f(\pi) = 0$. The maximum value is $3\sqrt{3}/2$ at $x = \pi/6 + n\pi$ and the minimum value is $-3\sqrt{3}/2$ at $x = 5\pi/6 + n\pi$, $n = 0, \pm 1, \pm 2, \cdots$.

29. $f'(x) = \begin{cases} 4, & x < 1 \\ 2x - 5, & x > 1 \end{cases}$ so $f'(x) = 0$ when $x = 5/2$, and $f'(x)$ does not exist when $x = 1$ because $\lim\limits_{x \to 1^-} f'(x) \neq \lim\limits_{x \to 1^+} f'(x)$ (see Theorem preceding Exercise 67, Section 3.3); $f(1/2) = 0$, $f(1) = 2$, $f(5/2) = -1/4$, $f(7/2) = 3/4$ so the maximum value is 2 and the minimum value is $-1/4$.

31. $f'(x) = p(x - a)^{p-1}$; critical point $x = a$

(a) if p is even then $p - 1$ is odd and $f'(x) < 0$ when $x < a$, $f'(x) > 0$ when $x > a$ so $f(a) = 0$ is a relative minimum.
(b) if p is odd then $p - 1$ is even and $f'(x)$ does not change sign at $x = a$ so f does not have relative extrema.

33. (a) $f'(x) = -\dfrac{64 \cos x}{\sin^2 x} + \dfrac{27 \sin x}{\cos^2 x} = \dfrac{-64 \cos^3 x + 27 \sin^3 x}{\sin^2 x \cos^2 x}$,

$f'(x) = 0$ when $27 \sin^3 x = 64 \cos^3 x$, $\tan^3 x = 64/27$, $\tan x = 4/3$ so the critical point is $x = x_0$ where $\tan x_0 = 4/3$ and $0 < x_0 < \pi/2$.

To test x_0 first rewrite $f'(x)$ as

$$f'(x) = \frac{27 \cos^3 x \, (\tan^3 x - 64/27)}{\sin^2 x \cos^2 x} = \frac{27 \cos x (\tan^3 x - 64/27)}{\sin^2 x};$$

if $x < x_0$ then $\tan x < 4/3$ and $f'(x) < 0$,

if $x > x_0$ then $\tan x > 4/3$ and $f'(x) > 0$

so $f(x_0)$ is the minimum value. f has no maximum because

$\lim\limits_{x \to 0^+} f(x) = +\infty$.

(b) If $\tan x_0 = 4/3$ then (see figure)

$\sin x_0 = 4/5$ and $\cos x_0 = 3/5$

so $f(x_0) = 64/\sin x_0 + 27/\cos x_0$

$= 64/(4/5) + 27/(3/5)$

$= 80 + 45 = 125$

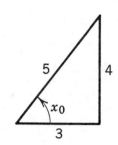

35. $(0,9)$ is on the graph so $f(0) = a_0 + a_1(0) + a_2(0)^2 = 9$, $a_0 = 9$

thus $f(x) = 9 + a_1 x + a_2 x^2$. $(2,1)$ is on the graph so
$f(2) = 9 + 2a_1 + 4a_2 = 1$,

$\qquad a_1 + 2a_2 = -4$ \hfill (i).

$f'(x) = a_1 + 2a_2 x$, but $f'(2) = 0$ because it is given that $f(2)$ is to be
an extreme value, so
$\qquad f'(2) = a_1 + 4a_2 = 0$ \hfill (ii).

Solve (i) and (ii) to get $a_1 = -8$ and $a_2 = 2$, thus $f(x) = 9 - 8x + 2x^2$.
As a check, we find that $f''(x) = 4 > 0$ so $f(2)$ is a minimum.

37. Let $f(x) = 1 - x^2/2 - \cos x$, then $f'(x) = -x + \sin x$ so $f'(x) = 0$ when
$\sin x = x$ which has no solution for $0 < x < 2\pi$ thus the maximum and
minimum values must occur at the endpoints of $[0, 2\pi]$. $f(0) = 0$,
$f(2\pi) = -2\pi^2$, so 0 is the maximum value, thus $1 - x^2/2 - \cos x \leq 0$,
$1 - x^2/2 \leq \cos x$ for all x in $[0, 2\pi]$.

39. By the quadratic formula, the roots are $x_1 = \dfrac{-b - \sqrt{b^2 - 4ac}}{2a}$ and

$x_2 = \dfrac{-b + \sqrt{b^2 - 4ac}}{2a}$. The midpoint is $\frac{1}{2}(x_1 + x_2) = -\dfrac{b}{2a}$. But

$f'(x) = 2ax + b$ so $f'(-\dfrac{b}{2a}) = 2a(-\dfrac{b}{2a}) + b = 0$.

41. $f'(x) = 2ax + b$; critical point is $x = -\dfrac{b}{2a}$

$f''(x) = 2a > 0$ so $f(-\dfrac{b}{2a})$ is the minimum value of f, but

$$f(-\frac{b}{2a}) = a(-\frac{b}{2a})^2 + b(-\frac{b}{2a}) + c = \frac{-b^2 + 4ac}{4a} \text{ thus } f(x) \geq 0 \text{ if and only}$$

$$\text{if } f(-\frac{b}{2a}) \geq 0, \frac{-b^2 + 4ac}{4a} \geq 0, -b^2 + 4ac \geq 0, b^2 - 4ac \leq 0$$

43. $f(\theta) = \sin^2\theta \cos\theta$, $f'(\theta) = \sin\theta (2\cos^2\theta - \sin^2\theta) = \sin\theta(3\cos^2\theta - 1)$;
      $f'(\theta) = 0$ for θ in $(0,\pi/2)$ if $\cos\theta = 1/\sqrt{3}$. $f(0) = 0$ and $f(\pi/2) = 0$, so
      the maximum value occurs when $\cos\theta = 1/\sqrt{3}$ where
      $f(\theta) = \sin^2\theta \cos\theta = (1 - \cos^2\theta)\cos\theta = (1 - 1/3)(1/\sqrt{3}) = 2/(3\sqrt{3})$.

EXERCISE SET 4.7

1.   Let x = one number, y = the other number, and $P = xy$ where $x + y = 10$.
     Thus $y = 10 - x$ so $P = x(10 - x) = 10x - x^2$ for x in $[0,10]$.
     $dP/dx = 10 - 2x$, $dP/dx = 0$ when $x = 5$.   If $x = 0$, 5, 10 then $P = 0$, 25,
     0 so P is maximum when $x = 5$ and, from $y = 10 - x$, when $y = 5$.

3.   If $y = x + 1/x$ for $1/2 \leq x \leq 3/2$ then $dy/dx = 1 - 1/x^2 = (x^2 - 1)/x^2$,
     $dy/dx = 0$ when $x = 1$.   If $x = 1/2$, 1, 3/2 then $y = 5/2$, 2, 13/6 so
     (a) y is as small as possible when $x = 1$.
     (b) y is as large as possible when $x = 1/2$.

5.   Let x and y be the dimensions shown
     in the figure and A the area, then
     $A = xy$ subject to the cost condition
     $3(2x) + 2(2y) = 6000$, or $y = 1500 - 3x/2$.

     Thus $A = x(1500 - 3x/2) = 1500x - 3x^2/2$
     for x in $[0,1000]$. $dA/dx = 1500 - 3x$,
     $dA/dx = 0$ when $x = 500$.   If $x = 0$ or
     1000 then $A = 0$, if $x = 500$ then
     $A = 375,000$ so the area is greatest when $x = 500$ ft and (from
     $y = 1500 - 3x/2$) when $y = 750$ ft.

7. $A = xy$ where $x^2 + y^2 = 20^2 = 400$ so

$y = \sqrt{400 - x^2}$ and $A = x\sqrt{400 - x^2}$ for

$0 \leq x \leq 20$; $dA/dx = 2(200 - x^2)/\sqrt{400 - x^2}$,

$dA/dx = 0$ when $x = \sqrt{200} = 10\sqrt{2}$. If

$x = 0$, $10\sqrt{2}$, 20 then $A = 0$, 200, 0 so

the area is maximum when $x = 10\sqrt{2}$ and

$y = \sqrt{400 - 200} = 10\sqrt{2}$.

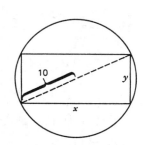

9. Let x, y, and z be as shown in the figure
    and A the area of the rectangle, then
    $A = xy$ and, by similar triangles, $z/10 = y/6$,
    $z = 5y/3$; also $x/10 = (8 - z)/8 = (8 - 5y/3)/8$
    thus $y = 24/5 - 12x/25$ so $A = x(24/5 - 12x/25) =$
    $24x/5 - 12x^2/25$ for x in $[0,10]$.
    $dA/dx = 24/5 - 24x/25$, $dA/dx = 0$ when
    $x = 5$. If $x = 0$, 5, 10 then $A = 0$, 12,
    0 so the area is greatest when $x = 5$ in.
    and $y = 12/5$ in.

11. $V = x(12 - 2x)^2$ for $0 \leq x \leq 6$;
    $dV/dx = 12(x - 2)(x - 6)$, $dV/dx = 0$
    when $x = 2$ for $0 < x < 6$. If $x = 0$,
    2, 6 then $V = 0$, 128, 0 so the volume
    is largest when $x = 2$ in.

13. Let x be the length of each side of a square, then
    $V = x(3 - 2x)(8 - 2x) = 4x^3 - 22x^2 + 24x$ for $0 \leq x \leq 3/2$;
    $dV/dx = 12x^2 - 44x + 24 = 4(3x - 2)(x - 3)$, $dV/dx = 0$ when $x = 2/3$ for

$0 < x < 3/2$. If $x = 0$, $2/3$, $3/2$ then $V = 0$, $200/27$, 0 so the maximum volume is $200/27$ ft^3.

15. With x, y, r, and s as shown in the figure, the sum of the enclosed areas is $A = \pi r^2 + s^2$ where $r = \dfrac{x}{2\pi}$ and $s = \dfrac{y}{4}$ because x is the circumference of the circle and y is the perimeter of the square, thus $A = \dfrac{x^2}{4\pi} + \dfrac{y^2}{16}$. But $x + y = 12$, so $y = 12 - x$ and

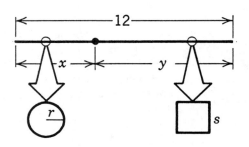

$$A = \frac{x^2}{4\pi} + \frac{(12 - x)^2}{16}$$

$$= \frac{\pi + 4}{16\pi} x^2 - \frac{3}{2} x + 9$$

for $0 \leq x \leq 12$. $\dfrac{dA}{dx} = \dfrac{\pi + 4}{8\pi} x - \dfrac{3}{2}$, $\dfrac{dA}{dx} = 0$ when $x = \dfrac{12\pi}{\pi + 4}$. If $x = 0$, $\dfrac{12\pi}{\pi + 4}$, 12 then $A = 9$, $\dfrac{36}{\pi + 4}$, $\dfrac{36}{\pi}$ so the sum of the enclosed areas is
   (a) a maximum when $x = 12$ in. (when all of the wire is used for the circle)
   (b) a minimum when $x = 12\pi/(\pi + 4)$ in.

17. The altitude of the triangle (see figure) is $\sqrt{y^2 - x^2/4}$ so $A = \dfrac{1}{2} x\sqrt{y^2 - x^2/4}$ where $x + 2y = 12$ thus $y = 6 - x/2$ and

$$A = \frac{1}{2} x\sqrt{(6 - x/2)^2 - x^2/4} = \frac{1}{2} x\sqrt{36 - 6x}$$

for $0 \leq x \leq 6$; $dA/dx = 9(4 - x)/(2\sqrt{36 - 6x})$, $dA/dx = 0$ when $x = 4$ for $0 < x < 6$. If $x = 0$, 4, 6 then $A = 0$, $4\sqrt{3}$, 0 so the area is maximum when $x = 4$ and $y = 6 - 4/2 = 4$, which is an equilateral triangle.

19. Let x be the length of each side of the squares and y the height of the frame, then the volume is $V = x^2 y$. The total length of the wire is L thus $8x + 4y = L$, $y = (L - 8x)/4$ so $V = x^2(L - 8x)/4 = (Lx^2 - 8x^3)/4$ for $0 \leq x \leq L/8$. $dV/dx = (2Lx - 24x^2)/4$, $dV/dx = 0$ for $0 < x < L/8$ when

x = L/12. If x = 0, L/12, L/8 then V = 0, $L^3/1728$, 0 so the volume is greatest when x = L/12 and y = L/12.

21. (a) The daily profit is
$$P = (\text{revenue}) - (\text{production cost})$$
$$= 100x - (100,000 + 50x + 0.0025x^2)$$
$$= -100,000 + 50x - 0.0025x^2$$
for $0 \le x \le 7000$, so dP/dx = 50 - 0.005x and dP/dx = 0 when x = 10,000. Because 10,000 is not in the interval [0,7000], the maximum profit must occur at an endpoint. When x = 0, P = -100,000; when x = 7000, P = 127,500 so 7000 units should be manufactured and sold daily.

(b) Yes, because dP/dx > 0 when x = 7000 so profit is increasing at this production level.

23. Let h and r be the dimensions shown in the figure, then the volume is $V = \frac{1}{3}\pi r^2 h$.

But $r^2 + h^2 = L^2$ thus $r^2 = L^2 - h^2$ so
$V = \frac{1}{3}\pi(L^2 - h^2)h = \frac{1}{3}\pi(L^2 h - h^3)$
for $0 \le h \le L$. $\frac{dV}{dh} = \frac{1}{3}\pi(L^2 - 3h^2)$.

$\frac{dV}{dh} = 0$ when $h = L/\sqrt{3}$. If h = 0, $L/\sqrt{3}$, 0

then V = 0, $\dfrac{2\pi}{9\sqrt{3}}L^3$, 0 so the volume is as

large as possible when $h = L/\sqrt{3}$ and $r = \sqrt{2/3}L$.

25. Let r and h be the dimensions shown in the figure, then the surface area is $S = 2\pi rh + 2\pi r^2$.

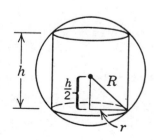

But $r^2 + (\frac{h}{2})^2 = R^2$ thus $h = 2\sqrt{R^2 - r^2}$ so

$S = 4\pi r\sqrt{R^2 - r^2} + 2\pi r^2$ for $0 \le r \le R$,
$\dfrac{dS}{dr} = \dfrac{4\pi(R^2 - 2r^2)}{\sqrt{R^2 - r^2}} + 4\pi r$. $\dfrac{dS}{dr} = 0$ when

$$\frac{R^2 - 2r^2}{\sqrt{R^2 - r^2}} = -r \qquad (i)$$

$$R^2 - 2r^2 = -r\sqrt{R^2 - r^2}$$

$$R^4 - 4R^2r^2 + 4r^4 = r^2(R^2 - r^2)$$

$$5r^2 - 5R^2r^2 + R^4 = 0$$

and using the quadratic formula $r^2 = \dfrac{5R^2 \pm \sqrt{25R^4 - 20R^4}}{10} = \dfrac{5 \pm \sqrt{5}}{10} R^2$,

$r = \sqrt{\dfrac{5 \pm \sqrt{5}}{10}} R$, of which only $r = \sqrt{\dfrac{5 + \sqrt{5}}{10}} R$ satisfies (i).

If $r = 0, \sqrt{\dfrac{5 + \sqrt{5}}{10}} R, 0$ then $S = 0, (5 + \sqrt{5})\pi R^2, 2\pi R^2$ so the surface area

is greatest when $r = \sqrt{\dfrac{5 + \sqrt{5}}{10}} R$ and, from $h = 2\sqrt{R^2 - r^2}$, $h = 2\sqrt{\dfrac{5 - \sqrt{5}}{10}} R$.

27. Let b and h be the dimensions shown in the figure, then the cross-sectional area is $A = \frac{1}{2} h(5 + b)$. But $h = 5\sin\theta$ and $b = 5 + 2(5\cos\theta) = 5 + 10\cos\theta$ so $A = \frac{5}{2}\sin\theta(10 + 10\cos\theta) =$ $25\sin\theta(1 + \cos\theta)$ for $0 \le \theta \le \pi/2$.

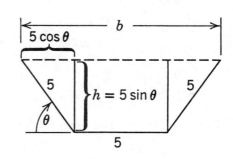

$$dA/d\theta = -25\sin^2\theta + 25\cos\theta(1 + \cos\theta)$$

$$= 25(-\sin^2\theta + \cos\theta + \cos^2\theta)$$

$$= 25(-1 + \cos^2\theta + \cos\theta + \cos^2\theta) = 25(2\cos^2\theta + \cos\theta - 1)$$

$$= 25(2\cos\theta - 1)(\cos\theta + 1).$$

$dA/d\theta = 0$ for $0 < \theta < \pi/2$ when $\cos\theta = 1/2$, $\theta = \pi/3$. If $\theta = 0, \pi/3, \pi/2$ then $A = 0, 75\sqrt{3}/4, 25$ so the cross-sectional area is greatest when $\theta = \pi/3$.

29. Let r and h be the radius and height of the cone (see figure). The slant height of any such cone will be R, the radius of the circular sheet. Refer to the solution of Exercise 23 to find that the largest volume is $\dfrac{2\pi}{9\sqrt{3}} R^3$.

31. (a) $C'(x) = 4 + 0.2x$, $C'(100) = 24$ (b) $C'(100) = 24$
     (c) $C(101) - C(100) = 24.1$
     (d) $R(x) = 10x$, $R'(x) = 10$;
         $P(x) = R(x) - C(x)$, $P'(x) = R'(x) - C'(x)$
         $\qquad\qquad\qquad\qquad\quad = 10 - (4 + 0.2x) = 6 - 0.2x$

33. Let $P(x,y)$ be a point on the curve $x^2 + y^2 = 1$. The distance between

     $P(x,y)$ and $P_0(2,0)$ is $D = \sqrt{(x-2)^2 + y^2}$, but $y^2 = 1 - x^2$ so

     $D = \sqrt{(x-2)^2 + 1 - x^2} = \sqrt{5 - 4x}$ for $-1 \leq x \leq 1$,
     $\dfrac{dD}{dx} = -\dfrac{2}{\sqrt{5-4x}}$ which has no critical points for $-1 < x < 1$.
     If $x = -1$, 1 then $D = 3$, 1 so the closest point occurs when $x = 1$ and
     $y = 0$.

35. The area of the window is $A = 2rh + \pi r^2/2$,
     the perimeter is $p = 2r + 2h + \pi r$ thus
     $h = \dfrac{1}{2}[p - (2 + \pi)r]$ so

     $A = r[p - (2 + \pi)r] + \pi r^2/2$.

     $\quad = pr - (2 + \pi/2)r^2$ for $0 \leq r \leq p/(2 + \pi)$,
     $dA/dr = p - (4 + \pi)r$, $dA/dr = 0$ when

     $r = p/(4 + \pi)$. $d^2A/dr^2 < 0$, so A is
     maximum when $r = p/(4 + \pi)$.

37. (a) Let x = diameter of the sphere, y = length of an edge of the cube.
         The combined volume is $V = \dfrac{1}{6}\pi x^3 + y^3$ and the surface area is

         $S = \pi x^2 + 6y^2 = $ constant. Thus $y = \dfrac{(S - \pi x^2)^{1/2}}{6^{1/2}}$ and

         $V = \dfrac{\pi}{6}x^3 + \dfrac{(S - \pi x^2)^{3/2}}{6^{3/2}}$ for $0 \leq x \leq \sqrt{\dfrac{S}{\pi}}$;

         $\dfrac{dV}{dx} = \dfrac{\pi}{2}x^2 - \dfrac{3\pi}{6^{3/2}}x(S - \pi x^2)^{1/2} = \dfrac{\pi}{2\sqrt{6}}x(\sqrt{6}x - \sqrt{S - \pi x^2})$.

         $\dfrac{dV}{dx} = 0$ when $x = 0$, or when $\sqrt{6}x = \sqrt{S - \pi x^2}$, $6x^2 = S - \pi x^2$,

$$x^2 = \frac{S}{6 + \pi}, \quad x = \sqrt{\frac{S}{6 + \pi}}. \quad \text{If } x = 0, \sqrt{\frac{S}{6 + \pi}}, \sqrt{\frac{S}{\pi}}, \text{ then } V = \frac{S^{3/2}}{6^{3/2}},$$

$$\frac{S^{3/2}}{6\sqrt{6 + \pi}}, \frac{S^{3/2}}{6\sqrt{\pi}} \text{ so that V is smallest when } x = \sqrt{\frac{S}{6 + \pi}}, \text{ and hence when}$$

$$y = \sqrt{\frac{S}{6 + \pi}}, \text{ thus } x = y.$$

(b) From part (a), the sum of the volumes is greatest when there is no cube.

39. (a) Let $y = x^2 - 3x + 2$ for $1 \leq x \leq 5/2$, then $dy/dx = 2x - 3$, $dy/dx = 0$ when $x = 3/2$. If $x = 1, 3/2, 5/2$ then $y = 0, -1/4, 3/4$ so

$-1/4 \leq x^2 - 3x + 2 \leq 3/4$ for $1 \leq x \leq 5/2$. Thus

$|x^2 - 3x + 2| \leq 3/4$ so M = 3/4.

(b) There are no critical values for $3/2 < x < 7/4$. If $x = 3/2, 7/4$ then $y = -1/4, -3/16$ so $-1/4 \leq x^2 - 3x + 2 \leq -3/16$. Thus

$|x^2 - 3x + 2| \geq 3/16$ so m = 3/16.

41. Let v = speed of light in the medium. The total time required for the light to travel from A to P to B is

$$t = (\text{total distance from A to P to B})/v$$

$$= \frac{1}{v} \left(\sqrt{(c - x)^2 + a^2} + \sqrt{x^2 + b^2} \right),$$

$$\frac{dt}{dx} = \frac{1}{v} \left[-\frac{c - x}{\sqrt{(c - x)^2 + a^2}} + \frac{x}{\sqrt{x^2 + b^2}} \right]$$

and $\frac{dt}{dx} = 0$ when $\dfrac{x}{\sqrt{x^2 + b^2}} = \dfrac{c - x}{\sqrt{(c - x)^2 + a^2}}$.

But $x/\sqrt{x^2 + b^2} = \sin\theta_2$ and $(c - x)/\sqrt{(c - x)^2 + a^2} = \sin\theta_1$ thus $dt/dx = 0$ when $\sin\theta_2 = \sin\theta_1$ so $\theta_2 = \theta_1$.

43. (a) The rate at which the farmer walks is
         analogous to the speed of light in
         Fermat's principle.
     (b) the best path occurs when $\theta_1 = \theta_2$
         (see figure).
     (c) by similar triangles,
         $x/(1/4) = (1 - x)/(3/4)$
         $$3x = 1 - x$$
         $$4x = 1$$
         $$x = 1/4.$$

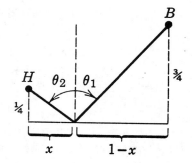

EXERCISE SET 4.8

1. Let x and y be two numbers, then their product is $P = xy$. But
    $x + y = 20$ thus $y = 20 - x$ so
    $$P = x(20 - x) = 20x - x^2 \text{ for } -\infty < x < +\infty,$$
    $dP/dx = 20 - 2x$, $dP/dx = 0$ when $x = 10$, $d^2P/dx^2 = -2$ so
    (a) P is a maximum when $x = 10$ and $y = 10$ and
    (b) P has no minimum.

3. Let x = length of each side that uses the $1 per foot fencing,
        y = length of each side that uses the $2 per foot fencing.
    The cost is $C = (1)(2x) + (2)(2y) = 2x + 4y$, but $A = xy = 3200$ thus
    $y = 3200/x$ so
    $$C = 2x + 12800/x \text{ for } x > 0,$$
    $dC/dx = 2 - 12800/x^2$, $dC/dx = 0$ when $x = 80$, $d^2C/dx^2 > 0$ so
    C is least when $x = 80$, $y = 40$.

5. Let x = length of each edge of base, y = height, $k = \$/cm^2$ for the
    sides. The cost is
    $$C = (2k)(2x^2) + (k)(4xy) = 4k(x^2 + xy),$$
    but $V = x^2y = 2000$ thus $y = 2000/x^2$ so
    $$C = 4k(x^2 + 2000/x) \text{ for } x > 0,$$
    $dC/dx = 4k(2x - 2000/x^2)$, $dC/dx = 0$ when $x = \sqrt[3]{1000} = 10$,
    $d^2C/dx^2 > 0$ so C is least when $x = 10$, $y = 20$.

7. Let x = height and width, y = length. The surface area is $S = 2x^2 + 3xy$ where $x^2y = V$, so $y = V/x^2$ and $S = 2x^2 + 3V/x$ for $x > 0$; $dS/dx = 4x - 3V/x^2$, $dS/dx = 0$ when $x = \sqrt[3]{3V/4}$, $d^2S/dx^2 > 0$ so S is minimum when $x = \sqrt[3]{\dfrac{3V}{4}}$, $y = \dfrac{4}{3}\sqrt[3]{\dfrac{3V}{4}}$.

9. The surface area is $S = \pi r^2 + 2\pi rh$ where $V = \pi r^2 h = 500$ so $h = 500/(\pi r^2)$ and $S = \pi r^2 + 1000/r$ for $r > 0$; $dS/dr = 2\pi r - 1000/r^2 = (2\pi r^3 - 1000)/r^2$, $dS/dr = 0$ when $r = \sqrt[3]{500/\pi}$, $d^2S/dr^2 > 0$ for $r > 0$ so S is minimum when $r = \sqrt[3]{500/\pi}$ and

$$h = \frac{500}{\pi r^2} = \frac{500}{\pi r^2} \, r = \frac{500}{\pi(500/\pi)}\sqrt[3]{500/\pi}$$

$$= \sqrt[3]{500/\pi}.$$

11. The area of the paper is

$A = \pi rL = \pi r\sqrt{r^2 + h^2}$, but

$V = \dfrac{1}{3}\pi r^2 h = 10$ thus $h = 30/(\pi r^2)$

so $A = \pi r\sqrt{r^2 + 900/(\pi^2 r^4)}$.

To simplify the computations let $S = A^2$,

$$S = \pi^2 r^2\left(r^2 + \frac{900}{\pi^2 r^4}\right) = \pi^2 r^4 + \frac{900}{r^2} \text{ for } r > 0,$$

$$\frac{dS}{dr} = 4\pi^2 r^3 - \frac{1800}{r^3} = \frac{4(\pi^2 r^6 - 450)}{r^3},$$

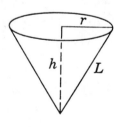

$dS/dr = 0$ when $r = \sqrt[6]{450/\pi^2}$, $d^2S/dr^2 > 0$, so S and hence A is least when $r = \sqrt[6]{450/\pi^2}$, $h = \dfrac{30}{\pi}\sqrt[3]{\pi^2/450}$.

13. Let (x,y) be a point on the curve, then the square of the distance between (x,y) and (0,2) is $S = x^2 + (y - 2)^2$ where $x^2 - y^2 = 1$,

$x^2 = y^2 + 1$ so

$$S = (y^2 + 1) + (y - 2)^2 = 2y^2 - 4y + 5 \text{ for any } y,$$
$$dS/dy = 4y - 4, \quad dS/dy = 0 \text{ when } y = 1,$$

$d^2S/dy^2 > 0$ so S is least when $y = 1$ and $x = \pm\sqrt{2}$.

15. The area of the triangle is $A = \frac{1}{2} ab$. Equate slopes to get

$\dfrac{b - 3}{0 - 1} = \dfrac{0 - 3}{a - 1}$, $b = \dfrac{3a}{a - 1}$ so

$$A = \frac{3}{2} \frac{a^2}{a - 1} \text{ for } a > 1,$$
$$\frac{dA}{da} = \frac{3a(a - 2)}{(a - 1)^2}, \quad \frac{dA}{da} = 0 \text{ for } a > 1 \text{ when } a = 2.$$

(a) there is no maximum for A because $\lim\limits_{a \to 1^+} A = +\infty$.

(b) by the first derivative test A is minimum when $a = 2$ so the slope is
$m = \dfrac{0 - 3}{2 - 1} = -3$.

17. The distance between the particles is

$D = \sqrt{(1 - t - t)^2 + (t - 2t)^2} = \sqrt{5t^2 - 4t + 1}$ for $t \geq 0$. For
convenience, we minimize D^2 instead, so $D^2 = 5t^2 - 4t + 1$,
$dD^2/dt = 10t - 4$, which is 0 when $t = 2/5$. $d^2D^2/dt^2 > 0$ so D^2 and hence
D is minimum when $t = 2/5$. The minimum distance is $D = 1/\sqrt{5}$.

19. If $P(x_0, y_0)$ is on the curve $y = 1/x^2$, then $y_0 = 1/x_0^2$. At P the slope of
the tangent line is $-2/x_0^3$ so its equation is

$y - \dfrac{1}{x_0^2} = -\dfrac{2}{x_0^3}(x - x_0)$, or $y = -\dfrac{2}{x_0^3} x + \dfrac{3}{x_0^2}$. The tangent line crosses

the y-axis at $\dfrac{3}{x_0^2}$, and the x-axis at $\dfrac{3}{2} x_0$. The length of the segment

then is $L = \sqrt{\dfrac{9}{x_0^4} + \dfrac{9}{4} x_0^2}$ for $x_0 > 0$. For convenience, we minimize L^2

instead, so $L^2 = \dfrac{9}{4x_0} + \dfrac{9}{4}x_0^2$,

$\dfrac{dL^2}{dx_0} = -\dfrac{36}{x_0^5} + \dfrac{9}{2}x_0 = \dfrac{9(x_0^6 - 8)}{2x_0^5}$, which is 0 when $x_0^6 = 8$, $x_0 = \sqrt{2}$.

$\dfrac{d^2L^2}{dx_0^2} > 0$ so L^2 and hence L is minimum when $x_0 = \sqrt{2}$, $y_0 = 1/2$.

21. At each point (x,y) on the curve the slope of the tangent line is

$m = \dfrac{dy}{dx} = -\dfrac{2x}{(1 + x^2)^2}$ for any x, $\dfrac{dm}{dx} = \dfrac{2(3x^2 - 1)}{(1 + x^2)^3}$, $\dfrac{dm}{dx} = 0$ when $x = \pm 1/\sqrt{3}$,

by the first derivative test the only relative maximum occurs at

$x = -1/\sqrt{3}$, which is the absolute maximum because $\lim\limits_{x\to\pm\infty} m = 0$.   The

tangent line has greatest slope at the point $(-1/\sqrt{3},\ 3/4)$.

23. The volume of the cone is $V = \dfrac{1}{3}\pi r^2 h$.

By similar triangles (see figure)

$\dfrac{r}{h} = \dfrac{R}{\sqrt{h^2 - 2Rh}}$, $r = \dfrac{Rh}{\sqrt{h^2 - 2Rh}}$ so

$V = \dfrac{1}{3}\pi R^2 \dfrac{h^3}{h^2 - 2Rh} = \dfrac{1}{3}\pi R^2 \dfrac{h^2}{h - 2R}$

for $h > 2R$, $\dfrac{dV}{dh} = \dfrac{1}{3}\pi R^2 \dfrac{h(h - 4R)}{(h - 2R)^2}$,

$\dfrac{dV}{dh} = 0$ for $h > 2R$ when $h = 4R$,

by the first derivative test V is

minimum when $h = 4R$.   If $h = 4R$ then $r = \sqrt{2}R$.

25. With x and y as shown in the figure, the maximum length of pipe will be the _smallest_ value of $L = x + y$. By similar triangles

$$\frac{y}{8} = \frac{x}{\sqrt{x^2 - 16}},$$

$$y = \frac{8x}{\sqrt{x^2 - 16}} \quad \text{so}$$

$$L = x + \frac{8x}{\sqrt{x^2 - 16}} \quad \text{for } x > 4,$$

$$\frac{dL}{dx} = 1 - \frac{128}{(x^2 - 16)^{3/2}},$$

$\frac{dL}{dx} = 0$ when

$$(x^2 - 16)^{3/2} = 128$$
$$x^2 - 16 = 128^{2/3} = 16(2^{2/3})$$
$$x^2 = 16(1 + 2^{2/3})$$
$$x = 4(1 + 2^{2/3})^{1/2},$$

$d^2L/dx^2 = 384x/(x^2 - 16)^{5/2} > 0$ if $x > 4$ so L is smallest when $x = 4(1 + 2^{2/3})^{1/2}$. For this value of x, $L = 4(1 + 2^{2/3})^{3/2}$.

27. Let x = distance from the weaker light source, I = the intensity at that point, and k the constant of proportionality. Then

$$I = \frac{kS}{x^2} + \frac{8kS}{(90 - x)^2} \quad \text{for } 0 < x < 90.$$

$$\frac{dI}{dx} = -\frac{2kS}{x^3} + \frac{16kS}{(90 - x)^3} = \frac{2kS[8x^3 - (90 - x)^3]}{x^3(90 - x)^3}, \text{ which is 0 when}$$

$8x^3 = (90 - x)^3$, $2x = 90 - x$, $x = 30$. $\frac{dI}{dx} < 0$ if $x < 30$, and $\frac{dI}{dx} > 0$ if $x > 30$, so the intensity is minimum at a distance of 30 cm from the weaker source.

29. Minimize $S = L^2 = (x - x_1)^2 + (y - y_1)^2$ where $ax + by + c = 0$. If $b \neq 0$ then $y = -\frac{a}{b}x - \frac{c}{b}$ so

$$S = (x - x_1)^2 + (-\frac{a}{b}x - \frac{c}{b} - y_1)^2 \text{ for all } x,$$

$$dS/dx = 2(x - x_1)^2 + 2(-\frac{a}{b}x - \frac{c}{b} - y_1)^2(-\frac{a}{b}), \ dS/dx = 0 \text{ when}$$

$$x = \frac{b^2 x_1 - aby_1 - ac}{a^2 + b^2}, \ d^2S/dx^2 = 2(1 + a^2/b^2) > 0$$

so S and hence L is minimum. Substitution of this value into the formula for S and simplification eventually gives

$$S = \frac{(ax_1 + by_1 + c)^2}{a^2 + b^2} \text{ so } L = \frac{|ax_1 + by_1 + c|}{\sqrt{a^2 + b^2}}.$$

The special case for b = 0 is treated in a similar way.

EXERCISE SET 4.9

1. $f(x) = x^2 - 2$

   $f'(x) = 2x$

   $x_{n+1} = \frac{x_n^2 + 2}{2x_n}$

   $x_1 = 1$

   $x_2 = 1.5$

   $x_3 = 1.4166667$

   $x_4 = 1.4142157$

   $x_5 = 1.4142136$

   $x_6 = 1.4142136$

3. $f(x) = x^3 - 6$

   $f'(x) = 3x^2$

   $x_{n+1} = \frac{2x_n^3 + 6}{3x_n^2}$

   $x_1 = 2$

   $x_2 = 1.8333333$

   $x_3 = 1.8172635$

   $x_4 = 1.8171206$

   $x_5 = 1.8171206$

5. $f(x) = x^3 - x + 3$

   $f'(x) = 3x^2 - 1$

   $x_{n+1} = \frac{2x_n^3 - 3}{3x_n^2 - 1}$

   $x_1 = -2$

7. $f(x) = x^5 + x^4 - 5$

   $f'(x) = 5x^4 + 4x^3$

   $x_{n+1} = \frac{4x_n^5 + 3x_n^4 + 5}{5x_n^4 + 4x_n^3}$

   $x_1 = 1$

$x_2 = -1.7272727$

$x_3 = -1.6736912$

$x_4 = -1.6717026$

$x_5 = -1.6716999$

$x_6 = -1.6716999$

$x_2 = 1.3333333$

$x_3 = 1.2394206$

$x_4 = 1.2247627$

$x_5 = 1.2244397$

$x_6 = 1.2244395$

$x_7 = 1.2244395$

9. $f(x) = 2x^2 + 4x - 3$

$f'(x) = 4x + 4$

$x_{n+1} = \dfrac{2x_n^2 + 3}{4x_n + 4}$

$x_1 = 1$

$x_2 = 0.625$

$x_3 = 0.5817307$

$x_4 = 0.5811389$

$x_5 = 0.5811388$

$x_6 = 0.5811388$

11. $f(x) = x^4 + x - 3$

$f'(x) = 4x^3 + 1$

$x_{n+1} = \dfrac{3x_n^4 + 3}{4x_n^3 + 1}$

$x_1 = -2$

$x_2 = -1.6451613$

$x_3 = -1.485724$

$x_4 = -1.4538064$

$x_5 = -1.4526284$

$x_6 = -1.4526269$

$x_7 = -1.4526269$

13. $f(x) = 2\sin x - x$

$f'(x) = 2\cos x - 1$

$x_{n+1} = \dfrac{2x_n \cos x_n - 2\sin x_n}{2\cos x_n - 1}$

$x_1 = 2$

$x_2 = 1.9009956$

$x_3 = 1.8955116$

$x_4 = 1.8954943$

$x_5 = 1.8954943$

15. $f(x) = x - \tan x$

$f'(x) = 1 - \sec^2 x$

$$x_{n+1} = \frac{\tan x_n - x_n \sec^2 x_n}{1 - \sec^2 x_n} = \frac{\sin x_n \cos x_n - x_n}{\cos^2 x_n - 1} = \frac{x_n - \sin x_n \cos x_n}{\sin^2 x_n}$$

$x_1 = 4.5$

$x_2 = 4.4936139$

$x_3 = 4.4934097$

$x_4 = 4.4934095$

$x_5 = 4.4934095$

17. (a) $f(x) = x^2 - a$
        $f'(x) = 2x$

        $x_{n+1} = \frac{1}{2}\left(x_n + \frac{a}{x_n}\right)$

(b) $a = 10,$
    $x_1 = 3$

    $x_2 = 3.1666667$

    $x_3 = 3.1622807$

    $x_4 = 3.1622777$

    $x_5 = 3.1622777$

19. At the point of intersection, $x^3 = 0.5x - 1$, $x^3 - 0.5x + 1 = 0$. Let $f(x) = x^3 - 0.5x + 1$. By graphing $y = x^3$ and $y = 0.5x - 1$ it is evident that there is only one point of intersection and it occurs in the interval $[-2,-1]$; note that $f(-2) < 0$ and $f(-1) > 0$. $f'(x) = 3x^2 - 0.5$ so

$$x_{n+1} = x_n - \frac{x_n^3 - 0.5x + 1}{3x_n^2 - 0.5}$$

$x_1 = -1$

$x_2 = -1.2$

$x_3 = -1.1664921$

$x_4 = -1.1653742$

$x_5 = -1.1653730$

$x_6 = -1.1653730$

EXERCISE SET 4.10

1. $f(2) = f(4) = 0$
$f'(x) = 2x - 6$
$2c - 6 = 0$
$c = 3$

3. $f(\pi/2) = f(3\pi/2) = 0$
$f'(x) = -\sin x$
$-\sin c = 0$
$c = \pi$

5. $f(0) = f(4) = 0$
$f'(x) = \dfrac{1}{2} - \dfrac{1}{2\sqrt{x}}$

$\dfrac{1}{2} - \dfrac{1}{2\sqrt{c}} = 0$
$c = 1$

7. $f(-4) = 12, \ f(6) = 42$
$f'(x) = 2x + 1$

$2c + 1 = \dfrac{42 - 12}{6 - (-4)} = 3$
$c = 1$

9. $f(0) = 1, \ f(3) = 2$
$f'(x) = \dfrac{1}{2\sqrt{x + 1}}$

$\dfrac{1}{2\sqrt{c + 1}} = \dfrac{2 - 1}{3 - 0} = \dfrac{1}{3}$

$\sqrt{c + 1} = 3/2$
$c + 1 = 9/4$
$c = 5/4$

11. $f(-5) = 0, \ f(3) = 4$
$f'(x) = -\dfrac{x}{\sqrt{25 - x^2}}$

$-\dfrac{c}{\sqrt{25 - c^2}} = \dfrac{4 - 0}{3 - (-5)} = \dfrac{1}{2}$

$-2c = \sqrt{25 - c^2}$
$4c^2 = 25 - c^2$
$c^2 = 5$
$c = -\sqrt{5}$

(we reject $c = \sqrt{5}$ because it does not satisfy the equation

$-2c = \sqrt{25 - c^2}$)

13.   (a) $f'(x) = \sec^2 x$, $\sec^2 c = 0$ has no solution
(b) $\tan x$ is not continuous on $[0, \pi]$

15.   Let $f(x) = \sin x$ and $x \neq y$. By the Mean-Value Theorem there is a number
c between x and y such that
$$\frac{\sin x - \sin y}{x - y} = \cos c$$
$$\frac{|\sin x - \sin y|}{|x - y|} = |\cos c| \leq 1$$
so $|\sin x - \sin y| \leq |x - y|$, which also holds when x = y.

17. Let $f(x) = \sqrt{x}$. By the Mean-Value Theorem there is a number c between x and y such that

$$\frac{\sqrt{y} - \sqrt{x}}{y - x} = \frac{1}{2\sqrt{c}} < \frac{1}{2\sqrt{x}} \text{ for } c \text{ in } (x,y),$$

thus $\sqrt{y} - \sqrt{x} < \dfrac{y - x}{2\sqrt{x}}$; multiply through and rearrange to get

$\sqrt{xy} < \dfrac{1}{2}(x + y)$.

19. $f'(x) = 2a_2x + a_1$,

$$2a_2c + a_1 = \frac{(a_2b^2 + a_1b + a_0) - (a_2a^2 + a_1a + a_0)}{b - a}$$

$$= \frac{a_2(b^2 - a^2) + a_1(b - a)}{b - a} = a_2(b + a) + a_1, \text{ so } c = \frac{1}{2}(b + a).$$

21. $f(0) = f(1) = 0$, $f'(x) = 3ax^2 + 2bx - (a + b)$, so there is at least one number c in $(0,1)$ where $f'(c) = 0$.

23. Let $f(x) = x^3 + 4x - 1$. Assume that $f(x) = 0$ has at least two distinct real solutions r_1 and r_2. Then $f(r_1) = f(r_2) = 0$ and so by Rolle's Theorem there is at least one number c between r_1 and r_2 where $f'(c) = 0$. But $f'(x) = 3x^2 + 4$ is never zero, so $f(x) = 0$ must have fewer than two distinct real solutions.

25. Assume that $f(x) = 0$ has at least four distinct real solutions $r_1 < r_2 < r_3 < r_4$, then by Rolle's Theorem there is at least one number in each of the intervals (r_1,r_2), (r_2,r_3) and (r_3,r_4) so that $f'(x) = 0$ at least three times. Apply Rolle's Theorem to $f'(x)$ to show that $f''(x) = 0$ at least twice; again to $f''(x)$ to show that $f'''(x) = 0$ at least once. But $f'''(x) = 60x^2 + 24ax + 6b$, and $f'''(x) = 0$ if $10x^2 + 4ax + b = 0$. Use the quadratic formula to get

$x = \dfrac{-4a \pm \sqrt{16a^2 - 40b}}{20}$, which has no real solutions if $16a^2 - 40b < 0$,

$16a^2 < 40b$, $2a^2 < 5b$.

27. $\dfrac{d}{dx} [f^2(x) + g^2(x)] = 2f(x)f'(x) + 2g(x)g'(x)$

$\qquad\qquad\qquad = 2f(x)g(x) + 2g(x)[-f(x)] = 0,$

so $f^2(x) + g^2(x)$ is constant.

29. $f'(x) = 3(x - 1)^2$, $g'(x) = (x^2 + 3) + 2x(x - 3) = 3x^2 - 6x + 3$

$= 3(x^2 - 2x + 1) = 3(x - 1)^2$, so $f'(x) = g'(x)$ and hence
$f(x) - g(x) = k$. Expand $f(x)$ and $g(x)$ to get

$f(x) - g(x) = (x^3 - 3x^2 + 3x - 1) - (x^3 - 3x^2 + 3x - 9) = 8.$

31. If $f'(x) = g'(x)$, then $f(x) = g(x) + k$. Let $x = 1$,

$f(1) = g(1) + k = (1)^3 - 4(1) + 6 + k = 3 + k = 2$, so $k = -1$.

$f(x) = x^3 - 4x + 5.$

33. Let $h = f - g$, then h is continuous on $[a,b]$, differentiable on (a,b),
and $h(a) = f(a) - g(a) = 0$, $h(b) = f(b) - g(b) = 0$. By Rolle's Theorem
there is some c in (a,b) where $h'(c) = 0$. But $h'(c) = f'(c) - g'(c)$ so
$f'(c) - g'(c) = 0$, $f'(c) = g'(c)$.

35. From Corollary 4.10.4, $f(x) - g(x) = k$ for x in (a,b). Also
$\lim\limits_{x \to a^+} [f(x) - g(x)] = k = f(a) - g(a)$ and

$\lim\limits_{x \to b^-} [f(x) - g(x)] = k = f(b) - g(b)$ because f and g, and hence $f - g$,
are continuous at a and b, so $f(x) - g(x) = k$ for all x in $[a,b]$.

37. From the Mean-Value Theorem there is a point c in (a,b) where

$f'(c) = \dfrac{f(b) - f(a)}{b - a} = \dfrac{0}{b - a} = 0.$

39. Similar to proof given in text; assume that $f(x) < 0$ and replace the
word "maximum" by "minimum".

41. (a) Let $x_1 < x_2$ where x_1 and x_2 are in $[a,b]$, then f is continuous on
$[x_1,x_2]$ and differentiable on (x_1,x_2) so for some c in (x_1,x_2)

$$\dfrac{f(x_2) - f(x_1)}{x_2 - x_1} = f'(c)$$

$f(x_2) - f(x_1) = (x_2 - x_1)f'(c) > 0$ because $f'(c) > 0$, so
$f(x_2) > f(x_1)$ which shows that f is increasing on $[a,b]$.

(b) Proceed as in part (a) using $f'(c) < 0$.

EXERCISE SET 4.11

1. $v = 3t^2 - 12t$, $a = 6t - 12$

t	s	v	|v|	a	direction; motion
1	-5	-9	9	-6	left; speeding up
2	-16	-12	12	0	left; neither
3	-27	-9	9	6	left; slowing down
4	-32	0	0	12	stopped
5	-25	15	15	18	right; speeding up

3. (a) $v = 10t - 22$, speed $= |v| = |10t - 22|$. $d|v|/dt$ does not exist at
    $t = 2.2$ which is the only critical point. If $t = 1, 2.2, 3$ then
    $|v| = 12, 0, 8$. The maximum speed is 12.

    (b) the distance from the origin is $|s| = |5t^2 - 22t| = |t(5t - 22)|$,
    but $t(5t - 22) < 0$ for $1 \leq t \leq 3$ so $|s| = -(5t^2 - 22t) = 22t - 5t^2$,
    $d|s|/dt = 22 - 10t$, thus the only critical point is $t = 2.2$.
    $d^2|s|/dt^2 < 0$ so the particle is farthest from the origin when
    $t = 2.2$. Its position is $s = 5(2.2)^2 - 22(2.2) = -24.2$.

5. $s = 1 + 6t - t^2$
    $v = 2(3 - t)$
    $a = -2$

7. $s = t^3 - 9t^2 + 24t$
    $v = 3(t - 2)(t - 4)$
    $a = 6(t - 3)$

9. $s = \begin{cases} \cos t, & 0 \leq t \leq 2\pi \\ 1, & t > 2\pi \end{cases}$

    $v = \begin{cases} -\sin t, & 0 \leq t \leq 2\pi \\ 0, & t > 2\pi \end{cases}$

    $a = \begin{cases} -\cos t, & 0 \leq t < 2\pi \\ 0, & t > 2\pi \end{cases}$

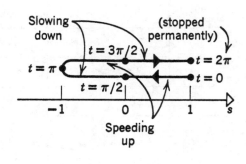

11. $s = 4t^{3/2} - 3t^2$, $v = 6t^{1/2} - 6t$, $a = 3t^{-1/2} - 6$.

    (a) $a = 0$ when $3t^{-1/2} = 6$, $t = 1/4$; $s = 5/16$, $v = 3/2$.

    (b) $v = 0$ when $6t^{1/2}(1 - t^{1/2}) = 0$ for $t > 0$, $t = 1$; $s = 1$, $a = -3$.

13. (a) $a = \dfrac{dv}{dt} = \dfrac{dv}{ds}\dfrac{ds}{dt} = v\dfrac{dv}{ds}$ because $v = \dfrac{ds}{dt}$.

    (b) $v = \dfrac{3}{2\sqrt{3t + 7}} = \dfrac{3}{2s}$; $\dfrac{dv}{ds} = -\dfrac{3}{2s^2}$; $a = -\dfrac{9}{4s^3} = -9/500$.

15. (a) $s_1 = s_2$ if they collide, so $\dfrac{1}{2}t^2 - t + 3 = -\dfrac{1}{4}t^2 + t + 1$,

    $\dfrac{3}{4}t^2 - 2t + 2 = 0$ which has no real solution.

    (b) Find the minimum value of $D = |s_1 - s_2| = \left|\dfrac{3}{4}t^2 - 2t + 2\right|$. From

    part (a), $\dfrac{3}{4}t^2 - 2t + 2$ is never zero, and for $t = 0$ it is positive,

    hence it is always positive, so $D = \dfrac{3}{4}t^2 - 2t + 2$. $\dfrac{dD}{dt} = \dfrac{3}{2}t - 2 = 0$

    when $t = \dfrac{4}{3}$. $\dfrac{d^2D}{dt^2} > 0$ so D is minimum when $t = \dfrac{4}{3}$, $D = \dfrac{2}{3}$.

(c) $v_1 = t - 1$, $v_2 = -\frac{1}{2} t + 1$. $v_1 < 0$ if $0 \leq t < 1$, $v_1 > 0$ if $t > 1$; $v_2 < 0$ if $t > 2$, $v_2 > 0$ if $0 \leq t < 2$. They are moving in opposite directions during the intervals $0 \leq t < 1$ and $t > 2$.

17. (a) negative direction because $v = ds/dt < 0$ at t_0.

(b) negative because $a = d^2s/dt^2$ and the curve is concave down at t_0 $(d^2s/dt^2 < 0)$.

(c) speeding up because v and a have the same sign.

(d) $v < 0$ and $a > 0$ at t_1 so the particle is slowing down because v and a have opposite signs.

19. (a) $\lim\limits_{t_1 \to t_0} v_{ave} = \lim\limits_{t_1 \to t_0} \dfrac{s(t_1) - s(t_0)}{t_1 - t_0} = s'(t_0) = v(t_0)$

(b) similar to part (a) with a in place of v.

SUPPLEMENTARY EXERCISES CHAPTER 4

1. $V = \pi R^2 h - \pi r^2 h = \pi(R^2 - r^2)h$

$dV/dt = \pi[(R^2 - r^2)dh/dt + h(2R \, dR/dt - 2r \, dr/dt)]$.
But $dR/dt = dr/dt = 2$, $dh/dt = -3$ so for $R = 7$, $r = 4$, and $h = 5$
$dV/dt = \pi[(49 - 16)(-3) + 5(14(2) - 8(2))] = -39\pi$.

The volume is decreasing at the rate of 39π m^3 sec.

3. By similar triangles
$x/48 = 10/y$, $x = 480/y$ so

$dx/dt = -(480/y^2)dy/dt$. But
$dy/dt = 32$ when $y = 16$ thus

$dx/dt = -(480/16^2)(32) = -60$.
The shadow is moving toward the pole
at the rate of 60 ft/sec.

5. $f'(x) = x^2(3 - 4x)$, critical points $x = 0$, $3/4$; $f(-1) = -2$, $f(0) = 0$,
    $f(3/4) = 27/256$, $f(3/2) = -27/16$. $m = -2$ at $x = -1$, $M = 27/256$ at
    $x = 3/4$.

7. $f'(x) = 2(3 - x^2)/(x^2 + 3)^2$, critical point $x = \sqrt{3}$; $f(\sqrt{3}) = \sqrt{3}/3$,
    $f(2) = 4/7$, $\lim_{x \to 0^+} f(x) = 0$. No minimum on $(0,2]$, $M = \sqrt{3}/3$ at $x = \sqrt{3}$.

9. $x^2 - 2x \geq 0$ when $x \leq 0$ or $x \geq 2$, $x^2 - 2x < 0$ when $0 < x < 2$
    $$f'(x) = \begin{cases} -2x + 2, & x < 0 \text{ or } x > 2 \\ 2x - 2, & 0 < x < 2 \end{cases}$$
    and $f'(x)$ does not exist when $x = 0$, 2. The only critical point in
    $(1,3)$ is $x = 2$; $f(1) = -1$, $f(2) = 0$, $f(3) = -3$, $m = -3$ at $x = 3$, $M = 0$
    at $x = 2$.

11. $f(x) = x^3 - 4x + 1$, $f'(x) = 3x^2 - 4$, $x_{n+1} = \dfrac{2x_n^3 - 1}{3x_n^2 - 4}$

$x_1 = -2$	$x_1 = 0$	$x_1 = 2$
$x_2 = -2.125$	$x_2 = 0.25$	$x_2 = 1.875$
$x_3 = -2.1149755$	$x_3 = 0.2540983$	$x_3 = 1.8609785$
$x_4 = -2.1149075$	$x_4 = 0.2541016$	$x_4 = 1.8608059$
$x_5 = -2.1149075$	$x_5 = 0.2541016$	$x_5 = 1.8608059$

13. $f'(x) = -\dfrac{2x}{(1 + x^2)^2}$

    $f''(x) = \dfrac{2(3x^2 - 1)}{(1 + x^2)^3}$

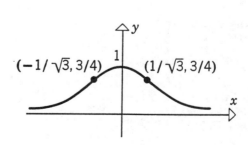

15. $f'(x) = \dfrac{2(x^3 + 1)}{x^2}$

$f''(x) = \dfrac{2(x^3 - 2)}{x^3}$

$f(x) = x^2 - \dfrac{2}{x}$ so $f(x)$ is

asymptotic to $y = x^2$ for $|x|$ large.

17. $f'(x) = -4\cos x \sin x$
     $\quad\quad = -2\sin 2x$
     $f''(x) = -4\cos 2x$

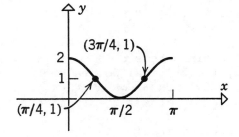

19. $f'(x) = \dfrac{3(8 - x)}{(x + 8)^3}$

$f''(x) = \dfrac{6(x - 16)}{(x + 8)^4}$

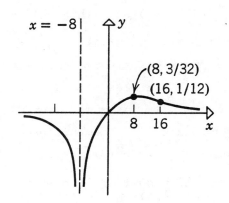

21. $f'(x) = 4x^3 - 18x^2 + 24x - 8$
     $f''(x) = 12x^2 - 36x + 24$
     $\quad\quad = 12(x - 1)(x - 2)$
     $f''(x) = 0$ when $x = 1, 2;\ f(1) = 2,\ f(2) = 3$. The inflection points are

(1,2) and (2,3) because the concavity changes at these points.
$f'(1) = 2$ so the tangent line at (1,2) is $y - 2 = 2(x - 1)$, $y = 2x$.
$f'(2) = 0$ so the tangent line at (2,3) is $y = 3$.

23. $f'(x) = 2\cos x + 2\sin 2x$
          $= 2\cos x + 4\sin x \cos x$
          $= 2\cos x(1 + 2\sin x)$;
  $f'(x) = 0$ when $\cos x = 0$ or $\sin x = -1/2$,
  critical points $x = \pi/2,\ 3\pi/2,\ 7\pi/6,\ 11\pi/6$
  relative max at $x = \pi/2,\ 3\pi/2$, relative min at $x = 7\pi/6,\ 11\pi/6$

25. $f'(x) = \dfrac{x - 9}{18x^{3/2}}$; critical point $x = 9$ (0 is not a critical point because

  it is not in the domain of f)
  $f''(x) = \dfrac{27 - x}{36x^{5/2}}$; $f''(9) > 0$, relative min at $x = 9$

27. $f'(x) = \sin x(2\cos x + 1)$; $f'(x) = 0$ when $\sin x = 0$ or $\cos x = -1/2$, in
  $(0,2\pi)$ the critical points are $x = \pi,\ 2\pi/3,\ 4\pi/3$
  $f''(x) = 2\cos 2x + \cos x$; $f''(\pi) > 0$, $f''(2\pi/3) < 0$, $f''(4\pi/3) < 0$
  relative max at $x = 2\pi/3,\ 4\pi/3$, relative min at $x = \pi$

29. Let (x,y) be a point in the first quadrant
  that is on the ellipse, then $A = (2x)(2y) = 4xy$.
  But, from the equation of the ellipse,

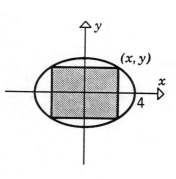

  $y^2 = \dfrac{9}{16}(16 - x^2)$ so with $S = A^2 = 16x^2y^2$

  $S = 9x^2(16 - x^2) = 9(16x^2 - x^4)$ for $0 < x < 4$,

  $dS/dx = 36x(8 - x^2)$, critical point at

  $x = \sqrt{8} = 2\sqrt{2}$. $d^2S/dx^2 > 0$ at $x = 2\sqrt{2}$ thus

  S and hence A is maximum there. If $x = 2\sqrt{2}$

  then $y = 3\sqrt{2}/2$. The dimensions of the

  rectangle are $4\sqrt{2}$ by $3\sqrt{2}$.

31. Let k be the amount of light admitted per unit area of clear glass. The
  total amount of light admitted by the entire window is

        $T = k \cdot$ (area of clear glass) $+ \dfrac{1}{2} k \cdot$ (area of blue glass)

          $= 2krh + \dfrac{1}{4}\pi kr^2$.

  But $P = 2h + 2r + \pi r$ which gives $h = \dfrac{1}{2}(P - 2r - \pi r)$ so

$$T = kr(P - 2r - \pi r) + \frac{1}{4}\pi k r^2 = k[Pr - (2 + \pi - \frac{\pi}{4})r^2]$$

$$= k[Pr - \frac{8 + 3\pi}{4} r^2] \text{ for } 0 < r < \frac{P}{2 + \pi},$$

$$\frac{dT}{dr} = k(P - \frac{8 + 3\pi}{2} r), \frac{dT}{dr} = 0 \text{ when } r = \frac{2P}{8 + 3\pi}.$$

This is the only critical point and $d^2T/dr^2 < 0$ there so the most light is admitted when $r = 2P/(8 + 3\pi)$ ft.

33. The total area of material used is

$$A = A_{top} + A_{bottom} + A_{side} = (2r)^2 + (2r)^2 + 2\pi rh = 8r^2 + 2\pi rh.$$

The volume is $V = \pi r^2 h$ thus $h = V/(\pi r^2)$ so $A = 8r^2 + 2V/r$ for $r > 0$,

$dA/dr = 16r - 2V/r^2 = 2(8r^3 - V)/r^2$, $dA/dr = 0$ when $r = \sqrt[3]{V}/2$.

This is the only critical point and $d^2A/dr^2 > 0$ there so the least

material is used when $r = \sqrt[3]{V}/2$,

$$\frac{r}{h} = \frac{r}{V/(\pi r^2)} = \frac{\pi}{V} r^3 \text{ and, for } r = \sqrt[3]{V}/2, \frac{r}{h} = \frac{\pi}{V} \frac{V}{8} = \frac{\pi}{8}.$$

35. f is continuous on $[-2,2]$,

$f'(x) = -x/\sqrt{4 - x^2}$ so f is differentiable on $(-2,2)$, $f(-2) = f(2) = 0$; hypotheses are satisfied. $f'(c) = 0$ for $c = 0$.

37. f is continuous on $[0,\sqrt{\pi}]$, $f'(x) = 2x \cos(x^2)$ so f is differentiable on $(0,\sqrt{\pi})$, $f(0) = f(\sqrt{\pi}) = 0$; hypotheses are satisfied. $f'(c) = 0$ when $2c \cos(c^2) = 0$ which yields $c = 0$, $\pm\sqrt{\pi/2}$ of which only $c = \sqrt{\pi/2}$ is in $(0,\sqrt{\pi})$.

39. f is continuous on $[0,4]$ and differentiable on $(0,4)$.

$$f'(c) = \frac{f(4) - f(0)}{4 - 0}, \frac{1}{2\sqrt{c}} = \frac{1}{2}, c = 1$$

41. By inspection, f is continuous on $[0,2]$ and differentiable on $(0,2)$ except perhaps at $x = 1$. For $x = 1$, $\lim_{x \to 1^-} f(x) = \lim_{x \to 1^+} f(x) = f(1)$ so f

is continuous at $x = 1$. $\lim_{x \to 1^-} f'(x) = \lim_{x \to 1^-} (-2x) = -2$,

$\lim_{x \to 1^+} f'(x) = \lim_{x \to 1^+} (-2/x^2) = -2$ so f is differentiable at $x = 1$ (see theorem preceding Exercise 67, Section 3.3).

$$f'(c) = \frac{f(2) - f(0)}{2 - 0} = \frac{1 - 3}{2} = -1$$

so $c \neq 1$. If $x < 1$ then $f'(x) = -2x$ thus $f'(c) = -1$ for $c = 1/2$. If $x > 1$ then $f'(x) = -2/x^2$ thus $f'(c) = -1$ for $c = \sqrt{2}$. The values of c are $1/2$, $\sqrt{2}$.

CHAPTER 5

EXERCISE SET 5.2

1. $x^9/9 + C$

3. $\frac{7}{12} x^{12/7} + C$

5. $4 \int t^{-1/2} dt = 8\sqrt{t} + C$

7. $\int x^{7/2} dx = \frac{2}{9} x^{9/2} + C$

9. $\int (x^{-3} + x^{1/2} - 3x^{1/4} + x^2) dx = -\frac{1}{2} x^{-2} + \frac{2}{3} x^{3/2} - \frac{12}{5} x^{5/4} + \frac{1}{3} x^3 + C$

11. $\int (7y^{-3/4} - y^{1/3} + 4y^{1/2}) dy = 28y^{1/4} - \frac{3}{4} y^{4/3} + \frac{8}{3} y^{3/2} + C$

13. $\int (x + x^4) dx = x^2/2 + x^5/5 + C$

15. $\int x^{1/3}(4 - 4x + x^2) dx = \int (4x^{1/3} - 4x^{4/3} + x^{7/3}) dx$
$$= 3x^{4/3} - \frac{12}{7} x^{7/3} + \frac{3}{10} x^{10/3} + C$$

17. $\int (x + 2x^{-2} - x^{-4}) dx = x^2/2 - 2/x + 1/(3x^3) + C$

19. $-4\cos x + 2\sin x + C$

21. $\int (\sec^2 x + \sec x \tan x) dx = \tan x + \sec x + C$

23. $\int (\sec x \tan x + 1) dx = \sec x + x + C$

25. $\int \sec x \tan x \, dx = \sec x + C$

27. $\int (1 + \sin \theta) d\theta = \theta - \cos \theta + C$

29. $\int (\cos \theta - 5\sec^2 \theta) d\theta = \sin \theta - 5\tan \theta + C$

31. $F(x) = \int x^{1/3}dx = \frac{3}{4} x^{4/3} + C$, $F(1) = \frac{3}{4} + C = 2$, $C = 5/4$;

$F(x) = \frac{3}{4} x^{4/3} + \frac{5}{4}$

33. $f'(x) = \frac{2}{3} x^{3/2} + C_1$; $f(x) = \frac{4}{15} x^{5/2} + C_1 x + C_2$

35. $f(x) = \frac{d}{dx} (5x^3 - 3x + C) = 15x^2 - 3$

37. $\int (\sec^2 x - 1)dx = \tan x - x + C$

39. $\frac{d}{dx} \left[\int f(x)dx - \int g(x)dx \right] = \frac{d}{dx} \left[\int f(x)dx \right] - \frac{d}{dx} \left[\int g(x)dx \right] = f(x) - g(x)$

EXERCISE SET 5.3

1. (a) $\int u^{23}du = u^{24}/24 + C = (x^2 + 1)^{24}/24 + C$

(b) $-\int u^3 du = -u^4/4 + C = -(\cos^4 x)/4 + C$

(c) $2 \int \sin u \, du = -2 \cos u + C = -2 \cos \sqrt{x} + C$

(d) $\frac{3}{8} \int u^{-1/2}du = \frac{3}{4} u^{1/2} + C = \frac{3}{4} \sqrt{4x^2 + 5} + C$

3. (a) $-\int u \, du = -\frac{1}{2} u^2 + C = -\frac{1}{2} \cot^2 x + C$

(b) $\int u^9 du = \frac{1}{10} u^{10} + C = \frac{1}{10} (1 + \sin t)^{10} + C$

(c) $\int (u - 1)^2 u^{1/2}du = \int (u^{5/2} - 2u^{3/2} + u^{1/2})du$

$= \frac{2}{7} u^{7/2} - \frac{4}{5} u^{5/2} + \frac{2}{3} u^{3/2} + C$

$= \frac{2}{7} (1 + x)^{7/2} - \frac{4}{5} (1 + x)^{5/2} + \frac{2}{3} (1 + x)^{3/2} + C$

(d) $\int \csc^2 u \, du = -\cot u + C = -\cot(\sin x) + C$

5. $u = 2 - x^2$, $du = -2x\,dx$; $-\dfrac{1}{2} \displaystyle\int u^3 du = -u^4/8 + C = -(2 - x^2)^4/8 + C$

7. $u = 8x$, $du = 8dx$; $\dfrac{1}{8} \displaystyle\int \cos u\,du = \dfrac{1}{8} \sin u + C = \dfrac{1}{8} \sin 8x + C$

9. $u = 4x$, $du = 4dx$; $\dfrac{1}{4} \displaystyle\int \sec u \tan u\,du = \dfrac{1}{4} \sec u + C = \dfrac{1}{4} \sec 4x + C$

11. $u = 7t^2 + 12$, $du = 14t\,dt$

$\dfrac{1}{14} \displaystyle\int u^{1/2} du = \dfrac{1}{21} u^{3/2} + C = \dfrac{1}{21} (7t^2 + 12)^{3/2} + C$

13. $u = x^3 + 1$, $du = 3x^2 dx$; $\dfrac{1}{3} \displaystyle\int u^{-1/2} du = \dfrac{2}{3} u^{1/2} + C = \dfrac{2}{3} \sqrt{x^3 + 1} + C$

15. $u = 4x^2 + 1$, $du = 8x\,dx$

$\dfrac{1}{8} \displaystyle\int u^{-3} du = -\dfrac{1}{16} u^{-2} + C = -\dfrac{1}{16} (4x^2 + 1)^{-2} + C$

17. $u = 5/x$, $du = -(5/x^2)dx$; $-\dfrac{1}{5} \displaystyle\int \sin u\,du = \dfrac{1}{5} \cos u + C = \dfrac{1}{5} \cos(5/x) + C$

19. $u = x^3$, $du = 3x^2 dx$; $\dfrac{1}{3} \displaystyle\int \sec^2 u\,du = \dfrac{1}{3} \tan u + C = \dfrac{1}{3} \tan(x^3) + C$

21. $u = \sin 3t$, $du = 3\cos 3t\,dt$; $\dfrac{1}{3} \displaystyle\int u^5 du = \dfrac{1}{18} u^6 + C = \dfrac{1}{18} \sin^6 3t + C$

23. $u = 2 - \sin 4\theta$, $du = -4\cos 4\theta\,d\theta$
$-\dfrac{1}{4} \displaystyle\int u^{1/2} du = -\dfrac{1}{6} u^{3/2} + C = -\dfrac{1}{6} (2 - \sin 4\theta)^{3/2} + C$

25. $u = \sec 2x$, $du = 2\sec 2x \tan 2x\,dx$
$\dfrac{1}{2} \displaystyle\int u^2 du = \dfrac{1}{6} u^3 + C = \dfrac{1}{6} \sec^3 2x + C$

27. $u = \cos 3\theta$, $du = -3\sin 3\theta\,d\theta$
$-\dfrac{1}{3} \displaystyle\int \sec^2 u\,du = -\dfrac{1}{3} \tan u + C = -\dfrac{1}{3} \tan(\cos 3\theta) + C$

29. $u = \sin(a + bx)$, $du = b\cos(a + bx)dx$

$\dfrac{1}{b} \displaystyle\int u^n du = \dfrac{1}{b(n + 1)} u^{n+1} + C = \dfrac{1}{b(n + 1)} \sin^{n+1}(a + bx) + C$

31. $u = x - 3$, $x = u + 3$, $dx = du$

$\displaystyle\int (u + 3)u^{1/2}du = \int (u^{3/2} + 3u^{1/2})du = \dfrac{2}{5} u^{5/2} + 2u^{3/2} + C$

$\qquad\qquad = \dfrac{2}{5} (x - 3)^{5/2} + 2(x - 3)^{3/2} + C$

33. $u = y + 1$, $y = u - 1$, $dy = du$

$\displaystyle\int \dfrac{u - 1}{u^{1/2}} du = \int (u^{1/2} - u^{-1/2})du = \dfrac{2}{3} u^{3/2} - 2u^{1/2} + C$

$\qquad\qquad = \dfrac{2}{3} (y + 1)^{3/2} - 2(y + 1)^{1/2} + C$

35. $u = 3\theta$, $du = 3\,d\theta$

$\dfrac{1}{3} \displaystyle\int \tan^2 u\,du = \dfrac{1}{3}\int (\sec^2 u - 1)du = \dfrac{1}{3} (\tan u - u) + C$

$\qquad\qquad = \dfrac{1}{3}(\tan 3\theta - 3\theta) + C$

37. $u = \sqrt{x - 1}$, $u^2 = x - 1$, $x = u^2 + 1$, $dx = 2u\,du$

$\displaystyle\int (u^2 + 1)^2 u(2u)du = 2\int (u^4 + 2u^2 + 1)u^2 du = 2\int (u^6 + 2u^4 + u^2)du$

$\qquad\qquad = \dfrac{2}{7} u^7 + \dfrac{4}{5} u^5 + \dfrac{2}{3} u^3 + C$

$\qquad\qquad = \dfrac{2}{7} (x - 1)^{7/2} + \dfrac{4}{5} (x - 1)^{5/2} + \dfrac{2}{3} (x - 1)^{3/2} + C$

39. (a) First method:

$\displaystyle\int (25x^2 - 10x + 1)dx = \dfrac{25}{3} x^3 - 5x^2 + x + C;$

second method:

$\dfrac{1}{5} \displaystyle\int u^2 du = \dfrac{1}{15} u^3 + C = \dfrac{1}{15} (5x - 1)^3 + C$

(b) $\dfrac{1}{15} (5x - 1)^3 + C = \dfrac{1}{15} (125x^3 - 75x^2 + 15x - 1) + C$

$\qquad\qquad = \dfrac{25}{3} x^3 - 5x^2 + x - \dfrac{1}{15} + C;$

the answers differ by a constant.

41. $f(x) = \int (6 - 5 \sin 2x)dx = 6x + \dfrac{5}{2} \cos 2x + C$,

$f(0) = \dfrac{5}{2} + C = 3$, $C = \dfrac{1}{2}$ so $f(x) = 6x + \dfrac{5}{2} \cos 2x + \dfrac{1}{2}$

43. $u = 3x + 2$, $du = 3dx$

$\dfrac{1}{3} \int f'(u)du = \dfrac{1}{3} f(u) + C = \dfrac{1}{3} f(3x + 2) + C$

45. $u = 2/x$, $du = -(2/x^2)dx$

$-\dfrac{1}{2} \int f'(u)du = -\dfrac{1}{2} f(u) + C = -\dfrac{1}{2} f(2/x) + C$

EXERCISE SET 5.4

1. (a) $1 + 8 + 27 = 36$ (b) $5 + 8 + 11 + 14 + 17 = 55$
    (c) $20 + 12 + 6 + 2 + 0 + 0 = 40$ (d) $1 + 1 + 1 + 1 + 1 + 1 = 6$

3. $\displaystyle\sum_{k=1}^{10} k$ 5. $\displaystyle\sum_{k=1}^{49} k(k + 1)$

7. $\displaystyle\sum_{k=1}^{10} 2k$ 9. $\displaystyle\sum_{k=1}^{6} (-1)^{k+1}(2k - 1)$

11. $\displaystyle\sum_{k=1}^{5} (-1)^k \dfrac{1}{k}$ 13. $\displaystyle\sum_{k=1}^{4} \sin \dfrac{(2k - 1)\pi}{8}$

15. $\displaystyle\sum_{k=1}^{5} \dfrac{k}{k + 1}$

17. (a) $\displaystyle\sum_{k=1}^{5} (-1)^{k+1}a_k$ (b) $\displaystyle\sum_{k=0}^{5} (-1)^{k+1}b_k$

(c) $\displaystyle\sum_{k=0}^{n} a_k x^k$ (d) $\displaystyle\sum_{k=0}^{5} a^{5-k} b^k$

19. $\displaystyle\sum_{k=1}^{100} k - \sum_{k=1}^{2} k = \frac{1}{2}(100)(100 + 1) - (1 + 2) = 5050 - 3 = 5047$

21. $\frac{1}{6}(20)(21)(41) = 2,870$

23. $4\displaystyle\sum_{k=1}^{6} k^3 - 2\sum_{k=1}^{6} k + \sum_{k=1}^{6} 1 = 4\left[\frac{1}{4}(6)^2(7)^2\right] - 2\left[\frac{1}{2}(6)(7)\right] + 6 = 1728$

25. $\displaystyle\sum_{k=1}^{30} k(k^2 - 4) = \sum_{k=1}^{30}(k^3 - 4k) = \sum_{k=1}^{30} k^3 - 4\sum_{k=1}^{30} k$

$$= \frac{1}{4}(30)^2(31)^2 - 4 \times \frac{1}{2}(30)(31) = 214,365$$

27. $(3^5 - 3^4) + (3^6 - 3^5) + \cdots + (3^{17} - 3^{16}) = 3^{17} - 3^4$

29. $(\frac{1}{2^2} - \frac{1}{1^2}) + (\frac{1}{3^2} - \frac{1}{2^2}) + \cdots + (\frac{1}{20^2} - \frac{1}{19^2}) = \frac{1}{20^2} - 1 = -\frac{399}{400}$

31. $(a_1 - a_0) + (a_2 - a_1) + \cdots + (a_n - a_{n-1}) = a_n - a_0$

33. (a) $n + n + \cdots + n = n^2$ (b) -3
        (n terms)

    (c) $x\displaystyle\sum_{k=1}^{n} k = \frac{1}{2}n(n + 1)x$

    (d) $c + c + \cdots + c = (n - m + 1)c$
        (n − m + 1 terms)

35. (a) $\displaystyle\sum_{k=0}^{14} (k + 4)(k + 1)$

(b) $\displaystyle\sum_{k=5}^{19} (k - 1)(k - 4)$

37. $\displaystyle\sum_{k=1}^{18} k \sin \frac{\pi}{k}$

39. both are valid

41. (a) $\displaystyle\sum_{k=0}^{19} 3^{k+1} = \sum_{k=0}^{19} 3(3^k) = \frac{3(1 - 3^{20})}{1 - 3} = \frac{3}{2} (3^{20} - 1)$

(b) $\displaystyle\sum_{k=0}^{25} 2^{k+5} = \sum_{k=0}^{25} 2^5 2^k = \frac{2^5(1 - 2^{26})}{1 - 2} = 2^{31} - 2^5$

(c) $\displaystyle\sum_{k=0}^{100} (-1)\left(-\frac{1}{2}\right)^k = \frac{(-1)(1 - (-1/2)^{101})}{1 - (-1/2)} = -\frac{2}{3} (1 + 1/2^{101})$

43. $\displaystyle\sum_{i=1}^{4} \left[\sum_{j=1}^{5} i + \sum_{j=1}^{5} j \right] = \sum_{i=1}^{4} \left[5i + \frac{1}{2} (5)(6)\right] = 5 \sum_{i=1}^{4} i + \sum_{i=1}^{4} 15$

$$= 5 \cdot \frac{1}{2} (4)(5) + (4)(15) = 110$$

45. $\displaystyle\sum_{k=1}^{n} (a_k - b_k) = (a_1 - b_1) + (a_2 - b_2) + \cdots + (a_n - b_n)$

$$= (a_1 + a_2 + \cdots + a_n) - (b_1 + b_2 + \cdots + b_n)$$

$$= \sum_{k=1}^{n} a_k - \sum_{k=1}^{n} b_k$$

EXERCISE SET 5.5

1. $\Delta x = \dfrac{6 - 2}{4} = 1$, $f(x) = 3x + 1$

   (a) $c_k = 2,3,4,5$; $\displaystyle\sum_{k=1}^{4} f(c_k)\Delta x = (7 + 10 + 13 + 16)(1) = 46$

   (b) $d_k = 3,4,5,6$; $\displaystyle\sum_{k=1}^{4} f(d_k)\Delta x = (10 + 13 + 16 + 19)(1) = 58$

3. $\Delta x = \pi/4$, $f(x) = \cos x$
   (a) $c_k = -\pi/2, \ -\pi/4, \ \pi/4, \ \pi/2$

   $\displaystyle\sum_{k=1}^{4} f(c_k)\Delta x = (0 + \sqrt{2}/2 + \sqrt{2}/2 + 0)(\pi/4) = \sqrt{2}\pi/4 \approx 1.111$

   (b) $d_k = -\pi/4, \ 0, \ 0, \ \pi/4$

   $\displaystyle\sum_{k=1}^{4} f(d_k)\Delta x = (\sqrt{2}/2 + 1 + 1 + \sqrt{2}/2)(\pi/4) = (2 + \sqrt{2})\pi/4 \approx 2.682$

5. $\Delta x = \dfrac{3}{n}$, $c_k = 1 + (k - 1)\dfrac{3}{n}$

   $f(c_k)\Delta x = \dfrac{1}{2}c_k\Delta x = \dfrac{1}{2}\left[1 + (k - 1)\dfrac{3}{n}\right]\dfrac{3}{n} = \dfrac{1}{2}\left[\dfrac{3}{n} + (k - 1)\dfrac{9}{n^2}\right]$

   $\displaystyle\sum_{k=1}^{n} f(c_k)\Delta x = \dfrac{1}{2}\left[\sum_{k=1}^{n}\dfrac{3}{n} + \dfrac{9}{n^2}\sum_{k=1}^{n}(k - 1)\right]$

   $= \dfrac{1}{2}\left[3 + \dfrac{9}{n^2}\cdot\dfrac{1}{2}(n - 1)n\right] = \dfrac{3}{2} + \dfrac{9}{4}\dfrac{n - 1}{n}$

   $A = \lim_{n\to+\infty}\left[\dfrac{3}{2} + \dfrac{9}{4}\left(1 - \dfrac{1}{n}\right)\right] = \dfrac{3}{2} + \dfrac{9}{4} = \dfrac{15}{4}$

7. $\Delta x = \dfrac{1}{n}$, $c_k = (k - 1)\dfrac{1}{n}$

   $f(c_k)\Delta x = c_k^2\Delta x = (k - 1)^2\dfrac{1}{n^2}\cdot\dfrac{1}{n} = \dfrac{1}{n^3}(k - 1)^2$

$$\sum_{k=1}^{n} f(c_k)\Delta x = \frac{1}{n^3} \sum_{k=1}^{n} (k-1)^2 = \frac{1}{n^3} \sum_{k=1}^{n-1} k^2$$

$$= \frac{1}{n^3} \cdot \frac{1}{6}(n-1)(n)(2n-1) = \frac{1}{6}\frac{(n-1)(2n-1)}{n^2}$$

$$A = \lim_{n\to+\infty} \frac{1}{6}(1-\frac{1}{n})(2-\frac{1}{n}) = \frac{1}{6}(1)(2) = \frac{1}{3}$$

9. $\Delta x = \frac{4}{n}$, $c_k = 2 + (k-1)\frac{4}{n}$

$$f(c_k)\Delta x = c_k^3 \Delta x = [2 + \frac{4}{n}(k-1)]^3 \frac{4}{n} = \frac{32}{n}[1 + \frac{2}{n}(k-1)]^3$$

$$= \frac{32}{n}\left[1 + \frac{6}{n}(k-1) + \frac{12}{n^2}(k-1)^2 + \frac{8}{n^3}(k-1)^3\right]$$

$$\sum_{k=1}^{n} f(c_k)\Delta x = \frac{32}{n}\left[\sum_{k=1}^{n} 1 + \frac{6}{n}\sum_{k=1}^{n}(k-1) + \frac{12}{n^2}\sum_{k=1}^{n}(k-1)^2 + \frac{8}{n^3}\sum_{k=1}^{n}(k-1)^3\right]$$

$$= \frac{32}{n}\left[n + \frac{6}{n}\sum_{k=1}^{n-1} k + \frac{12}{n^2}\sum_{k=1}^{n-1} k^2 + \frac{8}{n^3}\sum_{k=1}^{n-1} k^3\right]$$

$$= \frac{32}{n}\left[n + \frac{6}{n}\cdot\frac{1}{2}(n-1)(n) + \frac{12}{n^2}\cdot\frac{1}{6}(n-1)(n)(2n-1) + \frac{8}{n^3}\cdot\frac{1}{4}(n-1)^2 n^2\right]$$

$$= 32\left[1 + 3\frac{n-1}{n} + 2\frac{(n-1)(2n-1)}{n^2} + 2\frac{(n-1)^2}{n^2}\right]$$

$$A = \lim_{n\to+\infty} 32\left[1 + 3(1-\frac{1}{n}) + 2(1-\frac{1}{n})(2-\frac{1}{n}) + 2(1-\frac{1}{n})^2\right]$$

$$= 32[1 + 3(1) + 2(1)(2) + 2(1)^2] = 320$$

11. $\Delta x = \frac{3}{n}$, $d_k = 1 + \frac{3}{n}k$

$$f(d_k)\Delta x = \frac{1}{2}d_k\Delta x = \frac{1}{2}(1 + \frac{3}{n}k)\frac{3}{n} = \frac{3}{2}\left[\frac{1}{n} + \frac{3}{n^2}k\right]$$

$$\sum_{k=1}^{n} f(d_k)\Delta x = \frac{3}{2}\left[\sum_{k=1}^{n}\frac{1}{n} + \sum_{k=1}^{n}\frac{3}{n^2}k\right] = \frac{3}{2}\left[1 + \frac{3}{n^2}\cdot\frac{1}{2}n(n+1)\right]$$

$$= \frac{3}{2}\left[1 + \frac{3}{2}\frac{n+1}{n}\right]$$

$$A = \lim_{n\to+\infty}\frac{3}{2}\left[1 + \frac{3}{2}\left(1 + \frac{1}{n}\right)\right] = \frac{3}{2}\left(1 + \frac{3}{2}\right) = \frac{15}{4}$$

13. $\Delta x = \frac{1}{n}$, $d_k = \frac{k}{n}$

$$f(d_k)\Delta x = d_k^2\Delta x = \frac{k^2}{n^2}\frac{1}{n} = \frac{k^2}{n^3}$$

$$\sum_{k=1}^{n} f(d_k)\Delta x = \sum_{k=1}^{n}\frac{k^2}{n^3} = \frac{1}{n^3}\frac{1}{6}n(n+1)(2n+1) = \frac{1}{6}\frac{(n+1)(2n+1)}{n^2}$$

$$A = \lim_{n\to+\infty}\frac{1}{6}\left(1 + \frac{1}{n}\right)\left(2 + \frac{1}{n}\right) = \frac{1}{3}$$

15.   (a) Using inscribed rectangles, $\Delta x = \frac{b}{n}$, $c_k = \frac{b}{n}(k-1)$

$$f(c_k)\Delta x = c_k^3\Delta x = \frac{b^4}{n^4}(k-1)^3$$

$$\sum_{k=1}^{n} f(c_k)\Delta x = \frac{b^4}{n^4}\sum_{k=1}^{n}(k-1)^3 = \frac{b^4}{n^4}\sum_{k=1}^{n-1}k^3 = \frac{b^4}{4}\frac{(n-1)^2}{n^2}$$

$$A = \lim_{n\to+\infty}\frac{b^4}{4}\left(1 - \frac{1}{n}\right)^2 = b^4/4$$

(b) $\Delta x = \frac{b-a}{n}$, $c_k = a + \frac{b-a}{n}(k-1)$

$$f(c_k)\Delta x = c_k^3\Delta x = \left[a + \frac{b-a}{n}(k-1)\right]^2\frac{b-a}{n}$$

$$= \frac{b-a}{n}\left[a^3 + \frac{3a^2(b-a)}{n}(k-1) + \frac{3a(b-a)^2}{n^2}(k-1)^2\right.$$

$$\left. + \frac{(b-a)^3}{n^3}(k-1)^3\right]$$

$$\sum_{k=1}^{n} f(c_k)\Delta x = (b - a)\left[a^3 + \frac{3}{2}a^2(b - a)\frac{n - 1}{n} + \frac{1}{2}a(b - a)^2\right.$$

$$\left.\frac{(n - 1)(2n - 1)}{n^2} + \frac{1}{4}(b - a)^3\frac{(n - 1)^2}{n^2}\right]$$

$$A = \lim_{n \to +\infty} \sum_{k=1}^{n} f(c_k)\Delta x$$

$$= (b - a)[a^3 + \frac{3}{2}a^2(b - a) + a(b - a)^2 + \frac{1}{4}(b - a)^3]$$

$$= \frac{1}{4}(b^4 - a^4).$$

17. (a) $\Delta x = \dfrac{b - a}{n}$, $c_k = a + \dfrac{b - a}{n}(k - 1)$

$$f(c_k)\Delta x = \frac{b - a}{n}\left[a + \frac{b - a}{n}(k - 1)\right]$$

$$\sum_{k=1}^{n} f(c_k)\Delta x = (b - a)[a + \frac{b - a}{2} \cdot \frac{n - 1}{n}]$$

$$A = \lim_{n \to +\infty}(b - a)[a + \frac{b - a}{2}(1 - \frac{1}{n})] = \frac{1}{2}(b^2 - a^2)$$

(b) $\Delta x = \dfrac{b - a}{n}$, $d_k = a + \dfrac{b - a}{n}k$

$$f(d_k)\Delta x = \frac{b - a}{n}\left[a + \frac{b - a}{n}k\right]$$

$$\sum_{k=1}^{n} f(d_k)\Delta x = (b - a)[a + \frac{b - a}{2} \cdot \frac{n + 1}{n}]$$

$$A = \lim_{n \to +\infty}(b - a)[a + \frac{b - a}{2}(1 + \frac{1}{n})] = \frac{1}{2}(b^2 - a^2)$$

(c) The region is enclosed by a
trapezoid so its area is

$$A = \frac{1}{2}(b - a)(b + a)$$
$$= \frac{1}{2}(b^2 - a^2).$$

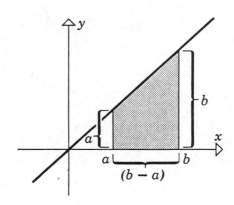

EXERCISE SET 5.6

1. (a) $(4/3)(1) + (5/2)(1) + (4)(2) = 71/6$
    (b) 2

3. (a) $(-9/4)(1) + (3)(2) + (63/16)(1) + (-5)(3) = -117/16$
    (b) ~~2~~ **3**

5. $\sum\limits_{k=1}^{4} f(c_k)\Delta x = (1/3 + 1/5 + 1/7 + 1/9)(2) = 496/315$

    $\sum\limits_{k=1}^{4} f(x_k^*)\Delta x = (1/2 + 1/4 + 1/6 + 1/8)(2) = 25/12$

    $\sum\limits_{k=1}^{4} f(d_k)\Delta x = (1 + 1/3 + 1/5 + 1/7)(2) = 352/105$

    $496/315 = 1984/1260$, $25/12 = 2625/1260$, $352/105 = 4224/1260$,
    $1984/1260 \leq 2625/1260 \leq 4224/1260$

7. $\displaystyle\int_{-3}^{3} 4x(1 - 3x)dx$

9. $\displaystyle\lim_{\max \Delta x_k \to 0} \sum_{k=1}^{n} 2x_k^* \Delta x_k$; $a = 1$, $b = 2$

11. $\displaystyle\lim_{\max \Delta x_k \to 0} \sum_{k=1}^{n} \frac{x_k^*}{x_k^* + 1} \Delta x_k$; $a = 0$, $b = 1$

13.   (a) 0.8   (b) −2.6   (c) −1.8   (d) −0.3

15.   (a)

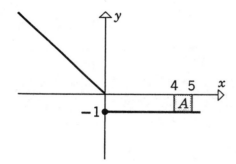

$-A = -(1)(1) = -1$

(b)

$A = \frac{1}{2} (1)(3 + 2) = \frac{5}{2}$

(c)

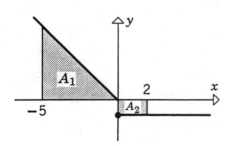

$A_1 - A_2 = \frac{25}{2} - 2 = 21/2$

(d)

$A_1 - A_2 = \frac{1}{2} k^2 - k$

17. $A_1 + A_2 = \frac{1}{2}(5)(5/2) + \frac{1}{2}(1)(1/2)$

$= 13/2$

19. $\frac{1}{2}[\pi(1)^2] = \pi/2$

21. $\sqrt{10x - x^2} = \sqrt{25 - (x - 5)^2}$; $\int_0^{10}\sqrt{10x - x^2}dx = \frac{1}{2}[\pi(5)^2] = 25\pi/2$

23. Let $\Delta x_k = \frac{1}{n}$ and $x_k^* = k\,\Delta x_k = \frac{k}{n}$, then

$$\sum_{k=1}^{n} f(x_k^*)\Delta x_k = \sum_{k=1}^{n}\frac{n^2}{k^2}\cdot\frac{1}{n} = n\sum_{k=1}^{n}\frac{1}{k^2} \geq n \text{ thus } \sum_{k=1}^{n} f(x_k^*)\Delta x_k \to +\infty \text{ as } n\to +\infty$$

(or as max $\Delta x_k \to 0$) so $\displaystyle\lim_{\max \Delta x_k \to 0}\sum_{k=1}^{n} f(x_k^*)\Delta x_k$ does not exist.

25. Let $S_n = \displaystyle\sum_{k=1}^{n} f(x_k^*)\Delta x_k$. From definition 5.7.5, for any $\epsilon > 0$ there are
numbers $\delta_1 > 0$ and $\delta_2 > 0$ such that

$$|S_n - L_1| < \epsilon \quad \text{for max } \Delta x_k < \delta_1 \text{ and}$$
$$|S_n - L_2| < \epsilon \quad \text{for max } \Delta x_k < \delta_2.$$

If $\delta = \min(\delta_1, \delta_2)$ then $|S_n - L_1| < \epsilon$ and $|S_n - L_2| < \epsilon$ for max $\Delta x_k < \delta$
thus

$$|L_1 - L_2| = |L_1 - S_n + S_n - L_2| = |(L_1 - S_n) + (S_n - L_2)|$$
$$\leq |S_n - L_1| + |S_n - L_2|$$
$$< \epsilon + \epsilon = 2\epsilon$$

so $|L_1 - L_2| < 2\epsilon$ for max $\Delta x_k < \delta$. Suppose $L_1 \neq L_2$ and let

$\epsilon = \frac{1}{2}|L_1 - L_2|$ then $|L_1 - L_2| < 2\epsilon$ yields $|L_1 - L_2| < |L_1 - L_2|$ which is false so $L_1 \neq L_2$ is impossible.

27. Let $R_n = \displaystyle\sum_{k=1}^{n} f(x_k^*)\Delta x_k$, $S_n = \displaystyle\sum_{k=1}^{n} g(x_k^*)\Delta x_k$, $T_n = \displaystyle\sum_{k=1}^{n} [f(x_k^*) + g(x_k^*)]\Delta x_k$,

$R = \displaystyle\int_a^b f(x)dx$, and $S = \displaystyle\int_a^b g(x)dx$ then $T_n = R_n + S_n$ and we want to prove

that $\displaystyle\lim_{\max \Delta x_k \to 0} T_n = R + S$.

$$|T_n - (R + S)| = |(R_n - R) + (S_n - S)|$$
$$\leq |R_n - R| + |S_n - S|$$

so for any $\epsilon > 0$
$$|T_n - (R + S)| < \epsilon \text{ if } |R_n - R| + |S_n - S| < \epsilon.$$
Because f and g are integrable on $[a,b]$, there are numbers δ_1 and δ_2 such that
$$|R_n - R| < \epsilon/2 \text{ for max } \Delta x_k < \delta_1 \text{ and}$$
$$|S_n - S| < \epsilon/2 \text{ for max } \Delta x_k < \delta_2.$$
If $\delta = \min(\delta_1, \delta_2)$ then $|R_n - R| < \epsilon/2$ and $|S_n - S| < \epsilon/2$ for
max $\Delta x_k < \delta$ thus $|R_n - R| + |S_n - S| < \epsilon$ and so $|T_n - (R + S)| < \epsilon$ for
max $\Delta x_k < \delta$ which shows that $\displaystyle\lim_{\max \Delta x_k \to 0} T_n = R + S$.

EXERCISE SET 5.7

1. $\left. \frac{1}{4} x^4 \right]_2^3 = 65/4$

3. $\left. \frac{1}{2} x^2 + \frac{1}{5} x^5 \right]_{-1}^2 = 81/10$

5. $\dfrac{1}{3}\,t^3 - t^2 + 8t\bigg]_{1}^{2} = 22/3$

7. $\displaystyle\int_{1}^{3} x^{-2}dx = -\dfrac{1}{x}\bigg]_{1}^{3} = 2/3$

9. $-\dfrac{1}{2x^2} + \dfrac{2}{x} - \dfrac{1}{3x^3}\bigg]_{1}^{2} = -1/3$

11. $\dfrac{2}{3}\,x^{3/2}\bigg]_{1}^{9} = 52/3$

13. $\dfrac{4}{5}\,y^{5/2}\bigg]_{4}^{9} = 844/5$

15. $6\sqrt{x} - \dfrac{10}{3}\,x^{3/2} + \dfrac{2}{\sqrt{x}}\bigg]_{1}^{4} = -55/3$

17. $-\cos\theta\bigg]_{-\pi/2}^{\pi/2} = 0$

19. $\displaystyle\int_{-\pi/4}^{\pi/4} \cos x\,dx = \sin x\bigg]_{-\pi/4}^{\pi/4} = \sqrt{2}$

21. $\dfrac{1}{2}\,x^2 - 2\cot x\bigg]_{\pi/6}^{\pi/2} = \pi^2/9 + 2\sqrt{3}$

23. $A = \displaystyle\int_{0}^{3} (x^2 + 1)dx = \dfrac{1}{3}\,x^3 + x\bigg]_{0}^{3} = 12$

25. $A = \displaystyle\int_{0}^{2\pi/3} 3\sin x\,dx = -3\cos x\bigg]_{0}^{2\pi/3} = 9/2$

27. $A_1 = \displaystyle\int_{-3}^{-2} (x^2 - 3x - 10)dx$

$= \dfrac{1}{3}\,x^3 - \dfrac{3}{2}\,x^2 - 10x\bigg]_{-3}^{-2} = 23/6,$

$A_2 = -\displaystyle\int_{-2}^{5} (x^2 - 3x - 10)dx = 343/6,$

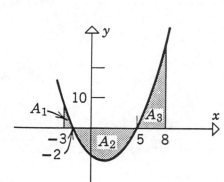

$$A_3 = \int_5^8 (x^2 - 3x - 10)dx = 243/6,$$

$$A = A_1 + A_2 + A_3 = 203/2$$

29.   (b), (c) are always valid. Use $f(x) = g(x) = 1$ to show that (a), (d), and (e) are false.

EXERCISE SET 5.8

1.   (a) $\int_1^3 u^7 du$ (b) $-\frac{1}{2} \int_7^4 u^{1/2} du$

      (c) $\frac{1}{\pi} \int_{-\pi}^{\pi} \sin u \, du$ (d) $\int_0^1 u^2 du$

      (e) $\frac{1}{2} \int_3^4 (u - 3)u^{1/2} du$ (f) $\int_2^0 (u + 5)u^{20} du$

3. $u = 2x + 1,$ 5. $u = 1 - 2x,$

    $\frac{1}{2} \int_1^3 u^4 du = \frac{1}{10} u^5 \Big]_1^3 = 121/5,$ $-\frac{1}{2} \int_3^1 u^3 du = -\frac{1}{8} u^4 \Big]_3^1 = 10,$

    or $\frac{1}{10} (2x + 1)^5 \Big]_0^1 = 121/5$ or $-\frac{1}{8} (1 - 2x)^4 \Big]_{-1}^0 = 10$

7. $u = 1 + x,$

    $\int_1^9 (u - 1)u^{1/2} du = \int_1^9 (u^{3/2} - u^{1/2})du = \frac{2}{5} u^{5/2} - \frac{2}{3} u^{3/2} \Big]_1^9 = 1192/15,$

    or $\frac{2}{5} (1 + x)^{5/2} - \frac{2}{3} (1 + x)^{3/2} \Big]_0^8 = 1192/15$

9. $u = x/2,$ $8\int_0^{\pi/4} \sin u \, du = -8 \cos u \Big]_0^{\pi/4} = 8 - 4\sqrt{2},$

    or $-8 \cos(x/2) \Big]_0^{\pi/2} = 8 - 4\sqrt{2}$

11. $u = x^2 + 2$,

$$\frac{1}{2} \int_6^3 u^{-3} du = -\left. \frac{1}{4u^2} \right]_6^3 = -1/48, \text{ or } -\frac{1}{4} \left. \frac{1}{(x^2 + 2)^2} \right]_{-2}^{-1} = -1/48$$

13. $\left. \frac{2}{3} (3u + 1)^{1/2} \right]_0^1 = 2/3$ 15. $\left. \frac{2}{3} (x^3 + 9)^{1/2} \right]_{-1}^1 = \frac{2}{3} (\sqrt{10} - 2\sqrt{2})$

17. $u = x^2 + 4x + 7$, $\frac{1}{2} \int_{12}^{28} u^{-1/2} du = \left. u^{1/2} \right]_{12}^{28} = \sqrt{28} - \sqrt{12} = 2(\sqrt{7} - \sqrt{3})$

19. $\left. \frac{1}{2} \sin^2 x \right]_{-3\pi/4}^{-\pi/4} = 0$ 21. $\left. \frac{5}{2} \sin(x^2) \right]_0^{\sqrt{\pi}} = 0$

23. $u = \sqrt{x}$, $2 \int_\pi^{2\pi} \sin u \, du = \left. -2 \cos u \right]_\pi^{2\pi} = -4$

25. $u = \sin 3x$, $\frac{1}{3} \int_0^{-1} u^2 du = \left. \frac{1}{9} u^3 \right]_0^{-1} = -1/9$

27. $u = 3\theta$, $\frac{1}{3} \int_{\pi/4}^{\pi/3} \sec^2 u \, du = \left. \frac{1}{3} \tan u \right]_{\pi/4}^{\pi/3} = (\sqrt{3} - 1)/3$

29. $u = 4 - 3y$, $y = \frac{1}{3}(4 - u)$, $dy = -\frac{1}{3} du$

$$-\frac{1}{27} \int_4^1 \frac{16 - 8u + u^2}{u^{1/2}} du = \frac{1}{27} \int_1^4 (16u^{-1/2} - 8u^{1/2} + u^{3/2}) du$$

$$= \frac{1}{27} \left[32u^{1/2} - \frac{16}{3} u^{3/2} + \frac{2}{5} u^{5/2} \right]_1^4 = 106/405$$

31. $A = \int_0^{\pi/8} 3 \cos 2x \, dx = \left. \frac{3}{2} \sin 2x \right]_0^{\pi/8} = 3\sqrt{2}/4$

33. $\int_{-2}^{2} \sqrt{4 - u^2}\,du = \frac{1}{2}\,[\pi(2)^2] = 2\pi$

35. $-\frac{1}{2}\int_{1}^{0} \sqrt{1 - u^2}\,du = \frac{1}{2}\int_{0}^{1} \sqrt{1 - u^2}\,du = \frac{1}{2} \cdot \frac{1}{4}\,[\pi(1)^2] = \pi/8$

37. $u = 1/x,\ -\int_{2}^{1} f(u)\,du = \int_{1}^{2} f(u)\,du = 3$

39. $\sin x = \cos(\pi/2 - x),$

$$\int_{0}^{\pi/2} \sin^n x\,dx = \int_{0}^{\pi/2} \cos^n(\pi/2 - x)\,dx = -\int_{\pi/2}^{0} \cos^n u\,du \quad (u = \pi/2 - x)$$

$$= \int_{0}^{\pi/2} \cos^n u\,du = \int_{0}^{\pi/2} \cos^n x\,dx \quad \text{(by replacing u by x)}$$

41. (a) $I = -\int_{a}^{0} \dfrac{f(a - u)}{f(a - u) + f(u)}\,du = \int_{0}^{a} \dfrac{f(a - u) + f(u) - f(u)}{f(a - u) + f(u)}\,du$

$$= \int_{0}^{a} du - \int_{0}^{a} \frac{f(u)}{f(a - u) + f(u)}\,du,$$

$I = a - I$ so $2I = a,\ I = a/2$

(b) 3/2   (c) $\pi/4$

EXERCISE SET 5.9

1. $\int_{0}^{3/2} (3 - 2x)\,dx + \int_{3/2}^{2} (2x - 3)\,dx = (3x - x^2)\Big]_{0}^{3/2} + (x^2 - 3x)\Big]_{3/2}^{2}$

$$= 9/4 + 1/4 = 5/2$$

3. $\int_{0}^{\pi/2} \cos x\,dx + \int_{\pi/2}^{3\pi/4} (-\cos x)\,dx = \sin x\Big]_{0}^{\pi/2} - \sin x\Big]_{\pi/2}^{3\pi/4} = 2 - \sqrt{2}/2$

5. $\int_{-2}^{0} x^2\,dx + \int_{0}^{3} (-x)\,dx = \frac{1}{3}\,x^3\Big]_{-2}^{0} - \frac{1}{2}\,x^2\Big]_{0}^{3} = -11/6$

7.   If $f(x) \le 0$ on $[a,b]$ then $-f(x) \ge 0$ so $\int_a^b [-f(x)]dx \ge 0$ from part (a) of

Theorem 5.9.2.   But $\int_a^b [-f(x)]dx = -\int_a^b f(x)dx \ge 0$ so $\int_a^b f(x)\,dx \le 0$.

9.   (a) $f_{ave} = \dfrac{1}{2-0} \int_0^2 x^2 dx = 4/3$ (c)

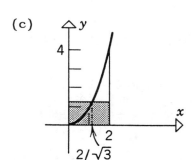

(b) $(x*)^2 = 4/3$, $x* = \pm 2/\sqrt{3}$,
     but only $2/\sqrt{3}$ is in $[0,2]$.

11. $f_{ave} = \dfrac{1}{9} \int_0^9 x^{1/2} dx = 2$; $\sqrt{x^*} = 2$, $x^* = 4$

13. $f_{ave} = \dfrac{1}{x_1 - x_0} \int_{x_0}^{x_1} (\alpha x + \beta)dx = \dfrac{1}{2}\alpha(x_0 + x_1) + \beta$;

$\alpha x^* + \beta = \dfrac{1}{2}\alpha(x_0 + x_1) + \beta$, $x^* = (x_0 + x_1)/2$ if $\alpha \ne 0$, x^* is any point
in $[x_0,x_1]$ if $\alpha = 0$.

15. $a_{ave} = \dfrac{v(t_1) - v(t_0)}{t_1 - t_0} = \dfrac{1}{t_1 - t_0} v(t) \Big]_{t_0}^{t_1} = \dfrac{1}{t_1 - t_0} \int_{t_0}^{t_1} a(t)dt$

17.
$$\text{time to fill tank} = (\text{volume of tank})/(\text{rate of filling})$$
$$= [\pi(3)^2 5]/(1) = 45\pi,$$
$$\text{force on bottom at time } t = \text{weight of water in tank at time } t$$
$$= (62.4)(\text{rate of filling})(\text{time})$$
$$= 62.4t,$$
$$\text{force}_{ave} = \dfrac{1}{45\pi} \int_0^{45\pi} 62.4t\,dt = 1404\pi \text{ lb.}$$

19. (a) $\displaystyle\int_a^b [f(x) - f_{ave}]dx = \int_a^b f(x)dx - \int_a^b f_{ave}dx$

$$= \int_a^b f(x)dx - f_{ave}(b - a) = 0$$

because $\displaystyle f_{ave}(b - a) = \int_a^b f(x)dx.$

(b) No, because if $\displaystyle\int_a^b [f(x) - c]dx = 0$ then $\displaystyle\int_a^b f(x)dx - c(b - a) = 0$ so

$c = \dfrac{1}{b - a}\displaystyle\int_a^b f(x)dx = f_{ave}$ is the only value.

21. (a) $x^3 + 1$

(b) $F(x) = \dfrac{1}{4} t^4 + t\Big]_1^x = \dfrac{1}{4} x^4 + x - \dfrac{5}{4}$; $F'(x) = x^3 + 1$

23. $\sin \sqrt{x}$ 25. $|x|$

27. $\displaystyle\int_2^x \dfrac{1}{t - 1} dt$ 29. $\displaystyle\int_0^x \dfrac{1}{t - 1} dt$

31. (a) $(0, +\infty)$ because f is continuous there and 1 is in $(0, +\infty)$.
     (b) at $x = 1$ because $F(1) = 0$

33. $F'(x) = \dfrac{\cos x}{x^2 + 3}$, $F''(x) = \dfrac{-(x^2 + 3)\sin x - 2x \cos x}{(x^2 + 3)^2}$

(a) 0   (b) 1/3   (c) 0

35. $x < 0$: $F(x) = \displaystyle\int_{-1}^x (-t)dt = -\dfrac{1}{2} t^2\Big]_{-1}^x = \dfrac{1}{2}(1 - x^2)$,

$x \geq 0$: $F(x) = \displaystyle\int_{-1}^0 (-t)dt + \int_0^x t\, dt = \dfrac{1}{2} + \dfrac{1}{2} x^2$;

$F(x) = \begin{cases} (1 - x^2)/2, & x < 0 \\ (1 + x^2)/2, & x \geq 0 \end{cases}$

37. $x \le 0$: $F(x) = \int_{-1}^{x} t^2 dt = \frac{1}{3} t^3 \Big]_{-1}^{x} = \frac{1}{3}(x^3 + 1)$,

   $x > 0$: $F(x) = \int_{-1}^{0} t^2 dt + \int_{0}^{x} 2t\, dt = \frac{1}{3} + x^2$;

   $F(x) = \begin{cases} (x^3 + 1)/3, & x \le 0 \\ x^2 + 1/3, & x > 0 \end{cases}$

39. $\frac{1}{x^3}(3x^2) = \frac{3}{x}$

41. $F'(x) = \dfrac{1}{1 + x^2} + \dfrac{1}{1 + (1/x)^2}(-1/x^2) = 0$ so F is constant on $(0, +\infty)$.

43. (a) $\sin^2(x^3)(3x^2) - \sin^2(x^2)(2x) = 3x^2 \sin^2(x^3) - 2x \sin^2(x^2)$

   (b) $\dfrac{1}{1+x}(1) - \dfrac{1}{1-x}(-1) = \dfrac{2}{1 - x^2}$

45. $\int_{x}^{b} f(t)dt = -\int_{b}^{x} f(t)dt$ so $\dfrac{d}{dx}\int_{x}^{b} f(t)dt = -\dfrac{d}{dx}\int_{b}^{x} f(t)dt = -f(x)$

SUPPLEMENTARY EXERCISES CHAPTER 5

1. $-x^{-2}/2 + 2\sqrt{x} + 5\cos x + C$

3. $u = \sqrt{x} + 2$, $2\int u^8 du = \frac{2}{9} u^9 + C = \frac{2}{9}(\sqrt{x} + 2)^9 + C$

5. $u = \sqrt{2x^2 - 5}$, $du = 2x/\sqrt{2x^2 - 5}\ dx$, $\frac{1}{2}\int \sin u\, du = -\frac{1}{2}\cos\sqrt{2x^2 - 5} + C$

7. $\int (3x^{1/2} + x^{11/6})dx = 2x^{3/2} + \frac{6}{17} x^{17/6} + C$

9. $u = \sin 5t$, $\frac{1}{5}\int \sec^2 u\, du = \frac{1}{5}\tan(\sin 5t) + C$

11.   (a) $\int (y^5 + 4y^3 + 4y)dy = \frac{1}{6} y^6 + y^4 + 2y^2 + C$

     (b) $\frac{1}{6} (y^2 + 2)^3 + C$

     [answer to (b)] − [answer to (a)]

$$= \frac{1}{6} (y^6 + 6y^4 + 12y^2 + 8) + C - (\frac{1}{6} y^6 + y^4 + 2y^2 + C) = 4/3$$

13. $\int_0^1 u^4 du = 1/5$

15. $u = x - 1$, $x = u + 1$,

$$\int_1^4 \frac{u - 1}{\sqrt{u}} du = \int_1^4 (u^{1/2} - u^{-1/2})du = \frac{2}{3} u^{3/2} - 2u^{1/2} \Big]_1^4 = 8/3$$

17. $\frac{4}{\pi} \int_{\pi/2}^{\pi} \cos u \, du = \frac{4}{\pi} \sin u \Big]_{\pi/2}^{\pi} = -4/\pi$

19. $\int_{-2}^{1/2} -(2x - 1)dx + \int_{1/2}^{2} (2x - 1)dx = (-x^2 + x) \Big]_{-2}^{1/2} + (x^2 - x) \Big]_{1/2}^{2} = 17/2$

21. $\int_0^x \frac{1}{(3t + 1)^2} dt = -\frac{1}{3(3t + 1)} \Big]_0^x = -\frac{1}{3(3x + 1)} + \frac{1}{3} = \frac{1}{6},$

     $3x + 1 = 2$, $x = 1/3$

23.   (a) $5 + 5 + 5 + 5 = 20$ (b) $2 + 2 + 2 + 2 = 8$
       (c) $n + n + n + n = 4n$ (d) $0 + 1/5 + 2/6 = 8/15$
       (e) $6/4 + 6/9 + 6/16 = 61/24$ (f) 9
       (g) $\sin(0) + \sin(\pi/4) + \sin(\pi/2) + \sin(3\pi/4) + \sin(\pi)$

$$= 0 + \sqrt{2}/2 + 1 + \sqrt{2}/2 + 0 = 1 + \sqrt{2}$$

       (h) $\sqrt{2}/2 + (\sqrt{2}/2)^2 + (\sqrt{2}/2)^3 + (\sqrt{2}/2)^4 = 3\sqrt{2}/4 + 3/4$

25.   (a) $\displaystyle\sum_{k=1}^{9} (-1)^{k+1} (\frac{k}{k + 1})^2 = \sum_{k=2}^{10} (-1)^k (\frac{k - 1}{k})^2$

       (b) $\displaystyle\sum_{k=1}^{11} (-1)^{k+1} \frac{\pi^{k+1}}{k} = \sum_{k=2}^{12} (-1)^k \frac{\pi^k}{k - 1}$

27. (a) $\Delta x = 4/n$, $c_k = 4k/n$

$$\sum_{k=1}^{n} f(c_k)\Delta x = \sum_{k=1}^{n} (16 - 16k^2/n^2)(4/n)$$

$$= \frac{64}{n} \sum_{k=1}^{n} 1 - \frac{64}{n^3} \sum_{k=1}^{n} k^2 = 64 - \frac{32}{3} \frac{(n + 1)(2n + 1)}{n^2}$$

(b) $d_k = 4(k - 1)/n$

$$\sum_{k=1}^{n} f(d_k)\Delta x = \sum_{k=1}^{n} (16 - 16(k - 1)^2/n^2)(4/n)$$

$$= \frac{64}{n} \sum_{k=1}^{n} 1 - \frac{64}{n^3} \sum_{k=1}^{n} (k - 1)^2$$

$$= 64 - \frac{32}{3} \frac{(n - 1)(2n - 1)}{n^2}$$

(c) area $= \lim_{n \to +\infty} [64 - \frac{32}{3}(1 + \frac{1}{n})(2 + \frac{1}{n})] = 128/3$

(d) $\int_{0}^{4} (16 - x^2)dx = 128/3$

29. (a) $\Delta x = 2/n$, because f is constant c_k can be chosen anywhere in the

k-th subinterval so $f(c_k) = 6$ and $\sum_{k=1}^{n} f(c_k)\Delta x = \sum_{k=1}^{n} (6)(2/n) = 12$

(b) same as for (a)   (c) area $= \lim_{n \to +\infty} 12 = 12$

(d) $\int_{-1}^{1} 6dx = 12$

31.   (a) $A_1 - A_2 = \frac{1}{2}(2)(1/2) - \frac{1}{2}(6)(3/2)$

           $= -4$

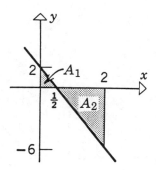

(b) $A = \frac{1}{4}\pi(3)^2 = 9\pi/4$

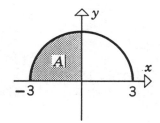

(c) $-A_1 + A_2 = -4 + 2 = -2$

(d) $-A_1 + A_2 = 0$ because
     $A_1 = A_2$ by symmetry

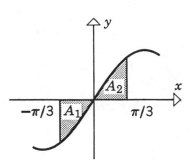

(e) $A_1 - A_2 = 0$ because
     $A_1 = A_2$ by symmetry

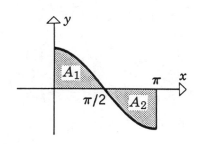

(f) $A = (6)(5) = 30$

(g) $-A_1 + A_2 + A_3 = -8 + 6 + 4 = 2$

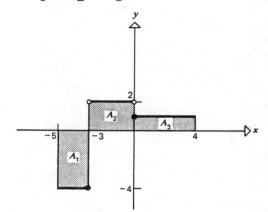

33. (a) $2\int_3^5 P(x)dx + \int_3^5 Q(x)dx = 2(3) + (4) = 10$

(b) $-\int_1^5 P(x)dx = -(-1) = 1$ (c) $-\int_3^5 Q(u)du = -\int_3^5 Q(x)dx = -4$

(d) $\int_3^5 P(x)dx + \int_5^1 P(x)dx = \int_3^5 P(x)dx - \int_1^5 P(x)dx = (3) - (-1) = 4$

35. $f_{ave} = \int_{-2}^{-1} 3x^2 dx = 7; \ 3(x^*)^2 = 7, \ x^* = \pm\sqrt{7/3}$ but only $-\sqrt{7/3}$ is in $[-2,-1]$

37. $f_{ave} = \dfrac{1}{4}\int_{-3}^1 (2 + |x|)dx = \dfrac{1}{4}\left[\int_{-3}^0 (2 - x)dx + \int_0^1 (2 + x)dx\right]$

$$= \frac{1}{4}[21/2 + 5/2] = 13/4;$$

$2 + |x^*| = 13/4, \ |x^*| = 5/4, \ x^* = \pm 5/4$ but only $-5/4$ is in $[-3,1]$

CHAPTER 6

EXERCISE SET 6.1

1. (a) $A = \int_0^4 (4x - x^2)dx = 32/3$

    (b) $A = \int_0^{16} (\sqrt{y} - y/4)dy = 32/3$

3. Eliminate x to get
    $$y^2 = y + 2,$$
    $$y^2 - y - 2 = 0,$$
    $$(y + 1)(y - 2) = 0,$$
    $y = -1$ and 2 with corresponding values of $x = 1/2$ and 2.

    (a) $A = \int_0^{1/2} [\sqrt{2x} - (-\sqrt{2x})]dx$

    $\qquad + \int_{1/2}^2 [\sqrt{2x} - (2x - 2)]dx$

    $\quad = \int_0^{1/2} 2\sqrt{2}x^{1/2}dx + \int_{1/2}^2 (\sqrt{2}x^{1/2} - 2x + 2)dx$

    $\quad = 2/3 + 19/12 = 9/4$

    (b) $A = \int_{-1}^2 [(y/2 + 1) - y^2/2]dy = 9/4$

173

5. $A = \int_{1/4}^{1} (\sqrt{x} - x^2)dx = 49/192$

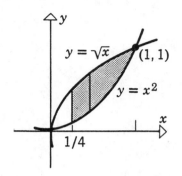

7. $A = \int_{\pi/4}^{\pi/2} (0 - \cos 2x)dx$

$= -\int_{\pi/4}^{\pi/2} \cos 2x \, dx = 1/2$

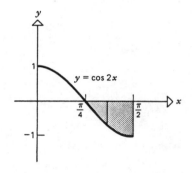

9. $A = \int_{0}^{4} [0 - (y^2 - 4y)]dy$

$= \int_{0}^{4} (4y - y^2)dy = 32/3$

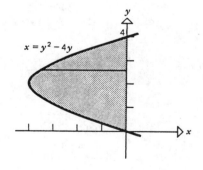

11. Equate $\sec^2 x$ and 2 to get $\sec^2 x = 2$,

$\sec x = \pm\sqrt{2}$, $x = \pm\pi/4$

$A = \int_{-\pi/4}^{\pi/4} (2 - \sec^2 x)dx = \pi - 2$

13. Eliminate y to get

$$6 - x = x^2 + 4$$
$$x^2 + x - 2 = 0$$
$$(x + 2)(x - 1) = 0$$

$x = -2$, 1 with corresponding values of $y = 8$, 5.

$A = \int_{-2}^{1} [(6 - x) - (x^2 + 4)]dx$

$= \int_{-2}^{1} (2 - x - x^2)dx = 9/2$

15. $A = \int_{-1}^{4} [(y + 6) - (-y^2)]dy$

$= \int_{-1}^{4} (y + 6 + y^2)dy$

$= 355/6$

17. $y = 2 + |x - 1| = \begin{cases} 3 - x, & x \le 1 \\ 1 + x, & x \ge 1 \end{cases}$,

$A = \int_{-5}^{1} [(-\frac{1}{5}x + 7) - (3 - x)]dx$

$\qquad + \int_{1}^{5} [(-\frac{1}{5}x + 7) - (1 + x)]dx$

$\qquad = \int_{-5}^{1} (\frac{4}{5}x + 4)dx + \int_{1}^{5} (6 - \frac{6}{5}x)dx$

$\qquad = 72/5 + 48/5 = 24$

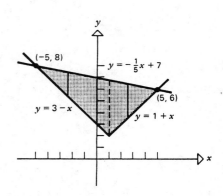

19. $A = \int_{-1}^{0} (y^3 - y)dy + \int_{0}^{1} -(y^3 - y)dy$

$\qquad = 1/2$

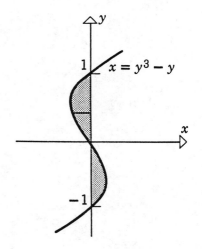

21. From the symmetry of the region

$A = 2\int_{\pi/4}^{5\pi/4} (\sin x - \cos x)dx = 4\sqrt{2}$

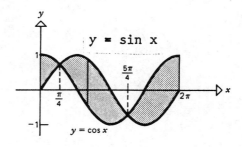

23. $A = \int_{1}^{4} (y - \frac{1}{\sqrt{y}}) dy = 11/2$

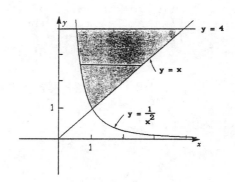

25. The tangent line at $(4,2)$ is
$y = \frac{1}{4} x + 1$;

$A = \int_{0}^{4} [(\frac{1}{4} x + 1) - \sqrt{x}] dx = 2/3$

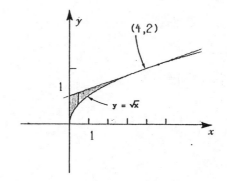

27. (a) $A = \int_{1}^{b} x^{-1/2} dx = 2(\sqrt{b} - 1)$ (b) $\lim\limits_{b \to +\infty} A = +\infty$

29. $\int_{0}^{k} 2\sqrt{y} \, dy = \int_{k}^{9} 2\sqrt{y} \, dy$

$\int_{0}^{k} y^{1/2} dy = \int_{k}^{9} y^{1/2} dy$

$\frac{2}{3} k^{3/2} = \frac{2}{3} (27 - k^{3/2})$

$k^{3/2} = 27/2$

$k = (27/2)^{2/3} = 9/\sqrt[3]{4}$

31. (a) It gives the area of the region that is between f and g when
    f(x) > g(x) <u>minus</u> the area of the region between f and g when
    f(x) < g(x), for a ≤ x ≤ b.
    (b) It gives the area of the region that is between f and g for
    a ≤ x ≤ b.

EXERCISE SET 6.2

1. $V = \pi \int_0^2 x^4 dx = 32\pi/5$

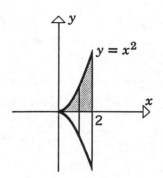

3. $V = \pi \int_1^2 (1 + x^3)^2 dx$

    $= \pi \int_1^2 (1 + 2x^3 + x^6) dx$

    $= 373\pi/14$

5. $V = \pi \int_{-3}^{3} (9 - x^2)^2 dx$

    $= \pi \int_{-3}^{3} (81 - 18x^2 + x^4) dx$

    $= 1296\pi/5$

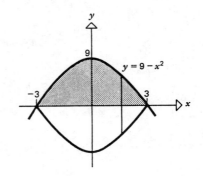

7. $V = \pi \int_{0}^{4} [(4x)^2 - (x^2)^2] dx$

    $= \pi \int_{0}^{4} (16x^2 - x^4) dx = 2048\pi/15$

9. $V = \pi \int_{0}^{\pi/4} (\cos^2 x - \sin^2 x) dx$

    $= \pi \int_{0}^{\pi/4} \cos 2x \, dx = \pi/2$

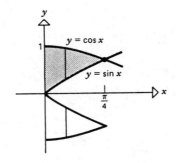

11. $V = \pi \int_0^1 [(\sqrt{x})^2 - x^2] dx$

   $= \pi \int_0^1 (x - x^2) dx = \pi/6$

13. $V = \pi \int_0^1 y^{2/3} dy = 3\pi/5$

15. $V = \pi \int_{-1}^3 (1 + y) dy = 8\pi$

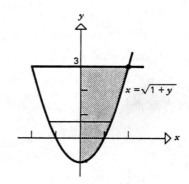

17. $V = \pi \int_{\pi/4}^{3\pi/4} \csc^2 y \, dy = 2\pi$

19. $V = \pi \int_{1}^{3} (9 - y^2) dy = 28\pi/3$

21. $V = \pi \int_{2}^{9} [(y - 1)^{2/3} - 1] dy$
     $= 58\pi/5$

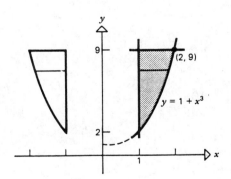

23. $V = \pi \int_{-1}^{2} [(y + 2)^2 - y^4] dy$

$= 72\pi/5$

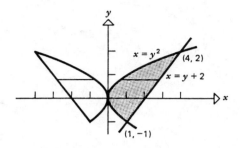

25. $V = \pi \int_{-4}^{4} [(25 - x^2) - 9] dx$

$= 2\pi \int_{0}^{4} (16 - x^2) dx$

$= 256\pi/3$

27. $V = \pi \int_{0}^{3} (9 - y^2)^2 dy$

$= \pi \int_{0}^{3} (81 - 18y^2 + y^4) dy$

$= 648\pi/5$

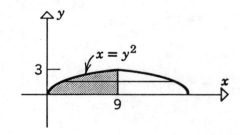

29. $V = \pi \int_0^1 [(\sqrt{x} + 1)^2 - (x + 1)^2]dx$

$= \pi \int_0^1 (2\sqrt{x} - x - x^2)dx = \pi/2$

31. $V = \pi \int_{-a}^a \frac{b^2}{a^2} (a^2 - x^2)dx$

$= 4\pi ab^2/3$

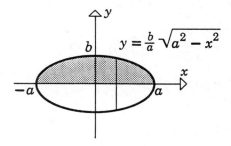

33. $V = \pi \int_{-1}^0 (x + 1)dx$

$+ \pi \int_0^1 [(x + 1) - 2x]dx$

$= \pi/2 + \pi/2 = \pi$

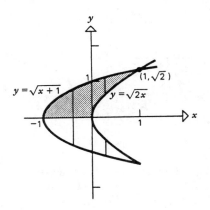

35. By similar triangles, $R/r = y/h$ so
     $R = ry/h$ and
     $A(y) = \pi r^2 y^2 / h^2$.
     $$V = (\pi r^2 / h^2) \int_0^h y^2 dy = \pi r^2 h / 3$$

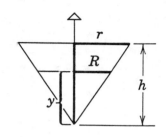

37. $$V = 2\pi \int_0^{L/2} [(r^2 - y^2) - (r^2 - L^2/4)] dy$$

     $$= 2\pi \int_0^{L/2} (L^2/4 - y^2) dy$$

     $$= \pi L^3 / 6$$

39. (a)   (b)

$0 \leqslant h < 2$ $2 \leqslant h \leqslant 4$

If the cherry is partially submerged then $0 \leq h < 2$ as shown in Figure
(a); if it is totally submerged then $2 \leq h \leq 4$ as shown in Figure (b).
The radius of the glass is 4 cm and that of the cherry is 1 cm so points
on the sections shown in the figures satisfy the equations $x^2 + y^2 = 16$
and $x^2 + (y + 3)^2 = 1$. We will find the volumes of the solids that are
generated when the shaded regions are revolved about the y-axis.

For $0 \leq h < 2$,

$$V = \pi \int_{-4}^{h-4} [(16 - y^2) - (1 - (y + 3)^2)]dy = 6\pi \int_{-4}^{h-4} (y + 4)dy = 3\pi h^2;$$

for $2 \le h \le 4$,

$$V = \pi \int_{-4}^{-2} [(16 - y^2) - (1 - (y + 3)^2)]dy + \pi \int_{-2}^{h-4} (16 - y^2)dy$$

$$= 6\pi \int_{-4}^{-2} (y + 4)dy + \pi \int_{-2}^{h-4} (16 - y^2)dy = 12\pi + \frac{1}{3}\pi(12h^2 - h^3 - 40)$$

$$= \frac{1}{3}\pi(12h^2 - h^3 - 4)$$

so

$$V = \begin{cases} 3\pi h^2 & \text{if } 0 \le h < 2 \\ \frac{1}{3}\pi(12h^2 - h^3 - 4) & \text{if } 2 \le h \le 4 \end{cases}.$$

41. $A(x) = \pi(x^2/4)^2 = \pi x^4/16$, $V = \int_0^{20} (\pi x^4/16)dx = 40,000\pi \text{ ft}^3$

43. With $y = \sqrt{9 - x^2}$, which is the upper half of the circle, $A(x)$ is the area of an equilateral triangle whose sides are each of length $2y$ so

$$A(x) = \frac{\sqrt{3}}{4}(2y)^2 = \sqrt{3}y^2$$

$$= \sqrt{3}(9 - x^2),$$

$$V = \int_{-3}^{3} \sqrt{3}(9 - x^2)dx$$

$$= 2\sqrt{3}\int_0^3 (9 - x^2)dx = 36\sqrt{3}.$$

45. $V = \int_{\pi/4}^{3\pi/4} \sin^2 x \, dx$

$\quad\ = \dfrac{1}{2} \int_{\pi/4}^{3\pi/4} (1 - \cos 2x) dx$

$\quad\ = (\pi + 2)/4$

47. $\tan \theta = h/x$ so $h = x \tan \theta$,

$\quad\quad A(y) = \dfrac{1}{2} hx = \dfrac{1}{2} x^2 \tan \theta$

$\quad\quad\quad\ = \dfrac{1}{2} (r^2 - y^2) \tan \theta$

because $x^2 = r^2 - y^2$,

$V = \dfrac{1}{2} \tan \theta \int_{-r}^{r} (r^2 - y^2) dy = \tan \theta \int_{0}^{r} (r^2 - y^2) dy = \dfrac{2}{3} r^3 \tan \theta$

49. Each cross section perpendicular to
the y-axis is a square so

$\quad\quad A(y) = x^2 = r^2 - y^2,$

$\quad\quad \dfrac{1}{8} V = \int_{0}^{r} (r^2 - y^2) dy$

$\quad\quad\quad V = 8(2r^3/3) = 16r^3/3$

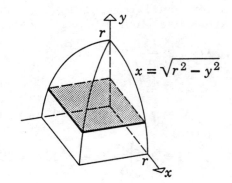

EXERCISE SET 6.3

1. $V = \int_0^1 2\pi(x)(x^3)dx$

$= 2\pi \int_0^1 x^4 dx = 2\pi/5$

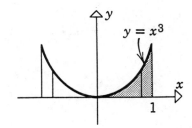

3. $V = \int_0^2 2\pi x(4 - x^2)^{1/3}dx$

$+ \int_2^4 2\pi x[-(4 - x^2)^{1/3}]dx$

$= 3\pi \sqrt[3]{4}(1 + 3 \sqrt[3]{3})$

5. $V = \int_1^2 2\pi x[(2x - 1) - (-2x + 3)]dx$

$= 8\pi \int_1^2 (x^2 - x)dx = 20\pi/3$

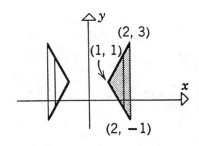

7. $V = \int_{1}^{3} 2\pi x(1/x)dx$

$= 2\pi \int_{1}^{3} dx = 4\pi$

9. $V = \int_{0}^{1} 2\pi y^{3}dy = \pi/2$

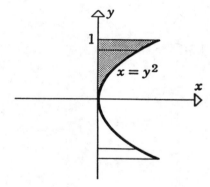

11. $V = \int_{0}^{1} 2\pi y(1 - \sqrt{y})dy$

$= 2\pi \int_{0}^{1} (y - y^{3/2})dy = \pi/5$

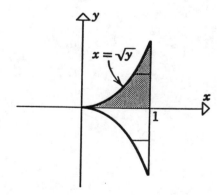

13. (a) $V = \int_0^1 2\pi x(x^3 - 3x^2 + 2x)dx = 7\pi/30$

(b) much easier; the method of
slicing would require that x
be expressed in terms of y.

15. $V = \int_0^1 2\pi(1 - y)y^{1/3}dy$

$= 2\pi\int_0^1 (y^{1/3} - y^{4/3})dy = 9\pi/14$

17. $x = \dfrac{h}{r}(r - y)$ is an equation of
line through $(0,r)$ and $(h,0)$ so

$V = \int_0^r 2\pi y\left[\dfrac{h}{r}(r - y)\right]dy$

$= \dfrac{2\pi h}{r}\int_0^r (ry - y^2)dy = \pi r^2 h/3$

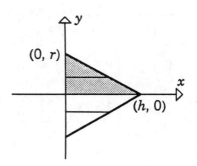

19. $V = \int_0^a 2\pi x (2\sqrt{r^2 - x^2})dx$

$= 4\pi \int_0^a x(r^2 - x^2)^{1/2}dx$

$= -\frac{4\pi}{3} (r^2 - x^2)^{3/2} \Big]_0^a$

$= \frac{4\pi}{3} [r^3 - (r^2 - a^2)^{3/2}]$

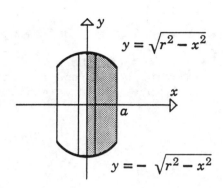

$y = \sqrt{r^2 - x^2}$

$y = -\sqrt{r^2 - x^2}$

21. $V_x = \pi \int_{1/2}^b \frac{1}{x^2} dx = \pi(2 - 1/b)$, $V_y = 2\pi \int_{1/2}^b dx = \pi(2b - 1)$;

$V_x = V_y$ if $2 - 1/b = 2b - 1$, $2b^2 - 3b + 1 = 0$, solve to get $b = 1/2$ (reject) or $b = 1$.

EXERCISE SET 6.4

1. (a) $L = \int_1^2 \sqrt{1 + 2^2} dx = \sqrt{5} \int_1^2 dx = \sqrt{5}$

(b) $L = \int_2^4 \sqrt{1 + (1/2)^2} dy = \frac{1}{2}\sqrt{5} \int_2^4 dy = \sqrt{5}$

(c) $L = \sqrt{(2 - 1)^2 + (4 - 2)^2} = \sqrt{5}$ ($y = 2x$ is a line).

3. $f'(x) = \frac{9}{2} x^{1/2}$, $1 + [f'(x)]^2 = 1 + \frac{81}{4} x$,

$L = \int_0^1 \sqrt{1 + 81x/4} \ dx = \frac{8}{243} (1 + \frac{81}{4} x)^{3/2} \Big]_0^1 = (85\sqrt{85} - 8)/243$

5. $\dfrac{dy}{dx} = \dfrac{2}{3} x^{-1/3}$, $1 + (\dfrac{dy}{dx})^2 = 1 + \dfrac{4}{9} x^{-2/3} = \dfrac{9x^{2/3} + 4}{9x^{2/3}}$,

$$L = \int_1^8 \frac{\sqrt{9x^{2/3} + 4}}{3x^{1/3}}\, dx = \frac{1}{18} \int_{13}^{40} u^{1/2}\,du, \quad u = 9x^{2/3} + 4$$

$$= \frac{1}{27} u^{3/2}\Big]_{13}^{40}$$

$$= \frac{1}{26}(40\sqrt{40} - 13\sqrt{13}) = \frac{1}{27}(80\sqrt{10} - 13\sqrt{13})$$

or (alternate solution)

$x = y^{3/2}$, $\dfrac{dx}{dy} = \dfrac{3}{2} y^{1/2}$, $1 + (\dfrac{dx}{dy})^2 = 1 + \dfrac{9}{4} y = \dfrac{4 + 9y}{4}$,

$$L = \frac{1}{2}\int_1^4 \sqrt{4 + 9y}\, dy = \frac{1}{18}\int_{13}^{40} u^{1/2}\,du, \quad u = 4 + 9y = \frac{1}{27}(80\sqrt{10} - 13\sqrt{13}).$$

7. $x = g(y) = \dfrac{1}{24} y^3 + 2y^{-1}$, $g'(y) = \dfrac{1}{8} y^2 - 2y^{-2}$,

$1 + [g'(y)]^2 = 1 + (\dfrac{1}{64} y^4 - \dfrac{1}{2} + 4y^{-4})$

$$= \frac{1}{64} y^4 + \frac{1}{2} + 4y^{-4} = (\frac{1}{8} y^2 + 2y^{-2})^2,$$

$$L = \int_2^4 (\frac{1}{8} y^2 + 2y^{-2})dy = 17/6$$

9. (a)

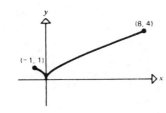

(b) dy/dx does not exist at x = 0

(c) $x = g(y) = y^{3/2}$, $g'(y) = \frac{3}{2} y^{1/2}$,

$$L = \int_0^1 \sqrt{1 + 9y/4} \; dy \quad \text{(portion for } -1 \le x \le 0)$$

$$+ \int_0^4 \sqrt{1 + 9y/4} \; dy \quad \text{(portion for } 0 \le x \le 8)$$

$$= \frac{8}{27} \left(\frac{13}{8} \sqrt{13} - 1\right) + \frac{8}{27} (10\sqrt{10} - 1) = (13\sqrt{13} + 80\sqrt{10} - 16)/27$$

11. (a) f' is continuous on [a,b] because f is smooth so by Theorem 4.6.4 f' has a maximum value M and a minimum value m on [a,b].

(b) From part (a)

$$m \le f'(x) \le M$$
$$m^2 \le [f'(x)]^2 \le M^2$$
$$1 + m^2 \le 1 + [f'(x)]^2 \le 1 + M^2$$
$$\sqrt{1 + m^2} \le \sqrt{1 + [f'(x)]^2} \le \sqrt{1 + M^2}$$
$$\int_a^b \sqrt{1 + m^2} \, dx \le \int_a^b \sqrt{1 + [f'(x)]^2} \, dx \le \int_a^b \sqrt{1 + M^2} \, dx$$
$$(b - a)\sqrt{1 + m^2} \le L \le (b - a)\sqrt{1 + M^2}$$

EXERCISE SET 6.5

1. $S = \int_0^1 2\pi(7x)\sqrt{1 + 49} \; dx = 70\pi\sqrt{2} \int_0^1 x \; dx = 35\pi\sqrt{2}$

3. $f'(x) = -x/\sqrt{4 - x^2}$, $1 + [f'(x)]^2 = 1 + \dfrac{x^2}{4 - x^2} = \dfrac{4}{4 - x^2}$,

$S = \int_{-1}^1 2\pi\sqrt{4 - x^2}(2/\sqrt{4 - x^2}) dx = 4\pi \int_{-1}^1 dx = 8\pi$

5. $f'(x) = \frac{1}{2} x^{-1/2} - \frac{1}{2} x^{1/2}$,

$1 + [f'(x)]^2 = 1 + \frac{1}{4} x^{-1} - \frac{1}{2} + \frac{1}{4} x = (\frac{1}{2} x^{-1} + \frac{1}{2} x)^2$,

$S = \int_1^3 2\pi(x^{1/2} - \frac{1}{3} x^{3/2})(\frac{1}{2} x^{-1} + \frac{1}{2} x)dx = \frac{\pi}{3} \int_1^3 (3 + 2x - x^2)dx = 16\pi/9$

7. $S = \int_0^2 2\pi(9y + 1)\sqrt{82} \, dy = 2\pi\sqrt{82} \int_0^2 (9y + 1)dy = 40\pi\sqrt{82}$

9. $g'(y) = -y/\sqrt{9 - y^2}$, $1 + [g'(y)]^2 = \frac{9}{9 - y^2}$,

$S = \int_{-2}^2 2\pi\sqrt{9 - y^2} \cdot \frac{3}{\sqrt{9 - y^2}} \, dy = 6\pi\int_{-2}^2 dy = 24\pi$

11. $x = g(y) = \frac{1}{4} y^4 + \frac{1}{8} y^{-2}$, $g'(y) = y^3 - \frac{1}{4} y^{-3}$,

$1 + [g'(y)]^2 = 1 + (y^6 - \frac{1}{2} + \frac{1}{16} y^{-6}) = (y^3 + \frac{1}{4} y^{-3})^2$

$S = \int_1^2 2\pi(\frac{1}{4} y^4 + \frac{1}{8} y^{-2})(y^3 + \frac{1}{4} y^{-3})dy = \frac{\pi}{16} \int_1^2 (8y^7 + 6y + y^{-5})dy$

$= 16,911\pi/1024$

13. Revolve the line segment joining the points $(0,0)$ and (h,r) about the x-axis. An equation of the line segment is $y = (r/h)x$ for $0 \leq x \leq h$ so

$S = \int_0^h 2\pi(r/h)x\sqrt{1 + r^2/h^2} \, dx = \frac{2\pi r}{h^2} \sqrt{r^2 + h^2} \int_0^h x \, dx = \pi r\sqrt{r^2 + h^2}$

15. $f(x) = \sqrt{r^2 - x^2}$, $f'(x) = -x/\sqrt{r^2 - x^2}$, $1 + [f'(x)]^2 = r^2/(r^2 - x^2)$,

$S = \int_a^{a+h} 2\pi\sqrt{r^2 - x^2}(r/\sqrt{r^2 - x^2})dx = 2\pi r\int_a^{a+h} dx = 2\pi rh$

17. (a) length of arc of sector = circumference of base of cone

$\ell\theta = 2\pi r$

$\theta = 2\pi r/\ell$

$S = $ area of sector $= \frac{1}{2} \ell^2(2\pi r/\ell) = \pi r\ell$

(b) $S = \pi r_2 \ell_2 - \pi r_1 \ell_1$

$\qquad = \pi r_2 (\ell_1 + \ell) - \pi r_1 \ell_1$

$\qquad = \pi [(r_2 - r_1)\ell_1 + r_2 \ell]$

Using similar triangles

$$\ell_2 / r_2 = \ell_1 / r_1$$

$$r_1 \ell_2 = r_2 \ell_1$$

$$r_1 (\ell_1 + \ell) = r_2 \ell_1$$

$$(r_2 - r_1)\ell_1 = r_1 \ell$$

so $S = \pi (r_1 \ell + r_2 \ell) = \pi (r_1 + r_2) \ell$

19.   (a) $1 \leq \sqrt{1 + [f'(x)]^2}$ so $2\pi f(x) \leq 2\pi f(x)\sqrt{1 + [f'(x)]^2}$. By part (b) of Theorem 5.8.3

$$\int_a^b 2\pi f(x)dx \leq \int_a^b 2\pi f(x)\sqrt{1 + [f'(x)]^2}\ dx$$

$$2\pi \int_a^b f(x)dx \leq S, \quad 2\pi A \leq S$$

(b) $2\pi A = S$ if $f'(x) = 0$ for all x in $[a,b]$ so $f(x)$ is constant on $[a,b]$.

EXERCISE SET 6.6

1. $s(t) = \int (2t - 3)dt = t^2 - 3t + C$, $s(1) = (1)^2 - 3(1) + C = 5$,

$C = 7$, $s(t) = t^2 - 3t + 7$.

3. $s(t) = \int (t^3 - 2t^2 + 1)dt = \frac{1}{4} t^4 - \frac{2}{3} t^3 + t + C$,

$s(0) = \frac{1}{4} (0)^4 - \frac{2}{3} (0)^3 + 0 + C = 1$, $C = 1$, $s(t) = \frac{1}{4} t^4 - \frac{2}{3} t^3 + t + 1$.

5. $v(t) = \int 4\,dt = 4t + C_1$, $v(0) = 4(0) + C_1 = 1$, $C_1 = 1$, $v(t) = 4t + 1$,

$s(t) = \int (4t + 1)dt = 2t^2 + t + C_2$, $s(0) = 2(0)^2 + 0 + C_2 = 0$, $C_2 = 0$,

$s(t) = 2t^2 + t$.

7. $v(t) = \int 4 \cos 2t \, dt = 2 \sin 2t + C_1$, $v(0) = 2 \sin 0 + C_1 = -1$, $C_1 = -1$,

$v(t) = 2 \sin 2t - 1$, $s(t) = \int (2 \sin 2t - 1) dt = -\cos 2t - t + C_2$,

$s(0) = -\cos 0 - 0 + C_2 = -3$, $C_2 = -2$, $s(t) = -\cos 2t - t - 2$.

9. (a) $s = \int \sin \frac{1}{2} \pi t \, dt = -\frac{2}{\pi} \cos \frac{1}{2} \pi t + C$

$s = 0$ when $t = 0$ which gives $C = \frac{2}{\pi}$ so $s = -\frac{2}{\pi} \cos \frac{1}{2} \pi t + \frac{2}{\pi}$.

$a = \frac{dv}{dt} = \frac{\pi}{2} \cos \frac{1}{2} \pi t$. When $t = 1$: $s = 2/\pi$, $v = 1$, $|v| = 1$, $a = 0$.

(b) $v = -3 \int t \, dt = -\frac{3}{2} t^2 + C_1$, $v = 0$ when $t = 0$ which gives $C_1 = 0$ so

$v = -\frac{3}{2} t^2$. $s = -\frac{3}{2} \int t^2 dt = -\frac{1}{2} t^3 + C_2$, $s = 1$ when $t = 0$ which

gives $C_2 = 1$ so $s = -\frac{1}{2} t^3 + 1$. When $t = 1$: $s = 1/2$, $v = -3/2$,

$|v| = 3/2$, $a = -3$.

11. If $a(t) = k$ then $v(t) = \int k \, dt = kt + C_1$, but $v(0) = 60$ mph $= 88$ ft/sec

so $k(0) + C_1 = 88$, $C_1 = 88$, $v(t) = kt + 88$;

$s(t) = \int (kt + 88) dt = \frac{1}{2} kt^2 + 88t + C_2$, $s(0) = 0$ so $C_2 = 0$,

$s(t) = \frac{1}{2} kt^2 + 88t$. $v(t) = 0$ at the instant when the car comes to a

stop so $kt + 88 = 0$, $t = -88/k$; $s(t) = 180$ at this instant so

$\frac{1}{2} k(-88/k)^2 + 88(-88/k) = 180$, $-3872/k = 180$, $k \approx -21.5$ ft/sec^2.

13. $s = 0$ and $v = 112$ when $t = 0$ so $v(t) = -32t + 112$, $s(t) = -16t^2 + 112t$.
(a) $v(3) = 16$ ft/sec, $v(5) = -48$ ft/sec.
(b) $v = 0$ when the projectile is at its maximum height so

$-32t + 112 = 0$, $t = 7/2$ sec, $s(7/2) = -16(7/2)^2 + 112(7/2) = 196$ ft.

(c) $s = 0$ when it reaches the ground so $-16t^2 + 112t = 0$,
$-16t(t - 7) = 0$, $t = 0$, 7 of which $t = 7$ is when it is at ground
level on its way down. $v(7) = -112$, $|v| = 112$ ft/sec.

15. (a) $s(t) = 0$ when it hits the ground,

$s(t) = -16t^2 + 16t = -16t(t - 1) = 0$ when $t = 1$ sec.

(b) The projectile moves upward until it gets to its highest point where
$v(t) = 0$, $v(t) = -32t + 16 = 0$ when $t = 1/2$ sec.

17. (a) $s(t) = 0$ when the package hits the ground,

   $s(t) = -16t^2 + 20t + 200 = 0$ when (use the quadratic formula)

   $t = (5 + 5\sqrt{33})/8$ sec.

 (b) $v(t) = -32t + 20$, $v[(5 + 5\sqrt{33})/8] = -20\sqrt{33}$, the speed at impact is

   $20\sqrt{33}$ ft/sec.

19. $s(t) = -4.9t^2 + 49t + 150$ and $v(t) = -9.8t + 49$.

 (a) The projectile reaches its maximum height when $v(t) = 0$,

   $-9.8t + 49 = 0$, $t = 5$ sec.

 (b) $s(5) = -4.9(5)^2 + 49(5) + 150 = 272.5$ m.

 (c) The projectile reaches its starting point when $s(t) = 150$,

   $-4.9t^2 + 49t + 150 = 150$, $-4.9t(t - 10) = 0$, $t = 10$ sec.

 (d) $v(10) = -9.8(10) + 49 = -49$ m/sec.

 (e) $s(t) = 0$ when the projectile hits the ground, $-4.9t^2 + 49t + 150 = 0$

   when (use the quadratic formula) $t \approx 12.46$ sec.

 (f) $v(12.46) = -9.8(12.46) + 49 \approx -73.1$, the speed at impact is about

   73.1 m/sec.

21. $s(t) = -16t^2 + v_0 t$, $v(t) = -32t + v_0$. $v = 0$ when it reaches its maximum

height so $-32t + v_0 = 0$, $t = v_0/32$ is the time it takes to get there,

thus $s(v_0/32) = -16(v_0/32)^2 + v_0(v_0/32) = 1,000$, $v_0^2 = 64,000$,

$v_0 = 80\sqrt{10}$ ft/sec (positive because fired upward).

23. $s = -16t^2 + v_0 t + s_0$, but $s_0 = 0$ so $s = -16t^2 + v_0 t$. $s = 0$ when $t = 8$

so $0 = -16(8)^2 + v_0(8)$, $v_0 = 128$ ft/sec. $v = 0$ at its highest point so

$-32t + 128 = 0$, $t = 4$, $s = -16(4)^2 + 128(4) = 256$ ft.

25. displacement $= \displaystyle\int_0^2 (t^2 + t - 2)dt = 2/3$

   distance $= \displaystyle\int_0^2 |t^2 + t - 2|dt$

     $= \displaystyle\int_0^1 -(t^2 + t - 2)dt + \int_1^2 (t^2 + t - 2)dt = 7/6 + 11/6 = 3$

27. displacement $= \displaystyle\int_0^{\pi} \cos t\, dt = 0$

distance $= \displaystyle\int_0^{\pi} |\cos t|\, dt = \int_0^{\pi/2} \cos t\, dt + \int_{\pi/2}^{\pi} -\cos t\, dt = 1 + 1 = 2$

29. $v(t) = t^3 - 3t^2 + 2t = t(t-1)(t-2)$

displacement $= \displaystyle\int_0^3 (t^3 - 3t^2 + 2t)\, dt = 9/4$

distance $= \displaystyle\int_0^3 |v(t)|\, dt = \int_0^1 v(t)\, dt + \int_1^2 -v(t)\, dt + \int_2^3 v(t)\, dt = 11/4$

31. $v(t) = \dfrac{1}{2} t^2 - 2t$

displacement $= \displaystyle\int_1^5 (\dfrac{1}{2} t^2 - 2t)\, dt = -10/3$

distance $= \displaystyle\int_1^5 |\dfrac{1}{2} t^2 - 2t|\, dt$

$= \displaystyle\int_1^4 -(\dfrac{1}{2} t^2 - 2t)\, dt + \int_4^5 (\dfrac{1}{2} t^2 - 2t)\, dt = 17/3$

33. $v(t) = \dfrac{2}{5} \sqrt{5t+1} + \dfrac{8}{5}$

displacement $= \displaystyle\int_0^3 (\dfrac{2}{5} \sqrt{5t+1} + \dfrac{8}{5})\, dt = \dfrac{4}{75} (5t+1)^{3/2} + \dfrac{8}{5} t\Big]_0^3 = 204/25$

distance $= \displaystyle\int_0^3 |v(t)|\, dt = \int_0^3 v(t)\, dt = 204/25$

EXERCISE SET 6.7

1. (a) $W = 30[5 - (-2)] = 210$ in-lb   (b) $W = \displaystyle\int_1^6 x^{-2} dx = 5/6$ in-lb

3. $F(x) = kx$, $F(5) = 5k = 1/2$, $k = 1/10$ ton/ft, $W = \int_0^6 (1/10)x\,dx = 9/5$

    ft-ton

5. $W = \int_0^1 kx\,dx = k/2 = 10$, $k = 20$ lb/ft

7. $W = \int_0^6 (9 - x)\rho(25\pi)dx = 900\pi\rho$ ft-lb

9. (a) $W = \int_0^9 (10 - x)62.4(300)dx$

    $= 18{,}720\int_0^9 (10 - x)dx$

    $= 926{,}640$ ft-lb

    (b) to empty the pool in one hour
    would require $926{,}640/3600 =$
    257.4 ft-lb of work per second
    so
    hp of motor $= 257.4/550 = 0.468$

11. $W = \int_0^{100} 15(100 - x)dx$

    $= 75{,}000$ ft-lb

13. (a) $F(4000) = k/(4000)^2 = 6000$, $k = 9.6 \times 10^{10}$

    (b) $W = \int_{4000}^{5000} 9.6 \times 10^{10}x^{-2}dx = 4{,}800{,}000$ mi-lb

EXERCISE SET 6.8

1. (a) F = ρhA = (62.4)(5)(9) = 2,808 lb
    (b) F = (40)(10)(9) = 3,600 lb

3. $F = \int_{1}^{3} 62.4x(4)dx$

    $= 249.6\int_{1}^{3} x\,dx$

    $= 998.4$ lb

5. by similar triangles

    $\dfrac{w(x)}{6} = \dfrac{10 - x}{8}$

    $w(x) = \dfrac{3}{4}(10 - x),$

    $F = \int_{2}^{10} 62.4x[\dfrac{3}{4}(10 - x)]dx$

    $= 46.8\int_{2}^{10}(10x - x^2)dx = 6988.8$ lb

7. $F = \int_{0}^{5} 62.4x(2\sqrt{25 - x^2})dx$

    $= 124.8\int_{0}^{5} x(25 - x^2)^{1/2}dx$

    $= 5200$ lb

9. Find the forces on the upper and
    lower halves and add them:

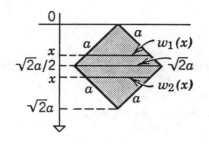

$$\frac{w_1(x)}{\sqrt{2}a} = \frac{x}{\sqrt{2}a/2}, \quad w_1(x) = 2x$$

$$F_1 = \int_0^{\sqrt{2}a/2} \rho x(2x)dx$$

$$= 2\rho \int_0^{\sqrt{2}a/2} x^2 dx = \sqrt{2}\rho a^3/6,$$

$$\frac{w_2(x)}{\sqrt{2}a} = \frac{\sqrt{2}a - x}{\sqrt{2}a/2}, \quad w_2(x) = 2(\sqrt{2}a - x)$$

$$F_2 = \int_{\sqrt{2}a/2}^{\sqrt{2}a} \rho x[2(\sqrt{2}a - x)]dx = 2\rho \int_{\sqrt{2}a/2}^{\sqrt{2}a} (\sqrt{2}ax - x^2)dx = \sqrt{2}\rho a^3/3,$$

$$F = F_1 + F_2 = \sqrt{2}\rho a^3/6 + \sqrt{2}\rho a^3/3 = \sqrt{2}\rho a^3/2$$

11. $\sqrt{16^2 + 4^2} = \sqrt{272} = 4\sqrt{17}$ is the
    other dimension of the bottom.

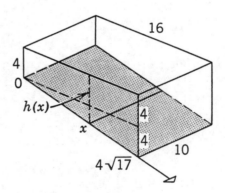

$(h(x) - 4)/4 = x/(4\sqrt{17})$

$h(x) = x/\sqrt{17} + 4,$

$$F = \int_0^{4\sqrt{17}} 62.4(x/\sqrt{17} + 4)10dx$$

$$= 624 \int_0^{4\sqrt{17}} (x/\sqrt{17} + 4)dx$$

$$= 14,976\sqrt{17} \text{ lb}$$

13. (a) From Exercise 12, $F = 4\rho_0(h + 1)$ so (assuming that ρ_0 is constant)
    $dF/dt = 4\rho_0(dh/dt)$ which is a positive constant if dh/dt is a
    positive constant.
    (b) If $dh/dt = 20$ then $dF/dt = 80\rho_0$ lb/min from part (a).

SUPPLEMENTARY EXERCISES CHAPTER 6

1.   (a) $\displaystyle\int_0^2 (x + 2 - x^2)dx$

(b) $\displaystyle\int_0^2 \sqrt{y}\ dy + \int_2^4 (\sqrt{y} - y + 2)dy$

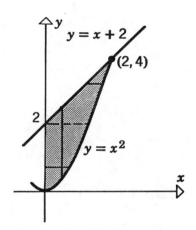

3.   (a) $\displaystyle\int_0^9 [\sqrt{x} - (-\sqrt{x})]dx = \int_0^9 2\sqrt{x}\ dx$

(b) $\displaystyle\int_{-3}^3 (9 - y^2)dy$

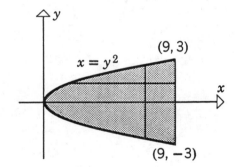

5.   (a) $\displaystyle\int_0^2 2\pi x(x + 2 - x^2)dx$

(b) $\displaystyle\int_0^2 \pi y\ dy + \int_2^4 \pi[y - (y - 2)^2]dy$

7.   (a) $\displaystyle\int_0^4 2\pi x[x/2 - (2 - \sqrt{4 - x})]dx$ (b) $\displaystyle\int_0^2 \pi[(4y - y^2)^2 - 4y^2]dy$

9.   (a) $\displaystyle\int_0^9 2\pi x(2\sqrt{x})dx = \int_0^9 4\pi x^{3/2}dx$ (b) $\displaystyle\int_{-3}^3 \pi(81 - y^4)dy$

11.   (a) $A = \int_0^4 \sqrt{4 - y}\ dy = 16/3$

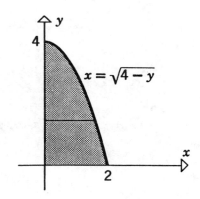

(b) $V = \int_0^4 \pi(4 - y)dy = 8\pi$

13. $A = \int_{-1}^0 (x^3 - x)dx + \int_0^1 (x - x^3)dx$

$\qquad + \int_1^2 (x^3 - x)dx$

$\qquad = 1/4 + 1/4 + 9/4 = 11/4$

15.   Let r be the radius of the semicircle shown in the figure, then by similar triangles

$\qquad 2r/a = (b - y)/b, \quad r = \dfrac{a}{2b}(b - y)$

so

$\qquad A(y) = \dfrac{1}{2}\pi r^2 = \dfrac{1}{2}\pi\ \dfrac{a^2}{4b^2}(b - y)^2$

$\qquad = \dfrac{\pi a^2}{8b^2}(b - y)^2$

$\qquad V = \int_0^b A(y)dy = \int_0^b \dfrac{\pi a^2}{8b^2}(b - y)^2 dy = \pi a^2 b/24$

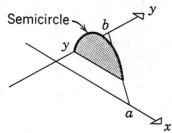

17. (a) $V = \int_0^4 2\pi(4 - x)\sqrt{x}\ dx$

$\qquad = 2\pi \int_0^4 (4x^{1/2} - x^{3/2})dx$

$\qquad = 256\pi/15$

(b) $V = \int_0^4 \pi[4 - (2 - \sqrt{x})^2]dx$

$\qquad = \pi \int_0^4 (4x^{1/2} - x)dx = 40\pi/3$

19. $y = \dfrac{x^{3/2}}{\sqrt{8}},\ 0 \le x \le 2;\ y' = \dfrac{3x^{1/2}}{2\sqrt{8}},\ L = \int_0^2 \sqrt{1 + \dfrac{9}{32}x}\ dx = 61/27$

21. $y' = \dfrac{1}{2}x^4 - \dfrac{1}{2}x^{-4}$,

$1 + (y')^2 = 1 + (\dfrac{1}{4}x^{16} - \dfrac{1}{2} + \dfrac{1}{4}x^{-16}) = (\dfrac{1}{2}x^4 + \dfrac{1}{2}x^{-4})^2$,

$L = \int_1^2 (\dfrac{1}{2}x^4 + \dfrac{1}{2}x^{-4})dx = 779/240$

23. $y' = 3x^2,\ 1 + (y')^2 = 1 + 9x^4$,

$S = \int_1^2 2\pi x^3\sqrt{1 + 9x^4}\ dx = \pi(145^{3/2} - 10^{3/2})/27$

25. $y' = x^{1/2} - \dfrac{1}{4}x^{-1/2}$,

$1 + (y')^2 = 1 + (x - \dfrac{1}{2} + \dfrac{1}{16}x^{-1}) = (x^{1/2} + \dfrac{1}{4}x^{-1/2})^2$,

$S = \int_0^9 2\pi x(x^{1/2} + \dfrac{1}{4}x^{-1/2})dx = 2\pi \int_0^9 (x^{3/2} + \dfrac{1}{4}x^{1/2})dx = 1017\pi/5$

27. $y' = \dfrac{1}{2}x^{-1/2} - \dfrac{1}{2}x^{1/2},\ 1 + (y')^2 = (\dfrac{1}{2}x^{-1/2} + \dfrac{1}{2}x^{1/2})^2$,

$S = \int_0^3 2\pi x(\dfrac{1}{2}x^{-1/2} + \dfrac{1}{2}x^{1/2})dx = \pi \int_0^3 (x^{1/2} + x^{3/2})dx = 28\pi\sqrt{3}/5$

29. $F(x) = kx$, $F(4) = 4k = 2$, $k = 1/2$, $W = \int_2^4 \frac{1}{2}\, x\, dx = 3$ in-lb

31. $F(x) = 250 + \dfrac{3}{4}\,(40 - x)$

$\qquad = 280 - \dfrac{3}{4}\,x,$

$W = \int_0^{40} (280 - \dfrac{3}{4}\,x)dx = 10{,}600$ ft-lb

40

¾ lb/ft

250 lb x

0

33. $A(y) = \pi x^2 = \pi(y/2 + 4)$,

$W = \int_{-8}^{0} 62.4(4 - y)[\pi(y/2 + 4)]dy$

$\qquad = 31.2\pi \int_{-8}^{0} (32 - 4y - y^2)dy$

$\qquad = 6656\pi$ ft-lb

35. By similar triangles
$$w(x)/4 = (x - 1)/2$$
$$w(x) = 2(x - 1)$$

$F = \int_1^3 \rho x[2(x - 1)]dx$

$\qquad = 2\rho \int_1^3 (x^2 - x)dx = 28\rho/3$ lb

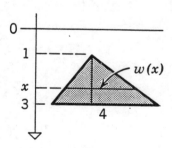

CHAPTER 7

EXERCISE SET 7.1

1. (a) $f(g(x)) = 4(x/4) = x$, $g(f(x)) = (4x)/4 = x$, f and g are inverse
        functions
    (b) $f(g(x)) = 3(3x - 1) + 1 = 9x - 2 \neq x$ so f and g are not inverse
        functions

    (c) $f(g(x)) = \sqrt[3]{(x^3 + 2) - 2} = x$, $g(f(x)) = (x - 2) + 2 = x$, f and g are
        inverse functions
    (d) $f(g(x)) = (x^{1/4})^4 = x$, $g(f(x)) = (x^4)^{1/4} = |x| \neq x$, f and g are not
        inverse functions

3. $f'(x) = 3$; f is increasing on $(-\infty, +\infty)$ so f has an inverse.

5. $f(x) = (2 + x)(1 - x)$; f does not have an inverse because f is not
    one-to-one, for example $f(-2) = f(1) = 0$.

7. f does not have an inverse because f is not one-to-one, for example
    $f(0) = f(1) = -1$.

9. $f(x) = (x - 1)^3$; f has an inverse because two different numbers cannot
    have the same cube so f is one-to-one.

11. $f'(x) = 10x^4 + 3x^2 + 3 \geq 3$ for $-\infty < x < +\infty$; f is increasing on $(-\infty, +\infty)$
    so f has an inverse.

13. $f'(x) = \cos x > 0$ for $-\pi/2 < x < \pi/2$; f is increasing on $(-\pi/2, \pi/2)$ so f
    has an inverse.

15. $y = f^{-1}(x)$, $x = f(y) = y^5$, $y = x^{1/5} = f^{-1}(x)$

17. $y = f^{-1}(x)$, $x = f(y) = 7y - 6$, $y = \frac{1}{7}(x + 6) = f^{-1}(x)$

19. $y = f^{-1}(x)$, $x = f(y) = 3y^3 - 5$, $y = \sqrt[3]{(x + 5)/3} = f^{-1}(x)$

21. $y = f^{-1}(x)$, $x = f(y) = \sqrt[3]{2y - 1}$, $y = (x^3 + 1)/2 = f^{-1}(x)$

23. $y = f^{-1}(x)$, $x = f(y) = 3/y^2$, $y = -\sqrt{3/x} = f^{-1}(x)$

25. $y = f^{-1}(x)$, $x = f(y) = \begin{cases} 5/2 - y, & y < 2 \\ 1/y, & y \geq 2 \end{cases}$,

$$y = f^{-1}(x) = \begin{cases} 5/2 - x, & x > 1/2 \\ 1/x, & x \leq 1/2 \end{cases}$$

27. $y = f^{-1}(x)$, $x = f(y) = 5y^3 + y - 7$, $\dfrac{dx}{dy} = 15y^2 + 1$, $\dfrac{dy}{dx} = \dfrac{1}{15y^2 + 1}$;

check: $1 = 15y^2 \dfrac{dy}{dx} + \dfrac{dy}{dx}$, $\dfrac{dy}{dx} = \dfrac{1}{15y^2 + 1}$.

29. $y = f^{-1}(x)$, $x = f(y) = \tan 2y$, $\dfrac{dx}{dy} = 2 \sec^2 2y$, $\dfrac{dy}{dx} = \dfrac{1}{2 \sec^2 2y}$;

check: $1 = (2 \sec^2 2y) \dfrac{dy}{dx}$, $\dfrac{dy}{dx} = \dfrac{1}{2 \sec^2 2y}$.

31. $y = f^{-1}(x)$, $x = f(y) = 2y^5 + y^3 + 1$, $\dfrac{dx}{dy} = 10y^4 + 3y^2$, $\dfrac{dy}{dx} = \dfrac{1}{10y^4 + 3y^2}$;

check: $1 = 10y^4 \dfrac{dy}{dx} + 3y^2 \dfrac{dy}{dx}$, $\dfrac{dy}{dx} = \dfrac{1}{10y^4 + 3y^2}$.

33. (a) $f(g(x)) = f(\sqrt{x})$
$$= (\sqrt{x})^2 = x, \ x > 1;$$
$$g(f(x)) = g(x^2)$$
$$= \sqrt{x^2} = x, \ x > 1.$$

(b)

(c) No, because $f(g(x)) = x$ for every x in the domain of g is not satisfied (the domain of g is $x > 0$).

35. $y = f^{-1}(x)$, $x = f(y) = (y + 2)^4$ for $y \geq 0$,

    $y = f^{-1}(x) = x^{1/4} - 2$ for $x \geq 16$.

37. $y = f^{-1}(x)$, $x = f(y) = -\sqrt{3 - 2y}$ for $y \leq 3/2$,

    $y = f^{-1}(x) = (3 - x^2)/2$ for $x \leq 0$.

39. $y = f^{-1}(x)$, $x = f(y) = y - 5y^2$ for $y \geq 1$, $5y^2 - y + x = 0$ for $y \geq 1$,

    $y = f^{-1}(x) = (1 + \sqrt{1 - 20x})/10$ for $x \leq -4$.

41. (a) $f(f(x)) = \dfrac{3 - \dfrac{3 - x}{1 - x}}{1 - \dfrac{3 - x}{1 - x}} = \dfrac{3 - 3x - 3 + x}{1 - x - 3 + x} = x$ so $f = f^{-1}$

    (b) symmetric about the line $y = x$

43. (a) $f(x) = x^3 - 3x^2 + 2x = x(x - 1)(x - 2)$ so $f(0) = f(1) = f(2) = 0$
    thus f is not one-to-one

    (b) $f'(x) = 3x^2 - 6x + 2$, $f'(x) = 0$ when $x = \dfrac{6 \pm \sqrt{36 - 24}}{6} = 1 \pm \sqrt{3}/3$.

    $f'(x) > 0$ (f is increasing) if $x < 1 - \sqrt{3}/3$, $f'(x) < 0$ (f is
    decreasing) if $1 - \sqrt{3}/3 < x < 1 + \sqrt{3}/3$, so $f(x)$ takes on values less
    than $f(1 - \sqrt{3}/3)$ on both sides of $1 - \sqrt{3}/3$ thus $1 - \sqrt{3}/3$ is the
    largest value of k.

45. (a) $f'(x) = \sqrt[3]{1 + x^2} > 0$ on $(-\infty, +\infty)$ so f is one-to-one there because f
    is increasing.

    (b) From (5), $(f^{-1})'(0) = 1/f'(f^{-1}(0))$ but $f^{-1}(0) = 1$ so

    $(f^{-1})'(0) = 1/\sqrt[3]{1 + 1} = 1/\sqrt[3]{2}$.

47.

49. $F'(x) = 2f'(2g(x))g'(x)$ so $F'(3) = 2f'(2g(3))g'(3)$. By inspection $f(1) = 3$, so $g(3) = f^{-1}(3) = 1$ and $g'(3) = (f^{-1})'(3) = 1/f'(f^{-1}(3)) = 1/f'(1) = 1/7$ because $f'(x) = 4x^3 + 3x^2$. Thus $F'(3) = 2f'(2)(1/7) = 2(44)(1/7) = 88/7$.

EXERCISE SET 7.2

1. (a) $\log_2 16 = \log_2(2^4) = 4$ (b) $\log_2(\frac{1}{32}) = \log_2(2^{-5}) = -5$

    (c) $\log_4 4 = 1$ (d) $\log_9 3 = \log_9(9^{1/2}) = 1/2$

    (e) $\log_{10}(0.001) = \log_{10}(10^{-3}) = -3$ (f) $\log_{10}(10^4) = 4$

    (g) $\ln(e^3) = 3$ (h) $\ln(\sqrt{e}) = \ln(e^{1/2}) = 1/2$

3. $\sqrt{x} = 10^{-1} = 0.1$, $x = 0.01$ 5. $1/x = e^{-2}$, $x = e^2$

7. $2x = 8$, $x = 4$ 9. $\log_{10}x = 5$, $x = 10^5$

11. $\ln 2x^2 = \ln 3$, $2x^2 = 3$, $x^2 = 3/2$, $x = \sqrt{3/2}$ (we discard $-\sqrt{3/2}$ because it does not satisfy the original equation).

13. (a) Let $y = \log_b x$, then $x = b^y = b^{\log_b x}$ for $x > 0$.

     (b) Let $y = b^x$, then $x = \log_b y = \log_b(b^x)$ for $-\infty < x < +\infty$.

15. Let $x = \log_b a$, then $b^x = a$ so $\log_a b^x = \log_a a = 1$, $x \log_a b = 1$,

     $x = \dfrac{1}{\log_a b}$.

17. (a) $\log_{10}(I/I_0) = 3$, $I/I_0 = 10^3 = 1000$, $I = 1000\ I_0$

     (b) $\log_{10}(I/I_0) = 8.2$, $I/I_0 = 10^{8.2}$, $I = 10^{8.2}\ I_0$

(c) If $R = 8$ then $I = 10^8 I_0$, if $R = 4$ then $I = 10^4 I_0$;

$(10^8 I_0)/(10^4 I_0) = 10^4 = 10,000$ so the earthquake is 10,000 times more intense.

EXERCISE SET 7.3

1. (a) $\ln 6 = \ln(2)(3) = \ln 2 + \ln 3 = r + s$
    (b) $\ln 1.5 = \ln \frac{3}{2} = \ln 3 - \ln 2 = s - r$
    (c) $\ln \frac{1}{2} = \ln 1 - \ln 2 = 0 - r = -r$ (d) $\ln 9 = \ln 3^2 = 2 \ln 3 = 2s$
    (e) $\ln 3^{1/5} = \frac{1}{5} \ln 3 = s/5$
    (f) $\ln \frac{1}{36} = \ln 1 - \ln 6^2 = -2 \ln 6 = -2(r + s)$

3. (a) if $\ln 3 \approx 1.1$ then $\ln \frac{1}{9} = \ln 3^{-2} = -2 \ln 3 \approx -2.2$,

    $\ln \frac{1}{3} = \ln 3^{-1} = -\ln 3 \approx -1.1$, $\ln 9 = \ln 3^2 = 2 \ln 3 \approx 2.2$
    (b) about 0.7   (c) about 2.7

5. $y = -\ln x$ 7. $y = \ln(x - 1)$

9. $\frac{1}{2x} (2) = 1/x$ 11. $2(\ln x) (\frac{1}{x}) = \frac{2 \ln x}{x}$

13. $\frac{1}{\tan x} (\sec^2 x) = \frac{\sec^2 x}{\tan x}$

15. $\dfrac{1}{x/(1+x^2)} \left[\dfrac{(1+x^2)(1) - x(2x)}{(1+x^2)^2} \right] = \dfrac{1-x^2}{x(1+x^2)}$

17. $\dfrac{3x^2 - 14x}{x^3 - 7x^2 - 3}$

19. $\dfrac{1}{2}(\ln x)^{-1/2} \left(\dfrac{1}{x}\right) = \dfrac{1}{2x\sqrt{\ln x}}$

21. $\cos(5/\ln x) \dfrac{d}{dx}[5(\ln x)^{-1}] = \cos(5/\ln x)[-5(\ln x)^{-2}(1/x)]$

$\qquad\qquad\qquad\qquad\qquad = -\dfrac{5\cos(5/\ln x)}{x(\ln x)^2}$

23. $-\dfrac{2x^3}{3-2x} + 3x^2\ln(3-2x)$

25. $2(x^2+1)[\ln(x^2+1)] \dfrac{2x}{x^2+1} + 2x[\ln(x^2+1)]^2$

$\qquad = 4x\ln(x^2+1) + 2x[\ln(x^2+1)]^2$

27. $\dfrac{(1+\ln x)(2x) - x^2(0 + 1/x)}{(1+\ln x)^2} = \dfrac{x(1 + 2\ln x)}{(1+\ln x)^2}$

29. $\dfrac{dy}{dx} + \dfrac{1}{xy}\left(x\dfrac{dy}{dx} + y\right) = 0, \ \dfrac{dy}{dx} = -\dfrac{y}{x(y+1)}$

31. $\dfrac{1}{2}\ln|x| + C$

33. $u = x^3 - 4, \ du = 3x^2 dx, \ \dfrac{1}{3}\displaystyle\int \dfrac{1}{u} du = \dfrac{1}{3}\ln|x^3 - 4| + C$

35. $u = \tan x, \ du = \sec^2 x\,dx, \ \displaystyle\int\dfrac{1}{u}du = \ln|\tan x| + C$

37. $u = 1 + \cos 3\theta, \ du = -3\sin 3\theta\,d\theta$

$\qquad -\dfrac{1}{3}\displaystyle\int\dfrac{1}{u}du = -\dfrac{1}{3}\ln|1 + \cos 3\theta| + C = -\dfrac{1}{3}\ln(1 + \cos 3\theta) + C$

$\qquad$ because $1 + \cos 3\theta \geq 0$

39. divide $x^2 + 1$ into x^3 to get

$$\int \frac{x^3}{x^2 + 1}\, dx = \int \left[x - \frac{x}{x^2 + 1}\right] dx = \int x\, dx - \int \frac{x}{x^2 + 1}\, dx$$
$$= \frac{1}{2} x^2 - \frac{1}{2} \ln(x^2 + 1) + C$$

41. $u = \ln y, \quad du = \frac{1}{y}\, dy, \quad \int u^3 du = \frac{1}{4}(\ln y)^4 + C$

43. $u = 3x + 2, \quad \frac{1}{3} \int_2^5 \frac{1}{u}\, du = \frac{1}{3} \ln|u|\Big]_2^5 = \frac{1}{3}(\ln 5 - \ln 2) = \frac{1}{3} \ln \frac{5}{2}$

45. $u = x^2 + 5, \quad \frac{1}{2} \int_6^5 \frac{1}{u}\, du = \frac{1}{2} \ln|u|\Big]_6^5 = \frac{1}{2}(\ln 5 - \ln 6) = \frac{1}{2} \ln \frac{5}{6}$

47. (a) all $x \neq 0$ \qquad\qquad (b) $x > 0$

49. $\frac{d}{dx}\left[\frac{1}{2}\ln x + \frac{1}{3}\ln(x + 3) + \frac{1}{5}\ln(3x - 2)\right] = \frac{1}{2x} + \frac{1}{3(x + 3)} + \frac{3}{5(3x - 2)}$

51. $\int \frac{1}{x}\ln(x^3)\, dx = 3 \int \ln x \left(\frac{1}{x}\right) dx = 3 \int u\, du = \frac{3}{2}(\ln x)^2 + C$

53. $\int \frac{\ln(1/x)}{x}\, dx = -\int \ln x \left(\frac{1}{x}\right) dx = -\frac{1}{2}(\ln x)^2 + C$

55. $\ln|y| = \ln|x| + \frac{1}{3}\ln|1 + x^2|, \quad \frac{dy}{dx} = x\sqrt[3]{1 + x^2}\left[\frac{1}{x} + \frac{2x}{3(1 + x^2)}\right]$

57. $\ln|y| = \frac{1}{3}\ln|x^2 - 8| + \frac{1}{2}\ln|x^3 + 1| - \ln|x^6 - 7x + 5|$

$$\frac{dy}{dx} = \frac{(x^2 - 8)^{1/3}\sqrt{x^3 + 1}}{x^6 - 7x + 5}\left[\frac{2x}{3(x^2 - 8)} + \frac{3x^2}{2(x^3 + 1)} - \frac{6x^5 - 7}{x^6 - 7x + 5}\right]$$

59. (a) $u = \frac{ab}{t}, \quad t = \frac{ab}{u}, \quad dt = -\frac{ab}{u^2}\, du$

$$\int_a^{ab} \frac{1}{t}\, dt = \int_b^1 \frac{u}{ab}\left(-\frac{ab}{u^2}\right) du = \int_1^b \frac{1}{u}\, du = \int_1^b \frac{1}{t}\, dt$$

(b) $\ln(ab) = \int_1^{ab} \frac{1}{t}\,dt = \int_1^a \frac{1}{t}\,dt + \int_a^{ab} \frac{1}{t}\,dt$

$$= \int_1^a \frac{1}{t}\,dt + \int_1^b \frac{1}{t}\,dt = \ln a + \ln b$$

61.   If $x > 1$ then $\ln x = \dfrac{1}{t_1^*}(x-1)$ for $1 \le t_1^* \le x$ so $\dfrac{1}{x} \le \dfrac{1}{t_1^*} \le 1$,

$\dfrac{x-1}{x} \le \dfrac{1}{t_1^*}(x-1) \le x-1,\ 1 - \dfrac{1}{x} \le \ln x \le x-1.$ If $0 < x < 1$ then

$\ln x = \int_1^x \frac{1}{t}\,dt = -\int_x^1 \frac{1}{t}\,dt = -\dfrac{1}{t_2^*}(1-x)$ for $x \le t_2^* \le 1$ so $1 \le \dfrac{1}{t_2^*} \le \dfrac{1}{x}$,

$-(1-x) \ge -\dfrac{1}{t_2^*}(1-x) \ge -\dfrac{1-x}{x},\ 1 - \dfrac{1}{x} \le \ln x \le x-1.$ In all cases

(including $x = 1$) $1 - \dfrac{1}{x} \le \ln x \le x - 1.$

63.   **(a)** If $x > 1$ then $0 < 1 - \dfrac{1}{x}$ and $x - 1 < x$ so

$\quad 0 < 1 - \dfrac{1}{x} \le \ln x \le x - 1 < x$, or simply $0 < \ln x < x$.

(b) If $x > 1$ then $\sqrt{x} > 1$ so $0 < \ln\sqrt{x} < \sqrt{x}$, $0 < \dfrac{1}{2}\ln x < \sqrt{x}$,

$\quad 0 < \ln x < 2\sqrt{x}$, $0 < \dfrac{\ln x}{x} < \dfrac{2}{\sqrt{x}}$

(c) $\lim\limits_{x \to +\infty} 0 = \lim\limits_{x \to +\infty} \dfrac{2}{\sqrt{x}} = 0$ so by the pinching theorem $\lim\limits_{x \to +\infty} \dfrac{\ln x}{x} = 0.$

(d) let $t = 1/x$ then $x = 1/t$ and

$\quad \lim\limits_{x \to 0^+} x \ln x = \lim\limits_{t \to +\infty} \dfrac{1}{t}\ln\dfrac{1}{t} = \lim\limits_{t \to +\infty}\left(-\dfrac{\ln t}{t}\right) = -\lim\limits_{t \to +\infty}\dfrac{\ln t}{t} = 0$

65. $p = c/v$ so the average pressure with respect to volume is

$P_{ave} = \dfrac{1}{v_1 - v_0}\int_{v_0}^{v_1} \dfrac{c}{v}\,dv = \dfrac{c}{v_1 - v_0}(\ln v_1 - \ln v_0) = \dfrac{c}{v_1 - v_0}\ln(v_1/v_0)$

67. Let $f(x) = x^2 - \ln x$, then $f'(x) = 2x - 1/x = (2x^2 - 1)/x$; $f'(x) = 0$ if $x = 1/\sqrt{2}$ at which there is a relative minimum and hence the absolute minimum on $(0, +\infty)$. The minimum value is $1/2 - \ln(1/\sqrt{2}) = (1 + \ln 2)/2$.

69. $V = \pi \displaystyle\int_1^4 \frac{1}{x}\, dx = \pi \ln|x|\Big]_1^4 = \pi \ln 4$

71. $A = \displaystyle\int_0^{\pi/3} \tan x\, dx = \int_0^{\pi/3} \frac{\sin x}{\cos x}\, dx = -\ln|\cos x|\Big]_0^{\pi/3} = \ln 2$

EXERCISE SET 7.4

1. (a) x^{-1}, $x > 0$

    (b) x^2, $x \neq 0$

    (c) $-x^2$, $-\infty < x < +\infty$

    (d) $-x$, $-\infty < x < +\infty$

    (e) x^3, $x > 0$

    (f) $\ln x + x$, $x > 0$

    (g) $x - \sqrt[3]{x}$, $-\infty < x < +\infty$

    (h) $\dfrac{e^x}{x}$, $x > 0$

3. (a) $\ln[(\sqrt{x})(x^{3/2})] = 1$, $\ln x^2 = 1$, but $\ln x^2 = 2\ln x$ because $x > 0$ so $2\ln x = 1$, $\ln x = 1/2$, $x = e^{1/2} = \sqrt{e}$

    (b) $2\pi x = \ln\sqrt{2}$, $x = \dfrac{\ln\sqrt{2}}{2\pi} = \dfrac{\ln 2}{4\pi}$ (c) $\cos x = e^0 = 1$, $x = 0$

    (d) $e^{2x} + 2e^x - 8 = 0$, $(e^x + 4)(e^x - 2) = 0$, so $e^x = -4$ (reject) or $e^x = 2$, $x = \ln 2$.

5. (a) $2^{1.7} = e^{1.7\ln 2} \approx e^{(1.7)(0.69)} \approx e^{1.2} \approx 3.3$

    (b) $5^{\sqrt{3}} = e^{\sqrt{3}\ln 5} \approx e^{(1.7)(1.6)} \approx e^{2.7} \approx 15$

    (c) $3^{\pi} = e^{\pi\ln 3} \approx e^{(3.1)(1.1)} \approx e^{3.4} \approx 30$

7. $-10xe^{-5x^2}$ 9. $x^3 e^x + 3x^2 e^x = x^2 e^x(x + 3)$

11. $\dfrac{dy}{dx} = \dfrac{(e^x + e^{-x})(e^x + e^{-x}) - (e^x - e^{-x})(e^x - e^{-x})}{(e^x + e^{-x})^2}$

$= \dfrac{(e^{2x} + 2 + e^{-2x}) - (e^{2x} - 2 + e^{-2x})}{(e^x + e^{-x})^2} = 4/(e^x + e^{-x})^2$

13. $(x\sec^2 x + \tan x)e^{x\tan x}$

15. $(1 - 3e^{3x})e^{(x-e^{3x})}$

17. $\dfrac{(x - 1)e^{-x}}{1 - xe^{-x}} = \dfrac{x - 1}{e^x - x}$

19. $e^{ax}(a\cos bx - b\sin bx)$

21. $y = e^{\ln(x^3+1)} = x^3 + 1, \ dy/dx = 3x^2$

23. $f'(x) = -3^{-x}\ln 3;$

$y = 3^{-x}, \ \ln y = -x\ln 3, \ \dfrac{1}{y}y' = -\ln 3, \ y' = -y\ln 3 = -3^{-x}\ln 3$

25. $f'(x) = \pi^{x\tan x}(\ln\pi)(x\sec^2 x + \tan x);$

$y = \pi^{x\tan x}, \ \ln y = (x\tan x)\ln\pi, \ \dfrac{1}{y}y' = (\ln\pi)(x\sec^2 x + \tan x)$

$y' = \pi^{x\tan x}(\ln\pi)(x\sec^2 x + \tan x)$

27. (a) Because x^x is not of the form a^x where a is constant.

(b) $y = x^x, \ \ln y = x\ln x, \ \dfrac{1}{y}y' = 1 + \ln x, \ y' = x^x(1 + \ln x)$

29. $\ln y = (\ln x)\ln(x^3 - 2x), \ \dfrac{1}{y}\dfrac{dy}{dx} = \dfrac{3x^2 - 2}{x^3 - 2x}\ln x + \dfrac{1}{x}\ln(x^3 - 2x),$

$\dfrac{dy}{dx} = (x^3 - 2x)^{\ln x}\left[\dfrac{3x^2 - 2}{x^3 - 2x}\ln x + \dfrac{1}{x}\ln(x^3 - 2x)\right]$

31. $\ln y = (\tan x)\ln(\ln x), \ \dfrac{1}{y}\dfrac{dy}{dx} = \dfrac{1}{x\ln x}\tan x + (\sec^2 x)\ln(\ln x),$

$\dfrac{dy}{dx} = (\ln x)^{\tan x}\left[\dfrac{\tan x}{x\ln x} + (\sec^2 x)\ln(\ln x)\right]$

33. $\ln y = e^x \ln x$, $\dfrac{1}{y}\dfrac{dy}{dx} = \dfrac{e^x}{x} + e^x \ln x$, $\dfrac{dy}{dx} = x^{(e^x)}\left[\dfrac{e^x}{x} + e^x \ln x\right]$

35. $y = Ae^{2x} + Be^{-4x}$, $y' = 2Ae^{2x} - 4Be^{-4x}$, $y'' = 4Ae^{2x} + 16Be^{-4x}$ so

$y'' + 2y' - 8y = (4Ae^{2x} + 16Be^{-4x}) + 2(2Ae^{2x} - 4Be^{-4x}) - 8(Ae^{2x} + Be^{-4x})$

$= 0.$

37.   (a) $f'(x) = ke^{kx}$, $f''(x) = k^2 e^{kx}$, $f'''(x) = k^3 e^{kx}$, ..., $f^{(n)}(x) = k^n e^{kx}$

(b) $f'(x) = -ke^{-kx}$, $f''(x) = k^2 e^{-kx}$, $f'''(x) = -k^3 e^{-kx}$, ...,

$f^{(n)}(x) = (-1)^n k^n e^{-kx}$

39. $f'(x) = \dfrac{1}{\sqrt{2\pi}\,\sigma}\exp\left[-\dfrac{1}{2}\left(\dfrac{x-\mu}{\sigma}\right)^2\right]\dfrac{d}{dx}\left[-\dfrac{1}{2}\left(\dfrac{x-\mu}{\sigma}\right)^2\right]$

$= \dfrac{1}{\sqrt{2\pi}\,\sigma}\exp\left[-\dfrac{1}{2}\left(\dfrac{x-\mu}{\sigma}\right)^2\right]\left[-\left(\dfrac{x-\mu}{\sigma}\right)\left(\dfrac{1}{\sigma}\right)\right]$

$= -\dfrac{1}{\sqrt{2\pi}\,\sigma^3}(x-\mu)\exp\left[-\dfrac{1}{2}\left(\dfrac{x-\mu}{\sigma}\right)^2\right]$

41. $-\dfrac{1}{5}e^{-5x} + C$ 43. $e^{\sin x} + C$

45. $-\dfrac{1}{6}\displaystyle\int e^{-2x^3}(-6x^2)dx = -\dfrac{1}{6}e^{-2x^3} + C$

47. $u = 1 + e^x$, $du = e^x dx$, $\displaystyle\int \dfrac{1}{u}\,du = \ln(1 + e^x) + C$

49. $u = 1 + e^{2t}$, $du = 2e^{2t}dt$, $\dfrac{1}{2}\displaystyle\int u^{1/2}du = \dfrac{1}{3}(1 + e^{2t})^{3/2} + C$

51. $\displaystyle\int e^{\sin x}\cos x\,dx = e^{\sin x} + C = \exp(\sin x) + C$

53. $u = 2 - e^{-x}$, $du = e^{-x}dx$, $\displaystyle\int \sec^2 u\,du = \tan(2 - e^{-x}) + C$

55. $\dfrac{\pi^{\sin x}}{\ln \pi} + C$ 57. $\dfrac{1}{2}x^2 \ln 3 - 4\pi e^2 \sin x + C$

59. $\ln(e^x) + \ln(e^{-x}) = \ln(e^x e^{-x}) = \ln 1 = 0$ so $\int [\ln(e^x) + \ln(e^{-x})]dx = C$

61. $u = \sqrt{y}$, $du = \dfrac{1}{2\sqrt{y}}\,dy$, $2\int e^u du = 2e^{\sqrt{y}} + C$

63. $u = 3 - 4e^x$, $du = -4e^x dx$, $u = -1$ when $x = 0$, $u = -17$ when $x = \ln 5$

$-\dfrac{1}{4}\int_{-1}^{-17} u\,du = -\dfrac{1}{8}u^2\Big]_{-1}^{-17} = -36$

65. $3x - e^x\Big]_1^2 = 3 + e - e^2$

67. $u = e^x + 4$, $du = e^x dx$, $u = e^{-\ln 3} + 4 = \dfrac{1}{3} + 4 = \dfrac{13}{3}$ when $x = -\ln 3$,

$u = e^{\ln 3} + 4 = 3 + 4 = 7$ when $x = \ln 3$

$\int_{13/3}^{7} \dfrac{1}{u}\,du = \ln u\Big]_{13/3}^{7} = \ln(7) - \ln(13/3) = \ln(21/13)$

69. $g(x) = e^{-x}f(x)$, $g'(x) = e^{-x}f'(x) - e^{-x}f(x) = e^{-x}[f'(x) - f(x)] = 0$ if $f'(x) = f(x)$ thus $g(x) = k$ because $g'(x) = 0$ so $e^{-x}f(x) = k$, $f(x) = ke^x$.

71. $2^x = 3^{x+1}$, $\ln(2^x) = \ln(3^{x+1})$, $x\ln 2 = (x + 1)\ln 3$, $x\ln 2 = x\ln 3 + \ln 3$, $x(\ln 2 - \ln 3) = \ln 3$, $x = \dfrac{\ln 3}{\ln 2 - \ln 3} = \dfrac{\ln 3}{\ln(2/3)}$

73. $f'(x) = ex^{e-1}$

75. $\dfrac{dk}{dT} = k_0 \exp\left[-\dfrac{q}{2}\dfrac{T - T_0}{T_0 T}\right]\left(-\dfrac{q}{2T^2}\right) = -\dfrac{qk_0}{2T^2}\exp\left[-\dfrac{q}{2}\dfrac{T - T_0}{T_0 T}\right]$

77. Divide $e^x + 3$ into e^{2x} to get $\dfrac{e^{2x}}{e^x + 3} = e^x - \dfrac{3e^x}{e^x + 3}$ so

$\int \dfrac{e^{2x}}{e^x + 3}\,dx = \int e^x dx - 3\int \dfrac{e^x}{e^x + 3}\,dx = e^x - 3\ln(e^x + 3) + C$

79. $y = f^{-1}(x)$, $x = f(y) = e^{1/y}$, $1/y = \ln x$, $y = \dfrac{1}{\ln x} = f^{-1}(x)$

81. $y = f^{-1}(x)$, $x = f(y) = 1 - \ln(3y)$, $\ln(3y) = 1 - x$, $y = \dfrac{1}{3} e^{1-x} = f^{-1}(x)$

83. (a) Let $f(x) = e^x$ and $x > 0$ then $\dfrac{f(x) - f(0)}{x - 0} = f'(c)$ where $0 < c < x$.

$\dfrac{e^x - 1}{x} = e^c > 1$ if $c > 0$, $e^x - 1 > x$, $e^x > 1 + x$

so $e^x \geq 1 + x$ if $x \geq 0$ (equality when $x = 0$)

(b) Let $f(x) = e^{-x}$ and $x > 0$ then $\dfrac{e^{-x} - 1}{x} = -e^{-c} > -1$ if $c > 0$.

$e^{-x} - 1 > -x$, $e^{-x} > 1 - x$ so $e^{-x} \geq 1 - x$ if $x \geq 0$ (equality when $x = 0$)

85. For any $\epsilon > 0$, $0 < e^x < \epsilon$ if $\ln e^x < \ln \epsilon$, $x < \ln \epsilon$; choose $x_0 = \ln \epsilon$.

87. $y = x^x$, $\ln y = x \ln x$, $y'/y = 1 + \ln x$, $y' = x^x(1 + \ln x)$; $y' = 0$ when $\ln x = -1$, $x = e^{-1}$ at which there is a relative minimum and hence the absolute minimum for $x > 0$. The minimum value is $e^{-1/e}$.

89. (a) The area under $1/t$ for $x \leq t \leq x + 1$ is less than the area of the rectangle with altitude $1/x$ and base 1, but greater than the area of the rectangle with altitude $1/(x + 1)$ and base 1.

(b) $\displaystyle\int_x^{x+1} \dfrac{1}{t}\, dt = \ln t \Big]_x^{x+1} = \ln(x + 1) - \ln x = \ln(1 + 1/x)$, so
$1/(x + 1) < \ln(1 + 1/x) < 1/x$ for $x > 0$.

(c) From part (b), $e^{1/(x+1)} < e^{\ln(1+1/x)} < e^{1/x}$,
$e^{1/(x+1)} < 1 + 1/x < e^{1/x}$, $e^{x/(x+1)} < (1 + 1/x)^x < e$; by the Pinching Theorem, $\displaystyle\lim_{x \to +\infty} (1 + 1/x)^x = e$.

(d) Use the inequality $e^{x/(x+1)} < (1 + 1/x)^x$ to get $e < (1 + 1/x)^{x+1}$ so $(1 + 1/x)^x < e < (1 + 1/x)^{x+1}$.

EXERCISE SET 7.5

1.   (a) $+\infty$ (b) 0   3.   (a) $+\infty$ (b) $+\infty$

5.   (a) 1   (b) 1

7. $f(x) = 4e^{3x} > 0$

  $f'(x) = 12e^{3x} > 0$

  $f''(x) = 36e^{3x} > 0$

9. $f(x) = e^{x} + e^{-x} = e^{-x}(e^{2x} + 1) > 0$

  $f'(x) = e^{x} - e^{-x} = e^{-x}(e^{2x} - 1)$

  $f''(x) = e^{x} + e^{-x} = f(x) > 0$

11. $f(x) = 1 - e^{-x^2}$

$f'(x) = 2xe^{-x^2}$

$f''(x) = 4(1/2 - x^2)e^{-x^2}$

$f''(x) = 0$ if $x = \pm 1/\sqrt{2}$
$\approx \pm 0.71$

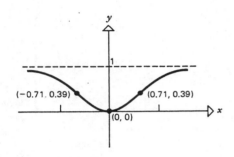

$\lim_{x \to \pm \infty} f(x) = 1$

13.   (a) yes, because $\lim_{x \to 0} f(x) = f(0)$

(b) no, because $\lim_{x \to 0^+} f'(x) = \lim_{x \to 0^+} e^x = 1$ and

$\lim_{x \to 0^-} f'(x) = \lim_{x \to 0^-} (-e^{-x}) = -1$ so $\lim_{x \to 0^+} f'(x) \neq \lim_{x \to 0^-} f'(x)$

(c) $f(x) = \begin{cases} e^x, & x \geq 0 \\ e^{-x}, & x < 0 \end{cases}$

15. $e^x \cos x = e^x$ when
    $\cos x = 1$
        $x = 0, \ \pm 2\pi, \ \pm 4\pi, \ \ldots$

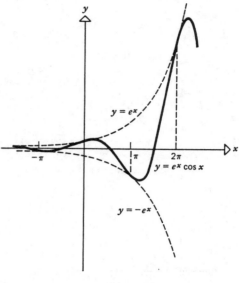

$y = e^x$

$y = e^x \cos x$

$y = -e^x$

$e^x \cos x = -e^x$ when
    $\cos x = -1$
        $x = \pm\pi, \ \pm 3\pi, \ \pm 5\pi, \ \ldots$

$e^x \cos x = 0$ when
    $\cos x = 0$
        $x = \pm\pi/2, \ \pm 3\pi/2, \ \pm 5\pi/2, \ \ldots$

$-1 \leq \cos x \leq 1$ thus
$-e^x \leq e^x \cos x \leq e^x$ and so
$\lim\limits_{x \to -\infty} e^x \cos x = 0$ by the pinching theorem.

$f(x) = e^x \cos x, \quad f'(x) = e^x(\cos x - \sin x),$
$f'(x) = 0$ when $\sin x = \cos x,$
$\tan x = 1, \quad x = \pi/4 + n\pi, \ n = 0, \ \pm 1, \ \pm 2, \ \ldots$
$f''(x) = -2e^x \sin x, \quad f''(x) = 0$ when $x = 0, \ \pm\pi, \ \pm 2\pi, \ \ldots$

17. $\lim\limits_{x \to \infty} \dfrac{e^x + e^{-x}}{e^x - e^{-x}} = \lim\limits_{x \to \infty} \dfrac{e^{2x} + 1}{e^{2x} - 1} = -1$

19. 0

21. $\dfrac{d}{dx}\left[e^x \right]\bigg|_{x=0} = \lim\limits_{h \to 0} \dfrac{e^{(0+h)} - e^0}{h} = \lim\limits_{h \to 0} \dfrac{e^h - 1}{h} = 1$

23. $\lim\limits_{x \to 0} \dfrac{1 - e^{-x}}{x} = \lim\limits_{x \to 0} -\dfrac{e^{-x} - e^0}{x - 0} = -\dfrac{d}{dx}\left[e^{-x} \right]\bigg|_{x=0} = e^{-x}\bigg|_{x=0} = 1$

25. let $h = 1/x$ then $x = 1/h$ and
    $\lim\limits_{x \to +\infty} x(e^{1/x} - 1) = \lim\limits_{h \to 0^+} \dfrac{e^h - 1}{h} = \dfrac{d}{dx}\left[e^x \right]\bigg|_{x=0} = 1$

27. (a) $\lim\limits_{x \to +\infty} xe^x = +\infty$; for $\lim\limits_{x \to -\infty} xe^x$, let $x = -t$ then

$\lim\limits_{x \to -\infty} xe^x = \lim\limits_{t \to +\infty} (-te^{-t}) = 0$.

(b) $y = xe^x$

$y' = (x + 1)e^x$

$y'' = (x + 2)e^x$

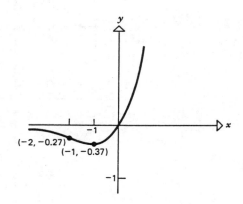

29. (a) $\lim\limits_{x \to +\infty} \dfrac{x^2}{e^{2x}} = \lim\limits_{x \to +\infty} x^2 e^{-2x} = \lim\limits_{x \to +\infty} (xe^{-x})^2 = (0)^2 = 0;$ $\lim\limits_{x \to -\infty} \dfrac{x^2}{e^{2x}} = +\infty.$

(b) $y = x^2/e^{2x} = x^2e^{-2x}$

$\quad y' = 2x(1 - x)e^{-2x}$

$\quad y'' = 2(2x^2 - 4x + 1)e^{-2x}$

$\quad y'' = 0$ if $2x^2 - 4x + 1 = 0$

$\quad x = \dfrac{4 \pm \sqrt{16 - 8}}{2},$

$\quad = 1 \pm \sqrt{2}/2 \approx 0.29,\ 1.71$

31. (a)

(b)

(c)

33. (a)

(b)

(c)

35. $m_{line} = \dfrac{1/e - 1}{1 - 0} = \dfrac{1 - e}{e}$, an equation

of the line is $y = \dfrac{1 - e}{e} x + 1$ so

$A = \displaystyle\int_0^1 (\dfrac{1 - e}{e} x + 1 - e^{-x}) dx$

$= \dfrac{1 - e}{2e} x^2 + x + e^{-x} \Big]_0^1 = \dfrac{3 - e}{2e}$

37. The volume of the solid obtained
     by revolving the region shown in
     the diagram about the y-axis is

     $$V = 2\pi\int_1^3 x \ln x \, dx \quad \text{so}$$

     $$\int_1^3 x \ln x \, dx = \frac{1}{2\pi} V.$$

     By the method of washers,

     $$V = \pi\int_0^{\ln 3}(9 - e^{2y})dy = \pi(9y - \frac{1}{2} e^{2y})\Big]_0^{\ln 3}$$

     $$= \pi(9 \ln 3 - 4)$$

     so $\int_1^3 x \ln x \, dx = (9 \ln 3 - 4)/2$

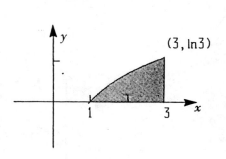

$(3, \ln 3)$

39. $y = \dfrac{\ln x}{x^2}$

     $y' = \dfrac{1 - 2 \ln x}{x^3}$

     $y'' = \dfrac{6 \ln x - 5}{x^4}$

     $y' = 0$ if $x = e^{1/2}$

     $y'' = 0$ if $x = e^{5/6}$

     $\lim\limits_{x\to+\infty} y = 0, \quad \lim\limits_{x\to0^+} y = -\infty$

$(e^{1/2} \cdot \frac{1}{2} e^{-1})$

41. If $\ln x < 2\sqrt{x}$, then $x < e^{2\sqrt{x}}$, $x^n < e^{2n\sqrt{x}}$, and $e^x/x^n > e^{x-2n\sqrt{x}}$. But

     $$\lim_{x\to+\infty} e^{x-2n\sqrt{x}} = \lim_{x\to+\infty} e^{\sqrt{x}(\sqrt{x}-2n)} = +\infty \text{ so } \lim_{x\to+\infty} e^x/x^n = +\infty.$$

EXERCISE SET 7.6

1.

	(a)	(b)	(c)	(d)	(e)	(f)
$\sinh x_0$	-2	$-3/4$	$-4/3$	$1/\sqrt{3}$	$8/15$	-1
$\cosh x_0$	$\sqrt{5}$	$5/4$	$5/3$	$2/\sqrt{3}$	$17/15$	$\sqrt{2}$
$\tanh x_0$	$-2/\sqrt{5}$	$-3/5$	$-4/5$	$1/2$	$8/17$	$-1/\sqrt{2}$
$\coth x_0$	$-\sqrt{5}/2$	$-5/3$	$-5/4$	2	$17/8$	$-\sqrt{2}$
$\operatorname{sech} x_0$	$1/\sqrt{5}$	$4/5$	$3/5$	$\sqrt{3}/2$	$15/17$	$1/\sqrt{2}$
$\operatorname{csch} x_0$	$-1/2$	$-4/3$	$-3/4$	$\sqrt{3}$	$15/8$	-1

(a) $\cosh^2 x_0 = 1 + \sinh^2 x_0 = 1 + (-2)^2 = 5$, $\cosh x_0 = \sqrt{5}$

(b) $\sinh^2 x_0 = \cosh^2 x_0 - 1 = \frac{25}{16} - 1 = \frac{9}{16}$, $\sinh x_0 = -\frac{3}{4}$ (because $x_0 < 0$)

(c) $\operatorname{sech}^2 x_0 = 1 - \tanh^2 x_0 = 1 - (-\frac{4}{5})^2 = 1 - \frac{16}{25} = \frac{9}{25}$,

$\operatorname{sech} x_0 = \frac{3}{5}$, $\cosh x_0 = \frac{1}{\operatorname{sech} x_0} = \frac{5}{3}$,

from $\dfrac{\sinh x_0}{\cosh x_0} = \tanh x_0$ we get $\sinh x_0 = (\frac{5}{3})(-\frac{4}{5}) = -\frac{4}{3}$

(d) $\operatorname{csch}^2 x_0 = \coth^2 x_0 - 1 = 4 - 1 = 3$, $\operatorname{csch} x_0 = \sqrt{3}$,

$\sinh x_0 = \dfrac{1}{\operatorname{csch} x_0} = \dfrac{1}{\sqrt{3}}$, from $\dfrac{\cosh x_0}{\sinh x_0} = \coth x_0$ we get

$\cosh x_0 = (\dfrac{1}{\sqrt{3}})(2) = \dfrac{2}{\sqrt{3}}$

(e) $\cosh x_0 = \dfrac{1}{\operatorname{sech} x_0} = \dfrac{17}{15}$, $\sinh^2 x_0 = \cosh^2 x_0 - 1 = \frac{289}{225} - 1 = \frac{64}{255}$,

$\sinh x_0 = \dfrac{8}{15}$ (because $x_0 > 0$)

(f) $\sinh x_0 = \dfrac{1}{\operatorname{csch} x_0} = -1$, $\cosh^2 x_0 = 1 + \sinh^2 x_0 = 2$, $\cosh x_0 = \sqrt{2}$

3. from (6b) and (1), $\cosh 2x = \cosh^2 x + \sinh^2 x$ and $\cosh^2 x = 1 + \sinh^2 x$ so $\cosh 2x = 2\sinh^2 x + 1$

5. $\cosh(-x) = \dfrac{1}{2}\left[e^{(-x)} + e^{-(-x)}\right] = \dfrac{1}{2}\left(e^{-x} + e^{x}\right) = \cosh x$

7. $\tanh(x + y) = \dfrac{\sinh(x + y)}{\cosh(x + y)} = \dfrac{\sinh x \cosh y + \cosh x \sinh y}{\cosh x \cosh y + \sinh x \sinh y}$

$$= \dfrac{\dfrac{\sinh x \cosh y}{\cosh x \cosh y} + \dfrac{\cosh x \sinh y}{\cosh x \cosh y}}{\dfrac{\cosh x \cosh y}{\cosh x \cosh y} + \dfrac{\sinh x \sinh y}{\cosh x \cosh y}} = \dfrac{\tanh x + \tanh y}{1 + \tanh x \tanh y}$$

9. $\tanh 2x = \dfrac{\sinh 2x}{\cosh 2x} = \dfrac{2\sinh x \cosh x}{\cosh^2 x + \sinh^2 x}$ (from (6a) and (6b))

$= \dfrac{2\tanh x}{1 + \tanh^2 x}$ (after dividing numerator and denominator by $\cosh^2 x$)

11. from (7a) with x replaced by $\dfrac{x}{2}$: $\cosh x = 2\sinh^2 \dfrac{x}{2} + 1$,

$2\sinh^2 \dfrac{x}{2} = \cosh x - 1$, $\sinh^2 \dfrac{x}{2} = \dfrac{1}{2}(\cosh x - 1)$,

$\sinh \dfrac{x}{2} = \pm\sqrt{\dfrac{1}{2}(\cosh x - 1)}$

13. add (4b) to (9b) then let $x = \dfrac{a + b}{2}$ and $y = \dfrac{a - b}{2}$.

15. (a) $\dfrac{d}{dx}(\sinh x) = \dfrac{d}{dx}\left[\dfrac{1}{2}(e^x - e^{-x})\right] = \dfrac{1}{2}(e^x + e^{-x}) = \cosh x$

(b) $\dfrac{d}{dx}(\coth x) = \dfrac{d}{dx}\left[\dfrac{e^x + e^{-x}}{e^x - e^{-x}}\right]$

$$= \dfrac{(e^x - e^{-x})(e^x - e^{-x}) - (e^x + e^{-x})(e^x + e^{-x})}{(e^x - e^{-x})^2}$$

$$= \frac{(e^{2x} - 2 + e^{-2x}) - (e^{2x} + 2 + e^{-2x})}{(e^x - e^{-x})^2}$$

$$= -\frac{4}{(e^x - e^{-x})^2} = -\operatorname{csch}^2 x$$

(c) $\frac{d}{dx} (\operatorname{sech} x) = \frac{d}{dx} \left[\frac{2}{e^x + e^{-x}} \right]$

$$= \frac{d}{dx} [2(e^x + e^{-x})^{-1}] = -2(e^x + e^{-x})^{-2}(e^x - e^{-x})$$

$$= -\frac{2}{(e^x + e^{-x})} \frac{e^x - e^{-x}}{e^x + e^{-x}} = -\operatorname{sech} x \tanh x$$

(d) proceed as in (c) using $\dfrac{2}{e^x - e^{-x}}$

17. $4\cosh(4x - 8)$

19. $-\dfrac{1}{x} \operatorname{csch}^2(\ln x)$

21. $\dfrac{1}{x^2} \operatorname{csch}(1/x)\coth(1/x)$

23. $\dfrac{2 + 5\cosh(5x)\sinh(5x)}{\sqrt{4x + \cosh^2(5x)}}$

25. $x^{5/2}\tanh(\sqrt{x})\operatorname{sech}^2(\sqrt{x}) + 3x^2\tanh^2(\sqrt{x})$

27. $\dfrac{1}{7} \sinh^7 x + C$

29. $\dfrac{2}{3} (\tanh x)^{3/2} + C$

31. $\ln(\cosh x) + C$

33. $-\dfrac{1}{3} \operatorname{sech}^3 x + C$

35. $2\sinh(\sqrt{x}) + C$

37. (a) $\dfrac{1}{2} (e^{\ln x} + e^{-\ln x}) = \dfrac{1}{2} (x + \dfrac{1}{x}) = \dfrac{x^2 + 1}{2x}, \quad x > 0$

(b) $\dfrac{1}{2} (e^{\ln x} - e^{-\ln x}) = \dfrac{1}{2} (x - \dfrac{1}{x}) = \dfrac{x^2 - 1}{2x}, \quad x > 0$

(c) $\dfrac{e^{2\ln x} - e^{-2\ln x}}{e^{2\ln x} + e^{-2\ln x}} = \dfrac{x^2 - 1/x^2}{x^2 + 1/x^2} = \dfrac{x^4 - 1}{x^4 + 1}, \quad x > 0$

(d) $\dfrac{1}{2} (e^{-\ln x} + e^{\ln x}) = \dfrac{1}{2} (\dfrac{1}{x} + x) = \dfrac{1 + x^2}{2x}, \quad x > 0$

39.   positive on $(0, +\infty)$, negative on $(-\infty, 0)$

increasing on $(-\infty, +\infty)$ $(\frac{dy}{dx} = \text{sech}^2 x > 0)$

concave up on $(-\infty, 0)$, concave down on $(0, +\infty)$ $(\frac{d^2 y}{dx^2} = -2 \text{ sech}^2 x \tanh x)$

41.   (a) $\cosh x = \sqrt{1 + \sinh^2 x} \geq \sqrt{1} = 1$

(b) $\cosh x \geq 1$ (part (a)) so $0 < \frac{1}{\cosh x} \leq 1$, $0 < \text{sech} x \leq 1$

43.   using $\sinh x + \cosh x = e^x$ (5a),

$(\sinh x + \cosh x)^n = (e^x)^n = e^{nx} = \sinh nx + \cosh nx$

45. $y' = \sinh x$, $1 + (y')^2 = 1 + \sinh^2 x = \cosh^2 x$

$L = \int_0^{\ln 2} \cosh x \, dx = \sinh x \Big]_0^{\ln 2} = \sinh(\ln 2) = \frac{1}{2}(e^{\ln 2} - e^{-\ln 2})$

$$= \frac{1}{2}(2 - \frac{1}{2}) = \frac{3}{4}$$

47. $V = \pi \int_0^5 (\cosh^2 2x - \sinh^2 2x) dx = \pi \int_0^5 dx = 5\pi$

EXERCISE SET 7.7

1. $\frac{1}{y} dy = \frac{1}{x} dx$, $\ln|y| = \ln|x| + C_1$, $\ln\left|\frac{y}{x}\right| = C_1$, $\left|\frac{y}{x}\right| = e^{C_1} = |C|$, $y = Cx$

3. $\frac{1}{1 + y} dy = -\frac{x}{\sqrt{1 + x^2}} dx$, $\ln|1 + y| = -\sqrt{1 + x^2} + C_1$, $y = Ce^{-\sqrt{1+x^2}} - 1$

5. $e^y dy = \frac{\sin x}{\cos^2 x} dx$, $e^y dy = \sec x \tan x \, dx$, $e^y = \sec x + C$, $y = \ln(\sec x + C)$

7. $\rho = e^{\int 3dx} = e^{3x}$, $e^{3x}y = \int e^{x}dx = e^{x} + C$, $y = e^{-2x} + Ce^{-3x}$

9. $\rho = e^{\int dx} = e^{x}$, $e^{x}y = \int e^{x}\cos(e^{x})dx = \sin(e^{x}) + C$, $y = e^{-x}\sin(e^{x}) + Ce^{-x}$

11. $y' + \dfrac{3}{x}y = -2x^3$, $\rho = e^{\int \frac{3}{x}dx} = e^{3\ln x} = x^3$,

   $x^3 y = \int -2x^6 dx = -\dfrac{2}{7}x^7 + C$, $y = -\dfrac{2}{7}x^4 + Cx^{-3}$

13. $\rho = e^{\int -x\,dx} = e^{-x^2/2}$, $e^{-x^2/2}y = \int xe^{-x^2/2}dx = -e^{-x^2/2} + C$,

   $y = -1 + Ce^{x^2/2}$, $3 = -1 + C$, $C = 4$, $y = -1 + 4e^{x^2/2}$

15. $\rho = e^{\int dt} = e^{t}$, $e^{t}y = \int 2e^{t}dt = 2e^{t} + C$, $y = 2 + Ce^{-t}$

   $1 = 2 + C$, $C = -1$, $y = 2 - e^{-t}$

17. $y^2 t \dfrac{dy}{dt} = t - 1$, $y^2 dy = (1 - \dfrac{1}{t})dt$, $\dfrac{1}{3}y^3 = t - \ln t + C$,

   $9 = 1 + C$, $C = 8$, $\dfrac{1}{3}y^3 = t - \ln t + 8$, $y = \sqrt[3]{3t - 3\ln t + 24}$

19. $\dfrac{dy}{dx} = 2 - \dfrac{y}{x}$, $\dfrac{dy}{dx} + \dfrac{1}{x}y = 2$, $\rho = e^{\int \frac{1}{x}dx} = e^{\ln x} = x$,

   $xy = \int 2x\,dx = x^2 + C$, $y = x + \dfrac{C}{x}$ but $y = 2$ when $x = 1$ so

   $2 = 1 + C$, $C = 1$, $y = x + \dfrac{1}{x}$

21. slope of normal $= \dfrac{y}{x - 3}$, $\dfrac{dy}{dx} = -\dfrac{x - 3}{y}$, $y\,dy = (3 - x)dx$,

   $\dfrac{1}{2}y^2 = 3x - \dfrac{1}{2}x^2 + C$, but $y = 2$ when $x = 3$ so $2 = 9 - \dfrac{9}{2} + C$, $C = -\dfrac{5}{2}$,

   $\dfrac{1}{2}y^2 = 3x - \dfrac{1}{2}x^2 - \dfrac{5}{2}$, $y^2 = 6x - x^2 - 5$

23. $\dfrac{dy}{dt} = $ rate in $-$ rate out, where y is the amount of salt at time t,

   $\dfrac{dy}{dt} = (4)(2) - (\dfrac{y}{50})(2) = 8 - \dfrac{1}{25}y$ so $\dfrac{dy}{dt} + \dfrac{1}{25}y = 8$ and $y(0) = 25$.

$$\rho = e^{\int \frac{1}{25} dt} = e^{t/25}, \quad e^{t/25}y = \int 8e^{t/25}dt = 200e^{t/25} + C,$$

$y = 200 + Ce^{-t/25}, \quad 25 = 200 + C, \quad C = -175,$

(a) $y = 200 - 175e^{-t/25}$

(b) when $t = 25$, $y = 200 - 175e^{-1} \approx 136$ lb

25. At time t there are $500 + (20 - 10)t = 500 + 10t$ gallons of brine in the tank so

$$\frac{dy}{dt} = 0 - \frac{y}{500 + 10t}(10) = -\frac{y}{50 + t}, \quad \frac{dy}{dt} + \frac{1}{50 + t}y = 0 \text{ and } y(0) = 50,$$

$$\rho = e^{\int 1/(50+t)dt} = e^{\ln(50+t)} = 50 + t, \quad (50 + t)y = C, \quad y = \frac{C}{50 + t},$$

$$50 = C/50, \quad C = 2500, \quad y = \frac{2500}{50 + t}$$

The tank reaches the point of overflowing when $500 + 10t = 1000$, $t = 50$ min so $y = 2500/(50 + 50) = 25$ lb.

27. (a) From (28), $k = -\frac{1}{T} \ln 2 = -\frac{1}{140} \ln 2 \approx -0.005$ so $y = 10e^{-0.005t}$
     (approximately).

(b) 10 weeks = 70 days so $y = 10e^{-0.35} \approx 7$ mg

29. $100e^{0.02t} = 5000$, $e^{0.02t} = 50$, $t = \frac{1}{0.02} \ln 50 \approx 196$ days

31. $y = y_0 e^{kt}$, but $y = 0.6y_0$ when $t = 5$ so $0.6y_0 = y_0 e^{5k}$, $e^{5k} = 0.6$

$k = \frac{1}{5} \ln 0.6$ so $T = -5 \frac{\ln 2}{\ln 0.6} \approx 6.8$ years

33. (a) $y = 10,000e^{kt}$, but $y = 12,000$ when $t = 10$ so $10,000e^{10k} = 12,000$,

$k = \frac{1}{10} \ln 1.2$. When $t = 20$, $y = 10,000e^{20k} = 10,000e^{2 \ln 1.2}$

$= 10,000(1.44) = 14,400$.

(b) From (27), $T = 10 \frac{\ln 2}{\ln 1.2} \approx 38$ years

35. $\frac{dT}{dt} = k(T - C)$, $k < 0$

$\frac{dT}{dt} - kT = -kC$, $\rho = e^{\int -kdt} = e^{-kt}$, $e^{-kt}T = \int -kCe^{-kt}dt = Ce^{-kt} + K,$

$T = C + Ke^{kt}$. But $T = T_0$ when $t = 0$ so $T_0 = C + K$, $K = T_0 - C$,

$T = C + (T_0 - C)e^{kt}$

37. (a) In t years the interest will be compounded nt times at an interest
rate of r/n each time. The value at the end of 1 interval is
$P + (r/n)P = P(1 + r/n)$, at the end of 2 intervals it is

$P(1 + r/n) + (r/n)P(1 + r/n) = P(1 + r/n)^2$, and continuing in this

fashion the value at the end of nt intervals is $P(1 + r/n)^{nt}$.
(b) Let $x = r/n$, then $n = r/x$ and

$$\lim_{n \to +\infty} P(1 + r/n)^{nt} = \lim_{x \to 0^+} P(1 + x)^{rt/x} = \lim_{x \to 0^+} P[(1 + x)^{1/x}]^{rt} = Pe^{rt}.$$

(c) The rate of increase is $dA/dt = rPe^{rt} = rA$.

39. Let $y = y_0 e^{kt}$ with $y = y_1$ when $t = t_1$ and $y = y_1/2$ when $t = t_1 + T$ then

$$y_0 e^{kt_1} = y_1 \qquad \text{(i)}$$

$$y_0 e^{k(t_1 + T)} = y_1/2 \qquad \text{(ii)}$$

Divide (i) by (ii) to get $e^{-kT} = 2$, $T = -\dfrac{1}{k} \ln 2$.

SUPPLEMENTARY EXERCISES CHAPTER 7

1. (a) $f(g(x)) = m(\dfrac{1}{mx}) = 1/x \neq x$; f and g are not inverses.

(b) $f(g(x)) = \dfrac{3}{(3 - x)/x + 1} = x$,

$g(f(x)) = \dfrac{3 - 3/(x + 1)}{3/(x + 1)} = x$; f and g are inverses.

(c) $f(g(x)) = (x^{1/3} + 2)^3 - 8 = x + 6x^{2/3} + 12x^{1/3} \neq x$; f and g are not
inverses.

(d) $f(g(x)) = x + 1 - 1 = x$,

$g(f(x)) = \sqrt[3]{x^3 - 1} + 1 = x$; f and g are inverses.

(e) $f(g(x)) = \sqrt{e^{2 \ln x}} = \sqrt{x^2} = x$ where $x > 0$,

   $g(f(x)) = 2 \ln\sqrt{e^x} = \ln e^x = x$; f and g are inverses.

3. $f(0) = f(2)$; f is not one-to-one so $f^{-1}(x)$ does not exist.

5. $y = f^{-1}(x)$, $x = f(y) = e^{2y} + 1$, $e^{2y} = x - 1$, $y = \frac{1}{2} \ln(x - 1) = f^{-1}(x)$.

7. f^{-1} will exist if and only if f is one-to-one. Let x_1, x_2 be any two distinct points in the domain of $y = f(x) = (ax + b)/(cx + d)$.
   $y_1 = f(x_1) = (ax_1 + b)/(cx_1 + d)$, $y_2 = f(x_2) = (ax_2 + b)/(cx_2 + d)$,

   $$y_2 - y_1 = \frac{(ax_2 + b)(cx_1 + d) - (ax_1 + b)(cx_2 + d)}{(cx_2 + d)(cx_1 + d)}$$

   $$= \frac{ad(x_2 - x_1) - bc(x_2 - x_1)}{(cx_2 + d)(cx_1 + d)} = \frac{(ad - bc)(x_2 - x_1)}{(cx_2 + d)(cx_2 + d)}$$

   f will be one-to-one if $y_1 \neq y_2$ (or equivalently $y_2 - y_1 \neq 0$) whenever $x_1 \neq x_2$, which occurs when $ad - bc \neq 0$. To find $f^{-1}(x)$ in this case, solve $y = (ax + b)/(cx + d)$ for x to get $x = (-dy + b)/(cy - a) = f^{-1}(y)$ so $f^{-1}(x) = (-dx + b)/(cx - a)$.

9. (a) $f(x) = \begin{cases} 2x - 5, & x \geq 5/2 \\ -2x + 5, & x < 5/2 \end{cases}$, $f'(x) = \begin{cases} 2, & x > 5/2 \\ -2, & x < 5/2 \end{cases}$
   and $f'(x)$ does not exist at $x = 5/2$. $f(x)$ is minimum when $x = 5/2$ and f is decreasing for $x < 5/2$ because $f'(x) < 0$, so f is one-to-one for x in the interval $(-\infty, 5/2)$.

   (b) $f'(x) = 2(x + 2)$, so f is decreasing for $x < -2$ and increasing for $x > -2$. f is one-to-one for x in $(-2, +\infty)$.

   (c) $f'(x) = -\sin(x - 2\pi/3)$, $f'(x) = 0$ when $x - 2\pi/3 = n\pi$, $x = 2\pi/3 + n\pi$ where n is an integer. $f'(-\pi/3) = f'(2\pi/3) = 0$ and $f'(x) > 0$ if $-\pi/3 < x < 2\pi/3$ so f is one-to-one for x in $(-\pi/3, 2\pi/3)$.

11. $y = f^{-1}(x)$, $x = f(y) = \dfrac{3}{y+1}$, $y = \dfrac{3}{x} - 1 = f^{-1}(x)$; $f'(x) = -\dfrac{3}{(x+1)^2}$,

    $f'(f^{-1}(x)) = -\dfrac{3}{(3/x)^2} = -\dfrac{x^2}{3}$, $(f^{-1})'(x) = -\dfrac{3}{x^2}$.

13. $y = f^{-1}(x)$, $x = f(y) = e^{y/2}$, $y = 2\ln x = f^{-1}(x)$; $f'(x) = \dfrac{1}{2} e^{x/2}$,

    $f'(f^{-1}(x)) = \dfrac{1}{2} e^{\ln x} = \dfrac{x}{2}$, $(f^{-1})'(x) = \dfrac{2}{x}$.

15.   (a) $\ln(1/12) = -\ln 12 = -\ln(2^2 \cdot 3) = -(2\ln 2 + \ln 3) = -(2r + s)$

    (b) $\ln(9/\sqrt{8}) = \ln(3^2 \cdot 2^{-3/2}) = 2\ln 3 - \dfrac{3}{2}\ln 2 = 2s - 3r/2$

    (c) $\ln(\sqrt[4]{8/3}) = \dfrac{1}{4}\ln(2^3/3) = \dfrac{1}{4}(3\ln 2 - \ln 3) = (3r - s)/4$

17.   (a) $25^x = 3^{1-x}$, $(5^2)^x = 3^{1-x}$, $5^{2x} = 3^{1-x}$, $\ln 5^{2x} = \ln 3^{1-x}$,

      $2x \ln 5 = (1-x)\ln 3$, $x = (\ln 3)/(2\ln 5 + \ln 3)$

    (b) $\sinh x = \dfrac{1}{4}\cosh x$, $\dfrac{1}{2}(e^x - e^{-x}) = \dfrac{1}{8}(e^x + e^{-x})$, $3e^x = 5e^{-x}$,

      $e^{2x} = 5/3$, $x = (\ln 5 - \ln 3)/2$

19.   (a) $\cosh x = (1 + \sinh^2 x)^{1/2} = (1 + 9/25)^{1/2} = \sqrt{34}/5$

    (b) $\tanh x = \sinh x/\cosh x = -3/\sqrt{34}$

    (c) $\sinh 2x = 2\sinh x \cosh x = -6\sqrt{34}/25$

21. $y = e^{-x/2}$, $dy/dx = -\dfrac{1}{2} e^{-x/2} = -1/(2\sqrt{e^x})$

23. $dy/dx = (\ln x - 1)/(\ln x)^2$ 25. $y = x/x = 1$, $dy/dx = 0$

27. $y = x\ln 10 - \ln \sin x$, $dy/dx = \ln 10 - \cot x$

29. $y = x^4 e^{\tan x}$, $dy/dx = x^3 e^{\tan x}(x\sec^2 x + 4)$

31. $dy/dx = \dfrac{1 + x/\sqrt{x^2 + a^2}}{x + \sqrt{x^2 + a^2}} = 1/\sqrt{x^2 + a^2}$

33. $y = \exp(3x^2)$, $dy/dx = 6x\exp(3x^2)$

35. $y = (\ln\sqrt{x})^{1/2}$, $dy/dx = \frac{1}{2}(\ln\sqrt{x})^{-1/2}(\frac{1}{2x}) = 1/(4x\sqrt{\ln\sqrt{x}})$

37. $dy/dx = \pi^x(\pi x^{\pi-1}) + x^\pi(\pi^x\ln\pi) = \pi^x x^{\pi-1}(\pi + x\ln\pi)$

39. $dy/dx = 5\cosh[\tanh(5x)]\text{sech}^2(5x)$

41. $y = e^{3x}(1 + 2e^{-x} + e^{-2x}) = e^{3x} + 2e^{2x} + e^x$, $dy/dx = 3e^{3x} + 4e^{2x} + e^x$.

43. $y = e^{ax}\sin bx$, $y' = e^{ax}[b\cos bx + a\sin bx]$,
    $y'' = e^{ax}[2ab\cos bx + (a^2 - b^2)\sin bx]$ so $y'' - 2ay' + (a^2 + b^2)y = 0$

45. (a) $dy = -e^{-x}dx$ (b) $dy = \dfrac{1}{1+x}dx$ (c) $dy = 2x(\ln 2)2^{x^2}dx$

47. (a) $\dfrac{1}{\sqrt{4 + e^{\ln x}}}(\frac{1}{x}) = \dfrac{1}{x\sqrt{4+x}}$

    (b) $\sqrt{\ln e^{5x} + e^{5x}}(5e^{5x}) = 5e^{5x}\sqrt{5x + e^{5x}}$

49. $Y = \ln y = \ln(Ce^{kt}) = \ln C + kt$ which is linear in t and Y so the graph is a straight line.

51. $u = 1 + e^x$, $\int\frac{1}{u}du = \ln(1 + e^x) + C$

53. $\dfrac{x^{e+1}}{e + 1} + C$ 55. $\int(\frac{4}{x} - \frac{3}{x^2})dx = 4\ln|x| + 3/x + C$

57. $u = 2\sec x - 1$, $\frac{1}{2}\int\frac{1}{u}du = \frac{1}{2}\ln|2\sec x - 1| + C$

59. $u = \sin 2x$, $\frac{1}{2}\int\exp(u)du = \frac{1}{2}\exp(\sin 2x) + C$

61. $u = \tanh x$, $\int u\,du = \frac{1}{2}\tanh^2 x + C$

63. $u = \ln x,$ $\displaystyle\int_1^2 \frac{1}{u}\,du = \ln 2$ 65. $u = \tan x,$ $\displaystyle\int_0^1 2^u du = \frac{2^u}{\ln 2}\Big]_0^1 = \frac{1}{\ln 2}$

67. $\dfrac{d}{dx}\left[\dfrac{e^{kx}}{k} + C\right] = e^{kx}$

69. $y = x^3 e^{-x}$
$y' = x^2(3 - x)e^{-x}$
$\quad y' = 0$ when $x = 0,\ 3$
$y'' = x(x^2 - 6x + 6)e^{-x}$
$\quad y'' = 0$ when $x = 0,\ 3 \pm\sqrt{3}$

71. $y'' = e^y y' + 2y' + 1.$ If x_0 is a critical point of y then when $x = x_0$, $y' = 0$ and $y'' = e^{y_0}(0) + 2(0) + 1 = 1 > 0$ so a relative minimum occurs at x_0.

73. $A = \displaystyle\int_0^b e^{-2x}dx = -\frac{1}{2}(e^{-2b} - 1) = \frac{1}{2} - \frac{1}{2}e^{-2b}.$ If $A = 1/4$ then
$1/2 - e^{-2b}/2 = 1/4,$ $e^{-2b} = 1/2,$ $b = (1/2)\ln 2.$ $\displaystyle\lim_{b\to+\infty} A = 1/2$

75. $y' = \sinh x$, $1 + (y')^2 = 1 + \sinh^2 x = \cosh^2 x$,

$$S = \int_0^1 2\pi \cosh x (\cosh x) dx = 2\pi \int_0^1 \cosh^2 x \, dx = \pi \int_0^1 (1 + \cosh 2x) dx$$

$$= \pi(x + \frac{1}{2} \sinh 2x) \Big]_0^1 = \pi(2 + \sinh 2)/2$$

77. Let y = population (in millions) t years after 1970, then
$y = 205e^{0.018t}$.

(a) $t = 2000 - 1970 = 30$ for the year 2000 so
$y = 205e^{(0.018)(30)} = 205e^{0.54} \approx 352$ million.

(b) 1 billion = 1000 million, $205e^{0.018t} = 1000$ when
$t = (1/0.018)\ln(1000/205) \approx 88$.   The population will reach one
billion in the year $1970 + 88 = 2058$.

EXERCISE SET 8.1

1. (a) $-\pi/2$ (b) π (c) $-\pi/4$
 (d) $\pi/4$ (e) 0 (f) $\pi/2$

3. $\theta = -\pi/3$; $\cos\theta = 1/2$, $\tan\theta = -\sqrt{3}$, $\cot\theta = -1/\sqrt{3}$,
    $\sec\theta = 2$, $\csc\theta = -2/\sqrt{3}$

5. $\tan\theta = 4/3$, $0 < \theta < \pi/2$; use the triangle shown to get $\sin\theta = 4/5$, $\cos\theta = 3/5$, $\cot\theta = 3/4$, $\sec\theta = 5/3$, $\csc\theta = 5/4$.

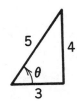

7. (a) $\pi/7$ (b) $\sin^{-1}(\sin\pi) = \sin^{-1}(\sin 0) = 0$
 (c) $\sin^{-1}(\sin(5\pi/7)) = \sin^{-1}(\sin(2\pi/7)) = 2\pi/7$
 (d) Note that $\pi/2 < 630 - 200\pi < \pi$ so
   $\sin(630) = \sin(630 - 200\pi)$
      $= \sin(\pi - (630 - 200\pi)) = \sin(201\pi - 630)$
   where $0 < 201\pi - 630 < \pi/2$;
   $\sin^{-1}(\sin 630) = \sin^{-1}(\sin(201\pi - 630)) = 201\pi - 630.$

9. (a) $0 \le x \le \pi$ (b) $-1 \le x \le 1$
 (c) $-\pi/2 < x < \pi/2$ (d) $-\infty < x < +\infty$
 (e) $0 < x \le \pi/2$ or $-\pi < x \le -\pi/2$ (f) $|x| \ge 1$

11. Let $\theta = \cos^{-1}(3/5)$,
  $\sin 2\theta = 2\sin\theta\cos\theta$
    $= 2(4/5)(3/5) = 24/25$

13. $\sin^{-1}(1) = \pi/2$

15. Let $\theta = \sec^{-1}(3/2)$,

$$\tan 2\theta = \frac{2 \tan \theta}{1 - \tan^2 \theta}$$

$$= \frac{2(\sqrt{5}/2)}{1 - 5/4} = -4\sqrt{5}$$

17. (a) $\tan^{-1} \frac{1}{2} + \tan^{-1} \frac{1}{3} = \tan^{-1} \dfrac{\frac{1}{2} + \frac{1}{3}}{1 - (\frac{1}{2})(\frac{1}{3})} = \tan^{-1} 1 = \pi/4$

(b) $2 \tan^{-1} \frac{1}{3} = \tan^{-1} \frac{1}{3} + \tan^{-1} \frac{1}{3} = \tan^{-1} \dfrac{\frac{1}{3} + \frac{1}{3}}{1 - (\frac{1}{3})(\frac{1}{3})} = \tan^{-1} \frac{3}{4}$,

$2 \tan^{-1} \frac{1}{3} + \tan^{-1} \frac{1}{7} = \tan^{-1} \frac{3}{4} + \tan^{-1} \frac{1}{7}$

$$= \tan^{-1} \dfrac{\frac{3}{4} + \frac{1}{7}}{1 - (\frac{3}{4})(\frac{1}{7})} = \tan^{-1} 1 = \pi/4$$

19. (a) $\cos(\tan^{-1} x) = \dfrac{x}{\sqrt{1 + x^2}}$ 　　　(b) $\tan(\cot^{-1} x) = \dfrac{1}{x}$

(c) $\sin(\sec^{-1}x) = \dfrac{\sqrt{x^2 - 1}}{x}$

(d) $\cot(\csc^{-1}x) = \sqrt{x^2 - 1}$

21. (a)   (b)

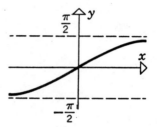

23. (a) let $\theta = \sin^{-1}(-x)$ then $\sin\theta = -x$, $-\pi/2 \le \theta \le \pi/2$. But
    $\sin(-\theta) = -\sin\theta$ and $-\pi/2 \le -\theta \le \pi/2$ so $\sin(-\theta) = -(-x) = x$,
    $-\theta = \sin^{-1}x$, $\theta = -\sin^{-1}x$.
    (b) proof is similar to that in part (a).
    (c) Let $\theta = \sec^{-1}(-x)$ for $x \ge 1$ then $\sec\theta = -x$ for $\pi \le \theta < 3\pi/2$. But
    $\sec(\theta - \pi) = -\sec\theta$ and $0 \le \theta - \pi < \pi/2$ so $\sec(\theta - \pi) = x$,
    $\theta - \pi = \sec^{-1}x$, $\theta = \pi + \sec^{-1}x$.

25. If $-1 \le x < 0$ then $0 < -x \le 1$ so $\sin^{-1}(-x) + \cos^{-1}(-x) = \pi/2$, but
    $\sin^{-1}(-x) = -\sin^{-1}x$ and $\cos^{-1}(-x) = \pi - \cos^{-1}x$ thus
    $-\sin^{-1}x + (\pi - \cos^{-1}x) = \pi/2$, $\sin^{-1}x + \cos^{-1}x = \pi/2$.

27. If $y = \sec^{-1}x$ for $-\infty < x \le -1$ or $1 \le x < +\infty$ then
    $\sec y = \sec(\sec^{-1}x) = x$. If $\sec y = x$ for $0 \le y < \pi/2$ or $\pi \le y < 3\pi/2$
    then $\sec^{-1}(\sec y) = \sec^{-1}x$, $y = \sec^{-1}x$.

29. (a) $\sin^{-1}X = \tan^{-1}\dfrac{X}{\sqrt{1 - X^2}}$

$= \text{ATN}\dfrac{X}{\sqrt{1 - X^2}}$

(b) $\sin^{-1}X + \cos^{-1}X = \pi/2$

$\cos^{-1}X = \pi/2 - \sin^{-1}X$

$= \pi/2 - \tan^{-1}\dfrac{X}{\sqrt{1 - X^2}}$

$\approx 1.5708 - \tan^{-1}\dfrac{X}{\sqrt{1 - X^2}}$

$= 1.5708 - \text{ATN}\dfrac{X}{\sqrt{1 - X^2}}$

EXERCISE SET 8.2

1. (a) $\dfrac{1}{\sqrt{1 - x^2/9}}(1/3) = 1/\sqrt{9 - x^2}$

    (b) $-2/\sqrt{1 - (2x + 1)^2}$

3. (a) $\dfrac{1}{x^7\sqrt{x^{14} - 1}}(7x^6) = \dfrac{7}{x\sqrt{x^{14} - 1}}$

    (b) $-1/\sqrt{e^{2x} - 1}$

5. (a) $\dfrac{1}{\sqrt{1 - 1/x^2}}(-1/x^2) = -\dfrac{1}{|x|\sqrt{x^2 - 1}}$

    (b) $\dfrac{\sin x}{\sqrt{1 - \cos^2 x}} = \dfrac{\sin x}{|\sin x|} = \begin{cases} 1, & \sin x > 0 \\ -1, & \sin x < 0 \end{cases}$

7. (a) $\dfrac{e^x}{x\sqrt{x^2 - 1}} + e^x\sec^{-1}x$

    (b) $\dfrac{3x^2(\sin^{-1}x)^2}{\sqrt{1 - x^2}} + 2x(\sin^{-1}x)^3$

9. (a) $\dfrac{1}{1 + (1 - x)^2/(1 + x)^2} \left[\dfrac{(1 + x)(-1) - (1 - x)(1)}{(1 + x)^2}\right]$

   $= -\dfrac{2}{(1 + x)^2 + (1 - x)^2} = -1/(x^2 + 1)$

   (b) $10(1 + x\ \csc^{-1}x)^9(-1/\sqrt{x^2 - 1} + \csc^{-1}x)$

11. (a) $\dfrac{1}{1 + (1 - x)/(1 + x)}\ \dfrac{1}{2}\ (\dfrac{1 - x}{1 + x})^{-1/2}\ \dfrac{(1 + x)(-1) - (1 - x)(1)}{(1 + x)^2}$

   $= -\dfrac{1}{2\sqrt{1 - x^2}}$

   (b) $\dfrac{x + 2x\ \ln x}{\sqrt{1 - x^4\ln^2 x}}$

13. $\sin^{-1}(xy) = \cos^{-1}(x - y)$,

   $\dfrac{1}{\sqrt{1 - x^2y^2}}\ (xy' + y) = -\dfrac{1}{\sqrt{1 - (x - y)^2}}\ (1 - y')$,

   $y' = \dfrac{y\sqrt{1 - (x - y)^2} + \sqrt{1 - x^2y^2}}{\sqrt{1 - x^2y^2} - x\sqrt{1 - (x - y)^2}}$

15. $\tan^{-1}x\ \Big]_{-1}^{1} = \tan^{-1}1 - \tan^{-1}(-1) = \pi/4 - (-\pi/4) = \pi/2$

17. $\sec^{-1}x\ \Big]_{-\sqrt{2}}^{-2/\sqrt{3}} = \sec^{-1}(-2/\sqrt{3}) - \sec^{-1}(-\sqrt{2}) = 7\pi/6 - 5\pi/4 = -\pi/12$

19. $u = 4x,\ \dfrac{1}{4} \displaystyle\int \dfrac{1}{1 + u^2}\ du = \dfrac{1}{4}\ \tan^{-1}4x + C$

21. $u = e^x,\ \displaystyle\int \dfrac{1}{1 + u^2}\ du = \tan^{-1}(e^x) + C$

23. $u = \sqrt{x}$, $2\int_{1}^{\sqrt{3}} \dfrac{1}{u^2 + 1}\, du = 2\,\tan^{-1}u\,\Big]_{1}^{\sqrt{3}} = 2(\tan^{-1}\sqrt{3} - \tan^{-1}1)$

$$= 2(\pi/3 - \pi/4) = \pi/6$$

25. $u = \tan x$, $\displaystyle\int \dfrac{1}{\sqrt{1 - u^2}}\, du = \sin^{-1}(\tan x) + C$

27. $u = \ln x$, $\displaystyle\int \dfrac{1}{\sqrt{1 - u^2}}\, du = \sin^{-1}(\ln x) + C$

29. (a) $\sin^{-1}(x/3) + C$ (b) $(1/\sqrt{5})\tan^{-1}(x/\sqrt{5}) + C$

    (c) $(1/\sqrt{\pi})\sec^{-1}(x/\sqrt{\pi}) + C$

31. $A = \displaystyle\int_{0}^{1/6} \dfrac{1}{\sqrt{1 - 9x^2}}\, dx = \dfrac{1}{3}\int_{0}^{1/2} \dfrac{1}{\sqrt{1 - u^2}}\, du = \dfrac{1}{3}\,\sin^{-1}u\,\Big]_{0}^{1/2} = \pi/18$

33. $V = \displaystyle\int_{-2}^{2} \pi\,\dfrac{1}{4 + x^2}\, dx = \dfrac{\pi}{2}\,\tan^{-1}(x/2)\,\Big]_{-2}^{2} = \pi^2/4$

35. $A = \displaystyle\int_{0}^{\pi/2} (1 - \sin y)\, dy$

$$= (y + \cos y)\,\Big]_{0}^{\pi/2} = \pi/2 - 1$$

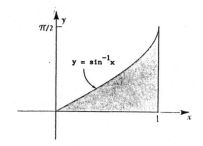

37. $\theta = \pi - (\alpha + \beta)$

$\quad = \pi - \cot^{-1}(x - 2) - \cot^{-1}\dfrac{5 - x}{4}$,

$\dfrac{d\theta}{dx} = \dfrac{1}{1 + (x - 2)^2} + \dfrac{-1/4}{1 + (5 - x)^2/16}$

$\quad = -\dfrac{3(x^2 - 2x - 7)}{[1 + (x - 2)^2][16 + (5 - x)^2]}$

$d\theta/dx = 0$ when $x = \dfrac{2 \pm \sqrt{4 + 28}}{2} = 1 \pm 2\sqrt{2}$,

only $1 + 2\sqrt{2}$ is in $[2,5]$.

$d\theta/dx > 0$ for x in $[2, 1 + 2\sqrt{2})$,

$d\theta/dx < 0$ for x in $(1 + 2\sqrt{2}, 5]$.

θ is maximum when $x = 1 + 2\sqrt{2}$.

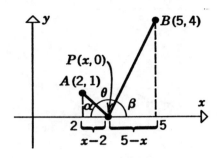

39. $\theta = \tan^{-1}(x/3)$

$\dfrac{d\theta}{dt} = \dfrac{3}{9 + x^2}\dfrac{dx}{dt}, \ \dfrac{dx}{dt} = \dfrac{9 + x^2}{3}\dfrac{d\theta}{dt}$

$\dfrac{dx}{dt}\bigg|_{x=2} = \dfrac{9 + 4}{3}(4\pi) = 52\pi/3 \text{ mi/min}$

41. $\theta = \alpha - \beta$

$\quad = \cot^{-1}(x/12) - \cot^{-1}(x/2)$

$\dfrac{d\theta}{dx} = -\dfrac{12}{144 + x^2} + \dfrac{2}{4 + x^2}$

$\quad = \dfrac{10(24 - x^2)}{(144 + x^2)(4 + x^2)}$

$d\theta/dx = 0$ when $x = \sqrt{24} = 2\sqrt{6}$, by the first derivative test θ is maximum there.

43. By the Mean-Value Theorem on the interval $[0, x]$,

$$\frac{\tan^{-1}x - \tan^{-1}0}{x - 0} = \frac{\tan^{-1}x}{x} = \frac{1}{1 + c^2} \text{ for c in } (0, x), \text{ but}$$

$$\frac{1}{1 + x^2} < \frac{1}{1 + c^2} < 1 \text{ for c in } (0, x) \text{ so}$$

$$\frac{1}{1 + x^2} < \frac{\tan^{-1}x}{x} < 1, \quad \frac{x}{1 + x^2} < \tan^{-1}x < x.$$

EXERCISE SET 8.3

1. (a) let $y = \cosh^{-1}x$, then $x = \cosh y = \frac{1}{2}(e^y + e^{-y})$, $e^y - 2x + e^{-y} = 0$,

    $e^{2y} - 2xe^y + 1 = 0$, $e^y = \dfrac{2x \pm \sqrt{4x^2 - 4}}{2} = x \pm \sqrt{x^2 - 1}$. To determine

    which sign to take, note that $y \geq 0$ so $e^{-y} \leq e^y$,

    $x = (e^y + e^{-y})/2 \leq (e^y + e^y)/2 = e^y$, hence $e^y \geq x$ thus

    $e^y = x + \sqrt{x^2 - 1}$, $y = \cosh^{-1}x = \ln(x + \sqrt{x^2 - 1})$.

    (b) $\dfrac{d}{dx}(\cosh^{-1}x) = \dfrac{1 + x/\sqrt{x^2 - 1}}{x + \sqrt{x^2 - 1}} = 1/\sqrt{x^2 - 1}$

3. (a) let $y = \text{sech}^{-1}x$ then $x = \text{sech}\,y = 1/\cosh y$, $\cosh y = 1/x$,

    $y = \cosh^{-1}(1/x)$; the proofs for the remaining two are similar.

    (b) $\dfrac{d}{dx}(\text{sech}^{-1}x) = \dfrac{d}{dx}(\cosh^{-1}(1/x)) = \dfrac{(-1/x^2)}{\sqrt{1/x^2 - 1}} = -\dfrac{1}{x\sqrt{1 - x^2}}$; the

    remaining two are done similarly.

    (c) $\text{sech}^{-1}x = \cosh^{-1}(1/x) = \ln\left[\dfrac{1}{x} + \sqrt{\dfrac{1}{x^2} - 1}\right] = \ln\left[\dfrac{1 + \sqrt{1 - x^2}}{x}\right]$; the

    remaining two are done similarly.

5. (a) $\ln(3 + \sqrt{8})$ (b) $\ln(\sqrt{5} - 2)$

7. (a) $\dfrac{1}{\sqrt{1 + x^2/9}}\,(\tfrac{1}{3}) = 1/\sqrt{9 + x^2}$ (b) $2/\sqrt{(2x + 1)^2 - 1}$

9. (a) $-\dfrac{7x^6}{x^7\sqrt{1 - x^{14}}} = -\dfrac{7}{x\sqrt{1 - x^{14}}}$ (b) $-1/\sqrt{1 + e^{2x}}$

11. (a) $\dfrac{1}{\sqrt{1 + 1/x^2}}\,(-1/x^2) = -\dfrac{1}{|x|\sqrt{x^2 + 1}}$

    (b) $\dfrac{\sinh x}{\sqrt{\cosh^2 x - 1}} = \dfrac{\sinh x}{|\sinh x|} = \begin{cases} 1, & x > 0 \\ -1, & x < 0 \end{cases}$

13. (a) $-\dfrac{e^x}{x\sqrt{1 - x^2}} + e^x\,\mathrm{sech}^{-1}x$

    (b) $3x^2(\sinh^{-1}x)^2/\sqrt{1 + x^2} + 2x(\sinh^{-1}x)^3$

15. (a) $\dfrac{1}{1 - (1 - x)^2/(1 + x)^2}\left[-\dfrac{2}{(1 + x)^2}\right] = -1/(2x)$

    (b) $10(1 + x\,\mathrm{csch}^{-1}x)^9\left(-\dfrac{x}{|x|\sqrt{1 + x^2}} + \mathrm{csch}^{-1}x\right)$

17. $x = \sqrt{2}u$, $\displaystyle\int \dfrac{\sqrt{2}}{\sqrt{2u^2 - 2}}\,du = \int \dfrac{1}{\sqrt{u^2 - 1}}\,du = \cosh^{-1}(x/\sqrt{2}) + C$, $x > \sqrt{2}$

19. $u = e^x$, $\displaystyle\int \dfrac{1}{u\sqrt{1 - u^2}}\,du = -\mathrm{sech}^{-1}(e^x) + C$

21. $u = x^3$, $\dfrac{1}{3}\displaystyle\int \dfrac{1}{u\sqrt{1 + u^2}}\,du = -\dfrac{1}{3}\,\mathrm{csch}^{-1}|x^3| + C$

23. $\left. \coth^{-1}x \right]_2^3 = \coth^{-1}3 - \coth^{-1}2$

$= \frac{1}{2} \ln \frac{3+1}{3-1} - \frac{1}{2} \ln \frac{2+1}{2-1} = \frac{1}{2} \ln 2 - \frac{1}{2} \ln 3$

$\approx \frac{1}{2} (0.6931 - 1.0986) \approx -0.2028$

25. (a)   (b)

          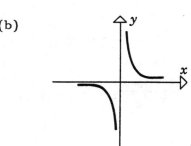

27. $\frac{d}{dx} (\mathrm{sech}^{-1}|x|) = \frac{d}{dx} (\mathrm{sech}^{-1}\sqrt{x^2})$

$= - \frac{1}{\sqrt{x^2} \sqrt{1 - x^2}} \frac{x}{\sqrt{x^2}} = - \frac{1}{x\sqrt{1 - x^2}}$

29. If $-1 < x < 1$ then $\frac{1+x}{1-x} > 0$, $\tanh^{-1}x = \frac{1}{2} \ln \frac{1+x}{1-x} = \frac{1}{2} \ln \left| \frac{1+x}{1-x} \right|$;

if $|x| > 1$ then $\frac{x+1}{x-1} > 0$, but $\frac{x+1}{x-1} = \left| \frac{1+x}{1-x} \right|$ so

$\coth^{-1}x = \frac{1}{2} \ln \frac{x+1}{x-1} = \frac{1}{2} \ln \left| \frac{1+x}{1-x} \right|$, $\int \frac{1}{1-x^2} dx = \frac{1}{2} \ln \left| \frac{1+x}{1-x} \right| + C$

31. Let $y = \sinh^{-1}x$ then $x = \sinh y$,

$\frac{dy}{dx} = \frac{1}{dx/dy} = \frac{1}{\cosh y} = \frac{1}{\sqrt{1 + \sinh^2 y}} = \frac{1}{\sqrt{1 + x^2}}$.

33. Let $u = -x$,

$\int \frac{1}{\sqrt{x^2 - 1}} dx = -\int \frac{1}{\sqrt{u^2 - 1}} du = -\cosh^{-1}u + C = -\cosh^{-1}(-x) + C$.

SUPPLEMENTARY EXERCISES CHAPTER 8

1. (a) $2\pi/3$ (b) $3/4$

    (c) $\cos[\sin^{-1}(4/5)] = \cos\theta = 3/5$

    (d) $\cos[\sin^{-1}(-4/5)] = \cos[-\sin^{-1}(4/5)]$
$$= \cos[\sin^{-1}(4/5)]$$
$$= 3/5$$

3. (a) $\pi/4$

    (b) $\sin^{-1}[\sin(5\pi/4)] = \sin^{-1}[\sin(-\pi/4)] = -\pi/4$

    (c) $\tan(\sec^{-1}5) = \tan\theta = 2\sqrt{6}$

    (d) $\tan^{-1}[\cot(\pi/6)] = \tan^{-1}(\sqrt{3}) = \pi/3$

5. (a) let $\alpha = \cos^{-1}(4/5)$, $\beta = \sin^{-1}(5/13)$
$$\cos(\alpha + \beta) = \cos\alpha\cos\beta - \sin\alpha\sin\beta$$
$$= (4/5)(12/13) - (3/5)(5/13) = 33/65$$

    (b) let $\alpha = \sin^{-1}(4/5)$, $\beta = \cos^{-1}(5/13)$
$$\sin(\alpha + \beta) = \sin\alpha\cos\beta + \cos\alpha\sin\beta$$
$$= (4/5)(5/13) + (3/5)(12/13) = 56/65$$

(c) let $\alpha = \tan^{-1}(1/3)$, $\beta = \tan^{-1}(2)$

$$\tan(\alpha + \beta) = \frac{\tan\alpha + \tan\beta}{1 - \tan\alpha\tan\beta} = \frac{1/3 + 2}{1 - (1/3)(2)} = 7$$

7. $\tanh u = -3/5$, $\operatorname{sech}^2 u = 1 - \tanh^2 u = 16/25$, $\operatorname{sech} u = 4/5$, $\cosh u = 5/4$, $\sinh u = (\tanh u)(\cosh u) = -3/4$, $\cosh 2u = \cosh^2 u + \sinh^2 u = 34/16$.

9.   (a)

(b)

$$f(x) = \cos^{-1}x + \sin^{-1}x = \pi/2$$

11. $-[\sec^{-1}(x^2)]^{-2} \dfrac{1}{x^2\sqrt{x^4 - 1}} (2x) = -\dfrac{2[\sec^{-1}(x^2)]^{-2}}{x\sqrt{x^4 - 1}}$

13. $(\sec x \tan x)/\sqrt{\sec^2 x - 1} = (\sec x \tan x)/|\tan x|$

15. $1/[1 - (\ln x)^2] + \tanh^{-1}(\ln x)$ 17. $3/[2\sqrt{\sin^{-1}3x}\ \sqrt{1 - 9x^2}]$

19. $\exp(\sec^{-1}x)/(x\sqrt{x^2 - 1})$ 21. $\pi^{\sin^{-1}x}(\ln\pi)/\sqrt{1 - x^2}$

23. $y = \tanh^{-1}(1/\coth x) = \tanh^{-1}(\tanh x)$ if $x \neq 0$, so $y = x$, $dy/dx = 1$ if $x \neq 0$.

25. $y = \tan^{-1}x$, $y' = 1/(1 + x^2)$, $y'' = -2x/(1 + x^2)^2$

$\sin y = x/\sqrt{1 + x^2}$, $\cos y = 1/\sqrt{1 + x^2}$

$-2 \sin y \cos^3 y = -2 \dfrac{x}{(1 + x^2)^{1/2}} \dfrac{1}{(1 + x^2)^{3/2}}$

$= -2x/(1 + x^2)^2 = y''$

27. $u = e^x$, $\displaystyle\int \frac{1}{1 - u^2}\, du = \frac{1}{2} \ln \left| \frac{1 + e^x}{1 - e^x} \right| + C$

29. $u = \ln x$, $\displaystyle\int \frac{1}{\sqrt{u^2 - 1}}\, du = \cosh^{-1}(\ln x) + C$, $x > e$

31. $u = 3x$, $\displaystyle\frac{1}{3} \int_{2/\sqrt{3}}^{2} \frac{1}{u\sqrt{u^2 - 1}}\, du = \frac{1}{3} \sec^{-1} u \Big]_{2/\sqrt{3}}^{2} = \frac{1}{3}(\pi/3 - \pi/6) = \pi/18$

33. $\displaystyle\int \frac{1}{x^{1/2} + x^{3/2}}\, dx = \int \frac{1}{x^{1/2}(1 + x)}\, dx$, $u = x^{1/2}$,

$2\displaystyle\int \frac{1}{1 + u^2}\, du = 2 \tan^{-1}\sqrt{x} + C$

35. $u = \sqrt{x}$, $2\displaystyle\int_{1/2}^{1/\sqrt{2}} \frac{1}{\sqrt{1 - u^2}}\, du = 2 \sin^{-1} u \Big]_{1/2}^{1/\sqrt{2}} = 2\left(\frac{\pi}{4} - \frac{\pi}{6}\right) = \pi/6$

37. $f'(x) = \dfrac{2}{1 + 4x^2} - \dfrac{1}{1 + x^2} = \dfrac{1 - 2x^2}{(1 + 4x^2)(1 + x^2)}$, the critical point is

$x = 1/\sqrt{2}$ for $x > 0$, $f'(x) > 0$ for $0 < x < 1/\sqrt{2}$ and $f'(x) < 0$ for

$x > 1/\sqrt{2}$ so $f(x)$ assumes its maximum value at $x = 1/\sqrt{2}$

CHAPTER 9

EXERCISE SET 9.2

1. $u = x$, $dv = e^{-x}dx$, $du = dx$, $v = -e^{-x}$
    $$\int xe^{-x}dx = -xe^{-x} + \int e^{-x}dx = -xe^{-x} - e^{-x} + C$$

3. $u = \ln(2x + 3)$, $dv = dx$, $du = \dfrac{2}{2x + 3}dx$, $v = x$
    $$\int \ln(2x + 3)dx = x\ln(2x + 3) - \int \frac{2x}{2x + 3}dx$$
    but $\displaystyle\int \frac{2x}{2x + 3}dx = \int(1 - \frac{3}{2x + 3})dx = x - \frac{3}{2}\ln(2x + 3) + C_1$ so
    $$\int \ln(2x + 3)dx = x\ln(2x + 3) - x + \frac{3}{2}\ln(2x + 3) + C$$

5. $u = \ln\sqrt{x} = \dfrac{1}{2}\ln x$, $dv = x\,dx$, $du = \dfrac{1}{2x}dx$, $v = \dfrac{1}{2}x^2$
    $$\int x\ln\sqrt{x}\,dx = \frac{1}{4}x^2\ln x - \frac{1}{4}\int x\,dx = \frac{1}{4}x^2\ln x - \frac{1}{8}x^2 + C$$

7. $u = \cos^{-1}(2x)$, $dv = dx$, $du = -\dfrac{2}{\sqrt{1 - 4x^2}}dx$, $v = x$
    $$\int \cos^{-1}(2x)dx = x\cos^{-1}(2x) + \int \frac{2x}{\sqrt{1 - 4x^2}}dx = x\cos^{-1}(2x) - \frac{1}{2}\sqrt{1 - 4x^2} + C$$

9. $u = x^2$, $dv = e^x dx$, $du = 2x\,dx$, $v = e^x$
    $$\int x^2 e^x dx = x^2 e^x - 2\int xe^x dx.$$
    For $\displaystyle\int xe^x dx$ use $u = x$, $dv = e^x dx$, $du = dx$, $v = e^x$ to get
    $$\int xe^x dx = xe^x - e^x + C_1 \text{ so } \int x^2 e^x dx = x^2 e^x - 2xe^x + 2e^x + C$$

11. $u = e^x$, $dv = \sin x\, dx$, $du = e^x dx$, $v = -\cos x$

$$\int e^x \sin x\, dx = -e^x \cos x + \int e^x \cos x\, dx.$$

For $\int e^x \cos x\, dx$ use $u = e^x$, $dv = \cos x\, dx$ to get

$$\int e^x \cos x = e^x \sin x - \int e^x \sin x\, dx \text{ so}$$

$$\int e^x \sin x\, dx = -e^x \cos x + e^x \sin x - \int e^x \sin x\, dx$$

$$2\int e^x \sin x\, dx = e^x(\sin x - \cos x) + C_1$$

$$\int e^x \sin x\, dx = \frac{1}{2} e^x(\sin x - \cos x) + C$$

13. Rewrite as $\int e^{-2\pi x} \cos 2\pi x\, dx$ then $u = e^{-2\pi x}$, $dv = \cos 2\pi x\, dx$,

$du = -2\pi e^{-2\pi x} dx$, $v = \frac{1}{2\pi} \sin 2\pi x$

$$\int e^{-2\pi x} \cos 2\pi x\, dx = \frac{1}{2\pi} e^{-2\pi x} \sin 2\pi x + \int e^{-2\pi x} \sin 2\pi x\, dx.$$

For $\int e^{-2\pi x} \sin 2\pi x\, dx$ use $u = e^{-2\pi x}$, $dv = \sin 2\pi x\, dx$ to get

$$\int e^{-2\pi x} \sin 2\pi x\, dx = -\frac{1}{2\pi} e^{-2\pi x} \cos 2\pi x - \int e^{-2\pi x} \cos 2\pi x\, dx \text{ so}$$

$$\int e^{-2\pi x} \cos 2\pi x\, dx = \frac{1}{2\pi} e^{-2\pi x} \sin 2\pi x - \frac{1}{2\pi} e^{-2\pi x} \cos 2\pi x - \int e^{-2\pi x} \cos 2\pi x\, dx$$

$$\int e^{-2\pi x} \cos 2\pi x\, dx = \frac{1}{4\pi} e^{-2\pi x}(\sin 2\pi x - \cos 2\pi x) + C$$

15. $u = e^{ax}$, $dv = \sin bx\, dx$, $du = ae^{ax} dx$, $v = -\frac{1}{b} \cos bx$

$$\int e^{ax} \sin bx\, dx = -\frac{1}{b} e^{ax} \cos bx + \frac{a}{b} \int e^{ax} \cos bx\, dx.$$

Use $u = e^{ax}$, $dv = \cos bx\, dx$ to get

$$\int e^{ax} \cos bx\, dx = \frac{1}{b} e^{ax} \sin bx - \frac{a}{b} \int e^{ax} \sin bx\, dx \text{ so}$$

$$\int e^{ax} \sin bx\, dx = -\frac{1}{b} e^{ax} \cos bx + \frac{a}{b^2} e^{ax} \sin bx - \frac{a^2}{b^2} \int e^{ax} \sin bx\, dx$$

$$\int e^{ax} \sin bx\, dx = \frac{e^{ax}}{a^2 + b^2} (a \sin bx - b \cos bx) + C$$

17. $u = x^2$, $dv = \cos x\,dx$, $du = 2x\,dx$, $v = \sin x$

$$\int x^2 \cos x\,dx = x^2 \sin x - 2 \int x \sin x\,dx$$

For $\int x \sin x\,dx$ use $u = x$, $dv = \sin x\,dx$ to get

$$\int x \sin x\,dx = -x \cos x + \sin x + C_1 \text{ so}$$

$$\int x^2 \cos x\,dx = x^2 \sin x + 2x \cos x - 2 \sin x + C$$

19. $u = x$, $dv = \sin(3x + 1)dx$, $du = dx$, $v = -\dfrac{1}{3}\cos(3x + 1)$

$$\int x \sin(3x + 1)dx = -\frac{1}{3} x \cos(3x + 1) + \frac{1}{3} \int \cos(3x + 1)dx$$

$$= -\frac{1}{3} x \cos(3x + 1) + \frac{1}{9} \sin(3x + 1) + C$$

21. $u = x$, $dv = \sec^2 x\,dx$, $du = dx$, $v = \tan x$

$$\int x \sec^2 x\,dx = x \tan x - \int \tan x\,dx$$

$$= x \tan x - \int \frac{\sin x}{\cos x}\,dx = x \tan x + \ln|\cos x| + C$$

23. $u = \cos(\ln x)$, $dv = dx$, $du = -\dfrac{1}{x}\sin(\ln x)dx$, $v = x$

$$\int \cos(\ln x)dx = x \cos(\ln x) + \int \sin(\ln x)dx.$$

Use $u = \sin(\ln x)$, $dv = dx$ to get

$$\int \sin(\ln x)dx = x \sin(\ln x) - \int \cos(\ln x)dx \text{ so}$$

$$\int \cos(\ln x)dx = x \cos(\ln x) + x \sin(\ln x) - \int \cos(\ln x)dx$$

$$\int \cos(\ln x)dx = \frac{1}{2} x[\cos(\ln x) + \sin(\ln x)] + C$$

25. $u = \sin(\ln x)$, $dv = dx$, $du = \dfrac{\cos(\ln x)}{x}\,dx$, $v = x$

$$\int \sin(\ln x)dx = x \sin(\ln x) - \int \cos(\ln x)dx.$$

Use $u = \cos(\ln x)$, $dv = dx$ to get

$$\int \cos(\ln x)dx = x \cos(\ln x) + \int \sin(\ln x)dx \text{ so}$$

$$\int \sin(\ln x)dx = x\sin(\ln x) - x\cos(\ln x) - \int \sin(\ln x)dx$$

$$\int \sin(\ln x)dx = (x/2)[\sin(\ln x) - \cos(\ln x)] + C$$

27. $u = x$, $dv = \tan^2 x\,dx = (\sec^2 x - 1)dx$, $du = dx$, $v = \tan x - x$

$$\int x\tan^2 x\,dx = x\tan x - x^2 - \int(\tan x - x)dx$$

$$= x\tan x - x^2 + \ln|\cos x| + \frac{1}{2}x^2 + C$$

$$= x\tan x - \frac{1}{2}x^2 + \ln|\cos x| + C$$

29. $u = x$, $dv = e^{-5x}dx$, $du = dx$, $v = -\frac{1}{5}e^{-5x}$

$$\int_0^1 xe^{-5x}dx = -\frac{1}{5}xe^{-5x}\Big]_0^1 + \frac{1}{5}\int_0^1 e^{-5x}dx$$

$$= -\frac{1}{5}e^{-5} - \frac{1}{25}e^{-5x}\Big]_0^1 = -\frac{1}{5}e^{-5} - \frac{1}{25}(e^{-5} - 1)$$

$$= (1 - 6e^{-5})/25$$

31. $u = \ln x$, $dv = x^2dx$, $du = \frac{1}{x}dx$, $v = \frac{1}{3}x^3$

$$\int_1^e x^2\ln x\,dx = \frac{1}{3}x^3\ln x\Big]_1^e - \frac{1}{3}\int_1^e x^2dx$$

$$= \frac{1}{3}e^3 - \frac{1}{9}x^3\Big]_1^e = \frac{1}{3}e^3 - \frac{1}{9}(e^3 - 1) = (2e^3 + 1)/9$$

33. $u = \ln(x + 3)$, $dv = dx$, $du = \dfrac{1}{x + 3}dx$, $v = x$

$$\int_{-2}^2 \ln(x + 3)dx = x\ln(x + 3)\Big]_{-2}^2 - \int_{-2}^2 \frac{x}{x + 3}dx$$

$$= 2\ln 5 + 2\ln 1 - \int_{-2}^2 \left[1 - \frac{3}{x + 3}\right]dx$$

$$= 2\ln 5 - \left[x - 3\ln(x + 3)\right]_{-2}^2$$

$$= 2\ln 5 - (2 - 3\ln 5) + (-2 - 3\ln 1) = 5\ln 5 - 4$$

35. $u = \sec^{-1}\sqrt{\theta}$, $dv = d\theta$, $du = \dfrac{1}{2\theta\sqrt{\theta - 1}}\, d\theta$, $v = \theta$

$$\int_2^4 \sec^{-1}\sqrt{\theta}\, d\theta = \theta \sec^{-1}\sqrt{\theta}\,\Big]_2^4 - \frac{1}{2}\int_2^4 \frac{1}{\sqrt{\theta - 1}}\, d\theta$$

$$= 4\sec^{-1}2 - 2\sec^{-1}\sqrt{2} - \sqrt{\theta - 1}\,\Big]_2^4$$

$$= 4\left(\frac{\pi}{3}\right) - 2\left(\frac{\pi}{4}\right) - \sqrt{3} + 1 = \frac{5\pi}{6} - \sqrt{3} + 1$$

37. $u = x$, $dv = \sin 4x\, dx$, $du = dx$, $v = -\dfrac{1}{4}\cos 4x$

$$\int_0^{\pi/2} x \sin 4x\, dx = -\frac{1}{4} x \cos 4x\,\Big]_0^{\pi/2} + \frac{1}{4}\int_0^{\pi/2} \cos 4x\, dx$$

$$= -\pi/8 + \frac{1}{16}\sin 4x\,\Big]_0^{\pi/2} = -\pi/8$$

39. $u = \tan^{-1}\sqrt{x}$, $dv = \sqrt{x}dx$, $du = \dfrac{1}{2\sqrt{x}(1 + x)}\, dx$, $v = \dfrac{2}{3} x^{3/2}$

$$\int_1^3 \sqrt{x}\tan^{-1}\sqrt{x}\, dx = \frac{2}{3} x^{3/2}\tan^{-1}\sqrt{x}\,\Big]_1^3 - \frac{1}{3}\int_1^3 \frac{x}{1 + x}\, dx$$

$$= \frac{2}{3} x^{3/2}\tan^{-1}\sqrt{x}\,\Big]_1^3 - \frac{1}{3}\int_1^3 \left[1 - \frac{1}{1 + x}\right]dx$$

$$= \left[\frac{2}{3} x^{3/2}\tan^{-1}\sqrt{x} - \frac{1}{3} x + \frac{1}{3}\ln|1 + x|\right]_1^3$$

$$= (2\sqrt{3}\pi - \pi/2 - 2 + \ln 2)/3$$

41. $u = x^2$, $dv = \dfrac{x}{\sqrt{x^2 + 1}}\, dx$, $du = 2x\, dx$, $v = \sqrt{x^2 + 1}$

$$\int_0^1 \frac{x^3}{\sqrt{x^2 + 1}}\, dx = x^2\sqrt{x^2 + 1}\,\Big]_0^1 - 2\int_0^1 x(x^2 + 1)^{1/2}dx$$

$$= \sqrt{2} - \frac{2}{3}(x^2 + 1)^{3/2}\,\Big]_0^1 = \sqrt{2} - \frac{2}{3}\left[2\sqrt{2} - 1\right] = (2 - \sqrt{2})/3$$

43.   (a) $A = \int_1^e \ln x \, dx = (x \ln x - x) \Big]_1^e = 1$

(b) $V = \pi \int_1^e (\ln x)^2 dx = \pi(x(\ln x)^2 - 2x \ln x + 2x) \Big]_1^e = \pi(e - 2)$

45. $V = 2\pi \int_0^\pi x \sin x \, dx = 2\pi(-x \cos x + \sin x) \Big]_0^\pi = 2\pi^2$

47.   (a) $\displaystyle\int \sin^3 x \, dx = -\frac{1}{3} \sin^2 x \cos x + \frac{2}{3} \int \sin x \, dx$

$\qquad\qquad = -\frac{1}{3} \sin^2 x \cos x - \frac{2}{3} \cos x + C$

(b) $\displaystyle\int \sin^4 x \, dx = -\frac{1}{4} \sin^3 x \cos x + \frac{3}{4} \int \sin^2 x \, dx$,

$\displaystyle\int \sin^2 x \, dx = -\frac{1}{2} \sin x \cos x + \frac{1}{2} x + C_1$ so

$\displaystyle\int_0^{\pi/4} \sin^4 x \, dx = -\frac{1}{4} \sin^3 x \cos x - \frac{3}{8} \sin x \cos x + \frac{3}{8} x \Big]_0^{\pi/4}$

$\qquad\qquad = -\frac{1}{4} (1/\sqrt{2})^3 (1/\sqrt{2}) - \frac{3}{8} (1/\sqrt{2})(1/\sqrt{2}) + 3\pi/32$

$\qquad\qquad = 3\pi/32 - 1/4$

49.   (a) $u = 5x$,

$\displaystyle\int \cos^3 5x \, dx = \frac{1}{5} \int \cos^3 u \, du$

$\qquad\qquad = \frac{1}{5} \left[\frac{1}{3} \cos^2 u \sin u + \frac{2}{3} \int \cos u \, du \right]$

$\qquad\qquad = \frac{1}{15} \cos^2 u \sin u + \frac{2}{15} \sin u + C$

$\qquad\qquad = \frac{1}{15} \cos^2 5x \sin 5x + \frac{2}{15} \sin 5x + C$

(b) $u = x^2$,

$\displaystyle\int x \cos^4(x^2) dx = \frac{1}{2} \int \cos^4 u \, du$

$\qquad\qquad = \frac{1}{2} \left[\frac{1}{4} \cos^3 u \sin u + \frac{3}{4} \int \cos^2 u \, du \right]$

$\qquad\qquad = \frac{1}{8} \cos^3 u \sin u + \frac{3}{8} \left[\frac{1}{2} \cos u \sin u + \frac{1}{2} \int du \right]$

$$= \frac{1}{8} \cos^3 u \sin u + \frac{3}{16} \cos u \sin u + \frac{3}{16} u + C$$

$$= \frac{1}{8} \cos^3(x^2)\sin(x^2) + \frac{3}{16} \cos(x^2)\sin(x^2) + \frac{3}{16} x^2 + C$$

51. $u = \sin^{n-1}x,\ dv = \sin x\,dx,\ du = (n-1)\sin^{n-2}x \cos x\,dx,\ v = -\cos x$

$$\int \sin^n x\,dx = -\sin^{n-1}x \cos x + (n-1)\int \sin^{n-2}x \cos^2 x\,dx$$

$$= -\sin^{n-1}x \cos x + (n-1)\int \sin^{n-2}x(1 - \sin^2 x)dx$$

$$= -\sin^{n-1}x \cos x + (n-1)\int \sin^{n-2}x\,dx - (n-1)\int \sin^n x\,dx$$

$$n\int \sin^n x\,dx = -\sin^{n-1}x \cos x + (n-1)\int \sin^{n-2}x\,dx$$

$$\int \sin^n x\,dx = -\frac{1}{n}\sin^{n-1}x \cos x + \frac{n-1}{n}\int \sin^{n-2}x\,dx$$

53. (a) $u = x^n,\ dv = e^x dx,\ du = nx^{n-1}dx,\ v = e^x$

$$\int x^n e^x dx = x^n e^x - n\int x^{n-1}e^x dx$$

(b) $\int x^3 e^x dx = x^3 e^x - 3\int x^2 e^x dx = x^3 e^x - 3\left[x^2 e^x - 2\int xe^x dx\right]$

$$= x^3 e^x - 3x^2 e^x + 6\left[xe^x - \int e^x dx\right]$$

$$= x^3 e^x - 3x^2 e^x + 6xe^x - 6e^x + C$$

55. $u = x,\ dv = f''(x)dx,\ du = dx,\ v = f'(x)$

$$\int_{-1}^{1} x f''(x)dx = x f'(x)\Big]_{-1}^{1} - \int_{-1}^{1} f'(x)dx$$

$$= f'(1) - f'(-1) - f(x)\Big]_{-1}^{1} = f'(1) - f'(-1) + f(-1) - f(1)$$

57. $du = -(1/x^2)dx,\ v = x$

$$\int \frac{1}{x}\,dx = 1 + \int \frac{1}{x}\,dx \text{ so } 0 = 1\ ???$$

The "obvious" cancellation of the indefinite integrals causes the problem. Instead, proceed as follows:

$$\int \frac{1}{x}\,dx - \int \frac{1}{x}\,dx = 1,\ \int(0)dx = 1,\ \text{but} \int(0)dx = C \text{ so } C = 1.$$

EXERCISE SET 9.3

1. $u = \cos x, \quad -\int u^5 du = -\frac{1}{6} \cos^6 x + C$

3. $u = \sin ax, \quad \frac{1}{a} \int u\, du = \frac{1}{2a} \sin^2 ax + C$

5. $\int \sin^2 5\theta\, d\theta = \frac{1}{2} \int (1 - \cos 10\theta) d\theta = \frac{1}{2}\theta - \frac{1}{20} \sin 10\theta + C$

7. $\displaystyle \int \cos^4(x/4) dx = \frac{1}{4} \int [1 + \cos(x/2)]^2 dx$

    $\displaystyle \qquad\qquad = \frac{1}{4} \int [1 + 2\cos(x/2) + \cos^2(x/2)] dx$

    $\displaystyle \qquad\qquad = \frac{1}{4} \int [1 + 2\cos(x/2) + \frac{1}{2}(1 + \cos x)] dx$

    $\displaystyle \qquad\qquad = \frac{1}{4} \int [\frac{3}{2} + 2\cos(x/2) + \frac{1}{2}\cos x] dx$

    $\displaystyle \qquad\qquad = \frac{3}{8} x + \sin(x/2) + \frac{1}{8}\sin x + C$

9. $\displaystyle \int \cos^5 \theta\, d\theta = \int (1 - \sin^2\theta)^2 \cos\theta\, d\theta$

    $\displaystyle \qquad\qquad = \int (1 - 2\sin^2\theta + \sin^4\theta)\cos\theta\, d\theta$

    $\displaystyle \qquad\qquad = \sin\theta - \frac{2}{3}\sin^3\theta + \frac{1}{5}\sin^5\theta + C$

11. $\displaystyle \int \sin^2 2t \cos^3 2t\, dt = \int \sin^2 2t(1 - \sin^2 2t)\cos 2t\, dt$

    $\displaystyle \qquad\qquad = \int (\sin^2 2t - \sin^4 2t)\cos 2t\, dt$

    $\displaystyle \qquad\qquad = \frac{1}{6}\sin^3 2t - \frac{1}{10}\sin^5 2t + C$

13. $\displaystyle \int \cos^4 x \sin^3 x\, dx = \int \cos^4 x(1 - \cos^2 x)\sin x\, dx$

    $\displaystyle \qquad = \int (\cos^4 x - \cos^6 x)\sin x\, dx = -\frac{1}{5}\cos^5 x + \frac{1}{7}\cos^7 x + C$

15. $\displaystyle\int \sin^5\theta \cos^4\theta\, d\theta = \int (1 - \cos^2\theta)^2\cos^4\theta \sin\theta\, d\theta$

$\displaystyle = \int (\cos^4\theta - 2\cos^6\theta + \cos^8\theta)\sin\theta\, d\theta$

$\displaystyle = -\frac{1}{5}\cos^5\theta + \frac{2}{7}\cos^7\theta - \frac{1}{9}\cos^9\theta + C$

17. $\displaystyle\int \sin^2 x \cos^2 x\, dx = \frac{1}{4}\int (1 - \cos 2x)(1 + \cos 2x)dx$

$\displaystyle = \frac{1}{4}\int (1 - \cos^2 2x)dx = \frac{1}{4}\int \sin^2 2x\, dx$

$\displaystyle = \frac{1}{8}\int (1 - \cos 4x)dx = \frac{1}{8}x - \frac{1}{32}\sin 4x + C$

19. $\displaystyle\int \sin x \cos 2x\, dx = \frac{1}{2}\int (\sin 3x - \sin x)dx = -\frac{1}{6}\cos 3x + \frac{1}{2}\cos x + C$

21. $\displaystyle\int \sin x \cos(x/2)dx = \frac{1}{2}\int [\sin(3x/2) + \sin(x/2)]dx$

$\displaystyle = -\frac{1}{3}\cos(3x/2) - \cos(x/2) + C$

23. $\displaystyle u = \cos x, \quad -\int u^{-8}du = 1/(7\cos^7 x) + C$

25. $\displaystyle\int_0^{\pi/4} \cos^3 x\, dx = \int_0^{\pi/4} (1 - \sin^2 x)\cos x\, dx$

$\displaystyle = \sin x - \frac{1}{3}\sin^3 x\Big]_0^{\pi/4} = (\sqrt{2}/2) - \frac{1}{3}(\sqrt{2}/2)^3 = 5\sqrt{2}/12$

27. $\displaystyle\int_0^{\pi/3} \sin^4 3x \cos^3 3x\, dx = \int_0^{\pi/3} \sin^4 3x(1 - \sin^2 3x)\cos 3x\, dx$

$\displaystyle = \frac{1}{15}\sin^5 3x - \frac{1}{21}\sin^7 3x\Big]_0^{\pi/3} = 0$

29. $\displaystyle\int_0^{\pi/6} \sin 2x \cos 4x\, dx = \frac{1}{2}\int_0^{\pi/6} (\sin 6x - \sin 2x)dx$

$\displaystyle = -\frac{1}{12}\cos 6x + \frac{1}{4}\cos 2x\Big]_0^{\pi/6}$

$\displaystyle = [(-1/12)(-1) + (1/4)(1/2)] - [-1/12 + 1/4] = 1/24$

31. (a) $\displaystyle\int_0^{2\pi} \sin mx \cos nx\, dx = \frac{1}{2} \int_0^{2\pi} [\sin(m + n)x + \sin(m - n)x]dx$

$$= -\left.\frac{\cos(m + n)x}{2(m + n)} - \frac{\cos(m - n)x}{2(m - n)}\right]_0^{2\pi},$$

m + n and m − n are integers so $\cos[(m + n)2\pi] = \cos[(m - n)2\pi] = 1$ thus

$$\int_0^{2\pi} \sin mx \cos nx\, dx$$

$$= \left[\left(-\frac{1}{2(m + n)} - \frac{1}{2(m - n)}\right) - \left(\frac{1}{2(m + n)} - \frac{1}{2(m - n)}\right)\right] = 0$$

(b) $\displaystyle\int_0^{2\pi} \cos mx \cos nx\, dx = \frac{1}{2} \int_0^{2\pi} [\cos(m + n)x + \cos(m - n)x]dx$

$$= \left.\frac{\sin(m + n)x}{2(m + n)} + \frac{\sin(m - n)x}{2(m - n)}\right]_0^{2\pi} = 0$$

(c) $\displaystyle\int_0^{2\pi} \sin mx \sin nx\, dx = \frac{1}{2} \int_0^{2\pi} [\cos(m - n)x - \cos(m + n)x]dx$

$$= \left.\frac{\sin(m - n)x}{2(m - n)} + \frac{\sin(m + n)x}{2(m + n)}\right]_0^{2\pi} = 0$$

33. $V = \pi \displaystyle\int_0^{\pi/4} (\cos^2 x - \sin^2 x)dx = \pi \int_0^{\pi/4} \cos 2x\, dx = \left.\frac{1}{2}\pi \sin 2x\right]_0^{\pi/4} = \pi/2$

35. (a) $\displaystyle\int_0^{\pi/2} \sin^3 x\, dx = \frac{2}{3}$

(b) $\displaystyle\int_0^{\pi/2} \sin^4 x\, dx = \frac{1 \cdot 3}{2 \cdot 4} \cdot \frac{\pi}{2} = 3\pi/16$

(c) $\displaystyle\int_0^{\pi/2} \sin^5 x\, dx = \frac{2 \cdot 4}{3 \cdot 5} = 8/15$

(d) $\displaystyle\int_0^{\pi/2} \sin^6 x\, dx = \frac{1 \cdot 3 \cdot 5}{2 \cdot 4 \cdot 6} \cdot \frac{\pi}{2} = 5\pi/32$

37. For (6), $\int \sin^3 x\,dx = -\frac{1}{3}\sin^2 x \cos x + \frac{2}{3}\int \sin x\,dx$

$= -\frac{1}{3}\sin^2 x \cos x - \frac{2}{3}\cos x + C,$

but $\sin^2 x = 1 - \cos^2 x$ so

$\int \sin^3 x\,dx = -\frac{1}{3}(1 - \cos^2 x)\cos x - \frac{2}{3}\cos x + C = -\cos x + \frac{1}{3}\cos^3 x + C$

The derivation of (7) is similar.

EXERCISE SET 9.4

1. $\frac{1}{3}\tan(3x + 1) + C$

3. $\frac{1}{2}\ln|\cos(e^{-2x})| + C$

5. $\frac{1}{2}\ln|\sec 2x + \tan 2x| + C$

7. $u = \tan x,\ \int u^2\,du = \frac{1}{3}\tan^3 x + C$

9. $\int \tan^3 4x(1 + \tan^2 4x)\sec^2 4x\,dx = \int (\tan^3 4x + \tan^5 4x)\sec^2 4x\,dx$

$= \frac{1}{16}\tan^4 4x + \frac{1}{24}\tan^6 4x + C$

11. $\int \sec^4 x(\sec^2 x - 1)\sec x \tan x\,dx = \int (\sec^6 x - \sec^4 x)\sec x \tan x\,dx$

$= \frac{1}{7}\sec^7 x - \frac{1}{5}\sec^5 x + C$

13. $\int (\sec^2 x - 1)^2 \sec x\,dx = \int (\sec^5 x - 2\sec^3 x + \sec x)dx$

$= \int \sec^5 x\,dx - 2\int \sec^3 x\,dx + \int \sec x\,dx$

$= \frac{1}{4}\sec^3 x \tan x + \frac{3}{4}\int \sec^3 x\,dx - 2\int \sec^3 x\,dx + \ln|\sec x + \tan x|$

$= \frac{1}{4}\sec^3 x \tan x - \frac{5}{4}\left[\frac{1}{2}\sec x \tan x + \frac{1}{2}\ln|\sec x + \tan x|\right]$

$+ \ln|\sec x + \tan x| + C$

$= \frac{1}{4}\sec^3 x \tan x - \frac{5}{8}\sec x \tan x + \frac{3}{8}\ln|\sec x + \tan x| + C$

15. $\int \sec^2 2t(\sec 2t \tan 2t)dt = \frac{1}{6}\sec^3 2t + C$

17. $\int \sec^4 x \, dx = \int (1 + \tan^2 x)\sec^2 x \, dx$

$= \int (\sec^2 x + \tan^2 x \sec^2 x) dx = \tan x + \frac{1}{3} \tan^3 x + C$

19. $u = \pi x$, use reduction formula (3) to get

$\frac{1}{\pi} \int \sec^6 u \, du = \frac{1}{\pi} \left[\frac{1}{5} \sec^4 u \tan u + \frac{4}{5} \int \sec^4 u \, du \right]$

$= \frac{1}{5\pi} \sec^4 u \tan u + \frac{4}{5\pi} \left[\frac{1}{3} \sec^2 u \tan u + \frac{2}{3} \tan u \right] + C$

$= \frac{1}{5\pi} \sec^4 \pi x \tan \pi x + \frac{4}{15\pi} \sec^2 \pi x \tan \pi x + \frac{8}{15\pi} \tan \pi x + C$

21. Use reduction formula (4) to get $\int \tan^4 x \, dx = \frac{1}{3} \tan^3 x - \tan x + x + C$

23. $u = \tan(x^2)$, $\frac{1}{2} \int u^2 du = \frac{1}{6} \tan^3(x^2) + C$

25. $\int (\csc^2 x - 1)\csc^2 x(\csc x \cot x) dx = \int (\csc^4 x - \csc^2 x)(\csc x \cot x) dx$

$= -\frac{1}{5} \csc^5 x + \frac{1}{3} \csc^3 x + C$

27. $\int (\csc^2 x - 1)\cot x \, dx = \int \csc x(\csc x \cot x) dt - \int \frac{\cos x}{\sin x} dx$

$= -\frac{1}{2} \csc^2 x - \ln|\sin x| + C$

29. $\int \tan^{1/2} x(1 + \tan^2 x)\sec^2 x \, dx = \frac{2}{3} \tan^{3/2} x + \frac{2}{7} \tan^{7/2} x + C$

31. $\int_0^{\pi/6} (\sec^2 2x - 1) dx = \frac{1}{2} \tan 2x - x \Big]_0^{\pi/6} = \sqrt{3}/2 - \pi/6$

33. $u = x/2$,

$2 \int_0^{\pi/4} \tan^5 u \, du = \frac{1}{2} \tan^4 u - \tan^2 u - 2 \ln|\cos u| \Big]_0^{\pi/4}$

$= 1/2 - 1 - 2 \ln(1/\sqrt{2}) = -1/2 + \ln 2$

35. $y' = \tan x$, $1 + (y')^2 = 1 + \tan^2 x = \sec^2 x$,

$L = \int_0^{\pi/4} \sqrt{\sec^2 x}\, dx = \int_0^{\pi/4} \sec x\, dx = \ln|\sec x + \tan x|\Big]_0^{\pi/4} = \ln(\sqrt{2} + 1)$

37. (a) $\int \csc x\, dx = \int \sec(\pi/2 - x)dx$

$= -\ln|\sec(\pi/2 - x) + \tan(\pi/2 - x)| + C$

$= -\ln|\csc x + \cot x| + C$

(b) $-\ln|\csc x + \cot x| = \ln \dfrac{1}{|\csc x + \cot x|}$

$= \ln \dfrac{|\csc x - \cot x|}{|\csc^2 x - \cot^2 x|} = \ln|\csc x - \cot x|.$

$-\ln|\csc x + \cot x| = -\ln\left|\dfrac{1}{\sin x} + \dfrac{\cos x}{\sin x}\right| = \ln\left|\dfrac{\sin x}{1 + \cos x}\right|$

$= \ln\left|\dfrac{2\sin(x/2)\cos(x/2)}{2\cos^2(x/2)}\right| = \ln|\tan(x/2)|$

39. $a\sin x + b\cos x = \sqrt{a^2 + b^2}\left[\dfrac{a}{\sqrt{a^2 + b^2}}\sin x + \dfrac{b}{\sqrt{a^2 + b^2}}\cos x\right]$

$= \sqrt{a^2 + b^2}(\sin x \cos\theta + \cos x \sin\theta)$

where $\cos\theta = a/\sqrt{a^2 + b^2}$ and $\sin\theta = b/\sqrt{a^2 + b^2}$ so

$a\sin x + b\cos x = \sqrt{a^2 + b^2}\sin(x + \theta)$ and

$\int \dfrac{dx}{a\sin x + b\cos x} = \dfrac{1}{\sqrt{a^2 + b^2}}\int \csc(x + \theta)dx$

$= -\dfrac{1}{\sqrt{a^2 + b^2}}\ln|\csc(x + \theta) + \cot(x + \theta)| + C$

EXERCISE SET 9.5

1. $x = 2 \sin \theta$, $dx = 2 \cos \theta \, d\theta$,

$$4 \int \cos^2 \theta \, d\theta = 2 \int (1 + \cos 2\theta) d\theta = 2\theta + \sin 2\theta + C$$

$$= 2\theta + 2 \sin \theta \cos \theta + C = 2 \sin^{-1}(x/2) + \frac{1}{2} x \sqrt{4 - x^2} + C$$

3. $x = 3 \sin \theta$, $dx = 3 \cos \theta \, d\theta$,

$$9 \int \sin^2 \theta \, d\theta = \frac{9}{2} \int (1 - \cos 2\theta) d\theta$$

$$= \frac{9}{2} \theta - \frac{9}{4} \sin 2\theta + C = \frac{9}{2} \theta - \frac{9}{2} \sin \theta \cos \theta + C$$

$$= \frac{9}{2} \sin^{-1}(x/3) - \frac{1}{2} x \sqrt{9 - x^2} + C$$

5. $x = 2 \tan \theta$, $dx = 2 \sec^2 \theta \, d\theta$,

$$\frac{1}{8} \int \frac{1}{\sec^2 \theta} \, d\theta = \frac{1}{8} \int \cos^2 \theta \, d\theta = \frac{1}{16} \int (1 + \cos 2\theta) d\theta = \frac{1}{16} \theta + \frac{1}{32} \sin 2\theta + C$$

$$= \frac{1}{16} \theta + \frac{1}{16} \sin \theta \cos \theta + C$$

$$= \frac{1}{16} \tan^{-1} \frac{x}{2} + \frac{x}{8(4 + x^2)} + C$$

7. $x = 3 \sec \theta$, $dx = 3 \sec \theta \tan \theta \, d\theta$,

$$3 \int \tan^2 \theta \, d\theta = 3 \int (\sec^2 \theta - 1) d\theta = 3 \tan \theta - 3\theta + C$$

$$= \sqrt{x^2 - 9} - 3 \sec^{-1} \frac{x}{3} + C$$

9. $x = \sqrt{2} \sin \theta$, $dx = \sqrt{2} \cos \theta \, d\theta$,

$$2\sqrt{2} \int \sin^3 \theta \, d\theta = 2\sqrt{2}(-\cos \theta + \frac{1}{3} \cos^3 \theta) + C = -2\sqrt{2 - x^2} + \frac{1}{3} (2 - x^2)^{3/2} + C$$

11. $x = \sqrt{3} \tan \theta$, $dx = \sqrt{3} \sec^2 \theta \, d\theta$,

$$\frac{1}{3} \int \frac{1}{\sec \theta} \, d\theta = \frac{1}{3} \int \cos \theta \, d\theta = \frac{1}{3} \sin \theta + C = \frac{x}{3\sqrt{3 + x^2}} + C$$

13. $x = \dfrac{3}{2} \sec \theta$, $dx = \dfrac{3}{2} \sec \theta \tan \theta \, d\theta$,

$$\dfrac{2}{9} \int \dfrac{1}{\sec \theta} \, d\theta = \dfrac{2}{9} \int \cos \theta \, d\theta = \dfrac{2}{9} \sin \theta + C = \dfrac{\sqrt{4x^2 - 9}}{9x} + C$$

15. $x = \sin \theta$, $dx = \cos \theta \, d\theta$,

$$\int \dfrac{1}{\cos^2 \theta} \, d\theta = \int \sec^2 \theta \, d\theta = \tan \theta + C = x/\sqrt{1 - x^2} + C$$

17. $\displaystyle\int \dfrac{x^2}{1 + x^2} \, dx = \int \dfrac{(x^2 + 1) - 1}{1 + x^2} \, dx = \int (1 - \dfrac{1}{1 + x^2}) dx = x - \tan^{-1}x + C$

19. $x = \dfrac{3}{2} \sin \theta$, $dx = \dfrac{3}{2} \cos \theta \, d\theta$,

$$\dfrac{2}{9} \int \dfrac{1}{\sin^2 \theta} \, d\theta = \dfrac{2}{9} \csc^2 \theta \, d\theta = -\dfrac{2}{9} \cot \theta + C = -\dfrac{\sqrt{9 - 4x^2}}{9x} + C$$

21. $x = \dfrac{1}{3} \sec \theta$, $dx = \dfrac{1}{3} \sec \theta \tan \theta \, d\theta$,

$$\dfrac{1}{3} \int \dfrac{\sec \theta}{\tan^2 \theta} \, d\theta = \dfrac{1}{3} \int \csc \theta \cot \theta \, d\theta = -\dfrac{1}{3} \csc \theta + C = -x/\sqrt{9x^2 - 1} + C$$

23. $e^x = \sin \theta$, $e^x dx = \cos \theta \, d\theta$,

$$\int \cos^2 \theta \, d\theta = \dfrac{1}{2} \int (1 + \cos 2\theta) d\theta = \dfrac{1}{2} \theta + \dfrac{1}{4} \sin 2\theta + C$$

$$= \dfrac{1}{2} \sin^{-1}(e^x) + \dfrac{1}{2} e^x \sqrt{1 - e^{2x}} + C$$

25. $x = 4 \sin \theta$, $dx = 4 \cos \theta \, d\theta$,

$$1024 \int_0^{\pi/2} \sin^3 \theta \cos^2 \theta \, d\theta = 1024 \left[-\dfrac{1}{3} \cos^3 \theta + \dfrac{1}{5} \cos^5 \theta \right]_0^{\pi/2}$$

$$= 1024(1/3 - 1/5) = 2048/15$$

27. $x = \sec \theta$, $dx = \sec \theta \tan \theta \, d\theta$,

$$\int_{\pi/4}^{\pi/3} \dfrac{1}{\sec \theta} \, d\theta = \int_{\pi/4}^{\pi/3} \cos \theta \, d\theta = \sin \theta \Big]_{\pi/4}^{\pi/3} = (\sqrt{3} - \sqrt{2})/2$$

29. $x = \sqrt{3} \tan \theta$, $dx = \sqrt{3} \sec^2 \theta \, d\theta$,

$$\frac{1}{9} \int_{\pi/6}^{\pi/3} \frac{\sec \theta}{\tan^4 \theta} \, d\theta = \frac{1}{9} \int_{\pi/6}^{\pi/3} \frac{\cos^3 \theta}{\sin^4 \theta} \, d\theta = \frac{1}{9} \int_{\pi/6}^{\pi/3} \frac{1 - \sin^2 \theta}{\sin^4 \theta} \cos \theta \, d\theta$$

$$= \frac{1}{9} \int_{1/2}^{\sqrt{3}/2} \frac{1 - u^2}{u^4} \, du, \quad u = \sin \theta$$

$$= \frac{1}{9} \int_{1/2}^{\sqrt{3}/2} (u^{-4} - u^{-2}) \, du$$

$$= \frac{1}{9} \left[-\frac{1}{3u^3} + \frac{1}{u} \right]_{1/2}^{\sqrt{3}/2} = \frac{10\sqrt{3} + 18}{243}$$

31. $u = x^2 + 4$, $du = 2x \, dx$, $\frac{1}{2} \int \frac{1}{u} \, du = \frac{1}{2} \ln|u| + C = \frac{1}{2} \ln(x^2 + 4) + C$.

$x = 2 \tan \theta$, $dx = 2 \sec^2 \theta \, d\theta$,

$$\int \tan \theta \, d\theta = \ln|\sec \theta| + C_1 = \ln \frac{\sqrt{x^2 + 4}}{2} + C_1$$

$$= \ln(x^2 + 4)^{1/2} - \ln 2 + C_1$$

$$= \frac{1}{2} \ln(x^2 + 4) + C \text{ with } C = C_1 - \ln 2$$

33. $y' = \frac{1}{x}$, $1 + (y')^2 = 1 + \frac{1}{x^2} = \frac{x^2 + 1}{x^2}$,

$$L = \int_1^2 \sqrt{\frac{x^2 + 1}{x^2}} \, dx = \int_1^2 \frac{\sqrt{x^2 + 1}}{x} \, dx; \quad x = \tan \theta, \quad dx = \sec^2 \theta \, d\theta,$$

$$L = \int_{\pi/4}^{\tan^{-1} 2} \frac{\sec^3 \theta}{\tan \theta} \, d\theta = \int_{\pi/4}^{\tan^{-1} 2} \frac{\tan^2 \theta + 1}{\tan \theta} \sec \theta \, d\theta$$

$$= \int_{\pi/4}^{\tan^{-1} 2} \left[\sec \theta \tan \theta + \frac{\sec \theta}{\tan \theta} \right] d\theta$$

$$= \int_{\pi/4}^{\tan^{-1} 2} (\sec \theta \tan \theta + \csc \theta) \, d\theta$$

$$= \sec \theta - \ln|\csc \theta + \cot \theta| \Big]_{\pi/4}^{\tan^{-1} 2}$$

$$= \left[\sqrt{5} - \ln\left|\frac{\sqrt{5}}{2} + \frac{1}{2}\right|\right] - \left[\sqrt{2} - \ln\left|\sqrt{2} + 1\right|\right]$$

$$= \sqrt{5} - \sqrt{2} + \ln\frac{2 + 2\sqrt{2}}{1 + \sqrt{5}}$$

35. $y' = 2x$, $1 + (y')^2 = 1 + 4x^2$,

$$S = 2\pi\int_0^1 x^2\sqrt{1 + 4x^2}\,dx; \quad x = \frac{1}{2}\tan\theta, \quad dx = \frac{1}{2}\sec^2\theta\,d\theta,$$

$$S = \frac{\pi}{4}\int_0^{\tan^{-1}2} \tan^2\theta\sec^3\theta\,d\theta = \frac{\pi}{4}\int_0^{\tan^{-1}2}(\sec^2\theta - 1)\sec^3\theta\,d\theta$$

$$= \frac{\pi}{4}\int_0^{\tan^{-1}2}(\sec^5\theta - \sec^3\theta)d\theta,$$

$$= \frac{\pi}{4}\left[\frac{1}{4}\sec^3\theta\tan\theta - \frac{1}{8}\sec\theta\tan\theta - \frac{1}{8}\ln\left|\sec\theta + \tan\theta\right|\right]_0^{\tan^{-1}2}$$

$$= \frac{1}{32}\pi[18\sqrt{5} - \ln(2 + \sqrt{5})]$$

37. (a) $x = 3\sinh u$, $dx = 3\cosh u\,du$,

$$\int du = u + C = \sinh^{-1}(x/3) + C$$

(b) $x = 3\tan\theta$, $dx = 3\sec^2\theta\,d\theta$,

$$\int \sec\theta\,d\theta = \ln\left|\sec\theta + \tan\theta\right| + C = \ln(\sqrt{x^2 + 9}/3 + x/3) + C$$

but $\sinh^{-1}(x/3) = \ln(x/3 + \sqrt{x^2/9 + 1}) = \ln(x/3 + \sqrt{x^2 + 9}/3)$
so the results in (a) and (b) agree.

39. If $-\pi/2 \le \theta \le \pi/2$ then

(a) $\cos\theta = \sqrt{1 - \sin^2\theta} = \sqrt{1 - x^2/a^2} = \sqrt{a^2 - x^2}/a$

(b) $\tan\theta = \dfrac{\sin\theta}{\cos\theta} = \dfrac{x/a}{\sqrt{a^2 - x^2}/a} = x/\sqrt{a^2 - x^2}$

EXERCISE SET 9.6

1. $\displaystyle\int \frac{1}{(x-2)^2+9}\, dx = \frac{1}{3}\,\tan^{-1}(\frac{x-2}{3}) + C$

3. $\displaystyle\int \frac{1}{\sqrt{9-(x-1)^2}}\, dx = \sin^{-1}(\frac{x-1}{3}) + C$

5. $\displaystyle\int \frac{1}{\sqrt{(x-3)^2+1}}\, dx = \sinh^{-1}(x-3) + C.$

   $-x^2 + 4x - 1$

   Alternate solution: let $x-3 = \tan\theta,$

   $\displaystyle\int \sec\theta\, d\theta = \ln|\sec\theta + \tan\theta| + C = \ln(\sqrt{x^2-6x+10} + x - 3) + C.$

7. $\displaystyle\int \sqrt{4-(x+1)^2}\, dx,$ let $x+1 = 2\sin\theta,$

   $4\displaystyle\int \cos^2\theta\, d\theta = 2\theta + \sin 2\theta + C = 2\theta + 2\sin\theta\cos\theta + C$

   $= 2\sin^{-1}(\frac{x+1}{2}) + \frac{1}{2}(x+1)\sqrt{3-2x-x^2} + C$

9. $\displaystyle\int \frac{1}{2(x+1)^2+5}\, dx = \frac{1}{2}\int \frac{1}{(x+1)^2+5/2}\, dx = \frac{1}{\sqrt{10}}\,\tan^{-1}\sqrt{2/5}(x+1) + C$

11. $\displaystyle\int \frac{2x+5}{(x+1)^2+4}\, dx,$ let $u = x+1,$

   $\displaystyle\int \frac{2u+3}{u^2+4}\, du = \int\left[\frac{2u}{u^2+4} + \frac{3}{u^2+4}\right] du = \ln(u^2+4) + \frac{3}{2}\,\tan^{-1}(u/2) + C$

   $= \ln(x^2+2x+5) + \frac{3}{2}\,\tan^{-1}(\frac{x+1}{2}) + C$

13. $\displaystyle\int \frac{x+3}{\sqrt{(x+1)^2+1}}\, dx,$ let $u = x+1,$

   $\displaystyle\int \frac{u+2}{\sqrt{u^2+1}}\, du = \int\left[u(u^2+1)^{-1/2} + \frac{2}{\sqrt{u^2+1}}\right] du = \sqrt{u^2+1} + 2\sinh^{-1}u + C$

   $= \sqrt{x^2+2x+2} + 2\sinh^{-1}(x+1) + C$

Alternate solution: let $x + 1 = \tan\theta$,

$$\int(\tan\theta + 2)\sec\theta\,d\theta = \int \sec\theta\tan\theta\,d\theta + 2\int \sec\theta\,d\theta$$

$$= \sec\theta + 2\ln|\sec\theta + \tan\theta| + C$$

$$= \sqrt{x^2 + 2x + 2} + 2\ln(\sqrt{x^2 + 2x + 2} + x + 1) + C.$$

15. $\displaystyle\int_0^1 \sqrt{4x - x^2}\,dx = \int_0^1 \sqrt{4 - (x - 2)^2}\,dx$, let $x - 2 = 2\sin\theta$,

$$4\int_{-\pi/2}^{-\pi/6} \cos^2\theta\,d\theta = 2\theta + \sin 2\theta\, \Big]_{-\pi/2}^{-\pi/6} = \frac{2\pi}{3} - \frac{\sqrt{3}}{2}$$

EXERCISE SET 9.7

1. $\displaystyle\frac{1}{(x + 4)(x - 1)} = \frac{A}{x + 4} + \frac{B}{x - 1}$; $A = -\frac{1}{5}$, $B = \frac{1}{5}$

$$-\frac{1}{5}\int \frac{1}{x + 4}\,dx + \frac{1}{5}\int \frac{1}{x - 1}\,dx = -\frac{1}{5}\ln|x + 4| + \frac{1}{5}\ln|x - 1| + C$$

$$= \frac{1}{5}\ln\left|\frac{x - 1}{x + 4}\right| + C$$

3. $\displaystyle\frac{x}{(x - 2)(x - 3)} = \frac{A}{x - 2} + \frac{B}{x - 3}$; $A = -2$, $B = 3$

$$-2\int \frac{1}{x - 2}\,dx + 3\int \frac{1}{x - 3}\,dx = -2\ln|x - 2| + 3\ln|x - 3| + C$$

5. $\displaystyle\frac{11x + 17}{(2x - 1)(x + 4)} = \frac{A}{2x - 1} + \frac{B}{x + 4}$; $A = 5$, $B = 3$

$$5\int \frac{1}{2x - 1}\,dx + 3\int \frac{1}{x + 4}\,dx = \frac{5}{2}\ln|2x - 1| + 3\ln|x + 4| + C$$

7. $\displaystyle\frac{1}{(x - 1)(x + 2)(x - 3)} = \frac{A}{x - 1} + \frac{B}{x + 2} + \frac{C}{x - 3}$; $A = -\frac{1}{6}$, $B = \frac{1}{15}$, $C = \frac{1}{10}$

$$-\frac{1}{6}\int \frac{1}{x - 1}\,dx + \frac{1}{15}\int \frac{1}{x + 2}\,dx + \frac{1}{10}\int \frac{1}{x - 3}\,dx$$

$$= -\frac{1}{6}\ln|x - 1| + \frac{1}{15}\ln|x + 2| + \frac{1}{10}\ln|x - 3| + C$$

9. $\dfrac{2x^2 - 9x - 9}{x(x + 3)(x - 3)} = \dfrac{A}{x} + \dfrac{B}{x + 3} + \dfrac{C}{x - 3}$; $A = 1$, $B = 2$, $C = -1$

$\displaystyle\int \frac{1}{x}\, dx + 2\int \frac{1}{x + 3}\, dx - \int \frac{1}{x - 3}\, dx = \ln|x| + 2\ln|x + 3| - \ln|x - 3| + C$

$$= \ln\left|\frac{x(x + 3)^2}{x - 3}\right| + C$$

11. $\dfrac{x^2 + 2}{x + 2} = x - 2 + \dfrac{6}{x + 2}$, $\displaystyle\int \left(x - 2 + \frac{6}{x + 2}\right) dx = \frac{1}{2}x^2 - 2x + 6\ln|x + 2| + C$

13. $\dfrac{3x^2 - 10}{x^2 - 4x + 4} = 3 + \dfrac{12x - 22}{x^2 - 4x + 4}$,

$\dfrac{12x - 22}{(x - 2)^2} = \dfrac{A}{x - 2} + \dfrac{B}{(x - 2)^2}$; $A = 12$, $B = 2$

$\displaystyle\int 3dx + 12\int \frac{1}{x - 2}\, dx + 2\int \frac{1}{(x - 2)^2}\, dx = 3x + 12\ln|x - 2| - 2/(x - 2) + C$

15. $\dfrac{x^3}{x^2 - 3x + 2} = x + 3 + \dfrac{7x - 6}{x^2 - 3x + 2}$,

$\dfrac{7x - 6}{(x - 1)(x - 2)} = \dfrac{A}{x - 1} + \dfrac{B}{x - 2}$; $A = -1$, $B = 8$

$\displaystyle\int (x + 3)dx - \int \frac{1}{x - 1}\, dx + 8\int \frac{1}{x - 2}\, dx$

$$= \frac{1}{2}x^2 + 3x - \ln|x - 1| + 8\ln|x - 2| + C$$

17. $\dfrac{x^5 + 2x^2 + 1}{x^3 - x} = x^2 + 1 + \dfrac{2x^2 + x + 1}{x^3 - x}$,

$\dfrac{2x^2 + x + 1}{x(x + 1)(x - 1)} = \dfrac{A}{x} + \dfrac{B}{x + 1} + \dfrac{C}{x - 1}$; $A = -1$, $B = 1$, $C = 2$

$\displaystyle\int (x^2 + 1)dx - \int \frac{1}{x}\, dx + \int \frac{1}{x + 1}\, dx + 2\int \frac{1}{x - 1}\, dx$

$$= \frac{1}{3}x^3 + x - \ln|x| + \ln|x + 1| + 2\ln|x - 1| + C$$

$$= \frac{1}{3}x^3 + x + \ln\left|\frac{(x + 1)(x - 1)^2}{x}\right| + C$$

19. $\dfrac{2x^2 + 3}{x(x - 1)^2} = \dfrac{A}{x} + \dfrac{B}{x - 1} + \dfrac{C}{(x - 1)^2}$; A = 3, B = -1, C = 5

$3\displaystyle\int \dfrac{1}{x}\ dx - \int \dfrac{1}{x - 1}\ dx + 5\int \dfrac{1}{(x - 1)^2}\ dx = 3\ \ln|x| - \ln|x - 1| - 5/(x - 1) + C$

21. $\dfrac{x^2 + x - 16}{(x + 1)(x - 3)^2} = \dfrac{A}{x + 1} + \dfrac{B}{x - 3} + \dfrac{C}{(x - 3)^2}$; A = -1, B = 2, C = -1

$-\displaystyle\int \dfrac{1}{x + 1}\ dx + 2\int \dfrac{1}{x - 3}\ dx - \int \dfrac{1}{(x - 3)^2}\ dx$

$= -\ln|x + 1| + 2\ \ln|x - 3| + \dfrac{1}{x - 3} + C = \ln \dfrac{(x - 3)^2}{|x + 1|} + \dfrac{1}{x - 3} + C$

23. $\dfrac{x^2}{(x + 2)^3} = \dfrac{A}{x + 2} + \dfrac{B}{(x + 2)^2} + \dfrac{C}{(x + 2)^3}$; A = 1, B = -4, C = 4

$\displaystyle\int \dfrac{1}{x + 2}\ dx - 4\int \dfrac{1}{(x + 2)^2}\ dx + 4\int \dfrac{1}{(x + 2)^3}\ dx$

$= \ln|x + 2| + \dfrac{4}{x + 2} - \dfrac{2}{(x + 2)^2} + C$

25. $\dfrac{2x^2 - 1}{(4x - 1)(x^2 + 1)} = \dfrac{A}{4x - 1} + \dfrac{Bx + C}{x^2 + 1}$; A = -14/17, B = 12/17, C = 3/17

$\displaystyle\int \dfrac{2x^2 - 1}{4x^3 - x^2 + 4x - 1}\ dx = -\dfrac{7}{34}\ \ln|4x - 1| + \dfrac{6}{17}\ \ln(x^2 + 1) + \dfrac{3}{17}\ \tan^{-1}x + C$

27. $\dfrac{1}{(x + 2)(x - 2)(x^2 + 4)} = \dfrac{A}{x + 2} + \dfrac{B}{x - 2} + \dfrac{Cx + D}{x^2 + 4}$;

A = -1/32, B = 1/32, C = 0, D = -1/8

$\displaystyle\int \dfrac{dx}{x^4 - 16} = \dfrac{1}{32}\ \ln\left|\dfrac{x - 2}{x + 2}\right| - \dfrac{1}{16}\ \tan^{-1}(x/2) + C$

29. $\dfrac{x^3 + 3x^2 + x + 9}{(x^2 + 1)(x^2 + 3)} = \dfrac{Ax + B}{x^2 + 1} + \dfrac{Cx + D}{x^2 + 3}$; A = 0, B = 3, C = 1, D = 0

$\displaystyle\int \dfrac{x^3 + 3x^2 + x + 9}{(x^2 + 1)(x^2 + 3)}\ dx = 3\ \tan^{-1}x + \dfrac{1}{2}\ \ln(x^2 + 3) + C$

31. $\dfrac{x^3 - 3x^2 + 2x - 3}{x^2 + 1} = x - 3 + \dfrac{x}{x^2 + 1}$,

$\displaystyle \int \dfrac{x^3 - 3x^2 + 2x - 3}{x^2 + 1}\, dx = \dfrac{1}{2} x^2 - 3x + \dfrac{1}{2} \ln(x^2 + 1) + C$

33. $\dfrac{x^2 + 1}{(x^2 + 2x + 3)^2} = \dfrac{Ax + B}{x^2 + 2x + 3} + \dfrac{Cx + D}{(x^2 + 2x + 3)^2}$; $A = 0$, $B = 1$, $C = D = -2$

$\displaystyle \int \dfrac{x^2 + 1}{(x^2 + 2x + 3)^2}\, dx = \int \dfrac{1}{(x + 1)^2 + 2}\, dx - \int \dfrac{2x + 2}{(x^2 + 2x + 3)^2}\, dx$

$\displaystyle = \dfrac{1}{\sqrt{2}} \tan^{-1} \dfrac{x + 1}{\sqrt{2}} + 1/(x^2 + 2x + 3) + C$

35. let $x = \sin \theta$ to get $\displaystyle \int \dfrac{1}{x^2 + 4x - 5}\, dx$,

$\dfrac{1}{(x + 5)(x - 1)} = \dfrac{A}{x + 5} + \dfrac{B}{x - 1}$; $A = -1/6$, $B = 1/6$

$\displaystyle -\dfrac{1}{6} \int \dfrac{1}{x + 5}\, dx + \dfrac{1}{6} \int \dfrac{1}{x - 1}\, dx = \dfrac{1}{6} \ln \left| \dfrac{x - 1}{x + 5} \right| + C = \dfrac{1}{6} \ln \left| \dfrac{\sin \theta - 1}{\sin \theta + 5} \right| + C$

37. let $u = e^x$ to get $\displaystyle \int \dfrac{dx}{1 + e^x} = \int \dfrac{e^x dx}{e^x(1 + e^x)} = \int \dfrac{du}{u(1 + u)}$,

$\dfrac{1}{u(1 + u)} = \dfrac{A}{u} + \dfrac{B}{1 + u}$; $A = 1$, $B = -1$

$\displaystyle \int \dfrac{du}{u(1 + u)} = \ln u - \ln(1 + u) + C = \ln \dfrac{e^x}{1 + e^x} + C$

39. (a) $x^4 + 1 = (x^4 + 2x^2 + 1) - 2x^2 = (x^2 + 1)^2 - 2x^2$

$= [(x^2 + 1) + \sqrt{2}x][(x^2 + 1) - \sqrt{2}x]$

$= (x^2 + \sqrt{2}x + 1)(x^2 - \sqrt{2}x + 1)$; $a = \sqrt{2}$, $b = -\sqrt{2}$

(b) $\dfrac{x}{(x^2 + \sqrt{2}x + 1)(x^2 - \sqrt{2}x + 1)} = \dfrac{Ax + B}{x^2 + \sqrt{2}x + 1} + \dfrac{Cx + D}{x^2 - \sqrt{2}x + 1}$;

$A = 0$, $B = -\dfrac{\sqrt{2}}{4}$, $C = 0$, $D = \dfrac{\sqrt{2}}{4}$ so

$$\int_0^1 \frac{x}{x^4 + 1}\, dx = -\frac{\sqrt{2}}{4} \int_0^1 \frac{1}{x^2 + \sqrt{2}x + 1}\, dx + \frac{\sqrt{2}}{4} \int_0^1 \frac{1}{x^2 - \sqrt{2}x + 1}\, dx$$

$$= -\frac{\sqrt{2}}{4} \int_0^1 \frac{1}{(x + \sqrt{2}/2)^2 + 1/2}\, dx + $$

$$\frac{\sqrt{2}}{4} \int_0^1 \frac{1}{(x - \sqrt{2}/2)^2 + 1/2}\, dx$$

$$= -\frac{\sqrt{2}}{4} \int_{\sqrt{2}/2}^{1+\sqrt{2}/2} \frac{1}{u^2 + 1/2}\, du + \frac{\sqrt{2}}{4} \int_{-\sqrt{2}/2}^{1-\sqrt{2}/2} \frac{1}{u^2 + 1/2}\, du$$

$$= -\frac{1}{2} \tan^{-1}\sqrt{2}u \Big]_{\sqrt{2}/2}^{1+\sqrt{2}/2} + \frac{1}{2} \tan^{-1}\sqrt{2}u \Big]_{-\sqrt{2}/2}^{1-\sqrt{2}/2}$$

$$= -\frac{1}{2} \tan^{-1}(\sqrt{2} + 1) + \frac{1}{2}\left(\frac{\pi}{4}\right) + \frac{1}{2} \tan^{-1}(\sqrt{2} - 1) - \frac{1}{2}\left(-\frac{\pi}{4}\right)$$

$$= \frac{\pi}{4} - \frac{1}{2}\left[\tan^{-1}(\sqrt{2} + 1) - \tan^{-1}(\sqrt{2} - 1)\right]$$

$$= \frac{\pi}{4} - \frac{1}{2}\left[\tan^{-1}(1 + \sqrt{2}) + \tan^{-1}(1 - \sqrt{2})\right]$$

$$= \frac{\pi}{4} - \frac{1}{2} \tan^{-1}\left[\frac{(1 + \sqrt{2}) + (1 - \sqrt{2})}{1 - (1 + \sqrt{2})(1 - \sqrt{2})}\right] \quad \text{(Exercise 16, 8.1)}$$

$$= \frac{\pi}{4} - \frac{1}{2} \tan^{-1}1 = \frac{\pi}{4} - \frac{1}{2}\left(\frac{\pi}{4}\right) = \frac{\pi}{8}$$

41. $V = \pi \int_0^2 \frac{x^4}{(9 - x^2)^2}\, dx, \quad \dfrac{x^4}{x^4 - 18x^2 + 81} = 1 + \dfrac{18x^2 - 81}{x^4 - 18x^2 + 81},$

$\dfrac{18x^2 - 81}{(9 - x^2)^2} = \dfrac{18x^2 - 81}{(x + 3)^2(x - 3)^2} = \dfrac{A}{x + 3} + \dfrac{B}{(x + 3)^2} + \dfrac{C}{x - 3} + \dfrac{D}{(x - 3)^2};$

     $A = -\dfrac{9}{4}, \ B = \dfrac{9}{4}, \ C = \dfrac{9}{4}, \ D = \dfrac{9}{4}$

$V = \pi\left[x - \dfrac{9}{4} \ln|x + 3| - \dfrac{9/4}{x + 3} + \dfrac{9}{4} \ln|x - 3| - \dfrac{9/4}{x - 3}\right]_0^2 = \pi\left(\dfrac{19}{5} - \dfrac{9}{4} \ln 5\right)$

43. $\displaystyle\int \frac{1}{y^2 - 5y + 6}\, dy = \int dx, \quad \dfrac{1}{(y - 3)(y - 2)} = \dfrac{1}{y - 3} - \dfrac{1}{y - 2},$

   $\ln|y - 3| - \ln|y - 2| = x + C_1,$

$$\ln\left|\frac{y-3}{y-2}\right| = x + C_1, \quad \frac{y-3}{y-2} = \pm e^{C_1}e^x = Ce^x,$$

$$y - 3 = Ce^x y - 2Ce^x, \quad y(1 - Ce^x) = 3 - 2Ce^x, \quad y = \frac{3 - 2Ce^x}{1 - Ce^x}.$$

45. $$\int \frac{1}{y^2 + y}\,dy = \int \frac{1}{t(t-1)}\,dt,$$

$$\frac{1}{y(y+1)} = \frac{1}{y} - \frac{1}{y+1} \text{ and } \frac{1}{t(t-1)} = -\frac{1}{t} + \frac{1}{t-1},$$

$$\ln|y| - \ln|y+1| = -\ln|t| + \ln|t-1| + C_1, \quad \ln\left|\frac{y}{y+1}\right| = \ln\left|\frac{t-1}{t}\right| + C_1,$$

$$\frac{y}{y+1} = C_2\frac{t-1}{t}; \text{ solve for } y \text{ to get}$$

$$y = \frac{C_2(t-1)/t}{1 - C_2(t-1)/t} = \frac{1}{Ct/(t-1) - 1} = \frac{t-1}{Ct - t + 1}.$$

47. (a) $x^3 - 6x^2 + 11x - 6 = (x-1)(x-2)(x-3)$

    (b) $x^3 - 3x^2 + x - 20 = (x-4)(x^2 + x + 5)$

    (c) $x^4 - 5x^3 + 7x^2 - 5x + 6 = (x-2)(x-3)(x^2+1)$

49. $x^4 - 3x^3 - 7x^2 + 27x - 18 = (x-1)(x-2)(x-3)(x+3),$

$$\frac{1}{(x-1)(x-2)(x-3)(x+3)} = \frac{A}{x-1} + \frac{B}{x-2} + \frac{C}{x-3} + \frac{D}{x+3};$$

$A = 1/8, \ B = -1/5, \ C = 1/12, \ D = -1/120$

$$\int \frac{dx}{x^4 - 3x^3 - 7x^2 + 27x - 18}$$

$$= \frac{1}{8}\ln|x-1| - \frac{1}{5}\ln|x-2| + \frac{1}{12}\ln|x-3| - \frac{1}{120}\ln|x+3| + C$$

51. (a) $\sqrt{2}$ is a positive root of $x^2 - 2 = 0$. The only possible positive rational roots of $x^2 - 2 = 0$ are the integers 1 and 2 neither of which satisfies the equation so $\sqrt{2}$ is not rational.

    (b) $\sqrt{a}$ is a positive root of $x^2 - a = 0$. The only possible positive rational roots of $x^2 - a = 0$ are the positive integers that divide a. If one of these satisfies the equation then $\sqrt{a}$ is an integer, otherwise $\sqrt{a}$ cannot be rational.

EXERCISE SET 9.8

1. $u = \sqrt{x - 2}$, $x = u^2 + 2$, $dx = 2u\,du$

$$\int 2u^2(u^2 + 2)du = 2\int(u^4 + 2u^2)du = \frac{2}{5}u^5 + \frac{4}{3}u^3 + C$$

$$= \frac{2}{5}(x - 2)^{5/2} + \frac{4}{3}(x - 2)^{3/2} + C$$

3. $u = \sqrt{x - 4}$, $x = u^2 + 4$, $dx = 2u\,du$

$$\int_0^2 \frac{2u^2}{u^2 + 4}\,du = 2\int_0^2\left[1 - \frac{4}{u^2 + 4}\right]du = 2u - 4\tan^{-1}(u/2)\,\Big]_0^2 = 4 - \pi$$

5. $u = 3 + \sqrt{x}$, $x = (u - 3)^2$, $dx = 2(u - 3)du$

$$\int_3^5 \frac{2(u - 3)}{u}\,du = 2\int_3^5 (1 - 3/u)du = 2u - 6\ln|u|\,\Big]_3^5 = 4 - 6\ln(5/3)$$

7. $u = \sqrt{x^3 + 1}$, $x^3 = u^2 - 1$, $3x^2 dx = 2u\,du$

$$\frac{2}{3}\int u^2(u^2 - 1)du = \frac{2}{3}\int(u^4 - u^2)du = \frac{2}{15}u^5 - \frac{2}{9}u^3 + C$$

$$= \frac{2}{15}(x^3 + 1)^{5/2} - \frac{2}{9}(x^3 + 1)^{3/2} + C$$

9. $u = x^{1/6}$, $x = u^6$, $dx = 6u^5 du$

$$\int \frac{6u^5}{u^3 + u^2}\,du = 6\int \frac{u^3}{u + 1}\,du = 6\int\left[u^2 - u + 1 - \frac{1}{u + 1}\right]du$$

$$= 2x^{1/2} - 3x^{1/3} + 6x^{1/6} - 6\ln(x^{1/6} + 1) + C$$

11. $u = v^{1/4}$, $v = u^4$, $dv = 4u^3 du$

$$4\int \frac{1}{u(1 - u)}\,du = 4\int\left[\frac{1}{u} + \frac{1}{1 - u}\right]du = 4\ln\frac{v^{1/4}}{|1 - v^{1/4}|} + C$$

13. $u = t^{1/6}$, $t = u^6$, $dt = 6u^5 du$

$$6\int \frac{u^3}{u - 1}\,du = 6\int\left[u^2 + u + 1 + \frac{1}{u - 1}\right]du$$

$$= 2t^{1/2} + 3t^{1/3} + 6t^{1/6} + 6\ln|t^{1/6} - 1| + C$$

15. $u = \sqrt{1 + x^2}$, $x^2 = u^2 - 1$, $2x\,dx = 2u\,du$, $x\,dx = u\,du$

$\int (u^2 - 1)du = \frac{1}{3}(1 + x^2)^{3/2} - (1 + x^2)^{1/2} + C$

17. $z = \sqrt{x}$, $x = z^2$, $dx = 2z\,dz$

$2\int z\sin z\,dz$; use integration by parts:

$u = z$, $dv = \sin z\,dz$, $du = dz$, $v = -\cos z$

$2\int z\sin z\,dz = 2(-z\cos z + \int \cos z\,dz) = -2z\cos z + 2\sin z + C$

so $\int \sin \sqrt{x}\,dx = -2\sqrt{x}\cos \sqrt{x} + 2\sin \sqrt{x} + C$

19. $u = \sqrt{e^x + 1}$, $e^x = u^2 - 1$, $x = \ln(u^2 - 1)$, $dx = \dfrac{2u}{u^2 - 1}\,du$

$\int \dfrac{2}{u^2 - 1}\,du = \int \left[\dfrac{1}{u - 1} - \dfrac{1}{u + 1}\right]du = \ln|u - 1| - \ln|u + 1| + C$

$$= \ln \frac{\sqrt{e^x + 1} - 1}{\sqrt{e^x + 1} + 1} + C$$

21. $\displaystyle\int \dfrac{1}{1 + \dfrac{2u}{1 + u^2} + \dfrac{1 - u^2}{1 + u^2}}\dfrac{2}{1 + u^2}\,du = \int \dfrac{1}{u + 1}\,du = \ln|\tan(x/2) + 1| + C$

23. $u = \tan(\theta/2)$, $\displaystyle\int \dfrac{d\theta}{1 - \cos \theta} = \int \dfrac{1}{u^2}\,du = -\dfrac{1}{u} + C = -\cot(\theta/2) + C$,

$\displaystyle\int_{\pi/2}^{\pi} \dfrac{d\theta}{1 - \cos \theta} = -\cot(\theta/2)\Big]_{\pi/2}^{\pi} = 1$

25. $u = \tan(x/2)$, $2\displaystyle\int \dfrac{1 - u^2}{(3u^2 + 1)(u^2 + 1)}\,du$

$\dfrac{1 - u^2}{(3u^2 + 1)(u^2 + 1)} = \dfrac{(0)u + 2}{3u^2 + 1} + \dfrac{(0)u - 1}{u^2 + 1} = \dfrac{2}{3u^2 + 1} - \dfrac{1}{u^2 + 1}$,

$\displaystyle\int \dfrac{\cos x}{2 - \cos x}\,dx = \dfrac{4}{\sqrt{3}}\tan^{-1}[\sqrt{3}\tan(x/2)] - x + C$

27. (a) $\displaystyle\int \sec x \, dx = \int \frac{1}{\cos x} \, dx = \int \frac{2}{1 - u^2} \, du$

$$= \ln\left|\frac{1 + u}{1 - u}\right| + C = \ln\left|\frac{1 + \tan(x/2)}{1 - \tan(x/2)}\right| + C$$

(b) $\displaystyle\frac{1 + \tan(x/2)}{1 - \tan(x/2)} = \frac{\tan(\pi/4) + \tan(x/2)}{1 - \tan(\pi/4)\tan(x/2)} = \tan(\pi/4 + x/2)$ (trig identity)

(c) $\displaystyle\frac{1 + \tan(x/2)}{1 - \tan(x/2)} = \frac{\cos(x/2) + \sin(x/2)}{\cos(x/2) - \sin(x/2)} = \frac{[\cos(x/2) + \sin(x/2)]^2}{\cos^2(x/2) - \sin^2(x/2)}$

$$= \frac{\cos^2(x/2) + 2\sin(x/2)\cos(x/2) + \sin^2(x/2)}{\cos x}$$

$$= \frac{1 + \sin x}{\cos x} = \sec x + \tan x$$

29. Let $u = \tanh(x/2)$ then

$$\cosh(x/2) = 1/\operatorname{sech}(x/2) = 1/\sqrt{1 - \tanh^2(x/2)} = 1/\sqrt{1 - u^2}$$

$$\sinh(x/2) = \tanh(x/2)\cosh(x/2) = u/\sqrt{1 - u^2}$$

so $\sinh x = 2\sinh(x/2)\cosh(x/2) = 2u/(1 - u^2)$,

$\cosh x = \cosh^2(x/2) + \sinh^2(x/2) = (1 + u^2)/(1 - u^2)$,

$x = 2\tanh^{-1}u, \quad dx = [1/(1 - u^2)]du$

$$\int \frac{dx}{2\cosh x + \sinh x} = \int \frac{1}{u^2 + u + 1} \, du$$

$$= \frac{2}{\sqrt{3}} \tan^{-1} \frac{2u + 1}{\sqrt{3}} + C = \frac{2}{\sqrt{3}} \tan^{-1} \frac{2\tanh(x/2) + 1}{\sqrt{3}} + C$$

31. $dx = -(1/u^2)du$, $\displaystyle -\int \frac{u}{\sqrt{3u^2 - 1}} \, du = -\frac{1}{3}\sqrt{3u^2 - 1} + C = -\frac{\sqrt{3 - x^2}}{3x} + C$

33. $dx = -(1/u^2)du$, $\displaystyle -\int u\sqrt{1 - 5u^2}\,du = \frac{1}{15}(1 - 5u^2)^{3/2} + C = \frac{(x^2 - 5)^{3/2}}{15x^3} + C$

EXERCISE SET 9.9

1. (a) $\frac{1}{6}$ (4.4193) $\approx$ 0.7366   (b) $\frac{1}{6}$ (3.9193) $\approx$ 0.6532

   (c) $\frac{1}{12}$ (8.3385) $\approx$ 0.6949   (d) $\frac{1}{18}$ (12.4771) $\approx$ 0.6932

3. (a) $\frac{\pi}{4}$ (2.4142) $\approx$ 1.8961   (b) $\frac{\pi}{4}$ (2.4142) $\approx$ 1.8961

   (c) $\frac{\pi}{8}$ (4.8284) $\approx$ 1.8961   (d) $\frac{\pi}{12}$ (7.6569) $\approx$ 2.0046

5. (a) $\frac{1}{2}$ (2.2521) $\approx$ 1.1261   (b) $\frac{1}{2}$ (1.2704) = 0.6352

   (c) $\frac{1}{4}$ (3.5225) $\approx$ 0.8806   (d) $\frac{1}{6}$ (5.2909) $\approx$ 0.8818

7. $\int_0^1 \frac{dx}{1 + x^2} \approx \frac{1}{30}$ (23.5618) $\approx$ 0.7854, $\pi/4 \approx$ 0.7854, $\pi \approx$ 3.1416

9. $\frac{1}{10}$ (6.9284) $\approx$ 0.6928

SUPPLEMENTARY EXERCISES CHAPTER 9

1. $u = x$, $dv = \cos 2x\, dx$, $du = dx$, $v = \frac{1}{2} \sin 2x$

   $\int x \cos 2x\, dx = \frac{1}{2} x \sin 2x + \frac{1}{4} \cos 2x + C$

3. $\int (\sec^2 x - 1)\sec x \tan x\, dx = \frac{1}{3} \sec^3 x - \sec x + C$

5. $u = \tan 3t$, $\frac{1}{3} \int u^2 du = \frac{1}{9} \tan^3 3t + C$

7. $\int \frac{1 - \cos^2 x}{1 + \cos x}\, dx = \int (1 - \cos x)dx = x - \sin x + C$

9. $\int x^2\cos^2 x\, dx = \frac{1}{2}\int x^2(1 + \cos 2x)dx = \frac{1}{2}\int x^2 dx + \frac{1}{2}\int x^2\cos 2x\, dx$,

use integration by parts twice to get

$\int x^2\cos 2x\, dx = \frac{1}{2}x^2\sin 2x + \frac{1}{2}x\cos 2x - \frac{1}{4}\sin 2x + C_1$

so $\int x^2\cos^2 x\, dx = \frac{1}{6}x^3 + \frac{1}{4}(x^2 - 1/2)\sin 2x + \frac{1}{4}x\cos 2x + C$

11. $\int \cos^{-5} x \sin x\, dx = \frac{1}{4}\cos^{-4} x + C = \frac{1}{4}\sec^4 x + C$

13. $\frac{1}{8}\int(1 - \cos 2x)^2(1 + \cos 2x)dx = \frac{1}{8}\int(1 - \cos 2x)\sin^2 2x\, dx$

$$= \frac{1}{8}\int \sin^2 2x\, dx - \frac{1}{8}\int \sin^2 2x \cos 2x\, dx$$

$$= \frac{1}{16}\int(1 - \cos 4x)dx - \frac{1}{48}\sin^3 2x$$

$$= \frac{1}{16}x - \frac{1}{64}\sin 4x - \frac{1}{48}\sin^3 2x + C$$

15. $\int_0^{\pi/4} \sin 5x \sin 3x\, dx = \frac{1}{2}\int_0^{\pi/4}(\cos 2x - \cos 8x)dx$

$$= \frac{1}{4}\sin 2x - \frac{1}{16}\sin 8x\Big]_0^{\pi/4} = 1/4$$

17. $\frac{1}{2}\int_0^1(1 - \cos 2\pi x)dx = \frac{1}{2}x - \frac{1}{4\pi}\sin 2\pi x\Big]_0^1 = 1/2$

19. $\frac{1}{2}\tan(x^2)\Big]_0^{\sqrt{\pi}/2} = 1/2$

21. $u = \cot^{-1}x$, $-\int \sin u\, du = \cos(\cot^{-1}x) + C = x/\sqrt{1 + x^2} + C$

23. $\ln|\sec(e^x) + \tan(e^x)| + C$

25. $u = e^{2x} + 1$, $\frac{1}{2}\int u^{-1/2}du = \sqrt{e^{2x} + 1} + C$

27. Use integration by parts with $u = e^{3x}$, $dv = \sin 2x\, dx$ to get
$$\int e^{3x}\sin 2x\, dx = -\frac{1}{2}e^{3x}\cos 2x + \frac{3}{2}\int e^{3x}\cos 2x\, dx \text{ and again with}$$
$u = e^{3x}$, $dv = \cos 2x\, dx$ to get
$$\int e^{3x}\cos 2x\, dx = \frac{1}{2}e^{3x}\sin 2x - \frac{3}{2}\int e^{3x}\sin 2x\, dx \text{ so, with}$$
$$I = \int e^{3x}\sin 2x\, dx, \quad I = -\frac{1}{2}e^{3x}\cos 2x + \frac{3}{4}e^{3x}\sin 2x - \frac{9}{4}I$$
$$I = \frac{1}{13}e^{3x}(3\sin 2x - 2\cos 2x) + C$$

29. $u = \sin^{-1}(x/2)$, $dv = dx$, $du = 1/\sqrt{4 - x^2}\, dx$, $v = x$
$$\int_1^2 \sin^{-1}(x/2)dx = x\sin^{-1}(x/2)\Big|_1^2 - \int_1^2 x(4 - x^2)^{-1/2}dx$$
$$= (2)(\pi/2) - (1)(\pi/6) + (4 - x^2)^{1/2}\Big|_1^2 = 5\pi/6 - \sqrt{3}$$

31. $u = \ln x$, $dv = (1/x^2)dx$, $du = dx/x$, $v = -1/x$
$$\int (\ln x)/x^2 dx = -(\ln x)/x - 1/x + C$$

33. $$\int x(x^2 - 9)^{-1/2}dx = \sqrt{x^2 - 9} + C$$

35. $x = 3\sin\theta$, $dx = 3\cos\theta\, d\theta$
$$3\int_{\sin^{-1}(1/3)}^{\pi/2} \frac{\cos^2\theta}{\sin\theta}\, d\theta = 3\int_{\sin^{-1}(1/3)}^{\pi/2} \frac{1 - \sin^2\theta}{\sin\theta}\, d\theta$$
$$= 3\int_{\sin^{-1}(1/3)}^{\pi/2} (\csc\theta - \sin\theta)d\theta$$
$$= -3\ln|\csc\theta + \cot\theta| + 3\cos\theta\Big|_{\sin^{-1}(1/3)}^{\pi/2}$$
$$= -3\ln(1) + 3\ln|3 + \sqrt{8}| - 3(\sqrt{8}/3)$$
$$= 3\ln(3 + \sqrt{8}) - \sqrt{8}$$

37. $u = \sqrt{2x + 3}$, $x = (u^2 - 3)/2$, $dx = u\,du$

$\dfrac{1}{4}\displaystyle\int (u^2 - 3)^2 du = \dfrac{1}{4}\int (u^4 - 6u^2 + 9)du$

$$= \dfrac{1}{4}\left(\dfrac{1}{5}u^5 - 2u^3 + 9u\right) + C = \dfrac{1}{20}u(u^4 - 10u^2 + 45) + C$$

$$= \dfrac{1}{20}\sqrt{2x + 3}\,(4x^2 + 12x + 9 - 20x - 30 + 45) + C$$

$$= \dfrac{1}{5}(x^2 - 2x + 6)\sqrt{2x + 3} + C$$

39. $\dfrac{1}{2}\displaystyle\int \dfrac{1}{\sqrt{1 - (t + 1/2)^2}}\,dt = \dfrac{1}{2}\sin^{-1}(t + 1/2) + C$

41. $x = a\sin\theta$, $dx = a\cos\theta\,d\theta$

$\dfrac{1}{a^2}\displaystyle\int \csc^2\theta\,d\theta = -\dfrac{1}{a^2}\cot\theta + C = -\dfrac{\sqrt{a^2 - x^2}}{a^2 x} + C$

43. $x = a\sin\theta$, $dx = a\cos\theta\,d\theta$

$a^2\displaystyle\int \cos^2\theta\,d\theta = \dfrac{1}{2}a^2\theta + \dfrac{1}{4}a^2\sin 2\theta + C$

$$= \dfrac{1}{2}a^2\sin^{-1}(x/a) + \dfrac{1}{2}x\sqrt{a^2 - x^2} + C$$

45. $\displaystyle\int \dfrac{x - 2}{\sqrt{4 - (x - 2)^2}}\,dx = \int \dfrac{u}{\sqrt{4 - u^2}}\,du$, $u = x - 2$

$$= -\sqrt{4 - u^2} + C = -\sqrt{4x - x^2} + C$$

47. $2x^2 + 3x + 1 = (2x + 1)(x + 1)$, $\dfrac{1}{(2x + 1)(x + 1)} = \dfrac{2}{2x + 1} - \dfrac{1}{x + 1}$

$\displaystyle\int \dfrac{dx}{(2x + 1)(x + 1)} = \ln\left|\dfrac{2x + 1}{x + 1}\right| + C$

49. $\dfrac{x + 1}{x(x + 3)(x - 2)} = \dfrac{-1/6}{x} + \dfrac{-2/15}{x + 3} + \dfrac{3/10}{x - 2}$

$\displaystyle\int \dfrac{x + 1}{x^3 + x^2 - 6x}\,dx = -\dfrac{1}{6}\ln|x| - \dfrac{2}{15}\ln|x + 3| + \dfrac{3}{10}\ln|x - 2| + C$

51. $u = x^3 - 3x$, $\dfrac{1}{3}\displaystyle\int \dfrac{1}{u}\,du = \dfrac{1}{3}\ln|x^3 - 3x| + C$

53. $x^4 - 1 = (x + 1)(x - 1)(x^2 + 1)$

$\int \dfrac{2x^2 + 5}{x^4 - 1} dx = \int \left[\dfrac{-7/4}{x + 1} + \dfrac{7/4}{x - 1} + \dfrac{-3/2}{x^2 + 1}\right] dx = \dfrac{7}{4} \ln\left|\dfrac{x - 1}{x + 1}\right| - \dfrac{3}{2} \tan^{-1}x + C$

55. $\int \dfrac{dx}{(x^2 + 4)(x - 3)} = \int \left[\dfrac{(-1/13)x - (3/13)}{x^2 + 4} + \dfrac{1/13}{x - 3}\right] dx$

$(3+x)(x+x)$

$= -\dfrac{1}{26} \ln(x^2 + 4) - \dfrac{3}{26} \tan^{-1} \dfrac{x}{2} + \dfrac{1}{13} \ln|x - 3| + C$

57. $\dfrac{3x^2 + 12x + 2}{(x^2 + 4)^2} = \dfrac{3}{x^2 + 4} + \dfrac{12x - 10}{(x^2 + 4)^2} = \dfrac{3}{x^2 + 4} + \dfrac{12x}{(x^2 + 4)^2} - \dfrac{10}{(x^2 + 4)^2}$

$\int \dfrac{1}{(x^2 + 4)^2} = \dfrac{1}{8} \int \cos^2\theta \, d\theta, \quad x = 2 \tan\theta$

$= \dfrac{1}{16} \theta + \dfrac{1}{32} \sin 2\theta + C_1 = \dfrac{1}{16} \tan^{-1}(x/2) + \dfrac{x}{8(x^2 + 4)} + C_1$

so $\int \dfrac{3x^2 + 12x + 2}{(x^2 + 4)^2} dx = \dfrac{3}{2} \tan^{-1} \dfrac{x}{2} - \dfrac{6}{x^2 + 4} - \dfrac{5}{8} \tan^{-1} \dfrac{x}{2} - \dfrac{5x}{4(x^2 + 4)} + C$

$= \dfrac{7}{8} \tan^{-1} \dfrac{x}{2} - \dfrac{5x + 24}{4(x^2 + 4)} + C$

59. $\int \dfrac{x}{(x + 1)^2 + 4} dx = \int \dfrac{u - 1}{u^2 + 4} du, \quad u = x + 1$

$= \dfrac{1}{2} \ln(u^2 + 4) - \dfrac{1}{2} \tan^{-1} \dfrac{u}{2} + C$

$= \dfrac{1}{2} \ln(x^2 + 2x + 5) - \dfrac{1}{2} \tan^{-1} \dfrac{x + 1}{2} + C$

61. $u = \sqrt{2}x, \quad \dfrac{1}{\sqrt{2}} \int \dfrac{1}{\sqrt{3 - u^2}} du = \dfrac{1}{\sqrt{2}} \sin^{-1}\sqrt{2/3}x + C$

63. $u = \sqrt{t}, \quad t = u^2, \quad dt = 2u \, du$

$\int \dfrac{2u^2}{u^2 + 1} du = 2\int \left[1 - \dfrac{1}{u^2 + 1}\right] du = 2\sqrt{t} - 2 \tan^{-1}\sqrt{t} + C$

65. $u = 1 + x^{1/3}$, $x = (u - 1)^3$, $dx = 3(u - 1)^2 du$

$$3\int_1^2 \frac{(u - 1)^4}{u}\, du = 3\int_1^2 \left[u^3 - 4u^2 + 6u - 4 + \frac{1}{u}\right] du$$

$$= 3\left[\frac{1}{4} u^4 - \frac{4}{3} u^3 + 3u^2 - 4u + \ln u\right]_1^2 = -7/4 + 3\ln 2$$

67. $u = \tan(x/2)$, $\tan x = \sin x/\cos x = 2u/(1 - u^2)$

$$\int \frac{dx}{1 - \tan x} = \int \frac{2u^2 - 2}{(u^2 + 1)(u^2 + 2u - 1)}\, du$$

$$\frac{2u^2 - 2}{(u^2 + 1)(u^2 + 2u - 1)} = \frac{u + 1}{u^2 + 1} + \frac{-u - 1}{u^2 + 2u - 1}$$

$$\int \frac{dx}{1 - \tan x} = \frac{1}{2}\ln(u^2 + 1) + \tan^{-1}u - \frac{1}{2}\ln|u^2 + 2u - 1| + C$$

$$= \tan^{-1}u - \frac{1}{2}\ln\left|\frac{u^2 + 2u - 1}{u^2 + 1}\right| + C$$

$$= \tan^{-1}u - \frac{1}{2}\ln\left|\frac{2u}{1 + u^2} - \frac{1 - u^2}{1 + u^2}\right| + C$$

$$= \frac{x}{2} - \frac{1}{2}\ln|\sin x - \cos x| + C$$

69. $u = \tan(x/2)$, $\tan x = 2u/(1 - u^2)$

$$\int \frac{dx}{\sin x - \tan x} = \frac{1}{2}\int \frac{u^2 - 1}{u^3}\, du = \frac{1}{2}\int (1/u - u^{-3})\, du$$

$$= \frac{1}{2}\ln\left|\tan \frac{x}{2}\right| + \frac{1}{4}\cot^2 \frac{x}{2} + C$$

71. $\displaystyle\int \frac{dx}{x^2 - a^2} = \int\left[\frac{1/(2a)}{x - a} + \frac{-1/(2a)}{x + a}\right] dx$

$$= \frac{1}{2a}\ln|x - a| - \frac{1}{2a}\ln|x + a| + C = \frac{1}{2a}\ln\left|\frac{x - a}{x + a}\right| + C$$

73.   (a) $\displaystyle A = \int_0^2 \frac{1}{4 + x^2}\, dx = \frac{1}{2}\tan^{-1}\frac{x}{2}\Big]_0^2 = \pi/8$

(b) $V = \pi \int_0^2 \dfrac{1}{(4 + x^2)^2} dx = \dfrac{\pi}{8} \int_0^{\pi/4} \cos^2\theta \, d\theta, \quad x = 2\tan\theta$

$$= \dfrac{\pi}{16} \left(\theta + \dfrac{1}{2}\sin 2\theta\right)\Big]_0^{\pi/4} = \pi(\pi + 2)/64$$

(c) $V = 2\pi \int_0^2 \dfrac{x}{4 + x^2} dx = \pi \ln(4 + x^2)\Big]_0^2 = \pi \ln 2$

75. (a) $\displaystyle\int x^3 e^{2x} dx = \dfrac{1}{2} x^3 e^{2x} - \dfrac{3}{2}\int x^2 e^{2x} dx = \dfrac{1}{2} x^3 e^{2x} - \dfrac{3}{2}\left[\dfrac{1}{2} x^2 e^{2x} - \int x e^{2x} dx\right]$

$\qquad = \dfrac{1}{2} x^3 e^{2x} - \dfrac{3}{4} x^2 e^{2x} + \dfrac{3}{2}\left[\dfrac{1}{2} x e^{2x} - \dfrac{1}{2}\int e^{2x} dx\right]$

$\qquad = \dfrac{1}{2} x^3 e^{2x} - \dfrac{3}{4} x^2 e^{2x} + \dfrac{3}{4} x e^{2x} - \dfrac{3}{8} e^{2x} + C$

(b) $\displaystyle\int_0^{\pi/10} x^2 \sin 5x \, dx = -\dfrac{1}{5} x^2 \cos 5x\Big]_0^{\pi/10} + \dfrac{2}{5}\int_0^{\pi/10} x \cos 5x \, dx$

$\qquad = 0 + \dfrac{2}{5}\left[\dfrac{1}{5} x \sin 5x\Big]_0^{\pi/10} - \dfrac{1}{5}\int_0^{\pi/10} \sin 5x \, dx\right]$

$\qquad = \dfrac{2}{25}(\pi/10) + \dfrac{2}{125}\cos 5x\Big]_0^{\pi/10}$

$\qquad = \pi/125 + \dfrac{2}{125}(0 - 1) = (\pi - 2)/125$

(c) $\displaystyle\int \sin^2 x \cos^4 x \, dx = \dfrac{1}{6}\sin^3 x \cos^3 x + \dfrac{1}{2}\int \sin^2 x \cos^2 x \, dx$

$\qquad = \dfrac{1}{6}\sin^3 x \cos^3 x + \dfrac{1}{2}\left[\dfrac{1}{4}\sin^3 x \cos x + \dfrac{1}{4}\int \sin^2 x \, dx\right]$

$\qquad = \dfrac{1}{6}\sin^3 x \cos^3 x + \dfrac{1}{8}\sin^3 x \cos x + \dfrac{1}{8}\left[-\dfrac{1}{2}\sin x \cos x + \dfrac{1}{2}\int dx\right]$

$\qquad = \dfrac{1}{6}\sin^3 x \cos^3 x + \dfrac{1}{8}\sin^3 x \cos x - \dfrac{1}{16}\sin x \cos x + \dfrac{x}{16} + C$

77. (a) $\displaystyle\int \dfrac{1 - \cos^2\theta}{\cos^5\theta}\sin\theta \, d\theta = \int(\cos^{-5}\theta - \cos^{-3}\theta)\sin\theta \, d\theta$

$$= \dfrac{1}{4}\cos^{-4}\theta - \dfrac{1}{2}\cos^{-2}\theta + C = \dfrac{1}{4}\sec^4\theta - \dfrac{1}{2}\sec^2\theta + C$$

(b) $\int \tan^3\theta \sec^2\theta \, d\theta = \frac{1}{4} \tan^4\theta + C$

but $\frac{1}{4} \tan^4\theta = \frac{1}{4} (\sec^2\theta - 1)^2 = \frac{1}{4} (\sec^4\theta - 2\sec^2\theta + 1)$
so the answers to (a) and (b) differ by 1/4.

79. (a) $\frac{6}{12} (117.8551) \approx 58.9276$ (b) $\frac{6}{18} (164.1985) \approx 54.7328$

81. (a) $\frac{1.2}{12} (18.2766) \approx 1.8277$ (b) $\frac{1.2}{18} (27.4178) \approx 1.8279$

83. 0.36972

85. $\int \frac{1}{e^{ax} + 1} \, dx = \int \frac{e^{-ax}}{1 + e^{-ax}} \, dx = -\frac{1}{a} \ln(1 + e^{-ax}) + C$

87. $\dfrac{\sqrt{1+x} + \sqrt{1-x}}{\sqrt{1+x} - \sqrt{1-x}} = \dfrac{(\sqrt{1+x} + \sqrt{1-x})^2}{2x} = \dfrac{1 + \sqrt{1-x^2}}{x}$

$\int \left[\frac{1}{x} + \frac{\sqrt{1-x^2}}{x} \right] dx = \int \frac{1}{x} \, dx + \int \frac{\sqrt{1-x^2}}{x} \, dx$

$\qquad = \ln|x| + \int \frac{u^2}{u^2 - 1} \, du, \quad u = \sqrt{1-x^2}$

$\qquad = \ln|x| + \int \left[1 + \frac{1}{u^2 - 1} \right] du$

$\qquad = \ln|x| + u + \frac{1}{2} \ln \frac{1-u}{1+u} + C$

$\qquad = \ln|x| + \sqrt{1-x^2} + \frac{1}{2} \ln \frac{1 - \sqrt{1-x^2}}{1 + \sqrt{1-x^2}} + C$

$\qquad = \sqrt{1-x^2} + \frac{1}{2} \ln(x^2) + \frac{1}{2} \ln \frac{(1 - \sqrt{1-x^2})^2}{x^2} + C$

$\qquad = \sqrt{1-x^2} + \ln(1 - \sqrt{1-x^2}) + C$

89. $x - 1 = 1/u$, $x = 1 + 1/u$, $dx = -(1/u^2)du$

$$\int \frac{\sqrt{x + 1}}{(x - 1)^{5/2}} \, dx = -\int \sqrt{2u + 1} \, du$$

$$= -\frac{1}{3}(2u + 1)^{3/2} + C = -\frac{1}{3}\left[\frac{x + 1}{x - 1}\right]^{3/2} + C$$

91. $\displaystyle\int \frac{1}{x^6(3 + 2x^{-5})} \, dx = -\frac{1}{10}\int \frac{1}{u} \, du$, $u = 3 + 2x^{-5}$

$$= -\frac{1}{10}\ln|u| + C = -\frac{1}{10}\ln|3 + 2x^{-5}| + C$$

93. $\displaystyle\int \sqrt{x - \sqrt{x^2 - 4}} \, dx = \frac{1}{\sqrt{2}}\int (\sqrt{x + 2} - \sqrt{x - 2}) \, dx$

$$= \frac{\sqrt{2}}{3}[(x + 2)^{3/2} - (x - 2)^{3/2}] + C$$

CHAPTER 10

EXERCISE SET 10.1

1. $\lim\limits_{\ell \to +\infty} (-e^{-x}) \Big]_0^\ell = \lim\limits_{\ell \to +\infty} (-e^{-\ell} + 1) = 1$

3. $\lim\limits_{\ell \to +\infty} 2\sqrt{x} \Big]_1^\ell = \lim\limits_{\ell \to +\infty} 2(\sqrt{\ell} - 1) = +\infty$, divergent

5. $\lim\limits_{\ell \to +\infty} \ln \dfrac{x-1}{x+1} \Big]_4^\ell = \lim\limits_{\ell \to +\infty} (\ln \dfrac{\ell-1}{\ell+1} - \ln \dfrac{3}{5}) = -\ln \dfrac{3}{5} = \ln \dfrac{5}{3}$

7. $\lim\limits_{\ell \to +\infty} -\dfrac{1}{2\ln^2 x} \Big]_e^\ell = \lim\limits_{\ell \to +\infty} \left[-\dfrac{1}{2\ln^2 \ell} + \dfrac{1}{2} \right] = \dfrac{1}{2}$

9. $\lim\limits_{\ell \to +\infty} -\dfrac{1}{2(x^2+1)} \Big]_a^\ell = \lim\limits_{\ell \to +\infty} \left[-\dfrac{1}{2(\ell^2+1)} + \dfrac{1}{2(a^2+1)} \right] = \dfrac{1}{2(a^2+1)}$

11. $\lim\limits_{\ell \to -\infty} -\dfrac{1}{4(2x-1)^2} \Big]_\ell^0 = \lim\limits_{\ell \to -\infty} \dfrac{1}{4} [-1 + 1/(2\ell-1)^2] = -1/4$

13. $\lim\limits_{\ell \to -\infty} \dfrac{1}{3} e^{3x} \Big]_\ell^0 = \lim\limits_{\ell \to -\infty} \left[\dfrac{1}{3} - \dfrac{1}{3} e^{3\ell} \right] = \dfrac{1}{3}$

15. $\int_{-\infty}^{+\infty} x^3 dx$ converges if $\int_{-\infty}^{0} x^3 dx$ and $\int_{0}^{+\infty} x^3 dx$ both converge; it diverges if either (or both) diverge.

$$\int_{0}^{+\infty} x^3 dx = \lim_{\ell \to +\infty} \frac{1}{4} x^4 \Big]_{0}^{\ell} = \lim_{\ell \to +\infty} \frac{1}{4} \ell^4 = +\infty \text{ so } \int_{-\infty}^{+\infty} x^3 dx \text{ is divergent.}$$

17. $\int_{0}^{+\infty} \frac{x}{(x^2 + 3)^2} dx = \lim_{\ell \to +\infty} - \frac{1}{2(x^2 + 3)} \Big]_{0}^{\ell} = \lim_{\ell \to +\infty} \frac{1}{2} [-1/(\ell^2 + 3) + 1/3] = \frac{1}{6}$,

similarly $\int_{-\infty}^{0} \frac{x}{(x^2 + 3)^2} dx = -1/6$ so $\int_{-\infty}^{\infty} \frac{x}{(x^2 + 3)^2} dx = 1/6 + (-1/6) = 0$

19. $\lim_{\ell \to 3^+} - \frac{1}{x - 3} \Big]_{\ell}^{4} = \lim_{\ell \to 3^+} [-1 + \frac{1}{\ell - 3}] = +\infty$, divergent

21. $\lim_{\ell \to \pi/2^-} - \ln(\cos x) \Big]_{0}^{\ell} = \lim_{\ell \to \pi/2^-} - \ln(\cos \ell) = +\infty$, divergent

23. $\lim_{\ell \to 1^-} \sin^{-1} x \Big]_{0}^{\ell} = \lim_{\ell \to 1^-} \sin^{-1} \ell = \pi/2$

25. $\lim_{\ell \to \pi/6^-} - \sqrt{1 - 2 \sin x} \Big]_{0}^{\ell} = \lim_{\ell \to \pi/6^-} (-\sqrt{1 - 2 \sin \ell} + 1) = 1$

27. $\int_{0}^{2} \frac{dx}{x - 2} = \lim_{\ell \to 2^-} \ln|x - 2| \Big]_{0}^{\ell} = \lim_{\ell \to 2^-} (\ln|\ell - 2| - \ln 2) = -\infty$, divergent

29. $\int_{0}^{8} x^{-1/3} dx = \lim_{\ell \to 0^+} \frac{3}{2} x^{2/3} \Big]_{\ell}^{8} = \lim_{\ell \to 0^+} \frac{3}{2} (4 - \ell^{2/3}) = 6$,

$\int_{-1}^{0} x^{-1/3} dx = \lim_{\ell \to 0^-} \frac{3}{2} x^{2/3} \Big]_{-1}^{\ell} = \lim_{\ell \to 0^-} \frac{3}{2} (\ell^{2/3} - 1) = -3/2$

so $\int_{-1}^{8} x^{-1/3} dx = 6 + (-3/2) = 9/2$

31. Define $\displaystyle\int_0^{+\infty} \frac{dx}{\sqrt{x}(x+4)} = \int_0^a \frac{dx}{\sqrt{x}(x+4)} + \int_a^{+\infty} \frac{dx}{\sqrt{x}(x+4)}$ where $a > 0$;

take $a = 1$ for convenience,

$$\int_0^1 \frac{dx}{\sqrt{x}(x+4)} = \lim_{\ell \to 0^+} \left. \tan^{-1}(\sqrt{x}/2)\right]_\ell^1 = \lim_{\ell \to 0^+} \left[\tan^{-1}(1/2) - \tan^{-1}(\sqrt{\ell}/2)\right]$$

$$= \tan^{-1}(1/2),$$

$$\int_1^{+\infty} \frac{dx}{\sqrt{x}(x+4)} = \lim_{\ell \to +\infty} \left. \tan^{-1}(\sqrt{x}/2)\right]_1^\ell = \lim_{\ell \to +\infty} \left[\tan^{-1}(\sqrt{\ell}/2) - \tan^{-1}(1/2)\right]$$

$$= \pi/2 - \tan^{-1}(1/2); \quad \int_0^{+\infty} \frac{dx}{\sqrt{x}(x+4)} = \pi/2$$

33. (a) $\displaystyle\int_0^{+\infty} e^{-ax}dx = \lim_{\ell \to +\infty} \left. -\frac{1}{a} e^{-ax}\right]_0^\ell = \lim_{\ell \to +\infty} \left[-\frac{1}{a} e^{-a\ell} + \frac{1}{a}\right] = \frac{1}{a} = 5, \ a = \frac{1}{5}$

(b) $\displaystyle\int_0^{+\infty} \frac{dx}{x^2 + a^2} = \lim_{\ell \to +\infty} \left. \frac{1}{a} \tan^{-1}(x/a)\right]_0^\ell$

$$= \lim_{\ell \to +\infty} \frac{1}{a} \tan^{-1}(\ell/a) = \frac{\pi}{2a} = 1, \ a = \pi/2$$

35. (a) $\displaystyle\int_0^{+\infty} \cos x\, dx = \lim_{\ell \to +\infty} \left. \sin x\right]_0^\ell = \lim_{\ell \to +\infty} \sin \ell$

which does not exist and does not become infinite.

(b) $\displaystyle\lim_{\ell \to +\infty} \int_0^\ell e^{-x}\cos x\, dx = \lim_{\ell \to +\infty} \left. \frac{1}{2} e^{-x}(\sin x - \cos x)\right]_0^\ell$

$$= \lim_{\ell \to +\infty} \frac{1}{2} \left[e^{-\ell}(\sin \ell - \cos \ell) + 1\right]$$

but both $e^{-\ell}\sin \ell$ and $e^{-\ell}\cos \ell \to 0$ as $\ell \to +\infty$ (by the pinching theorem because $-e^{-\ell} \le e^{-\ell}\sin \ell \le e^{-\ell}$ and $-e^{-\ell} \le e^{-\ell}\cos \ell \le e^{-\ell}$) so

$$\int_0^{+\infty} e^{-x}\cos x\, dx = 1/2.$$

37. If $p = 1$, $\displaystyle\int_0^1 \frac{dx}{x} = \lim_{\ell \to 0^+} \ln x \Big]_\ell^1 = +\infty$;

if $p \neq 1$, $\displaystyle\int_0^1 \frac{dx}{x^p} = \lim_{\ell \to 0^+} \frac{x^{1-p}}{1-p}\Big]_\ell^1 = \lim_{\ell \to 0^+}[(1 - \ell^{1-p})/(1-p)]$

$$= \begin{cases} 1/(1-p), & p < 1 \\ +\infty, & p > 1 \end{cases}$$

39. $A = \displaystyle\int_0^4 (1-x)^{-2}dx = \int_0^1 (1-x)^{-2}dx + \int_1^4 (1-x)^{-2}dx$,

$\displaystyle\int_0^1 (1-x)^{-2}dx = \lim_{\ell \to 1^-}(1-x)^{-1}\Big]_0^\ell = +\infty$,

similarly $\displaystyle\int_1^4 (1-x)^{-2}dx = +\infty$ so $A = +\infty$

41. (a) $\displaystyle\int_2^{+\infty} \frac{x}{x^5 + 1}\, dx \leq \int_2^{+\infty} \frac{dx}{x^4} = \lim_{\ell \to +\infty} -\frac{1}{3x^3}\Big]_2^\ell = 1/24$

(b) $\displaystyle\int_1^{+\infty} e^{-x^2} dx \leq \int_1^{+\infty} e^{-x}dx = \lim_{\ell \to +\infty} -e^{-x}\Big]_1^\ell = e^{-1}$

43. (a) $V = \displaystyle\int_1^{+\infty} \frac{\pi}{x^2}\, dx = \lim_{\ell \to +\infty} -\pi/x \Big]_1^\ell = \pi$,

(b) $S = \displaystyle\int_1^{+\infty} 2\pi(1/x)\sqrt{1 + 1/x^4}\, dx$, but

$\sqrt{1 + 1/x^4} \geq 1$ if $x \geq 1$ so $(2\pi/x)\sqrt{1 + 1/x^4} \geq 2\pi/x$,

$\displaystyle\lim_{\ell \to +\infty} \int_1^\ell (2\pi/x)dx = +\infty$; S is infinite.

45. $\displaystyle\int \frac{dx}{(r^2 + x^2)^{3/2}} = \frac{1}{r^2}\int \cos\theta\, d\theta$, $x = r\tan\theta$

$$= \frac{1}{r^2}\sin\theta + C = \frac{x}{r^2\sqrt{r^2 + x^2}} + C$$

$$\text{so } u = \frac{2\pi NIr}{k} \lim_{\ell \to +\infty} \frac{x}{r^2\sqrt{r^2 + x^2}}\Bigg]_a^\ell = \frac{2\pi NI}{kr} \lim_{\ell \to +\infty} (\ell/\sqrt{r^2 + \ell^2} - a/\sqrt{r^2 + a^2})$$

$$= \frac{2\pi NI}{kr} (1 - a/\sqrt{r^2 + a^2})$$

47. (a) $\displaystyle\int_{4000}^{4000+\ell} 9.6 \times 10^{10}x^{-2}dx$

(b) $\displaystyle\int_{4000}^{+\infty} 9.6 \times 10^{10}x^{-2}dx = \lim_{\ell \to +\infty} -9.6 \times 10^{10}/x\Bigg]_{4000}^\ell = 2.4 \times 10^7$

EXERCISE SET 10.2

1. $\displaystyle\lim_{x \to 1} \frac{1/x}{1} = 1$

3. $\displaystyle\lim_{x \to 0} \frac{e^x}{\cos x} = 1$

5. $\displaystyle\lim_{\theta \to 0} \frac{\sec^2\theta}{1} = 1$

7. $\displaystyle\lim_{x \to 1} \frac{1/x}{\pi \sec^2 \pi x} = 1/\pi$

9. $\displaystyle\lim_{x \to \pi^+} \frac{\cos x}{1} = -1$

11. $\displaystyle\lim_{x \to \pi/2^-} \frac{-\sin x}{-(1/2)(\pi/2 - x)^{-1/2}} = \lim_{x \to \pi/2^-} 2(\sin x)\sqrt{\pi/2 - x} = 0$

13. $\displaystyle\lim_{x \to 0} \frac{e^x - e^{-x}}{2\sin 2x} = \lim_{x \to 0} \frac{e^x + e^{-x}}{4\cos 2x} = 1/2$

15. $\displaystyle\lim_{x \to 0} \frac{3\cos 3x}{2\cosh 2x} = 3/2$

17. $\displaystyle\lim_{x \to \pi/2} \frac{2\cos 2x}{8x} = -\frac{1}{2\pi}$

19. $\lim\limits_{x\to 0} \dfrac{1 - 1/(x + 1)}{2 \sin 2x} = \lim\limits_{x\to 0} \dfrac{x}{2(x + 1)\sin 2x}$

$\qquad\qquad = \lim\limits_{x\to 0} \dfrac{1}{4(x + 1)\cos 2x + 2 \sin 2x} = 1/4$

21. $\lim\limits_{x\to 0} \dfrac{-2x + 2\sin x}{4x^3} = \lim\limits_{x\to 0} \dfrac{-2 + 2\cos x}{12x^2} = \lim\limits_{x\to 0} - \dfrac{\sin x}{12x} = -1/12$

23. $\lim\limits_{x\to 0} \dfrac{x - \sin x}{x^3} = \lim\limits_{x\to 0} \dfrac{1 - \cos x}{3x^2} = \lim\limits_{x\to 0} \dfrac{\sin x}{6x} = 1/6$

25. $\lim\limits_{x\to 0} \dfrac{ae^{ax} - be^{bx}}{1} = a - b$

27. $\lim\limits_{x\to 0} \dfrac{1 - \sec^2 x}{\cos x - 1} = \lim\limits_{x\to 0} \dfrac{\cos^2 x - 1}{\cos^2 x(\cos x - 1)} = \lim\limits_{x\to 0} \dfrac{\cos x + 1}{\cos^2 x} = 2$

29. $\lim\limits_{x\to +\infty} \dfrac{-\dfrac{1}{1 + x^2}}{\dfrac{1}{1 + 1/x^2}\left(-\dfrac{2}{x^3}\right)} = \lim\limits_{x\to +\infty} \dfrac{x}{2} = +\infty$

31. $\lim\limits_{x\to 0^+} \dfrac{-\sin x/\cos x}{-3 \sin 3x/\cos 3x} = \lim\limits_{x\to 0^+} \dfrac{\tan x}{3 \tan 3x} = \lim\limits_{x\to 0^+} \dfrac{\sec^2 x}{9 \sec^2 3x} = 1/9$

33. (a) L'Hôpital's rule does not apply to the problem $\lim\limits_{x\to 1} \dfrac{3x^2 - 2x + 1}{3x^2 - 2x}$

   because it is not a $\dfrac{0}{0}$ form

   (b) $\lim\limits_{x\to 1} \dfrac{3x^2 - 2x + 1}{3x^2 - 2x} = 2$

35. $\lim\limits_{x \to 0} \dfrac{k + \cos \ell x}{x^2}$ does not exist if $k \neq -1$ so suppose $k = -1$, then

$\lim\limits_{x \to 0} \dfrac{-1 + \cos \ell x}{x^2} = \lim\limits_{x \to 0} \dfrac{-\ell \sin \ell x}{2x} = \lim\limits_{x \to 0} \dfrac{-\ell^2 \cos \ell x}{2} = -\ell^2/2 = -4$

if $\ell^2 = 8$, $\ell = \pm 2\sqrt{2}$

37. $\lim\limits_{x \to +\infty} \dfrac{\ln(\frac{x+1}{x-1})}{1/x} = \lim\limits_{x \to +\infty} \dfrac{-2/(x^2-1)}{-1/x^2} = \lim\limits_{x \to +\infty} \dfrac{2x^2}{x^2-1} = \lim\limits_{x \to +\infty} \dfrac{4x}{2x} = 2$

39. $\lim\limits_{x \to 0^+} \dfrac{\sin(1/x)}{(\sin x)/x}$, $\lim\limits_{x \to 0^+} \dfrac{\sin x}{x} = 1$ but $\lim\limits_{x \to 0^+} \sin(1/x)$ does not exist because

$\sin(1/x)$ oscillates between -1 and 1 as $x \to +\infty$, so $\lim\limits_{x \to 0^+} \dfrac{x \sin(1/x)}{\sin x}$ does

not exist.

41. $T(\theta) = \dfrac{1}{2} \sin \theta (1 - \cos \theta)$,

$S(\theta) = \dfrac{1}{2}(\theta - \sin \theta)$,

$\lim\limits_{\theta \to 0^+} \dfrac{T(\theta)}{S(\theta)} = \lim\limits_{\theta \to 0^+} \dfrac{\sin \theta - \sin \theta \cos \theta}{\theta - \sin \theta}$

$= \lim\limits_{\theta \to 0^+} \dfrac{\cos \theta + \sin^2 \theta - \cos^2 \theta}{1 - \cos \theta}$

$= \lim\limits_{\theta \to 0^+} \dfrac{\cos \theta - \cos 2\theta}{1 - \cos \theta}$

$= \lim\limits_{\theta \to 0^+} \dfrac{-\sin \theta + 2 \sin 2\theta}{\sin \theta}$

$= \lim\limits_{\theta \to 0^+} \dfrac{-\cos \theta + 4 \cos 2\theta}{\cos \theta} = 3$

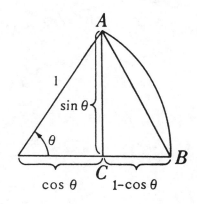

EXERCISE SET 10.3

1. $\lim\limits_{x \to +\infty} \dfrac{1/x}{1} = 0$

3. $\lim\limits_{x\to 0^+} \dfrac{-\csc^2 x}{1/x} = \lim\limits_{x\to 0^+} \dfrac{-x}{\sin^2 x} = \lim\limits_{x\to 0^+} \dfrac{-1}{2\sin x \cos x} = -\infty$

5. $\lim\limits_{x\to +\infty} \dfrac{1 + \ln x}{1 + 1/x} = +\infty$

7. $\lim\limits_{x\to +\infty} \dfrac{100 x^{99}}{e^x} = \lim\limits_{x\to +\infty} \dfrac{(100)(99) x^{98}}{e^x} = \cdots = \lim\limits_{x\to +\infty} \dfrac{(100)(99)(98)\cdots(1)}{e^x} = 0$

9. $\lim\limits_{x\to +\infty} x e^{-x} = \lim\limits_{x\to +\infty} \dfrac{x}{e^x} = \lim\limits_{x\to +\infty} \dfrac{1}{e^x} = 0$

11. $\lim\limits_{x\to +\infty} x \sin(\pi/x) = \lim\limits_{x\to +\infty} \dfrac{\sin(\pi/x)}{1/x} = \lim\limits_{x\to +\infty} \dfrac{(-\pi/x^2)\cos(\pi/x)}{-1/x^2}$

    $= \lim\limits_{x\to +\infty} \pi \cos(\pi/x) = \pi$

13. $\lim\limits_{x\to +\infty} x(e^{\sin(2/x)} - 1) = \lim\limits_{x\to +\infty} \dfrac{e^{\sin(2/x)} - 1}{1/x}$

    $= \lim\limits_{x\to +\infty} \dfrac{(-2/x^2) e^{\sin(2/x)}}{-1/x^2} = \lim\limits_{x\to +\infty} 2 e^{\sin(2/x)} = 2$

15. $y = (1 - 3/x)^x$

    $\lim\limits_{x\to +\infty} \ln y = \lim\limits_{x\to +\infty} \dfrac{\ln(1 - 3/x)}{1/x} = \lim\limits_{x\to +\infty} \dfrac{-3}{1 - 3/x} = -3, \quad \lim\limits_{x\to +\infty} y = e^{-3}$

17. $y = (e^x + x)^{1/x}$

    $\lim\limits_{x\to 0} \ln y = \lim\limits_{x\to 0} \dfrac{\ln(e^x + x)}{x} = \lim\limits_{x\to 0} \dfrac{e^x + 1}{e^x + x} = 2, \quad \lim\limits_{x\to 0} y = e^2$

19. $y = (1 + 1/x^2)^x$

    $\lim\limits_{x\to +\infty} \ln y = \lim\limits_{x\to +\infty} \dfrac{\ln(1 + 1/x^2)}{1/x} = \lim\limits_{x\to +\infty} \dfrac{2x}{x^2 + 1} = \lim\limits_{x\to +\infty} \dfrac{1}{x} = 0, \quad \lim\limits_{x\to +\infty} y = e^0 = 1$

21. $y = (1 + 1/x)^{x^2}$

$$\lim_{x \to +\infty} \ln y = \lim_{x \to +\infty} \frac{\ln(1 + 1/x)}{1/x^2} = \lim_{x \to +\infty} \frac{x^2}{2(x + 1)} = \lim_{x \to +\infty} x = +\infty, \quad \lim_{x \to +\infty} y = +\infty$$

23. $y = (2 - x)^{\tan(\pi x/2)}$

$$\lim_{x \to 1} \ln y = \lim_{x \to 1} \frac{\ln(2 - x)}{\cot(\pi x/2)} = \lim_{x \to 1} \frac{2 \sin^2(\pi x/2)}{\pi(2 - x)} = 2/\pi, \quad \lim_{x \to 1} y = e^{2/\pi}$$

25. $y = x^{\sin x}$, $\displaystyle \lim_{x \to 0^+} \ln y = \lim_{x \to 0^+} \frac{\ln x}{\csc x}$

$$= \lim_{x \to 0^+} \frac{1/x}{-\csc x \cot x} = \lim_{x \to 0^+} (- \frac{\sin x}{x}) \tan x = 0,$$

$\displaystyle \lim_{x \to 0^+} y = e^0 = 1$

27. $y = (\sin x)^{3/\ln x}$

$$\lim_{x \to 0^+} \ln y = \lim_{x \to 0^+} \frac{3 \ln \sin x}{\ln x} = \lim_{x \to 0^+} (3 \cos x) \frac{x}{\sin x} = 3, \quad \lim_{x \to 0^+} y = e^3$$

29. $y = (\tan x)^{\cos x}$,

$$\lim_{x \to \pi/2^-} \ln y = \lim_{x \to \pi/2^-} \frac{\ln \tan x}{\sec x} = \lim_{x \to \pi/2^-} \frac{\sec^2 x/\tan x}{\sec x \tan x} = \lim_{x \to \pi/2^-} \frac{\cos x}{\sin^2 x} = 0,$$

$\displaystyle \lim_{x \to \pi/2^-} y = e^0 = 1$

31. $y = (1 + x^2)^{1/\ln x}$

$$\lim_{x \to +\infty} \ln y = \lim_{x \to +\infty} \frac{\ln(1 + x^2)}{\ln x} = \lim_{x \to +\infty} \frac{2x^2}{1 + x^2} = \lim_{x \to +\infty} 2 = 2, \quad \lim_{x \to +\infty} y = e^2$$

33. $\displaystyle \lim_{\theta \to 0} (\frac{1 + \cos \theta}{1 - \cos^2\theta} - \frac{2}{\sin^2\theta}) = \lim_{\theta \to 0} \frac{\cos \theta - 1}{\sin^2\theta} = \lim_{\theta \to 0} \frac{-\sin \theta}{2 \sin \theta \cos \theta} = -1/2$

35. $\lim\limits_{x\to 0}(\dfrac{1}{\sin x} - \dfrac{1}{x}) = \lim\limits_{x\to 0} \dfrac{x - \sin x}{x \sin x} = \lim\limits_{x\to 0} \dfrac{1 - \cos x}{x \cos x + \sin x}$

$= \lim\limits_{x\to 0} \dfrac{\sin x}{2 \cos x - x \sin x} = 0$

37. $\lim\limits_{x\to 0} (\dfrac{\cos x}{\sin x} - \dfrac{1}{\sin x}) = \lim\limits_{x\to 0} \dfrac{\cos x - 1}{\sin x} = \lim\limits_{x\to 0} \dfrac{-\sin x}{\cos x} = 0$

39. $\lim\limits_{x\to +\infty}[x - \ln(x^2 + 1)] = \lim\limits_{x\to +\infty}[\ln e^x - \ln(x^2 + 1)] = \lim\limits_{x\to +\infty} \ln \dfrac{e^x}{x^2 + 1},$

$\lim\limits_{x\to +\infty} \dfrac{e^x}{x^2 + 1} = \lim\limits_{x\to +\infty} \dfrac{e^x}{2x} = \lim\limits_{x\to +\infty} \dfrac{e^x}{2} = +\infty$ so $\lim\limits_{x\to +\infty}[x - \ln(x^2 + 1)] = +\infty$

41. $\lim\limits_{x\to 0^+} \dfrac{\cot x}{\cot 2x} = \lim\limits_{x\to 0^+} \dfrac{\tan 2x}{\tan x} = \lim\limits_{x\to 0^+} \dfrac{2 \sec^2 2x}{\sec^2 x} = 2$

43. In each case, apply L'Hôpital's rule n times.

   (a) $\lim\limits_{x\to +\infty} \dfrac{x^n}{e^x} = \cdots = \lim\limits_{x\to +\infty} \dfrac{n(n - 1)(n - 2)\cdots(1)}{e^x} = 0$

   (b) $\lim\limits_{x\to +\infty} \dfrac{e^x}{x^n} = \cdots = \lim\limits_{x\to +\infty} \dfrac{e^x}{n(n - 1)(n - 2)\cdots(1)} = +\infty$

45. (a) 0   (b) $+\infty$ (c) 0   (d) $-\infty$
    (e) $+\infty$ (f) $+\infty$ (g) $-\infty$ (h) $-\infty$

47. Let $y = x^x$, $\lim\limits_{x\to +\infty} y = +\infty$ and

   $\lim\limits_{x\to 0^+} y = 1$.   Use logarithmic differ-

   entiation to get $dy/dx = x^x(1 + \ln x)$
   so $dy/dx = 0$ when $x = e^{-1}$, $dy/dx < 0$
   if $x < e^{-1}$, $dy/dx > 0$ if $x > e^{-1}$,
   and $dy/dx \to -\infty$ as $x\to 0^+$.   Also,
   $d^2y/dx^2 = x^x[1/x + (1 + \ln x)^2] > 0$
   for $x > 0$ so the curve is always
   concave up.

49. $\int \ln x \, dx = x \ln x - x + C$ (integration by parts),

$$\int_0^1 \ln x \, dx = \lim_{\ell \to 0^+} \int_\ell^1 \ln x \, dx = \lim_{\ell \to 0^+} (x \ln x - x) \Big]_\ell^1$$
$$= \lim_{\ell \to 0^+} (-1 - \ell \ln \ell + \ell),$$

but $\lim_{\ell \to 0^+} \ell \ln \ell = \lim_{\ell \to 0^+} \dfrac{\ln \ell}{1/\ell} = \lim_{\ell \to 0^+} (-\ell) = 0$ so $\int_0^1 \ln x \, dx = -1$

51. $\int xe^{-3x} dx = -\dfrac{1}{3} xe^{-3x} - \dfrac{1}{9} e^{-3x} + C$ (integration by parts),

$$\int_0^{+\infty} xe^{-3x} dx = \lim_{\ell \to +\infty} \int_0^\ell xe^{-3x} dx = \lim_{\ell \to +\infty} (-\dfrac{1}{3} xe^{-3x} - \dfrac{1}{9} e^{-3x}) \Big]_0^\ell$$
$$= \lim_{\ell \to +\infty} (-\dfrac{1}{3} \ell e^{-3\ell} - \dfrac{1}{9} e^{-3\ell} + \dfrac{1}{9}),$$

$\dfrac{1}{\ell} \dfrac{\ell - 1}{\ell^2}$

$1 - \dfrac{1}{\ell}$

but $\lim_{\ell \to +\infty} \ell e^{-3\ell} = \lim_{\ell \to +\infty} \dfrac{\ell}{e^{3\ell}} = \lim_{\ell \to +\infty} \dfrac{1}{3e^{3\ell}} = 0$ so $\int_0^{+\infty} xe^{-3x} dx = 1/9$

53. (a) $\displaystyle\int_0^\ell \sqrt{1 + t^3} \, dt \geq \int_0^\ell t^{3/2} dt = \dfrac{2}{5} t^{5/2} \Big]_0^\ell = \dfrac{2}{5} \ell^{5/2},$

$$\lim_{\ell \to +\infty} \int_0^\ell t^{3/2} dt = \lim_{\ell \to +\infty} \dfrac{2}{5} \ell^{5/2} = +\infty \text{ so } \int_0^{+\infty} \sqrt{1 + t^3} \, dt = +\infty$$

(b) $\lim_{x \to +\infty} \dfrac{2\sqrt{1 + 8x^3}}{(5/2)x^{3/2}} = \lim_{x \to +\infty} \dfrac{4}{5} \sqrt{1/x^3 + 8} = 8\sqrt{2}/5$

55. Use of L'Hôpital's rule leads to $\lim_{x \to +\infty} (2 + x \cos x + \sin x)$ which does not exist, nor is it $\pm\infty$, so the original limit cannot be determined by this method. $\lim_{x \to +\infty} \dfrac{x(2 + \sin x)}{x + 1} = \lim_{x \to +\infty} \dfrac{2 + \sin x}{1 + 1/x}$, which does not exist because $\sin x$ oscillates between -1 and 1 as $x \to +\infty$.

57. (a) $y = x^{\frac{\ln a}{1 + \ln x}}$,

$$\lim_{x \to 0^+} \ln y = \lim_{x \to 0^+} \frac{(\ln a)\ln x}{1 + \ln x} = \lim_{x \to 0^+} \frac{(\ln a)/x}{1/x} = \lim_{x \to 0^+} \ln a = \ln a,$$

$$\lim_{x \to 0^+} y = e^{\ln a} = a$$

(b) same as part (a) with $x \to +\infty$

(c) $y = (x + 1)^{\frac{\ln a}{x}}$, $\lim_{x \to 0} \ln y = \lim_{x \to 0} \frac{(\ln a)\ln(x + 1)}{x} = \lim_{x \to 0} \frac{\ln a}{x + 1} = \ln a,$

$$\lim_{x \to 0} y = e^{\ln a} = a$$

SUPPLEMENTARY EXERCISES CHAPTER 10

1. $\displaystyle\int_0^{+\infty} \frac{dx}{x^2 + 4} = \lim_{\ell \to +\infty} \frac{1}{2} \tan^{-1}(x/2)\Big]_0^{\ell} = \lim_{\ell \to +\infty} \frac{1}{2} \tan^{-1}(\ell/2) = \pi/4,$

$\displaystyle\int_\ell^0 \frac{dx}{x^2 + 4} = \lim_{\ell \to -\infty} \frac{1}{2} \tan^{-1}(x/2)\Big]_\ell^0 = \pi/4$ so $\displaystyle\int_{-\infty}^{+\infty} \frac{dx}{x^2 + 4} = \pi/2$

3. $\displaystyle\lim_{\ell \to 1^-} -\sqrt{1 - x^2}\,\Big]_0^{\ell} = \lim_{\ell \to 1^-} (-\sqrt{1 - \ell^2} + 1) = 1$

5. $\displaystyle\int_0^1 x^{-2/3} dx = \lim_{\ell \to 0^+} 3x^{1/3}\Big]_\ell^1 = 3,$

$\displaystyle\int_{-1}^0 x^{-2/3} dx = \lim_{\ell \to 0^-} 3x^{1/3}\Big]_{-1}^{\ell} = 3,\ \int_{-1}^1 x^{-2/3} dx = 6$

7. $\lim\limits_{\ell \to +\infty} -\frac{1}{2} e^{-x^2}\Big]_0^{\ell} = \lim\limits_{\ell \to +\infty} \frac{1}{2}(-e^{-\ell^2} + 1) = 1/2,$

$\lim\limits_{\ell \to -\infty} -\frac{1}{2} e^{-x^2}\Big]_{\ell}^0 = \lim\limits_{\ell \to -\infty} \frac{1}{2}(-1 + e^{-\ell^2}) = -1/2,$

so $\displaystyle\int_{-\infty}^{+\infty} xe^{-x^2}dx = 1/2 - 1/2 = 0$

9. $\lim\limits_{\ell \to 0^+} \ln|\sin x|\,\Big]_{\ell}^{\pi/2} = \lim\limits_{\ell \to 0^+} -\ln|\sin \ell| = +\infty,$ diverges

11. $\lim\limits_{\ell \to +\infty} -1/\ln x\,\Big]_e^{\ell} = \lim\limits_{\ell \to +\infty} (-1/\ln \ell + 1) = 1$

13. $\lim\limits_{\ell \to +\infty} \tan^{-1}(x + 1)\,\Big]_e^{\ell} = \lim\limits_{\ell \to +\infty} [\tan^{-1}(\ell + 1) - \tan^{-1}(1)] = \pi/2 - \pi/4 = \pi/4.$

15. If $n = -1$ then $\displaystyle\int_0^1 \frac{\ln x}{x} dx = \lim\limits_{\ell \to 0^+} \frac{1}{2}(\ln x)^2\,\Big]_{\ell}^1 = -\infty$

so the integral diverges.
If $n \neq -1$ then (using integration by parts)

$$\int_0^1 x^n \ln x\, dx = \lim\limits_{\ell \to 0^+} \left[\frac{x^{n+1}}{n + 1}\ln x - \frac{x^{n+1}}{(n + 1)^2}\right]_{\ell}^1$$

$$= \lim\limits_{\ell \to 0^+} \left[-\frac{1}{(n + 1)^2} - \frac{\ell^{n+1}}{n + 1}\ln \ell + \frac{\ell^{n+1}}{(n + 1)^2}\right].$$

If $n < -1$ then $n + 1 < 0$, $\lim\limits_{\ell \to 0^+} \ell^{n+1}\ln \ell = -\infty$ and $\lim\limits_{\ell \to 0^+} \ell^{n+1} = +\infty$ so the integral diverges. If $n > -1$ then

$$\lim\limits_{\ell \to 0^+} \ell^{n+1}\ln \ell = \lim\limits_{\ell \to 0^+} \frac{\ln \ell}{\ell^{-(n+1)}} = \lim\limits_{\ell \to 0^+} -\frac{\ell^{n+1}}{n + 1} = 0 \text{ and}$$

$\lim\limits_{\ell \to 0^+} \ell^{n+1} = 0$ so the integral converges to $-1/(n + 1)^2$.

17. $\displaystyle\lim_{x\to 0} \frac{3xe^{3x} + e^{3x} - 1}{2\sin 2x} = \lim_{x\to 0} \frac{9xe^{3x} + 6e^{3x}}{4\cos 2x} = 3/2$

19. $\displaystyle\lim_{x\to 0^+} \frac{e^{1/x}}{1/x^2} = \lim_{x\to 0^+} \frac{(-1/x^2)e^{1/x}}{-2/x^3} = \lim_{x\to 0^+} \frac{e^{1/x}}{2/x} = \lim_{x\to 0^+} \frac{(-1/x^2)e^{1/x}}{-2/x^2}$

$\displaystyle = \lim_{x\to 0^+} (1/2)e^{1/x} = +\infty$

21. $\displaystyle\lim_{x\to 0^-} x^2 e^{1/x} = (0)(0) = 0$

23. $\displaystyle\lim_{\theta\to 0}\left(\frac{1}{\theta\sin\theta} - \frac{1}{\theta^2}\right) = \lim_{\theta\to 0} \frac{\theta - \sin\theta}{\theta^2\sin\theta} = \lim_{\theta\to 0} \frac{1 - \cos\theta}{\theta^2\cos\theta + 2\theta\sin\theta}$

$\displaystyle = \lim_{\theta\to 0} \frac{\sin\theta}{-\theta^2\sin\theta + 4\theta\cos\theta + 2\sin\theta}$

$\displaystyle = \lim_{\theta\to 0} \frac{\cos\theta}{-\theta^2\cos\theta - 6\theta\sin\theta + 6\cos\theta} = 1/6$

25. $\displaystyle\lim_{x\to 2} \frac{1 - e^{x-2}}{2\pi\sin 2\pi x} = \lim_{x\to 2} \frac{-e^{x-2}}{4\pi^2\cos 2\pi x} = -1/(4\pi^2)$

27. $\displaystyle\lim_{x\to 0} \frac{\sin(x^2)}{2x\cos(x^2)} = \lim_{x\to 0} \frac{2x\cos(x^2)}{-4x^2\sin(x^2) + 2\cos(x^2)} = 0$

29. $\displaystyle\lim_{x\to +\infty} \frac{3(\ln x)^2}{x} = \lim_{x\to +\infty} \frac{6\ln x}{x} = \lim_{x\to +\infty} \frac{6}{x} = 0$

31. $y = (1 + x)^{\ln x}$, $\displaystyle\lim_{x\to 0^+} \ln y = \lim_{x\to 0^+} \ln x \ln(1 + x) = \lim_{x\to 0^+} \frac{\ln(1 + x)}{1/\ln x}$

$\displaystyle = \lim_{x\to 0^+} \frac{1/(1 + x)}{-1/[x(\ln x)^2]} = \lim_{x\to 0^+} \frac{-x(\ln x)^2}{1 + x},$

but $\lim\limits_{x\to 0^+} x(\ln x)^2 = \lim\limits_{x\to 0^+} \dfrac{(\ln x)^2}{1/x} = \lim\limits_{x\to 0^+} \dfrac{(2\ln x)/x}{-1/x^2}$

$\qquad\qquad\qquad\qquad = \lim\limits_{x\to 0^+} \dfrac{2\ln x}{-1/x} = \lim\limits_{x\to 0^+} \dfrac{2/x}{-1/x^2} = \lim\limits_{x\to 0^+}(-2x) = 0$

so $\lim\limits_{x\to 0^+} \dfrac{-x(\ln x)^2}{1+x} = \dfrac{0}{1} = 0$ and $\lim\limits_{x\to 0^+} y = e^0 = 1$

33. (a) $A = \displaystyle\int_8^{+\infty} x^{-2/3}\,dx = \lim\limits_{\ell\to +\infty} 3x^{1/3}\Big]_8^{\ell} = +\infty$

(b) $V = \displaystyle\int_8^{+\infty} \pi x^{-4/3}\,dx = \lim\limits_{\ell\to +\infty} -3\pi x^{-1/3}\Big]_8^{\ell} = 3\pi/2$

35. $f(x) = \sqrt{x - x^2} - \sin^{-1}\sqrt{x} = \sqrt{x(1-x)} - \sin^{-1}\sqrt{x}$, the domain of f is $0 \le x \le 1$,

$f'(x) = \dfrac{1}{2}(x - x^2)^{-1/2}(1 - 2x) - \dfrac{1}{\sqrt{1-x}}\left(\dfrac{1}{2}x^{-1/2}\right) = -\dfrac{x}{\sqrt{x-x^2}}$,

$1 + [f'(x)]^2 = 1 + \dfrac{x^2}{x - x^2} = \dfrac{1}{1-x}$,

$L = \displaystyle\int_0^1 \dfrac{1}{\sqrt{1-x}}\,dx = \lim\limits_{\ell\to 1^-} -2\sqrt{1-x}\Big]_0^{\ell} = \lim\limits_{\ell\to 1^-} 2(-\sqrt{1-\ell}+1) = 2$

CHAPTER 11

EXERCISE SET 11.1

1. 1/3, 2/4, 3/5, 4/6, 5/7,...; $\lim\limits_{n\to+\infty} \dfrac{n}{n+2} = 1$, converges

3. 2, 2, 2, 2, 2,...; $\lim\limits_{n\to+\infty} 2 = 2$, converges

5. $\dfrac{\ln 1}{1}$, $\dfrac{\ln 2}{2}$, $\dfrac{\ln 3}{3}$, $\dfrac{\ln 4}{4}$, $\dfrac{\ln 5}{5}$, ...; $\lim\limits_{n\to+\infty} \dfrac{\ln n}{n} = \lim\limits_{n\to+\infty} \dfrac{1}{n} = 0$, converges

7. 0, 2, 0, 2, 0,...; diverges

9. −1, 16/9, −54/28, 128/65, −250/126,...; diverges because odd-numbered terms approach −2, even-numbered terms approach 2.

11. 6/2, 12/8, 20/18, 30/32, 42/50,...; $\lim\limits_{n\to+\infty} \dfrac{1}{2}(1 + 1/n)(1 + 2/n) = 1/2$, converges

13. cos(3), cos(3/2), cos(1), cos(3/4), cos(3/5), ...; $\lim\limits_{n\to+\infty} \cos(3/n) = 1$, converges

15. e^{-1}, $4e^{-2}$, $9e^{-3}$, $16e^{-4}$, $25e^{-5}$, ...; $\lim\limits_{x\to+\infty} x^2 e^{-x} = \lim\limits_{x\to+\infty} \dfrac{x^2}{e^x} = 0$, so $\lim\limits_{n\to+\infty} n^2 e^{-n} = 0$, converges

17.   2, $(5/3)^2$, $(6/4)^3$, $(7/5)^4$, $(8/6)^5$, ...; let $y = \left[\dfrac{x + 3}{x + 1}\right]^x$,

$$\lim_{x\to+\infty} \ln y = \lim_{x\to+\infty} \frac{\ln \dfrac{x + 3}{x + 1}}{1/x} = \lim_{x\to+\infty} \frac{2x^2}{(x + 1)(x + 3)} = 2,$$

so $\lim\limits_{n\to+\infty} \left[\dfrac{n + 3}{n + 1}\right]^n = e^2$

$$\frac{x + 1}{x + 3}$$

19. $\left\{\dfrac{2n - 1}{2n}\right\}_{n=1}^{+\infty}$; $\lim\limits_{n\to+\infty} \dfrac{2n - 1}{2n} = 1$, converges

21. $\{1/3^n\}_{n=1}^{+\infty}$; $\lim\limits_{n\to+\infty} 1/3^n = 0$, converges

23. $\left\{\dfrac{1}{n} - \dfrac{1}{n + 1}\right\}_{n=1}^{+\infty}$; $\lim\limits_{n\to+\infty} \left(\dfrac{1}{n} - \dfrac{1}{n + 1}\right) = 0$, converges

25. $\{\sqrt{n + 1} - \sqrt{n + 2}\}_{n=1}^{+\infty}$;

$$\lim_{n\to+\infty} (\sqrt{n + 1} - \sqrt{n + 2}) = \lim_{n\to+\infty} \frac{(n + 1) - (n + 2)}{\sqrt{n + 1} + \sqrt{n + 2}} = \lim_{n\to+\infty} \frac{-1}{\sqrt{n + 1} + \sqrt{n + 2}}$$
$$= 0, \text{ converges}$$

27.   (a) 3, 6, 12, 24, 48, 96, 192, 384, $\cdots$
      (b) 1, 1, 2, 3, 5, 8, 13, 21, $\cdots$

29.   (a) 1, $\dfrac{1}{4} + \dfrac{2}{4}$, $\dfrac{1}{9} + \dfrac{2}{9} + \dfrac{3}{9}$, $\dfrac{1}{16} + \dfrac{2}{16} + \dfrac{3}{16} + \dfrac{4}{16}$, $\cdots$ = 1, 3/4, 2/3, 5/8, $\cdots$

      (b) $a_n = \dfrac{1}{n^2} (1 + 2 + \cdots + n) = \dfrac{1}{n^2} \dfrac{1}{2} n(n + 1) = \dfrac{1}{2} \dfrac{n + 1}{n}$

         $\lim\limits_{n\to+\infty} a_n = \lim\limits_{n\to+\infty} \dfrac{1}{2} (1 + 1/n) = 1/2$

31. $\left|\dfrac{1}{n} - 0\right| = \dfrac{1}{n} < \epsilon$ if $n > 1/\epsilon$
      (a) $1/\epsilon = 1/0.5$ = 2,   N = 3
      (b) $1/\epsilon = 1/0.1$ = 10,   N = 11
      (c) $1/\epsilon = 1/0.001$ = 1000, N = 1001

33. (a) $\left|\dfrac{1}{n} - 0\right| = \dfrac{1}{n} < \epsilon$ if $n > 1/\epsilon$, choose any $N > 1/\epsilon$.

(b) $\left|\dfrac{n}{n+1} - 1\right| = \dfrac{1}{n+1} < \epsilon$ if $n > 1/\epsilon - 1$, choose any $N > 1/\epsilon - 1$.

35. (a) Let s_n denote the length of each of the n sides of the polygon, then

$$\sin(\pi/n) = \dfrac{s_n/2}{r}$$

$$s_n = 2r\sin(\pi/n)$$

$$P_n = n\, s_n = 2rn\sin(\pi/n)$$

(b) $\displaystyle\lim_{n\to+\infty} 2rn\sin(\pi/n) = \lim_{n\to+\infty} \dfrac{2r\sin(\pi/n)}{1/n}$

$$= \lim_{n\to+\infty} 2\pi r\cos(\pi/n) = 2\pi r$$

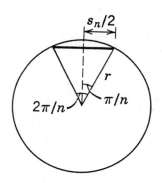

37. Let $y = (2^x + 3^x)^{1/x}$,

$$\lim_{x\to+\infty} \ln y = \lim_{x\to+\infty} \dfrac{\ln(2^x + 3^x)}{x} = \lim_{x\to+\infty} \dfrac{2^x\ln 2 + 3^x\ln 3}{2^x + 3^x}$$

$$= \lim_{x\to+\infty} \dfrac{(2/3)^x\ln 2 + \ln 3}{(2/3)^x + 1} = \ln 3 \text{ so } \lim_{n\to+\infty} (2^n + 3^n)^{1/n} = e^{\ln 3} = 3$$

EXERCISE SET 11.2

1. $a_n - a_{n+1} = \dfrac{1}{n} - \dfrac{1}{n+1} = \dfrac{1}{n(n+1)} > 0$ for $n \geq 1$, so decreasing.

3. $a_n - a_{n+1} = \dfrac{n}{2n+1} - \dfrac{n+1}{2n+3} = -\dfrac{1}{(2n+1)(2n+3)} < 0$ for $n \geq 1$, so increasing.

5. $a_n - a_{n+1} = (n - 2^n) - (n+1 - 2^{n+1}) = 2^n - 1 > 0$ for $n \geq 1$, so decreasing.

7. $\dfrac{a_{n+1}}{a_n} = \dfrac{(n+1)/(2n+3)}{n/(2n+1)} = \dfrac{(n+1)(2n+1)}{n(2n+3)} = \dfrac{2n^2+3n+1}{2n^2+3n} > 1$

   for $n \geq 1$, so increasing.

9. $\dfrac{a_{n+1}}{a_n} = \dfrac{(n+1)e^{-(n+1)}}{ne^{-n}} = (1+1/n)e^{-1} < 1$ for $n \geq 1$, so decreasing.

11. $\dfrac{a_{n+1}}{a_n} = \dfrac{2^{n+1}/(n+1)!}{2^n/n!} = \dfrac{2^{n+1}}{2^n} \cdot \dfrac{n!}{(n+1)!} = \dfrac{2}{n+1} \leq 1$ for $n \geq 1$, so

    nonincreasing.

13. $\dfrac{a_{n+1}}{a_n} = \dfrac{(n+1)!}{3^{n+1}} \cdot \dfrac{3^n}{n!} = \dfrac{n+1}{3}$, $\dfrac{n+1}{3} \leq 1$ for $n = 1$ and 2, $\dfrac{n+1}{3} > 1$ for

    $n \geq 3$, so not monotone.

15. $\dfrac{a_{n+1}}{a_n} = \dfrac{10^{n+1}}{(2n+2)!} \cdot \dfrac{(2n)!}{10^n} = \dfrac{10}{(2n+2)(2n+1)} < 1$ for $n \geq 1$,

    so decreasing.

17. $\dfrac{a_{n+1}}{a_n} = \dfrac{(n+1)^{n+1}}{(n+1)!} \cdot \dfrac{n!}{n^n} = \dfrac{(n+1)^n}{n^n} = (1+1/n)^n > 1$ for $n \geq 1$,

    so increasing.

19. $f(x) = x/(2x+1)$, $f'(x) = 1/(2x+1)^2 > 0$ for $x \geq 1$, so increasing.

21. $f(x) = 1/(x + \ln x)$, $f'(x) = -\dfrac{1 + 1/x}{(x + \ln x)^2} < 0$ for $x \geq 1$, so decreasing.

23. $f(x) = \dfrac{\ln(x+2)}{x+2}$, $f'(x) = \dfrac{1 - \ln(x+2)}{(x+2)^2} < 0$ for $x \geq 1$, so decreasing.

25. $a_{n+1}/a_n = (n+1)/(5n) < 1$ for $n \geq 1$ so decreasing; 0 is a lower bound

    of $n/5^n$ for $n \geq 1$ so the sequence converges.

27. Let $f(x) = x - 1/x$, then $f'(x) = 1 + 1/x^2 > 0$ so increasing; $n - 1/n$ has
    no upper bound so the sequence diverges.

29. Let $f(x) = 2 + 1/x$, then $f'(x) = -1/x^2 < 0$ for $x \geq 1$ so decreasing; 2 is a lower bound because $2 + 1/n \geq 2$ for $n \geq 1$ so the sequence converges.

31. (a) We can discard the first two terms without affecting the convergence or the limit. This leaves the sequence $1, \frac{1}{2}, \frac{1}{3}, \ldots, \frac{1}{n}, \ldots$.

   Because $\lim\limits_{n \to +\infty} \frac{1}{n} = 0$, this is also the limit of the given sequence.

   (b) Discard the first four terms, leaving $1, 2, 3, \ldots, n, \ldots$;

   $\lim\limits_{n \to +\infty} n = +\infty$ so the limit does not exist.

33. $\dfrac{a_{n+1}}{a_n} = \dfrac{3^{n+1}(1 + 3^{2n})}{3^n(1 + 3^{2n+2})} = \dfrac{3(1 + 3^{2n})}{1 + 3^{2n+2}} < \dfrac{3(1 + 3^{2n})}{3^{2n+2}} = \dfrac{1}{3^{2n+1}} + \dfrac{1}{3} < 1$

   for $n \geq 1$ so decreasing.

35. (a) If $\{a_n\}_{n=1}^{+\infty}$ is nonincreasing then $a_n \geq a_{n+1}$ for all n, thus

   $-a_n \leq -a_{n+1}$ so $\{-a_n\}_{n=1}^{+\infty}$ is nondecreasing.

   (b) From part (a), $\{-a_n\}_{n=1}^{+\infty}$ is nondecreasing so Theorem 11.2.2 applies. Suppose there is a finite constant M_1 such that

   $-a_n \leq M_1$ and $\lim\limits_{n \to +\infty} (-a_n) = L_1 \leq M_1$, then $a_n \geq -M_1$, and

   $\lim\limits_{n \to +\infty} a_n = -L_1 \geq -M_1$; part (a) of Theorem 11.2.3 follows by

   letting $M = -M_1$ and $L = -L_1$. Suppose no such constant exists,

   then $\lim\limits_{n \to +\infty} (-a_n) = +\infty$ so $\lim\limits_{n \to +\infty} a_n = -\infty$.

37. (a) $\dfrac{a_{n+1}}{a_n} = \dfrac{(n + 1)^{n+1}}{(n + 1)!e^{n+1}} \cdot \dfrac{n!e^n}{n^n} = \dfrac{(1 + 1/n)^n}{e} < 1$ for $n \geq 1$ because

   $(1 + 1/n)^n < e$ for $n \geq 1$.

   (b) Yes, because it is decreasing and bounded below by 0.

EXERCISE SET 11.3

1. (a) $s_1 = 2$, $s_2 = 12/5$, $s_3 = 62/25$, $s_4 = 312/125$

$$s_n = \frac{2 - 2(1/5)^n}{1 - 1/5} = \frac{5}{2} - \frac{5}{2}(1/5)^n, \quad \lim_{n \to +\infty} s_n = 5/2, \text{ converges}$$

(b) $\frac{1}{(k + 1)(k + 2)} = \frac{1}{k + 1} - \frac{1}{k + 2}$, $s_1 = 1/6$, $s_2 = 1/4$, $s_3 = 3/10$, $s_4 = 1/3$

$$s_n = \frac{1}{2} - \frac{1}{n + 2}, \quad \lim_{n \to +\infty} s_n = 1/2, \text{ converges}$$

(c) $s_1 = 1/4$, $s_2 = 3/4$, $s_3 = 7/4$, $s_4 = 15/4$

$$s_n = \frac{(1/4) - (1/4)2^n}{1 - 2} = -\frac{1}{4} + \frac{1}{4}(2^n), \quad \lim_{n \to +\infty} s_n = +\infty, \text{ diverges}$$

3. geometric, $a = 1$, $r = -3/4$, sum $= \dfrac{1}{1 - (-3/4)} = 4/7$

5. geometric, $a = 7$, $r = -1/6$, sum $= \dfrac{7}{1 + 1/6} = 6$

7. geometric, $r = -3/2$, diverges

9. $s_n = \displaystyle\sum_{k=1}^{n} \left(\frac{1}{k + 2} - \frac{1}{k + 3}\right) = \frac{1}{3} - \frac{1}{n + 3}$, $\displaystyle\lim_{n \to +\infty} s_n = 1/3$

11. $s_n = \displaystyle\sum_{k=1}^{n} \left(\frac{1/3}{3k - 1} - \frac{1/3}{3k + 2}\right) = \frac{1}{6} - \frac{1/3}{3n + 2}$, $\displaystyle\lim_{n \to +\infty} s_n = 1/6$

13. $\displaystyle\sum_{k=1}^{\infty} \frac{4^{k+2}}{7^{k-1}} = \sum_{k=1}^{\infty} 64\left(\frac{4}{7}\right)^{k-1}$; geometric, $a = 64$, $r = 4/7$,

sum $= \dfrac{64}{1 - 4/7} = 448/3$

15. geometric, $a = -1/2$, $r = -1/2$, sum $= \dfrac{-1/2}{1 + 1/2} = -1/3$

17. $0.4444\cdots = 0.4 + 0.04 + 0.004 + \cdots = \dfrac{0.4}{1 - 0.1} = 4/9$

19. $5.373737\cdots = 5 + 0.37 + 0.0037 + 0.000037 + \cdots$

$$= 5 + \dfrac{0.37}{1 - 0.01} = 5 + 37/99 = 532/99$$

21. $0.782178217821\cdots = 0.7821 + 0.00007821 + 0.000000007821 + \cdots$

$$= \dfrac{0.7821}{1 - 0.0001} = 7821/9999 = 869/1111$$

23. $s_n = \ln \dfrac{1}{2} + \ln \dfrac{2}{3} + \ln \dfrac{3}{4} + \cdots + \ln \dfrac{n}{n + 1}$

$$= \ln(\dfrac{1}{2} \cdot \dfrac{2}{3} \cdot \dfrac{3}{4} \cdots \dfrac{n}{n + 1}) = \ln \dfrac{1}{n + 1} = -\ln(n + 1),$$

$\lim\limits_{n \to +\infty} s_n = -\infty$, series diverges.

25. $\ln(1 - 1/k^2) = \ln \dfrac{k^2 - 1}{k^2} = \ln \dfrac{(k - 1)(k + 1)}{k^2}$

$$= \ln \dfrac{k - 1}{k} + \ln \dfrac{k + 1}{k} = \ln \dfrac{k - 1}{k} - \ln \dfrac{k}{k + 1},$$

$$s_n = \sum_{k=2}^{n+1} \left[\ln \dfrac{k - 1}{k} - \ln \dfrac{k}{k + 1} \right]$$

$$= (\ln \dfrac{1}{2} - \ln \dfrac{2}{3}) + (\ln \dfrac{2}{3} - \ln \dfrac{3}{4}) + (\ln \dfrac{3}{4} - \ln \dfrac{4}{5}) + \cdots$$

$$+ (\ln \dfrac{n}{n + 1} - \ln \dfrac{n + 1}{n + 2})$$

$$= \ln \dfrac{1}{2} - \ln \dfrac{n + 1}{n + 2}, \ \lim\limits_{n \to +\infty} s_n = \ln \dfrac{1}{2} = -\ln 2$$

27. $s_n = (1 - 1/3) + (1/2 - 1/4) + (1/3 - 1/5) + (1/4 - 1/6) + \cdots$

$$+ [1/n - 1/(n + 2)]$$

$$= (1 + 1/2 + 1/3 + \cdots + 1/n) - (1/3 + 1/4 + 1/5 + \cdots + 1/(n + 2))$$

$$= 3/2 - 1/(n + 1) - 1/(n + 2), \ \lim\limits_{n \to +\infty} s_n = 3/2$$

29. (a) $\displaystyle\sum_{k=0}^{\infty} (-1)^k x^k = 1 - x + x^2 - x^3 + \cdots = \dfrac{1}{1 - (-x)} = \dfrac{1}{1 + x}$

if $|-x| < 1, \ |x| < 1, \ -1 < x < 1.$

(b) $\displaystyle\sum_{k=0}^{\infty} (x - 3)^k = 1 + (x - 3) + (x - 3)^2 + \cdots = \dfrac{1}{1 - (x - 3)} = \dfrac{1}{4 - x}$

if $|x - 3| < 1$, $2 < x < 4$.

(c) $\displaystyle\sum_{k=0}^{\infty} (-1)^k x^{2k} = 1 - x^2 + x^4 - x^6 + \cdots = \dfrac{1}{1 - (-x^2)} = \dfrac{1}{1 + x^2}$

if $|-x^2| < 1$, $|x| < 1$, $-1 < x < 1$.

31. Geometric series, $a = 1/x^2$, $r = 2/x$. Converges for $|2/x| < 1$, $|x| > 2$;

$S = \dfrac{1/x^2}{1 - 1/x} = \dfrac{1}{x^2 - 1}$.

33. Geometric series, $a = \sin x$, $r = -\dfrac{1}{2}\sin x$. Converges for

$\left|-\dfrac{1}{2}\sin x\right| < 1$, $|\sin x| < 2$, so converges for all values of x.

$S = \dfrac{\sin x}{1 + \dfrac{1}{2}\sin x} = \dfrac{2\sin x}{2 + \sin x}$.

35. $a_2 = \dfrac{1}{2} a_1 + \dfrac{1}{2}$

$a_3 = \dfrac{1}{2} a_2 + \dfrac{1}{2} = \dfrac{1}{2^2} a_1 + \dfrac{1}{2^2} + \dfrac{1}{2}$

$a_4 = \dfrac{1}{2} a_3 + \dfrac{1}{2} = \dfrac{1}{2^3} a_1 + \dfrac{1}{2^3} + \dfrac{1}{2^2} + \dfrac{1}{2}$

$a_5 = \dfrac{1}{2} a_4 + \dfrac{1}{2} = \dfrac{1}{2^4} a_1 + \dfrac{1}{2^4} + \dfrac{1}{2^3} + \dfrac{1}{2^2} + \dfrac{1}{2}$

$\vdots$

$a_n = \dfrac{1}{2^{n-1}} a_1 + \dfrac{1}{2^{n-1}} + \dfrac{1}{2^{n-2}} + \cdots + \dfrac{1}{2}$

$\displaystyle\lim_{n\to+\infty} a_n = \lim_{n\to+\infty} \dfrac{a_1}{2^{n-1}} + \sum_{n=1}^{\infty} \left(\dfrac{1}{2}\right)^n = 0 + \dfrac{1/2}{1 - 1/2} = 1$

EXERCISE SET 11.4

1. $\displaystyle\sum_{k=1}^{\infty} \frac{1}{2^k} = \frac{1/2}{1 - 1/2} = 1;$ $\displaystyle\sum_{k=1}^{\infty} \frac{1}{4^k} = \frac{1/4}{1 - 1/4} = 1/3;$

$\displaystyle\sum_{k=1}^{\infty} \left(\frac{1}{2^k} + \frac{1}{4^k}\right) = 1 + 1/3 = 4/3$

3. $\displaystyle\sum_{k=2}^{\infty} \frac{1}{k^2 - 1} = \sum_{k=2}^{\infty} \left[\frac{1/2}{k - 1} - \frac{1/2}{k + 1}\right],$

$s_n = \frac{1}{2}\left[(1 - \frac{1}{3}) + (\frac{1}{2} - \frac{1}{4}) + (\frac{1}{3} - \frac{1}{5}) + \cdots + (\frac{1}{n} - \frac{1}{n + 2})\right]$

$ = \frac{1}{2}\left[(1 + \frac{1}{2} + \frac{1}{3} + \cdots + \frac{1}{n}) - (\frac{1}{3} + \frac{1}{4} + \frac{1}{5} + \cdots + \frac{1}{n + 2})\right]$

$ = \frac{3}{4} - \frac{1}{2}(\frac{1}{n + 1} + \frac{1}{n + 2}),$

$\displaystyle\sum_{k=2}^{\infty} \frac{1}{k^2 - 1} = \lim_{n \to +\infty} s_n = 3/4;$ $\displaystyle\sum_{k=2}^{\infty} \frac{7}{10^{k-1}} = \frac{7/10}{1 - 1/10} = 7/9;$

so $\displaystyle\sum_{k=2}^{\infty} \left[\frac{1}{k^2 - 1} - \frac{7}{10^{k-1}}\right] = 3/4 - 7/9 = -1/36$

5. (a) $p = 3$, converges   (b) $p = 1/2$, diverges
    (c) $p = 1$, diverges   (d) $p = 2/3$, diverges
    (e) $p = 4/3$, converges   (f) $p = 1/4$, diverges
    (g) $p = 5/3$, converges   (h) $p = \pi$, converges

7. (a) $\displaystyle\lim_{k \to +\infty} \frac{k^2 + k + 3}{2k^2 + 1} = \frac{1}{2}$ (b) $\displaystyle\lim_{k \to +\infty} (1 + \frac{1}{k})^k = e$

9. $\displaystyle\sum_{k=1}^{\infty} \frac{1}{k + 6} = \sum_{k=7}^{\infty} \frac{1}{k},$ diverges because the harmonic series diverges.

11. $\displaystyle\int_1^{+\infty} \frac{1}{5x + 2} = \lim_{\ell\to+\infty} \frac{1}{5} \ln(5x + 2)\Big]_1^\ell = +\infty$, the series diverges by the integral test.

13. $\displaystyle\int_1^{+\infty} \frac{1}{1 + 9x^2}\, dx = \lim_{\ell\to+\infty} \frac{1}{3} \tan^{-1}3x\Big]_1^\ell = \frac{1}{3}\,(\pi/2 - \tan^{-1}3)$, the series converges by the integral test.

15. $\displaystyle\sum_{k=1}^{\infty} \frac{1}{\sqrt{k + 5}} = \sum_{k=6}^{\infty} \frac{1}{\sqrt{k}}$, diverges because the p-series with $p = 1/2 \leq 1$ diverges.

17. $\displaystyle\int_1^{+\infty} (2x - 1)^{-1/3} dx = \lim_{\ell\to+\infty} \frac{3}{4}\,(2x - 1)^{2/3}\Big]_1^\ell = +\infty$, the series diverges by the integral test.

19. $\displaystyle\lim_{k\to+\infty} \frac{k}{\ln(k + 1)} = \lim_{k\to+\infty} \frac{1}{1/(k + 1)} = +\infty$, the series diverges because

   $\displaystyle\lim_{k\to+\infty} u_k \neq 0.$

21. $\displaystyle\int_1^{+\infty} \frac{[\ln(x + 1)]^{-2}}{x + 1}\, dx = \lim_{\ell\to+\infty} -1/\ln(x + 1)\Big]_1^\ell = 1/\ln 2$, the series converges by the integral test.

23. $\displaystyle\lim_{k\to+\infty} (1 + 1/k)^k = e \neq 0$, the series diverges.

25. $\displaystyle\int_1^{+\infty} \frac{\tan^{-1}x}{1 + x^2}\, dx = \lim_{\ell\to+\infty} \frac{1}{2}\,(\tan^{-1}x)^2\Big]_1^\ell = 3\pi^2/32$, the series converges by the integral test.

27. $\displaystyle\sum_{k=5}^{\infty} 7k^{-p} = \sum_{k=5}^{\infty} 7\Big(\frac{1}{k^p}\Big)$, converges because a p-series with $p > 1$ converges.

29. $\lim\limits_{k\to+\infty} k^2\sin^2(1/k) = 1 \neq 0$, the series diverges.

31.   Use the integral test with $\displaystyle\int_2^{+\infty} \frac{dx}{x(\ln x)^p}$ to get $\lim\limits_{\ell\to+\infty} \ln(\ln x)]_2^\ell = +\infty$ if

$p = 1$, $\lim\limits_{\ell\to+\infty} \dfrac{(\ln x)^{1-p}}{1 - p}\Big]_2^\ell = \begin{cases} +\infty & \text{if} & p < 1 \\[2mm] \dfrac{1}{(p - 1)(\ln 2)^{p-1}} & \text{if } p > 1 \end{cases}$

33.   Suppose $\Sigma(u_k + v_k)$ converges then so does $\Sigma[(u_k + v_k)-u_k]$, but $\Sigma[(u_k + v_k)-u_k] = \Sigma v_k$ so Σv_k converges which contradicts the assumption that Σv_k diverges. Suppose $\Sigma(u_k - v_k)$ converges then so does $\Sigma[u_k - (u_k - v_k)] = \Sigma v_k$ which leads to the same contradiction as before.

35.   (a) diverges because $\displaystyle\sum_{k=1}^{\infty} (2/3)^{k-1}$ converges and $\displaystyle\sum_{k=1}^{\infty} 1/k$ diverges.

(b) diverges because $\displaystyle\sum_{k=1}^{\infty} \frac{k^2}{1 + k^2}$ diverges and $\displaystyle\sum_{k=1}^{\infty} \frac{1}{k(k + 1)}$ converges.

(c) diverges because $\displaystyle\sum_{k=1}^{\infty} 1/(3k + 2)$ diverges and $\displaystyle\sum_{k=1}^{\infty} 1/k^{3/2}$ converges.

(d) converges because both $\displaystyle\sum_{k=2}^{\infty} \frac{1}{k(\ln k)^2}$ and $\displaystyle\sum_{k=2}^{\infty} 1/k^2$ converge.

37.   (a) If $S = \displaystyle\sum_{k=1}^{\infty} u_k$ and $s_n = \displaystyle\sum_{k=1}^{n} u_k$, then $S - s_n = \displaystyle\sum_{k=n+1}^{\infty} u_k < \int_n^{+\infty} f(x)dx$ by interpreting u_{n+1}, u_{n+2}, $\cdots$ as the areas of inscribed rectangles for $x \geq n$.

(b) $\int_{10}^{+\infty} \frac{1}{x^3} dx = \lim_{\ell \to +\infty} - \frac{1}{2x^2} \Big]_{10}^{\ell} = \lim_{\ell \to +\infty} \Big[- \frac{1}{2\ell^2} + \frac{1}{200} \Big] = 0.005$

(c) $\int_{n}^{+\infty} \frac{1}{x^4} dx = \lim_{\ell \to +\infty} - \frac{1}{3x^4} \Big]_{n}^{\ell} = \lim_{\ell \to +\infty} \Big[- \frac{1}{3\ell^4} + \frac{1}{3n^4} \Big] = \frac{1}{3n^4},$

$\frac{1}{3n^4} \leq 10^{-5}, \; n^4 \geq \frac{10^5}{3}, \; n \geq 14; \text{ take } n = 14.$

EXERCISE SET 11.5

1. $\rho = \lim_{k \to +\infty} \frac{3^{k+1}/(k + 1)!}{3^k/k!} = \lim_{k \to +\infty} \frac{3}{k + 1} = 0$, the series converges.

3. $\rho = \lim_{k \to +\infty} \frac{k}{k + 1} = 1$, the result is inconclusive.

5. $\rho = \lim_{k \to +\infty} \frac{(k + 1)!/(k + 1)^3}{k!/k^3} = \lim_{k \to +\infty} \frac{k^3}{(k + 1)^2} = +\infty$, the series diverges.

7. $\rho = \lim_{k \to +\infty} \frac{3k + 2}{2k - 1} = 3/2$, the series diverges.

9. $\rho = \lim_{k \to +\infty} \frac{k^{1/k}}{5} = 1/5$, the series converges.

11. ratio test, $\rho = \lim_{k \to +\infty} \frac{2k^3}{(k + 1)^3} = 2$, diverges

13. ratio test, $\rho = \lim_{k \to +\infty} 7/(k + 1) = 0$, converges

15. ratio test, $\rho = \lim_{k \to +\infty} \frac{(k + 1)^2}{5k^2} = 1/5$, converges

17. ratio test, $\rho = \lim\limits_{k \to +\infty} e^{-1}(k + 1)^{50}/k^{50} = e^{-1} < 1$, converges

19. root test, $\rho = \lim\limits_{k \to +\infty} k^{1/k}(2/3) = 2/3$, converges

21. diverges (integral test)

23. root test, $\rho = \lim\limits_{k \to +\infty} 4/(7k - 1) = 0$, converges

25. ratio test, $\rho = \lim\limits_{k \to +\infty} \dfrac{(k + 1)^2}{(2k + 2)(2k + 1)} = 1/4$, converges

27. integral test, $\displaystyle\int_1^{+\infty} \dfrac{dx}{1 + \sqrt{x}} = \lim\limits_{\ell \to +\infty} 2[\sqrt{x} - \ln(1 + \sqrt{x})]_1^{\ell} = +\infty$, diverges

29. ratio test, $\rho = \lim\limits_{k \to +\infty} \dfrac{\ln(k + 1)}{e \ln k} = \lim\limits_{k \to +\infty} \dfrac{k}{e(k + 1)} = 1/e < 1$, converges

31. ratio test, $\rho = \lim\limits_{k \to +\infty} \dfrac{k + 5}{4(k + 1)} = 1/4$, converges

33. $u_k = \dfrac{k!}{1 \cdot 3 \cdot 5 \cdots (2k - 1)}$, by the ratio test

$\rho = \lim\limits_{k \to +\infty} \dfrac{1 \cdot 3 \cdot 5 \cdots (2k - 1)(k + 1)!}{1 \cdot 3 \cdot 5 \cdots (2k - 1)(2k + 1)k!} = \lim\limits_{k \to +\infty} \dfrac{k + 1}{2k + 1} = 1/2$,

the series converges

35. (a) $\lim\limits_{x \to +\infty} \ln y = \lim\limits_{x \to +\infty} \dfrac{\ln(\ln x)}{x} = \lim\limits_{x \to +\infty} \dfrac{1}{x \ln x} = 0$

so $\lim\limits_{x \to +\infty} y = e^0 = 1 = \lim\limits_{k \to +\infty} (\ln k)^{1/k}$

(b) $\rho = \lim\limits_{k \to +\infty} \dfrac{1}{3} (\ln k)^{1/k} = 1/3$

(c) $\rho = \lim\limits_{k \to +\infty} \dfrac{\ln(k + 1)}{3 \ln k} = \lim\limits_{k \to +\infty} \dfrac{k}{3(k + 1)} = 1/3$

37. The result follows trivially if $a = 0$ so suppose $a \neq 0$ and consider the series $\sum\limits_{k=1}^{\infty} a^k/k!$. By the ratio test, $\rho = \lim\limits_{k \to +\infty} \dfrac{a}{k + 1} = 0$ so the series converges for every real number a and hence $\lim\limits_{k \to +\infty} a^k/k! = 0$.

EXERCISE SET 11.6

1. $\dfrac{1}{3^k + 5} < \dfrac{1}{3^k}, \quad \sum\limits_{k=1}^{\infty} \dfrac{1}{3^k}$ converges

3. $\dfrac{1}{5k^2 - k} \leq \dfrac{1}{5k^2 - k^2} = \dfrac{1}{4k^2}, \quad \sum\limits_{k=1}^{\infty} \dfrac{1}{4k^2}$ converges

5. $\dfrac{2^k - 1}{3^k + 2k} < \dfrac{2^k}{3^k} = (2/3)^k, \quad \sum\limits_{k=1}^{\infty} (2/3)^k$ converges

7. $\dfrac{3}{k - 1/4} > \dfrac{3}{k}, \quad \sum\limits_{k=1}^{\infty} 3/k$ diverges

9. $\dfrac{9}{\sqrt{k} + 1} \geq \dfrac{9}{\sqrt{k} + \sqrt{k}} = \dfrac{9}{2\sqrt{k}}, \quad \sum\limits_{k=1}^{\infty} \dfrac{9}{2\sqrt{k}}$ diverges

11. $\dfrac{k^{4/3}}{8k^2 + 5k + 1} \geq \dfrac{k^{4/3}}{8k^2 + 5k^2 + k^2} = \dfrac{1}{14k^{2/3}}, \quad \sum\limits_{k=1}^{\infty} \dfrac{1}{14k^{2/3}}$ diverges

13. compare with the convergent series $\displaystyle\sum_{k=1}^{\infty} 1/k^5$,

$\rho = \displaystyle\lim_{k\to\infty} \frac{4k^7 - 2k^6 + 6k^5}{8k^7 + k - 8} = 1/2$, converges

15. compare with the convergent series $\displaystyle\sum_{k=1}^{\infty} 5/3^k$, $\rho = \displaystyle\lim_{k\to\infty} \frac{3^k}{3^k + 1} = 1$,

converges

17. compare with the divergent series $\displaystyle\sum_{k=1}^{\infty} \frac{1}{k^{2/3}}$,

$\rho = \displaystyle\lim_{k\to\infty} \frac{k^{2/3}}{(8k^2 - 3k)^{1/3}} = \displaystyle\lim_{k\to\infty} \frac{1}{(8 - 3/k)^{1/3}} = 1/2$, diverges

19. $\dfrac{1}{k^3 + 2k + 1} < \dfrac{1}{k^3}$, $\displaystyle\sum_{k=1}^{\infty} 1/k^3$ converges so $\displaystyle\sum_{k=1}^{\infty} \frac{1}{k^3 + 2k + 1}$ converges by the comparison test

21. $\dfrac{1}{9k - 2} > \dfrac{1}{9k}$, $\displaystyle\sum_{k=1}^{\infty} \frac{1}{9k}$ diverges so $\displaystyle\sum_{k=1}^{\infty} \frac{1}{9k - 2}$ diverges by the comparison test

23. limit comparison test, compare with the convergent series $\displaystyle\sum_{k=1}^{\infty} 1/k^{5/2}$,

$\rho = \displaystyle\lim_{k\to\infty} \frac{k^3}{k^3 + 1} = 1$, converges

25. limit comparison test, compare with the divergent series $\displaystyle\sum_{k=1}^{\infty} 1/k$,

$$\rho = \lim_{k\to+\infty} \frac{k}{\sqrt{k^2 + k}} = 1, \text{ diverges}$$

27. limit comparison test, compare with the convergent series $\displaystyle\sum_{k=1}^{\infty} 1/k^{5/2}$,

$$\rho = \lim_{k\to+\infty} \frac{k^3 + 2k^{5/2}}{k^3 + 3k^2 + 3k} = 1, \text{ converges}$$

29. diverges because $\displaystyle\lim_{k\to+\infty} \frac{1}{4 + 2^{-k}} = 1/4 \neq 0$

31. $\dfrac{\tan^{-1}k}{k^2} < \dfrac{\pi/2}{k^2}$, $\displaystyle\sum_{k=1}^{\infty} \frac{\pi/2}{k^2}$ converges so $\displaystyle\sum_{k=1}^{\infty} \frac{\tan^{-1}k}{k^2}$ converges

33. $\displaystyle\sum_{k=1}^{\infty} \frac{\ln k}{k\sqrt{k}} = \sum_{k=2}^{\infty} \frac{\ln k}{k\sqrt{k}}$ because $\ln 1 = 0$,

$$\int_{2}^{+\infty} \frac{\ln x}{x^{3/2}}\, dx = \lim_{\ell\to+\infty} \left[-\frac{2\ln x}{x^{1/2}} - \frac{4}{x^{1/2}} \right]_{2}^{\ell} = \sqrt{2}(\ln 2 + 2)$$

so $\displaystyle\sum_{k=2}^{\infty} \frac{\ln k}{k^{3/2}}$ converges.

35. $\rho = \displaystyle\lim_{k\to+\infty} \frac{1 - \cos(1/k)}{1/k^2}$, but

$$\lim_{x\to+\infty} \frac{1 - \cos(1/x)}{1/x^2} = \lim_{x\to+\infty} \frac{(-1/x^2)\sin(1/x)}{-2/x^3} = \lim_{x\to+\infty} \frac{\sin(1/x)}{2(1/x)} = 1/2,$$

so $\rho = 1/2$; the series converges.

37. $\displaystyle\sum_{k=1}^{\infty} \frac{\ln k}{k^2} = \sum_{k=2}^{\infty} \frac{\ln k}{k^2}$, $\dfrac{\ln k}{k^2} < \dfrac{2\sqrt{k}}{k^2} = \dfrac{2}{k^{3/2}}$ for $k \geq 2$, $\displaystyle\sum_{k=2}^{\infty} \frac{2}{k^{3/2}}$ converges so

   $\displaystyle\sum_{k=1}^{\infty} \frac{\ln k}{k^2}$ converges

39. limit comparison test, compare with $\displaystyle\sum_{k=1}^{\infty} 1/k^p$,

   $\rho = \displaystyle\lim_{k \to +\infty} \frac{k^p}{(a + bk)^p} = \lim_{k \to +\infty} \frac{1}{(a/k + b)^p} = 1/b^p$, converges for $p > 1$.

41. Compare with the convergent series $\displaystyle\sum_{k=1}^{\infty} 1/k!$,

   $\rho = \displaystyle\lim_{k \to +\infty} \frac{(k + 1)^2 k!}{(k + 2)!} = \lim_{k \to +\infty} \frac{k + 1}{k + 2} = 1$, converges

43. $k! = k(k - 1)(k - 2)\cdots(2)(1) \geq 2 \cdot 2 \cdot 2 \cdots 2 \cdot 1 = 2^{k-1}$,

   $1/k! \leq 1/2^{k-1}$, $\displaystyle\sum_{k=1}^{\infty} 1/2^{k-1}$ converges so $\displaystyle\sum_{k=1}^{\infty} 1/k!$ converges

EXERCISE SET 11.7

1. converges

3. diverges because $\displaystyle\lim_{k \to +\infty} a_k = \lim_{k \to +\infty} \frac{k + 1}{3k + 1} = 1/3 \neq 0$

5. converges

7. $\rho = \lim\limits_{k \to +\infty} \dfrac{(3/5)^{k+1}}{(3/5)^k} = 3/5$, converges absolutely

9. $\rho = \lim\limits_{k \to +\infty} \dfrac{3k^2}{(k+1)^2} = 3$, diverges

11. $\rho = \lim\limits_{k \to +\infty} \dfrac{(k+1)^3}{ek^3} = 1/e$, converges absolutely

13. conditionally convergent, $\displaystyle\sum_{k=1}^{\infty} \dfrac{(-1)^{k+1}}{3k}$ converges by the alternating

    series test but $\displaystyle\sum_{k=1}^{\infty} \dfrac{1}{3k}$, diverges

15. divergent, $\lim\limits_{k \to +\infty} a_k \neq 0$

17. $\displaystyle\sum_{k=1}^{\infty} \dfrac{\cos k\pi}{k} = \displaystyle\sum_{k=1}^{\infty} \dfrac{(-1)^k}{k}$ is conditionally convergent, $\displaystyle\sum_{k=1}^{\infty} \dfrac{(-1)^k}{k}$ converges

    by the alternating series test but $\displaystyle\sum_{k=1}^{\infty} 1/k$ diverges.

19. absolutely convergent, $\displaystyle\sum_{k=1}^{\infty} \left[\dfrac{k+2}{3k-1}\right]^k$ converges by the root test.

21. conditionally convergent, $\displaystyle\sum_{k=1}^{\infty} (-1)^{k+1} \frac{k+2}{k(k+3)}$ converges by the

alternating series test but $\displaystyle\sum_{k=1}^{\infty} \frac{k+2}{k(k+3)}$ diverges (limit comparison test

with $\Sigma\ 1/k$)

23. $\displaystyle\sum_{k=1}^{\infty} \sin(k\pi/2) = 1 + 0 - 1 + 0 + 1 + 0 - 1 + 0 + \cdots$, divergent

($\displaystyle\lim_{k\to+\infty} \sin(k\pi/2)$ does not exist)

25. conditionally convergent, $\displaystyle\sum_{k=2}^{\infty} \frac{(-1)^k}{k \ln k}$ converges by the alternating series

test but $\displaystyle\sum_{k=2}^{\infty} \frac{1}{k \ln k}$ diverges (integral test)

27. absolutely convergent, $\displaystyle\sum_{k=2}^{\infty} (1/\ln k)^k$ converges by the root test

29. conditionally convergent, let $f(x) = \dfrac{x^2 + 1}{x^3 + 2}$ then

$f'(x) = \dfrac{x(4 - 3x - x^3)}{(x^3 + 2)^2} \leq 0$ for $x \geq 2$ so $\{a_k\}_{k=2}^{+\infty} = \left\{\dfrac{k^2 + 1}{k^3 + 2}\right\}_{k=2}^{+\infty}$ is

nonincreasing, $\displaystyle\lim_{k\to+\infty} a_k = 0$; the series converges by the alternating

series test but $\displaystyle\sum_{k=2}^{\infty} \frac{k^2 + 1}{k^3 + 2}$ diverges (limit comparison test with $\Sigma\ 1/k$)

31. $|\text{error}| \le a_8 = 1/8 = 0.125$ 33. $|\text{error}| \le a_{100} = 1/\sqrt{100} = 0.1$

35. $|\text{error}| < 0.0001$ if $a_{n+1} < 0.0001$, $1/(n + 1) < 0.0001$,
     $n + 1 > 10{,}000$, $n > 9{,}999$; $n = 10{,}000$

37. $|\text{error}| < 0.005$ if $a_{n+1} < 0.005$, $1/\sqrt{n + 1} < 0.005$, $\sqrt{n + 1} > 200$,
     $n + 1 > 40{,}000$, $n > 39{,}999$; $n = 40{,}000$

39. Suppose $\Sigma |a_k|$ converges, then $\lim\limits_{k \to +\infty} |a_k| = 0$ so $|a_k| < 1$ for $k \ge K$ and
     thus $|a_k|^2 < |a_k|$, $a_k^2 < |a_k|$ hence Σa_k^2 converges by the comparison
     test.

41. Suppose $a_1 - a_2 + a_3 - a_4 + \cdots + (-1)^{k+1} a_k + \cdots$ satisfies (a) and
     (b) of Theorem 11.7.1, then the series converges and by Theorem 11.4.3b
     so does $-(a_1 - a_2 + a_3 - a_4 + \cdots + (-1)^{k+1} a_k + \cdots)$ which equals
     $-a_1 + a_2 - a_3 + a_4 - \cdots + (-1)^k a_k + \cdots$

43. $(1 - \frac{1}{2} - \frac{1}{4}) + (\frac{1}{3} - \frac{1}{6} - \frac{1}{8}) + (\frac{1}{5} - \frac{1}{10} - \frac{1}{12}) + \cdots$
     $= (\frac{1}{2} - \frac{1}{4}) + (\frac{1}{6} - \frac{1}{8}) + (\frac{1}{10} - \frac{1}{12}) + \cdots$
     $= \frac{1}{2}(1 - \frac{1}{2} + \frac{1}{3} - \frac{1}{4} + \frac{1}{5} - \frac{1}{6} + \cdots) = S/2$

EXERCISE SET 11.8

1. $\rho = \lim\limits_{k \to +\infty} \dfrac{k + 1}{k + 2} |x| = |x|$, the series converges if $|x| < 1$ and diverges
    if $|x| > 1$. If $x = -1$, $\sum\limits_{k=0}^{\infty} \dfrac{(-1)^k}{k + 1}$ converges by the alternating series

test; if $x = 1$, $\displaystyle\sum_{k=0}^{\infty} \frac{1}{k+1}$ diverges. The radius of convergence is 1, the interval of convergence is $[-1,1)$.

3. $\rho = \displaystyle\lim_{k\to\infty} \frac{|x|}{k+1} = 0$, the radius of convergence is $+\infty$, the interval is $(-\infty, +\infty)$.

5. $\rho = \displaystyle\lim_{k\to\infty} \frac{5k^2|x|}{(k+1)^2} = 5|x|$, converges if $|x| < 1/5$ and diverges if $|x| > 1/5$. If $x = -1/5$, $\displaystyle\sum_{k=1}^{\infty} \frac{(-1)^k}{k^2}$ converges; if $x = 1/5$, $\displaystyle\sum_{k=1}^{\infty} 1/k^2$ converges. Radius of convergence is $1/5$, interval of convergence is $[-1/5, 1/5]$.

7. $\rho = \displaystyle\lim_{k\to\infty} \frac{k|x|}{k+2} = |x|$, converges if $|x| < 1$, diverges if $|x| > 1$. If $x = -1$, $\displaystyle\sum_{k=1}^{\infty} \frac{(-1)^k}{k(k+1)}$ converges; if $x = 1$, $\displaystyle\sum_{k=1}^{\infty} \frac{1}{k(k+1)}$ converges. Radius of convergence is 1, interval of convergence is $[-1,1]$.

9. $\rho = \displaystyle\lim_{k\to\infty} \frac{\sqrt{k}}{\sqrt{k+1}} |x| = |x|$, converges if $|x| < 1$, diverges if $|x| > 1$. If $x = -1$, $\displaystyle\sum_{k=1}^{\infty} \frac{-1}{\sqrt{k}}$ diverges; if $x = 1$, $\displaystyle\sum_{k=1}^{\infty} \frac{(-1)^{k-1}}{\sqrt{k}}$ converges. Radius of convergence is 1, interval of convergence is $(-1,1]$.

11. $\rho = \displaystyle\lim_{k\to\infty} \frac{|x|^2}{(2k+3)(2k+2)} = 0$, radius of convergence is $+\infty$, interval of convergence is $(-\infty, +\infty)$.

13. $\rho = \lim\limits_{k \to +\infty} \dfrac{3|x|}{k + 1} = 0$, radius of convergence is $+\infty$, interval of convergence is $(-\infty, +\infty)$.

15. $\rho = \lim\limits_{k \to +\infty} \dfrac{1 + k^2}{1 + (k + 1)^2} |x| = |x|$, converges if $|x| < 1$, diverges if

   $|x| > 1$. If $x = -1$, $\sum\limits_{k=0}^{\infty} \dfrac{(-1)^k}{1 + k^2}$ converges; if $x = 1$, $\sum\limits_{k=0}^{\infty} \dfrac{1}{1 + k^2}$

   converges. Radius of convergence is 1, interval of convergence is $[-1, 1]$.

17. $\rho = \lim\limits_{k \to +\infty} \dfrac{k|x + 1|}{k + 1} = |x + 1|$, converges if $|x + 1| < 1$, diverges if

   $|x + 1| > 1$. If $x = -2$, $\sum\limits_{k=1}^{\infty} \dfrac{-1}{k}$ diverges; if $x = 0$, $\sum\limits_{k=1}^{\infty} \dfrac{(-1)^{k+1}}{k}$

   converges. Radius of convergence is 1, interval of convergence is $(-2, 0]$.

19. $\rho = \lim\limits_{k \to +\infty} (3/4)|x + 5| = \dfrac{3}{4}|x + 5|$, convergence if $|x + 5| < 4/3$,

   diverges if $|x + 5| > 4/3$. If $x = -19/3$, $\sum\limits_{k=0}^{\infty} (-1)^k$ diverges; if

   $x = -11/3$, $\sum\limits_{k=0}^{\infty} 1$ diverges. Radius of convergence is 4/3, interval of
   convergence is $(-19/3, -11/3)$.

21. $\rho = \lim\limits_{k \to +\infty} \dfrac{k^2 + 4}{(k + 1)^2 + 4}|x + 1|^2 = |x + 1|^2$, converges if $|x + 1| < 1$,

   diverges if $|x + 1| > 1$. If $x = -2$, $\sum\limits_{k=1}^{\infty} \dfrac{(-1)^{3k+1}}{k^2 + 4}$ converges; if $x = 0$,

$$\sum_{k=1}^{\infty} \frac{(-1)^k}{k^2 + 4}$$ converges. Radius of convergence is 1, interval of convergence is $[-2,0]$.

23. $\rho = \lim_{k \to +\infty} \frac{\pi|x - 1|^2}{(2k + 3)(2k + 2)} = 0$, radius of convergence $+\infty$, interval of convergence $(-\infty, +\infty)$.

25. $\rho = \lim_{k \to +\infty} \sqrt[k]{|u_k|} = \lim_{k \to +\infty} \frac{|x|}{\ln k} = 0$, the series converges absolutely for all x so the interval of convergence is $(-\infty, +\infty)$.

27. By the ratio test for absolute convergence, $\rho = \lim_{k \to +\infty} \frac{|x - a|}{b} = \frac{|x - a|}{b}$; converges if $|x - a| < b$, diverges if $|x - a| > b$. If $x = a - b$, $\sum_{k=0}^{\infty} (-1)^k$ diverges; if $x = a + b$, $\sum_{k=0}^{\infty} 1$ diverges. The interval of convergence is $(a - b, a + b)$.

29. By the ratio test for absolute convergence,
$$\rho = \lim_{k \to +\infty} \frac{(k + 1 + p)!k!(k + q)!}{(k + p)!(k + 1)!(k + 1 + q)!} |x|$$
$$= \lim_{k \to +\infty} \frac{k + 1 + p}{(k + 1)(k + 1 + q)} |x| = 0,$$
radius of convergence is $+\infty$.

31. By assumption $\sum_{k=0}^{\infty} c_k x^k$ converges if $|x| < R$ so $\sum_{k=0}^{\infty} c_k x^{2k} = \sum_{k=0}^{\infty} c_k (x^2)^k$ converges if $|x^2| < R$, $|x| < \sqrt{R}$. Thus $\sum_{k=0}^{\infty} c_k x^{2k}$ has radius of convergence $\sqrt{R}$.

EXERCISE SET 11.9

1. $1 - 2x + 2x^2 - \frac{4}{3}x^3 + \frac{2}{3}x^4$ 3. $2x - \frac{4}{3}x^3$

5. $x + \frac{1}{3}x^3$ 7. $x + x^2 + \frac{1}{2}x^3 + \frac{1}{6}x^4$

9. $1 + \frac{1}{2}x^2 + \frac{5}{24}x^4$ 11. $\ln 3 + \frac{2}{3}x - \frac{2}{9}x^2 + \frac{8}{81}x^3 - \frac{4}{81}x^4$

13. $e + e(x - 1) + \frac{e}{2}(x - 1)^2 + \frac{e}{6}(x - 1)^3$

15. $2 + \frac{1}{4}(x - 4) - \frac{1}{64}(x - 4)^2 + \frac{1}{512}(x - 4)^3$

17. $\frac{\sqrt{2}}{2} - \frac{\sqrt{2}}{2}(x - \pi/4) - \frac{\sqrt{2}}{4}(x - \pi/4)^2 + \frac{\sqrt{2}}{12}(x - \pi/4)^3$

19. $-\frac{\sqrt{3}}{2} + \frac{\pi}{2}(x + 1/3) + \frac{\sqrt{3}\pi^2}{4}(x + 1/3)^2 - \frac{\pi^3}{12}(x + 1/3)^3$

21. $\frac{\pi}{4} + \frac{1}{2}(x - 1) - \frac{1}{4}(x - 1)^2 + \frac{1}{12}(x - 1)^3$

23. $f^{(k)}(x) = (-1)^k e^{-x}$, $f^{(k)}(0) = (-1)^k$; $\displaystyle\sum_{k=0}^{\infty} \frac{(-1)^k}{k!}x^k$

25. $f^{(k)}(x) = \frac{(-1)^k k!}{(1 + x)^{k+1}}$, $f^{(k)}(0) = (-1)^k k!$; $\displaystyle\sum_{k=0}^{\infty}(-1)^k x^k$

27. $f^{(0)}(0) = 0$; for $k \geq 1$, $f^{(k)}(x) = \frac{(-1)^{k+1}(k - 1)!}{(1 + x)^k}$,

$f^{(k)}(0) = (-1)^{k+1}(k - 1)!$; $\displaystyle\sum_{k=1}^{\infty} \frac{(-1)^{k+1}}{k}x^k$

29. $f^{(k)}(0) = 0$ if k is odd, $f^{(k)}(0)$ is alternately $1/2^k$ and $-1/2^k$ if k is

even. $\displaystyle\sum_{k=0}^{\infty} \frac{f^{(k)}(0)}{k!} x^k = 1 - \frac{1}{2^2 2!} x^2 + \frac{1}{2^4 4!} x^4 - \cdots = \sum_{k=0}^{\infty} \frac{(-1)^k}{2^{2k}(2k)!} x^{2k}$

31. $f^{(k)}(0) = 0$ if k is odd, $f^{(k)}(0) = 1$ if k is even;

$\displaystyle\sum_{k=0}^{\infty} \frac{f^{(k)}(0)}{k!} x^k = 1 + \frac{1}{2!} x^2 + \frac{1}{4!} x^4 + \cdots = \sum_{k=0}^{\infty} \frac{1}{(2k)!} x^{2k}$

33. $f^{(k)}(x) = \dfrac{(-1)^k k!}{x^{k+1}}$, $f^{(k)}(-1) = -k!$; $\displaystyle\sum_{k=0}^{\infty} (-1)(x + 1)^k$

35. $f^{(0)}(1) = 0$; for $k \geq 1$, $f^{(k)}(x) = \dfrac{(-1)^{k+1}(k-1)!}{x^k}$,

$f^{(k)}(1) = (-1)^{k+1}(k-1)!$; $\displaystyle\sum_{k=1}^{\infty} \frac{(-1)^{k+1}}{k} (x-1)^k$

37. $f^{(k)}(1/2) = 0$ if k is odd, $f^{(k)}(1/2)$ is alternately π^k and $-\pi^k$ is k is
even;

$\displaystyle\sum_{k=0}^{\infty} \frac{f^{(k)}(1/2)}{k!} (x - 1/2)^k = 1 - \frac{\pi^2}{2!} (x - 1/2)^2 + \frac{\pi^4}{4!} (x - 1/2)^4 - \cdots$

$\displaystyle = \sum_{k=0}^{\infty} \frac{(-1)^k \pi^{2k}}{(2k)!} (x - 1/2)^{2k}$

39. $f^{(k)}(\ln 4) = 15/8$ for k even, $f^{(k)}(\ln 4) = 17/8$ for k odd, which can be

written as $f^{(k)}(\ln 4) = \dfrac{16 - (-1)^k}{8}$; $\displaystyle\sum_{k=0}^{\infty} \frac{16 - (-1)^k}{8k!} (x - \ln 4)^k$

EXERCISE SET 11.10

1. $f^{(6)}(x) = 2^6 e^{2x}$, $R_5(x) = \dfrac{2^6 e^{2c}}{6!} x^6$

3. $f^{(5)}(x) = -5!/(x + 1)^6$, $R_4(x) = -\dfrac{1}{(c + 1)^6} x^5$

5. $f^{(4)}(x) = (4 + x)e^x$, $R_3(x) = \dfrac{(4 + c)e^c}{4!} x^4$

7. $f^{(3)}(x) = -\dfrac{2(1 - 3x^2)}{(1 + x^2)^3}$, $R_2(x) = -\dfrac{1 - 3c^2}{3(1 + c^2)^3} x^3$

9. $f^{(4)}(x) = -\dfrac{15}{16} x^{-7/2}$, $R_3(x) = -\dfrac{5}{128c^{7/2}} (x - 4)^4$

11. $f^{(5)}(x) = \cos x$, $R_4(x) = \dfrac{\cos c}{5!} (x - \pi/6)^5$

13. $f^{(6)}(x) = 7!/(1 + x)^8$, $R_5(x) = \dfrac{7}{(1 + c)^8} (x + 2)^6$

15. $f^{(n+1)}(x) = (n + 1)!/(1 - x)^{n+2}$, $R_n(x) = \dfrac{1}{(1 - c)^{n+2}} x^{n+1}$

17. $f^{(n+1)}(x) = 2^{n+1} e^{2x}$, $R_n(x) = \dfrac{2^{n+1} e^{2c}}{(n + 1)!} x^{n+1}$

19. $f(x) = \cos x$, $f^{(n+1)}(x) = \pm\sin x$ or $\pm\cos x$, $|f^{(n+1)}(x)| \leq 1$,

$|R_n(x)| = \dfrac{|f^{(n+1)}(c)|}{(n + 1)!} |x|^{n+1} \leq \dfrac{|x|^{n+1}}{(n + 1)!}$, $\displaystyle\lim_{n \to +\infty} \dfrac{|x|^{n+1}}{(n + 1)!} = 0$, by the

pinching theorem $\displaystyle\lim_{n \to +\infty} |R_n(x)| = 0$ so $\displaystyle\lim_{n \to +\infty} R_n(x) = 0$ for all x.

21. $f(x) = e^x$, $f^{(n+1)}(x) = e^x$, $|R_n(x)| = \dfrac{e^c}{(n+1)!}\,|x-1|^{n+1}$.

If $x \geq 1$, $e^c \leq e^x$, $|R_n(x)| \leq e^x\,\dfrac{|x-1|^{n+1}}{(n+1)!}$,

$\lim\limits_{n\to+\infty} e^x\,\dfrac{|x-1|^{n+1}}{(n+1)!} = e^x(0) = 0$ so $\lim\limits_{n\to+\infty} R_n(x) = 0$. If $x < 1$, $e^c < e$,

$|R_n(x)| < e\,\dfrac{|x-1|^{n+1}}{(n+1)!}$ and again $\lim\limits_{n\to+\infty} R_n(x) = 0$.

23. $f(x) = e^x$, $f^{(n+1)}(x) = e^x$, $|R_n(x)| = \dfrac{e^c}{(n+1)!}\,|x-a|^{n+1}$. If $x \geq a$,

$e^c \leq e^x$, $|R_n(x)| \leq e^x\,\dfrac{|x-a|^{n+1}}{(n+1)!}$, $\lim\limits_{n\to+\infty} e^x\,\dfrac{|x-a|^{n+1}}{(n+1)!} = e^x(0) = 0$

so $\lim\limits_{n\to+\infty} R_n(x) = 0$. If $x < a$, $e^c < e^a$, $|R_n(x)| < e^a\,\dfrac{|x-a|^{n+1}}{(n+1)!}$ and

again $\lim\limits_{n\to+\infty} R_n(x) = 0$.

25. $f(x) = \cos x$, $f^{(n+1)}(x) = \pm\sin x$ or $\pm\cos x$, $|f^{(n+1)}(x)| \leq 1$,

$|R_n(x)| = \dfrac{|f^{(n+1)}(c)|}{(n+1)!}\,|x-a|^{n+1} \leq \dfrac{|x-a|^{n+1}}{(n+1)!}$, $\lim\limits_{n\to+\infty}\dfrac{|x-a|^{n+1}}{(n+1)!} = 0$

so $\lim\limits_{n\to+\infty} R_n(x) = 0$ for all x.

27. $e^{-2x} = \sum\limits_{k=0}^{\infty} \dfrac{(-2x)^k}{k!} = \sum\limits_{k=0}^{\infty} \dfrac{(-1)^k 2^k}{k!}\,x^k$, valid on $(-\infty, +\infty)$

29. $\dfrac{1}{1+x} = \dfrac{1}{1-(-x)} = \sum\limits_{k=0}^{\infty} (-x)^k = \sum\limits_{k=0}^{\infty} (-1)^k x^k$,

valid if $-1 < -x < 1$, $-1 < x < 1$

31. $\dfrac{x^2}{1 + 3x} = x^2 \left[\dfrac{1}{1 - (-3x)}\right] = x^2 \displaystyle\sum_{k=0}^{\infty} (-3x)^k = \sum_{k=0}^{\infty} (-1)^k 3^k x^{k+2}$,

   valid if $-1 < -3x < 1$, $-1/3 < x < 1/3$

33. $\cos^2 x = \dfrac{1}{2}(1 + \cos 2x) = \dfrac{1}{2}\left[1 + \displaystyle\sum_{k=0}^{\infty} (-1)^k \dfrac{(2x)^{2k}}{(2k)!}\right]$

   $= \dfrac{1}{2}\left[1 + 1 + \displaystyle\sum_{k=1}^{\infty} \dfrac{(-1)^k 2^{2k}}{(2k)!} x^{2k}\right]$

   $= 1 + \displaystyle\sum_{k=1}^{\infty} \dfrac{(-1)^k 2^{2k-1}}{(2k)!} x^{2k}$, valid on $(-\infty, +\infty)$

35. $\sin 2x = \displaystyle\sum_{k=0}^{\infty} (-1)^k \dfrac{(2x)^{2k+1}}{(2k + 1)!} = \sum_{k=0}^{\infty} \dfrac{(-1)^k 2^{2k+1}}{(2k + 1)!}$, valid on $(-\infty, +\infty)$

37. $\cos(x^2) = \displaystyle\sum_{k=0}^{\infty} (-1)^k \dfrac{(x^2)^{2k}}{(2k)!} = \sum_{k=0}^{\infty} \dfrac{(-1)^k}{(2k)!} x^{4k}$, valid on $(-\infty, +\infty)$

39. $\dfrac{1}{x} = \dfrac{1}{1 + (x - 1)} = \dfrac{1}{1 - [-(x - 1)]} = \displaystyle\sum_{k=0}^{\infty} [-(x - 1)]^k = \sum_{k=0}^{\infty} (-1)^k (x - 1)^k$

   which is valid for $-1 < -(x - 1) < 1$, $0 < x < 2$

41. $\sin \pi = 0$ 43. $e^{-\ln 3} = 1/3$

45. (a) $\cos 2x = 1 - \dfrac{(2x)^2}{2!} + \dfrac{(2x)^4}{4!} - \dfrac{(2x)^6}{6!} + \cdots$

$\qquad = 1 - 2x^2 + \dfrac{2}{3}x^4 - \dfrac{4}{45}x^6 + \cdots,$

$\qquad x^2\cos 2x = x^2 - 2x^4 + \dfrac{2}{3}x^6 - \dfrac{4}{45}x^8 + \cdots$

(b) $\dfrac{f^{(5)}(0)}{5!} = 0$ so $f^{(5)}(0) = 0$

EXERCISE SET 11.11

1. (a) $3/(n+1)! < 0.5 \times 10^{-5}$, $n = 9$

   (b) $3/(n+1)! < 0.5 \times 10^{-10}$, $n = 13$

3. $|R_n(1/2)| = \dfrac{e^c}{(n+1)!}(1/2)^{n+1} < \dfrac{2}{(n+1)!}\dfrac{1}{2^{n+1}} = \dfrac{1}{2^n(n+1)!} < 0.5 \times 10^{-4}$

   if $n = 5$ so $\sqrt{e} = e^{0.5} \approx 1 + 0.5 + \dfrac{(0.5)^2}{2!} + \dfrac{(0.5)^3}{3!} + \dfrac{(0.5)^4}{4!} + \dfrac{(0.5)^5}{5!}$,

   $\sqrt{e} \approx 1.6487$

5. $|R_n(\pi/20)| \leq \dfrac{(\pi/20)^{n+1}}{(n+1)!} < 0.5 \times 10^{-4}$ if $n = 3$,

   $\cos(\pi/20) \approx 1 - \dfrac{(\pi/20)^2}{2!} \approx 0.9877$

7. Expand about $\pi/3$ to get $\cos x = \dfrac{1}{2} - \dfrac{\sqrt{3}}{2}(x - \pi/3) - \dfrac{1}{4}(x - \pi/3)^2 + \cdots,$

   $58^\circ = 29\pi/90$ radians, $|R_n(x)| \leq \dfrac{|x - \pi/3|^{n+1}}{(n+1)!}$,

   $|R_n(29\pi/90)| \leq \dfrac{|29\pi/90 - \pi/3|^{n+1}}{(n+1)!} = \dfrac{(\pi/90)^{n+1}}{(n+1)!} < 0.5 \times 10^{-4}$ if $n = 2$,

   $\cos 58^\circ \approx \dfrac{1}{2} - \dfrac{\sqrt{3}}{2}(-\pi/90) - \dfrac{1}{4}(-\pi/90)^2 \approx 0.5299$

9. Let $x = 1/9$ in series (16) to get $\ln 1.25 \approx 0.223$

11. $(0.1)^3/3 < 0.5 \times 10^{-3}$ so $\tan^{-1}(0.1) \approx 0.100$ to three decimal place accuracy.

13. $|R_n(0.1)| = \dfrac{|f^{(n+1)}(c)|}{(n+1)!}(0.1)^{n+1}$ where $f^{(n+1)}(c) = \sinh c$ or $\cosh c$ for

$0 < c < 0.1$, but $\sinh c < \cosh c < \cosh 0.1$, and
$$\cosh 0.1 = \frac{1}{2}(e^{0.1} + e^{-0.1}) < \frac{1}{2}(2+1) = 1.5$$

so $|R_n(0.1)| < \dfrac{1.5(0.1)^{n+1}}{(n+1)!} < 0.5 \times 10^{-4}$ if $n = 3$,

$\cosh 0.1 \approx 1 + \dfrac{(0.1)^2}{2!} \approx 1.0050$

15. $\sin x = x - \dfrac{x^3}{3!} + (0)x^4 + R_4(x)$, $|R_4(x)| \le \dfrac{|x|^5}{5!} < 0.5 \times 10^{-3}$

if $|x|^5 < 0.06$, $|x| < (0.06)^{1/5} \approx 0.569$

17. $|R_2(x)| = \dfrac{e^c}{3!}|x|^3$. If $x \ge 0$ then $e^c \le e^x$, $|R_2(x)| \le \dfrac{e^x x^3}{3!} \le 0.0005$,

$e^x x^3 \le 0.003$ if $0 \le x < 0.137$ (by trial and error with a hand

calculator). If $x < 0$ then $e^c < 1$, $|R_2(x)| < \dfrac{|x|^3}{3!} \le 0.0005$,

$|x|^3 \le 0.003$, $x^3 \ge -0.003$, $x \ge -(0.003)^{1/3} \approx -0.144$.

19. (a) $\ln 2 = 1 - 1/2 + 1/3 - 1/4 + \cdots$, $|\text{error}| \le 1/(n+1) < 0.5 \times 10^{-6}$
     if $n + 1 > 2 \times 10^6$, $n > 1,999,999$; $n = 2,000,000$
     (b) Let $f(x) = \ln \dfrac{1+x}{1-x} = \ln(1+x) - \ln(1-x)$,

$$f^{(n+1)}(c) = n!\left[\frac{(-1)^n}{(1+c)^{n+1}} + \frac{1}{(1-c)^{n+1}}\right]$$

$$|f^{n+1}(c)| \le n!\left[\frac{1}{(1+c)^{n+1}} + \frac{1}{(1-c)^{n+1}}\right]$$

$$< n!\left[\frac{1}{(1+0)^{n+1}} + \frac{1}{(1-1/3)^{n+1}}\right] \text{ for } 0 < c < 1/3$$

$$= n!\,[1 + 1/(2/3)^{n+1}],$$

$$|R_n(1/3)| < [1 + 1/(2/3)^{n+1}] \frac{(1/3)^{n+1}}{n+1}$$

$$= (1/3^{n+1} + 1/2^{n+1})/(n+1) < 0.5 \times 10^{-6}$$

if $n \geq 16$ so 8 <u>terms</u> in series (16) are sufficient to assure six decimal place accuracy (even powers of x have zero coefficients).

21. $(1/2)^9/9! < 0.5 \times 10^{-3}$ and $(1/3)^7/7! < 0.5 \times 10^{-3}$ so

$$\tan^{-1}1/2 \approx 1/2 - \frac{(1/2)^3}{3} + \frac{(1/2)^5}{5} - \frac{(1/2)^7}{7} \approx 0.463$$

$$\tan^{-1}1/3 \approx 1/3 - \frac{(1/3)^3}{3} + \frac{(1/3)^5}{5} \approx 0.322, \quad \pi \approx 4(0.463 + 0.322) = 3.140$$

EXERCISE SET 11.12

1. (a) $\frac{d}{dx}(1 + x + x^2/2! + x^3/3! + \cdots) = 1 + x + x^2/2! + \cdots = e^x$

    (b) $\int(1 + x + x^2/2! + \cdots)dx = (x + x^2/2! + x^3/3! + \cdots) + C_1$

    $$= (1 + x + x^2/2! + x^3/3! + \cdots) + C_1 - 1$$

    $$= e^x + C$$

3. (a) $\frac{d}{dx}(x + x^3/3! + x^5/5! + \cdots) = 1 + x^2/2! + x^4/4! + \cdots = \cosh x$

    (b) $\int(x + x^3/3! + x^5/5! + \cdots)dx = (x^2/2! + x^4/4! + x^6/6! + \cdots) + C_1$

    $$= (1 + x^2/2! + x^4/4! + x^6/6! + \cdots)$$
    $$+ C_1 - 1$$

    $$= \cosh x + C$$

5. $1/(1+x)^2 = \frac{d}{dx}\left[-\frac{1}{1+x}\right] = \frac{d}{dx}\left[-\sum_{k=0}^{\infty}(-1)^k x^k\right]$

    $$= \frac{d}{dx}\left[\sum_{k=0}^{\infty}(-1)^{k+1}x^k\right] = \sum_{k=1}^{\infty}(-1)^{k+1}kx^{k-1}$$

7. $\ln \dfrac{1}{1 - x} = -\ln(1 - x) = \displaystyle\int_0^x \dfrac{1}{1 - t}\, dt = \int_0^x \left(\sum_{k=0}^{\infty} t^k \right) dt$

$= \displaystyle\sum_{k=0}^{\infty} \dfrac{t^{k+1}}{k + 1}\Bigg]_0^x = \sum_{k=0}^{\infty} \dfrac{x^{k+1}}{k + 1} = \sum_{k=1}^{\infty} \dfrac{x^k}{k}$

9. $x = 1/4$, $S = \ln \dfrac{1}{1 - 1/4} = \ln \dfrac{4}{3}$

11. $f(x) = \dfrac{1}{1 - x} = \displaystyle\sum_{k=0}^{\infty} x^k$, $f'(x) = \dfrac{1}{(1 - x)^2} = \displaystyle\sum_{k=1}^{\infty} kx^{k-1}$,

$f''(x) = \dfrac{2}{(1 - x)^3} = \displaystyle\sum_{k=2}^{\infty} k(k - 1)x^{k-2} = 2 + 6x + 12x^2 + 20x^3 + \cdots$

13. (a) $\rho = \displaystyle\lim_{k \to +\infty} \dfrac{|x|^{k+2}}{k + 2} \cdot \dfrac{k + 1}{|x|^{k+1}} = |x| \lim_{k \to +\infty} \dfrac{k + 1}{k + 2} = |x|$;

converges if $|x| < 1$.

(b) $f'(x) = \displaystyle\sum_{k=0}^{\infty} (-1)^k x^k$, converges on $(-1,1)$

(c) $f'(x) = \dfrac{1}{1 + x}$ so $f(x) = \ln(1 + x) + C$ for x in $(-1,1)$, let $x = 0$ to

find that $C = 0$, thus $f(x) = \ln(1 + x)$.

15. $(1 - x)^{1/3} = 1 + \dfrac{1}{3}(-x) + \dfrac{(1/3)(-2/3)}{2!}(-x)^2$

$+ \dfrac{(1/3)(-2/3)(-5/3)}{3!}(-x)^3 + \cdots$

$= 1 - \dfrac{1}{3}x - \dfrac{1}{9}x^2 - \dfrac{5}{81}x^4 - \cdots$; $R = 1$

17. $x(1 - 2x)^{-1/3} = x[1 + (-1/3)(-2x) + \dfrac{(-1/3)(-4/3)}{2!} (-2x)^2$

$$+ \dfrac{(-1/3)(-4/3)(-7/3)}{3!} (-2x)^3 + \cdots]$$

$$= x + \dfrac{2}{3} x^2 + \dfrac{8}{9} x^3 + \dfrac{112}{81} x^4 + \cdots ;$$

converges if $|-2x| < 1$, $|x| < 1/2$ so $R = 1/2$

19. $(1 - 4x^2)^{1/2} = 1 + \dfrac{1}{2} (-4x^2) + \dfrac{(1/2)(-1/2)}{2!} (-4x^2)^2$

$$+ \dfrac{(1/2)(-1/2)(-3/2)}{3!} (-4x^2)^3 + \cdots$$

$$= 1 - 2x^2 - 2x^4 - 4x^6 - \cdots ;$$

converges if $|-4x^2| < 1$, $|x|^2 < 1/4$, $|x| < 1/2$ so $R = 1/2$

21. $\displaystyle\int_0^1 \cos \sqrt{x}\, dx = \int_0^1 (1 - x/2! + x^2/4! - x^3/6! + \cdots)dx$

$$= x - \dfrac{1}{2 \cdot 2!} x^2 + \dfrac{1}{3 \cdot 4!} x^3 - \dfrac{1}{4 \cdot 6!} x^4 + \cdots \Big]_0^1$$

$$= 1 - \dfrac{1}{2 \cdot 2!} + \dfrac{1}{3 \cdot 4!} - \dfrac{1}{4 \cdot 6!} + \cdots ,$$

but $\dfrac{1}{4 \cdot 6!} < 0.5 \times 10^{-3}$ so $\displaystyle\int_0^1 \cos \sqrt{x}\, dx \approx 1 - \dfrac{1}{2 \cdot 2!} + \dfrac{1}{3 \cdot 4!} \approx 0.764$

23. $\displaystyle\int_0^{1/2} \dfrac{dx}{1 + x^4} = \int_0^{1/2} (1 - x^4 + x^8 - \cdots)dx$

$$= x - \dfrac{1}{5} x^5 + \dfrac{1}{9} x^9 - \cdots \Big]_0^{1/2} = 1/2 - \dfrac{(1/2)^5}{5} + \dfrac{(1/2)^9}{9} - \cdots ,$$

but $\dfrac{(1/2)^9}{9} < 0.5 \times 10^{-3}$ so $\displaystyle\int_0^{1/2} \dfrac{dx}{1 + x^4} \approx 1/2 - \dfrac{(1/2)^5}{5} \approx 0.494$

25. $\displaystyle\int_0^{0.1} e^{-x^3} dx = \int_0^{0.1} (1 - x^3 + x^6/2! - \cdots)dx$

$$= x - \dfrac{1}{4} x^4 + \dfrac{1}{7 \cdot 2!} x^7 - \cdots \Big]_0^{0.1}$$

$$= 0.1 - \frac{1}{4} (0.1)^4 + \frac{1}{7 \cdot 2!} (0.1)^7 - \cdots,$$

but $\frac{1}{4} (0.1)^4 < 0.5 \times 10^{-3}$ so $\int_0^{0.1} e^{-x^3} dx \approx 0.100$

27. $\displaystyle\int_0^{1/2} (1 + x^2)^{-1/4} dx = \int_0^{1/2} (1 - \frac{1}{4} x^2 + \frac{5}{32} x^4 - \frac{15}{128} x^6 + \cdots) dx$

$$= x - \frac{1}{12} x^3 + \frac{1}{32} x^5 - \frac{15}{896} x^7 + \cdots \Big]_0^{1/2}$$

$$= 1/2 - \frac{1}{12} (1/2)^3 + \frac{1}{32} (1/2)^5 - \frac{15}{896} (1/2)^7 + \cdots,$$

but $\frac{15}{896} (1/2)^7 < 0.5 \times 10^{-3}$ so

$$\int_0^{1/2} (1 + x^2)^{-1/4} dx \approx 1/2 - \frac{1}{12} (1/2)^3 + \frac{1}{32} (1/2)^5 \approx 0.491$$

29. $e^{-x^2} \cos x = (1 - x^2 + \frac{x^4}{2!} - \frac{x^6}{3!} + \cdots)(1 - \frac{x^2}{2!} + \frac{x^4}{4!} - \frac{x^6}{6!} + \cdots)$

$$= 1 - \frac{3}{2} x^2 + \frac{25}{24} x^4 - \frac{331}{720} x^6 + \cdots$$

31. $\dfrac{\sin x}{e^x} = e^{-x} \sin x = (1 - x + \frac{x^2}{2!} - \frac{x^3}{3!} + \frac{x^4}{4!} - \cdots)(x - \frac{x^3}{3!} + \frac{x^5}{5!} - \cdots)$

$$= x - x^2 + \frac{1}{3} x^3 - \frac{1}{30} x^5 + \cdots$$

33. $x \ln(1 - x^2) = x(-x^2 - \frac{1}{2} x^4 - \frac{1}{3} x^6 - \frac{1}{4} x^8 - \cdots)$

$$= -x^3 - \frac{1}{2} x^5 - \frac{1}{3} x^7 - \frac{1}{4} x^9 - \cdots$$

35. $x^2 e^{4x} \sqrt{1 + x} = x^2(1 + 4x + 8x^2 + \frac{32}{3} x^3 + \cdots)$

$$(1 + \frac{1}{2} x - \frac{1}{8} x^2 + \frac{1}{16} x^3 + \cdots)$$

$$= x^2(1 + \frac{9}{2} x + \frac{79}{8} x^2 + \frac{683}{48} x^3 + \cdots)$$

$$= x^2 + \frac{9}{2} x^3 + \frac{79}{8} x^4 + \frac{683}{48} x^5 + \cdots$$

37. (a) $\dfrac{1 - \cos x}{\sin x} = \dfrac{1 - (1 - x^2/2! + x^4/4! - x^6/6! + \cdots)}{x - x^3/3! + x^5/5! - \cdots}$

$\qquad = \dfrac{x^2/2! - x^4/4! + x^6/6! - \cdots}{x - x^3/3! + x^5/5! - \cdots}$

$\qquad = \dfrac{x/2! - x^3/4! + x^5/6! - \cdots}{1 - x^2/3! + x^4/5! - \cdots}, \qquad x \neq 0$

$\displaystyle\lim_{x \to 0} \dfrac{1 - \cos x}{\sin x} = \dfrac{0}{1} = 0$

(b) $\ln\sqrt{1 + x} - \sin 2x = \dfrac{1}{2} \ln(1 + x) - \sin 2x$

$\qquad = \dfrac{1}{2}(x - \dfrac{1}{2} x^2 + \dfrac{1}{3} x^3 - \cdots) - (2x - \dfrac{4}{3} x^3 + \dfrac{4}{15} x^5 - \cdots)$

$\qquad = -\dfrac{3}{2} x - \dfrac{1}{4} x^2 + \dfrac{3}{2} x^3 + \cdots,$

$\displaystyle\lim_{x \to 0} \dfrac{\ln\sqrt{1 + x} - \sin 2x}{x} = \lim_{x \to 0} (-\dfrac{3}{2} - \dfrac{1}{4} x + \dfrac{3}{2} x^2 + \cdots) = -3/2$

39. (a) $\sinh^{-1} x = \displaystyle\int (1 + x^2)^{-1/2} dx - C$

$\qquad = \displaystyle\int (1 - \dfrac{1}{2} x^2 + \dfrac{3}{8} x^4 - \dfrac{5}{16} x^6 + \cdots) dx - C$

$\qquad = (x - \dfrac{1}{6} x^3 + \dfrac{3}{40} x^5 - \dfrac{5}{112} x^7 + \cdots) - C,$

$\sinh^{-1} 0 = 0 \quad$ so $\quad C = 0$

(b) $(1 + x^2)^{-1/2} = 1 + \displaystyle\sum_{k=1}^{\infty} \dfrac{(-1/2)(-3/2)(-5/2)\cdots(-1/2 - k + 1)}{k!} (x^2)^k$

$\qquad = 1 + \displaystyle\sum_{k=1}^{\infty} (-1)^k \dfrac{1 \cdot 3 \cdot 5 \cdots (2k - 1)}{2^k k!} x^{2k},$

$\sinh^{-1} x = x + \displaystyle\sum_{k=1}^{\infty} (-1)^k \dfrac{1 \cdot 3 \cdot 5 \cdots (2k - 1)}{2^k k!(2k + 1)} x^{2k+1}$

(c) $R = 1$

41. $f^{(0)}(x) = (1 + x)^m$, $f^{(k)}(x) = m(m - 1)(m - 2)\cdots(m - k + 1)(1 + x)^{m-k}$

for $k \geq 1$, $f^{(0)}(0) = 1$, $f^{(k)}(0) = m(m - 1)(m - 2)\cdots(m - k + 1)$, so the

Maclaurin series for $(1 + x)^m$ is
$$1 + mx + \frac{m(m - 1)}{2!} x^2 + \cdots + \frac{m(m - 1)(m - 2)\cdots(m - k + 1)}{k!} x^k + \cdots$$

43. $(1 + x)^m = \binom{m}{0} + \sum_{k=1}^{\infty} \binom{m}{k} x^k = \sum_{k=0}^{\infty} \binom{m}{k} x^k$

45. $(1 - x)^{1/3} = (1 + (-x))^{1/3} = \sum_{k=0}^{\infty} \binom{1/3}{k}(-x)^k = \sum_{k=0}^{\infty} (-1)^k \binom{1/3}{k} x^k$, $R = 1$

47. $(1 + x^2)^{2/3} = \sum_{k=0}^{\infty} \binom{2/3}{k}(x^2)^k = \sum_{k=0}^{\infty} \binom{2/3}{k} x^{2k}$, $R = 1$

49. $x(1 + x)^{-1/2} = x \sum_{k=0}^{\infty} \binom{-1/2}{k} x^k = \sum_{k=0}^{\infty} \binom{-1/2}{k} x^{k+1}$, $R = 1$

SUPPLEMENTARY EXERCISES CHAPTER 11

1. $L = 0$ 3. $L = 0 - 0 = 0$

5. $\sin[(2n - 1)\pi/2]$ is alternately 1 and -1 so the limit does not exist

7. $a_n = (-1)^n/e^n$ is alternating,

   $a_n = e^{1/n}$ is decreasing;

   $a_n = \dfrac{1}{\sqrt{n}} - \dfrac{1}{\sqrt{n + 1}} = \dfrac{\sqrt{n + 1} - \sqrt{n}}{\sqrt{n}\,\sqrt{n + 1}} = \dfrac{1}{\sqrt{n}\,\sqrt{n + 1}\,(\sqrt{n + 1} + \sqrt{n})}$ is decreasing;

   $a_n = \sin \pi n = 0$ is nondecreasing;

   $a_n = \sin[(2n - 1)\pi/2]$ is alternating;

$$a_n = \frac{n + 1}{n(n + 2)}, \text{ let } f(x) = \frac{x + 1}{x^2 + 2x} \text{ then } f'(x) = -\frac{x^2 + 2x + 2}{(x^2 + 2x)^2} < 0$$

if $x \geq 1$ so a_n is decreasing.

9. (a) $\displaystyle\sum_{k=0}^{\infty} (\pi/q^2)^k$ is a geometric series which converges for $\pi/q^2 < 1$,

$q^2 > \pi$, $|q| > \sqrt{\pi}$

(b) $\displaystyle\sum_{k=1}^{\infty} 1/k^{3q}$ is a p-series with $p = 3q$, converges for $3q > 1$, $q > 1/3$

(c) $\displaystyle\sum_{k=2}^{\infty} 1/(k \ln q) = \sum_{k=2}^{\infty} (1/\ln q)(1/k)$ diverges for all q because

$\displaystyle\sum_{k=2}^{\infty} 1/k$ diverges

(d) $\displaystyle\sum_{k=2}^{\infty} (1/\ln q)^k$ is a geometric series which converges

for $|1/\ln q| < 1$, $|\ln q| > 1$, $q > e$ or $0 < q < e^{-1}$

11. (a) $1.3636\cdots = 1 + \displaystyle\sum_{k=1}^{\infty} 36(0.01)^k$

(b) $1.3636\cdots = 1 + \dfrac{0.36}{1 - 0.01} = 1 + 36/99 = 1 + 4/11 = 15/11$

13. (a) $\dfrac{1}{6}\left[\displaystyle\sum_{k=1}^{\infty} (\tfrac{1}{3})^k + \sum_{k=1}^{\infty} (\tfrac{1}{2})^k\right] = \dfrac{1}{6}\left[\dfrac{1/3}{1 - 1/3} + \dfrac{1/2}{1 - 1/2}\right] = 1/4$

(b) $\displaystyle\sum_{k=2}^{\infty} \ln \frac{k+1}{k} = \sum_{k=2}^{\infty} [\ln(k+1) - \ln k],$

$s_n = [\ln 3 - \ln 2] + [\ln 4 - \ln 3] + \cdots + [\ln(n+2) - \ln(n+1)]$

$\quad = \ln(n+2) - \ln 2, \quad \lim\limits_{n\to+\infty} s_n = +\infty,$ diverges

(c) $s_n = [1^{-1/2} - 2^{-1/2}] + [2^{-1/2} - 3^{-1/2}] + \cdots + [n^{-1/2} - (n+1)^{-1/2}]$

$\quad = 1 - (n+1)^{-1/2}, \quad \lim\limits_{n\to+\infty} s_n = 1$

15. converges (integral test, $\displaystyle\int_1^{\infty} x e^{-x^2} dx$ converges)

17. converges (comparison test, $\dfrac{\sqrt{k}}{k^2 + 7} < \dfrac{\sqrt{k}}{k^2} = \dfrac{1}{k^{3/2}}$)

19. converges (ratio test, $\rho = 0$)

21. diverges (root test, $\rho = (5/2)^3 > 1$)

23. absolutely convergent (comparison test, $2^k/(3^k + 1) < 2^k/3^k = (2/3)^k$, $\Sigma(2/3)^k$ is a convergent geometric series)

25. diverges ($\lim\limits_{k\to+\infty} |u_k| = \lim\limits_{k\to+\infty} \frac{1}{2}(3/2)^k = +\infty$)

27. $\rho = \lim\limits_{k\to+\infty} \dfrac{k^{3/2}|x-1|}{(k+1)^{3/2}} = |x-1|$, converges if $|x-1| < 1$, diverges if

$|x-1| > 1$. If $x = 0$, $\displaystyle\sum_{k=1}^{\infty} \dfrac{(-1)^k}{k^{3/2}}$ converges;

if $x = 2$, $\displaystyle\sum_{k=1}^{\infty} \dfrac{1}{k^{3/2}}$ converges. $R = 1$, interval of convergence $[0,2]$.

29. $\rho = \lim_{k \to +\infty} \dfrac{k|1 - x|^2}{4(k + 1)} = \dfrac{1}{4} |1 - x|^2$,

converges if $|x - 1|^2 < 4$, $|x - 1| < 2$; diverges if $|x - 1| > 2$.

If $x = -1$, $\displaystyle\sum_{k=1}^{\infty} 1/k$ diverges; if $x = 3$, $\displaystyle\sum_{k=1}^{\infty} 1/k$ diverges. $R = 2$,

interval of convergence $(-1,3)$.

31. $\rho = \lim_{k \to +\infty} \dfrac{1}{5} (k + 1)|x - 1| = +\infty$, $R = 0$, converges only for $x = 1$.

33. (a) $(x - 2) - \dfrac{1}{2} (x - 2)^2 + \dfrac{1}{3} (x - 2)^3$

(b) $R_3(x) = - \dfrac{(x - 2)^4}{4(c - 1)^4}$, c between 2 and x

(c) $|R_3(x)| = \dfrac{|x - 2|^4}{4|c - 1|^4} < \dfrac{|3/2 - 2|^4}{4|3/2 - 1|^4} = 1/4$

35. (a) $1 + \dfrac{1}{2} (x - 1) - \dfrac{1}{8} (x - 1)^2$

(b) $R_2(x) = \dfrac{(x - 1)^3}{16c^{5/2}}$, c between 1 and x

(c) $|R_2(x)| = \dfrac{|x - 1|^3}{16c^{5/2}} < \dfrac{|4/9 - 1|^3}{16(4/9)^{5/2}} = \dfrac{(5/9)^3}{16(2/3)^5} < 0.0814$

37. $\ln(a + x) = \ln a(1 + x/a) = \ln a + \ln(1 + x/a)$

$$= \ln a + \sum_{k=0}^{\infty} (-1)^k \dfrac{(x/a)^{k+1}}{k + 1},$$

converges if $|x/a| < 1$, $|x| < |a| = a$ so $R = a$

39. $1/(9 + x)^{1/2} = \dfrac{1}{3} (1 + x/9)^{-1/2}$

$$= \dfrac{1}{3} \left[1 + \sum_{k=1}^{\infty} (-1)^k \dfrac{1 \cdot 3 \cdot 5 \cdots (2k - 1)}{2^k k!} (x/9)^k \right],$$

converges if $|x/9| < 1$, $|x| < 9$ so $R = 9$

41. $\sec x = 1/\cos x = 1/(1 - x^2/2! + x^4/4! - \cdots) = 1 + x^2/2 + 5x^4/24 + \cdots$

43. $[\cos x]^{1/2} = [1 - x^2/2! + x^4/4! - \cdots]^{1/2}$

$$= [1 + (-x^2/2! + x^4/4! - \cdots)]^{1/2}$$

$$= 1 + \frac{1}{2}(-x^2/2! + x^4/4! - \cdots) - \frac{1}{8}(-x^2/2! + x^4/4! - \cdots)^2 + \cdots$$

$$= 1 - x^2/4 - x^4/96 + \cdots$$

45. $f(x) = \ln(1 + \sin x)$, $f'(x) = \dfrac{\cos x}{1 + \sin x}$, $f''(x) = -\dfrac{1}{1 + \sin x}$,

$f''' = \dfrac{\cos x}{(1 + \sin x)^2}$; $f(0) = 0$, $f'(0) = 1$, $f''(0) = -1$, $f'''(0) = 1$;

$\ln(1 + \sin x) = x - \dfrac{1}{2} x^2 + \dfrac{1}{6} x^3 + \cdots$

47. $\dfrac{\ln(1 - 2x)}{x} = \dfrac{1}{x}\left(-2x - 2x^2 - \dfrac{8}{3} x^3 - \cdots\right) = -2 - 2x - \dfrac{8}{3} x^2 - \cdots$,

$\displaystyle\lim_{x \to 0} \dfrac{\ln(1 - 2x)}{x} = -2$

49. $\sin x = x - x^3/3! + x^5/5! + (0)x^6 + R_6(x)$,

$|R_6(x)| \le \dfrac{|x|^7}{7!} < 6 \times 10^{-4}$ if $|x|^7 < 3.024$, $|x| < (3.024)^{1/7} \approx 1.17$

51. $\displaystyle\int_0^1 \dfrac{1 - e^{-t/2}}{t}\, dt = \int_0^1 \dfrac{\left[1 - \left(1 - \dfrac{t}{2} + \dfrac{t^2}{8} - \dfrac{t^3}{48} + \dfrac{t^4}{384} - \dfrac{t^5}{3840} + \cdots\right)\right]}{t}\, dt$

$$= \int_0^1 \left(\dfrac{1}{2} - \dfrac{t}{8} + \dfrac{t^2}{48} - \dfrac{t^3}{384} + \dfrac{t^4}{3840} - \cdots\right) dt$$

$$= \dfrac{t}{2} - \dfrac{t^2}{16} + \dfrac{t^3}{144} - \dfrac{t^4}{1436} + \dfrac{t^5}{19200} - \cdots \Bigg]_0^1$$

$$= 1/2 - 1/16 + 1/144 - 1/1436 + 1/19200 - \cdots,$$

but $1/19200 < 0.5 \times 10^{-3}$ so

$$\int_0^1 \dfrac{1 - e^{-t/2}}{t}\, dt \approx 1/2 - 1/16 + 1/144 - 1/1436 \approx 0.444$$

53. $y' = \displaystyle\sum_{n=1}^{\infty} \frac{k^n x^{n-1}}{(n-1)!} = \sum_{n=0}^{\infty} \frac{k^{n+1} x^k}{n!} = k \sum_{n=0}^{\infty} \frac{k^n x^k}{n!} = ky$, so $y' - ky = 0$

CHAPTER 12

EXERCISE SET 12.2

1.

3.

5.

$y^2 = (12/5)x$

7.

9.

11.

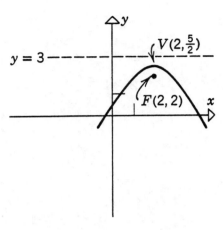

$$(x - 2)^2 = -2(y - 5/2)$$

13.

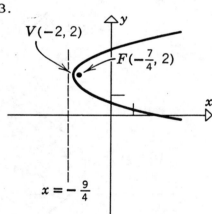

$$(y - 2)^2 = x + 2$$

15.

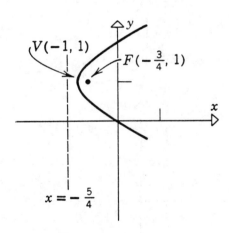

$$(y - 1)^2 = x + 1$$

17. $y^2 = 4px$, $p = 3$, $y^2 = 12x$

19. $y^2 = 4px$, $p = -7$, $y^2 = -28x$

21. $y^2 = 4px$, $2^2 = 4p(2)$, $4p = 2$, $y^2 = 2x$

23. $x^2 = 4py$, $p = -3$, $x^2 = -12y$

25. $y^2 = 4p(x - h)$, $4 = 4p(3 - h)$ and $9 = 4p(2 - h)$, solve simultaneously to get $h = 19/5$, $4p = -5$ so $y^2 = -5(x - 19/5)$

27. The vertex is half way between the focus and directrix so the vertex is at $(3/2, 0)$ and $p = 3/2$, $y^2 = 6(x - 3/2)$.

29. The vertex is 3 units above the directrix so $p = 3$,
$(x - 1)^2 = 12(y - 1)$.

31. $(x - 5)^2 = 4p(y + 3)$, $(9 - 5)^2 = 4p(5 + 3)$ so $4p = 2$,
$(x - 5)^2 = 2(y + 3)$

33. (a) $y = Ax^2 + Bx + C$; use $(0, 3)$, $(2, 0)$ and $(3, 2)$ to get the system of equations
$$C = 3$$
$$9A + 3B + C = 2$$
$$4A + 2B + C = 0$$
which when solved yields $A = 7/6$, $B = -23/6$, $C = 3$ so
$$y = \frac{7}{6}x^2 - \frac{23}{6}x + 3.$$

(b) $x = Ay^2 + By + C$;
$$9A + 3B + C = 0$$
$$4A + 2B + C = 3$$
$$C = 2$$
$A = -7/6$, $B = 17/6$, $C = 2$ so $x = -\frac{7}{6}y^2 + \frac{17}{6}y + 2$

35. Complete the square to get $(x + \frac{B}{2A})^2 = \frac{1}{A}(y - C + \frac{B^2}{4A})$ so the vertex is
at $(-\frac{B}{2A}, \frac{4AC - B^2}{4A})$, the focus is at $(-\frac{B}{2A}, \frac{4AC - B^2 + 1}{4A})$, and the
directrix is $y = \frac{4AC - B^2 - 1}{4A}$

37. $y = ax^2 + b$, $(20,0)$ and $(10,12)$ are on the curve so $400a + b = 0$ and $100a + b = 12$. Solve for b to get $b = 16 =$ height of arch.

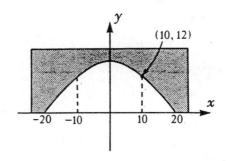

39.   Let (x_0, y_0) be a point on the parabola $y^2 = 4px$ for $p > 0$ then

$$PF = \sqrt{(x_0 - p)^2 + y_0^2} = \sqrt{x_0^2 - 2px_0 + p^2 + 4px_0} = \sqrt{(x_0 + p)^2}$$

so $PF = x_0 + p$ where $x_0 \geq 0$ and PF is a minimum when $x_0 = 0$ (the vertex).

41.   Use an xy-coordinate system so that $y^2 = 4px$ is an equation of the parabola, then $(1, 1/2)$ is a point on the curve so $(1/2)^2 = 4p(1)$, $p = 1/16$. The light source should be placed at the focus which is 1/16 ft. from the vertex.

43.   Similar to proof in text.

EXERCISE SET 12.3

1.

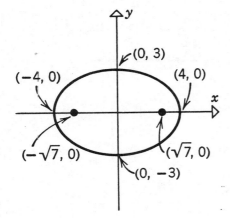

$$c^2 = 16 - 9 = 7, \quad c = \sqrt{7}$$

3.

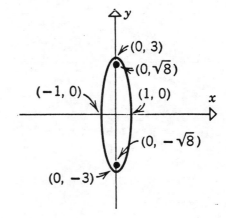

$$\frac{x^2}{1} + \frac{y^2}{9} = 1$$
$$c^2 = 9 - 1 = 8, \quad c = \sqrt{8}$$

5.

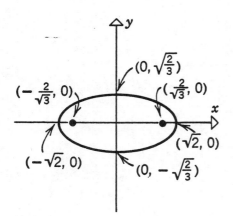

$$\frac{x^2}{2} + \frac{y^2}{2/3} = 1$$
$$c^2 = 2 - 2/3 = 4/3, \quad c = 2/\sqrt{3}$$

7.

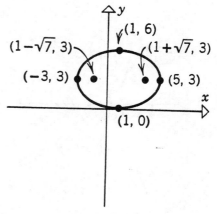

$$\frac{(x - 1)^2}{16} + \frac{(y - 3)^2}{9} = 1$$
$$c^2 = 16 - 9 = 7, \quad c = \sqrt{7}$$

9.

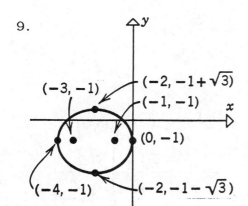

$$\frac{(x + 2)^2}{4} + \frac{(y + 1)^2}{3} = 1$$

$$c^2 = 4 - 3 = 1, \ c = 1$$

11.

$$\frac{(x + 1)^2}{9} + \frac{(y - 1)^2}{1} = 1$$

$$c^2 = 9 - 1 = 8, \ c = \sqrt{8}$$

$$\frac{(x + 1)^2}{4} + \frac{(y - 5)^2}{16} = 1$$

$$c^2 = 16 - 4 = 12, \ c = 2\sqrt{3}$$

13.

15. $x^2/9 + y^2/4 = 1$

17. $a = 26/2 = 13, \ c = 5, \ b^2 = a^2 - c^2 = 169 - 25 = 144;$
      $x^2/169 + y^2/144 = 1$

19. $c = 1, \ a^2 = b^2 + c^2 = 2 + 1 = 3; \ x^2/3 + y^2/2 = 1$

21. $b^2 = 16 - 12 = 4$; $x^2/16 + y^2/4 = 1$ and $x^2/4 + y^2/16 = 1$

23. $a = 6$, $(2,3)$ satisfies $x^2/36 + y^2/b^2 = 1$ so $4/36 + 9/b^2 = 1$, $b^2 = 81/8$; $x^2/36 + y^2/(81/8) = 1$

25. The center is midway between the foci so it is at $(1,3)$ thus $c = 1$, $b = 1$, $a^2 = 1 + 1 = 2$; $(x - 1)^2 + (y - 3)^2/2 = 1$

27. $(4,1)$ and $(4,5)$ are the foci so the center is at $(4,3)$ thus $c = 2$, $a = 12/2 = 6$, $b^2 = 36 - 4 = 32$; $(x - 4)^2/32 + (y - 3)^2/36 = 1$

29. Substitute $x = 8 - 2y$ into $x^2 + 4y^2 = 40$ to get $y^2 - 4y + 3 = 0$ which yields $y = 1, 3$. Substitute these into $x = 8 - 2y$ to get $x = 6, 2$ so the points of intersection are $(2,3)$ and $(6,1)$.

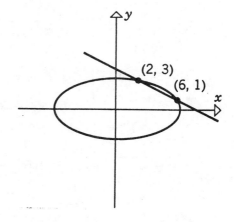

31. Substitute $x^2 = 20 - y^2$ into $x^2 + 9y^2 = 36$ to get $y^2 = 2$, $y = \pm\sqrt{2}$. Use $x^2 = 20 - y^2$ to get $x = \pm 3\sqrt{2}$ when $y = \sqrt{2}$ or $-\sqrt{2}$. The points of intersection are $(3\sqrt{2},\sqrt{2})$, $(3\sqrt{2},-\sqrt{2})$, $(-3\sqrt{2},\sqrt{2})$, and $(-3\sqrt{2},-\sqrt{2})$.

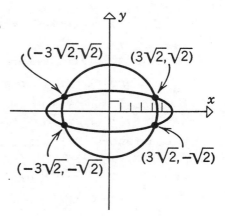

33. Use implicit differentiation on $x^2 + 4y^2 = 8$ to get

$dy/dx \Big|_{(x_0, y_0)} = - \dfrac{x_0}{4y_0}$ where (x_0, y_0) is the point of tangency, but

$-x_0/(4y_0) = -1/2$ because the slope of the line is $-1/2$ so $x_0 = 2y_0$.

(x_0, y_0) is on the ellipse so $x_0^2 + 4y_0^2 = 8$ which when solved with

$x_0 = 2y_0$ yields the points of tangency $(2, 1)$ and $(-2, -1)$. Substitute

these into the equation of the line to get $k = \pm 4$.

35. $y = (b/a)\sqrt{a^2 - x^2}$ is the upper half of the ellipse,

$A = 4 \displaystyle\int_0^a \frac{b}{a} \sqrt{a^2 - x^2} \, dx = \frac{4b}{a} \int_0^a \sqrt{a^2 - x^2} \, dx = \frac{4b}{a} (\tfrac{1}{4} \pi a^2) = \pi ab$

37. The vertex in the first quadrant is at the point where $y = x > 0$ so

$x^2/a^2 + x^2/b^2 = 1$, $x^2 = a^2 b^2/(a^2 + b^2)$, $x = ab/\sqrt{a^2 + b^2}$.

$A = (2x)^2 = 4x^2 = 4a^2 b^2/(a^2 + b^2)$.

39. $\sqrt{(x - 4)^2 + y^2} = \frac{4}{5} \left| \frac{25}{4} - x \right|$, $x^2 - 8x + 16 + y^2 = \frac{16}{25} (\frac{625}{16} - \frac{25}{2} x + x^2)$,

$9x^2 + 25y^2 = 225$, $x^2/25 + y^2/9 = 1$

center: $(0, 0)$, major axis: 10, minor axis: 6

41. Let R be the radius of a circle C that
is tangent to both C_1 and C_2. The distances
between the center of C and the centers
of C_1 and C_2 are, respectively,
$r_1 + R$ and $r_2 - R$ (see accompanying
diagram). Their sum is
$(r_1 + R) + (r_2 - R) = r_1 + r_2$ which is
a constant so the centers lie on an
ellipse with foci at the centers of C_1
and C_2. The length of the major axis is
$2a = r_1 + r_2$; the center of the ellipse

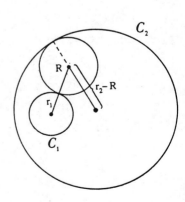

is at the midpoint of the line segment that joins the centers of C_1 and C_2.

43.   (a) $0 < c < a$ so $0 < c/a < 1$, $0 < e < 1$.

(b) Keeping a fixed, c approaches 0 if e approaches 0 so $b = \sqrt{a^2 - c^2}$ approaches a; the ellipse approaches a circle of radius a.

c approaches a if e approaches 1 so $b = \sqrt{a^2 - c^2}$ approaches 0; the ellipse approaches the major axis.

45.   Similar to derivation in text.

EXERCISE SET 12.4

1.

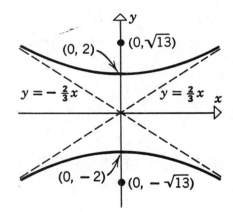

$c^2 = a^2 + b^2 = 16 + 4 = 20$,
$c = 2\sqrt{5}$

3.

$y^2/4 - x^2/9 = 1$
$c^2 = 4 + 9 = 13$, $c = \sqrt{13}$

5.

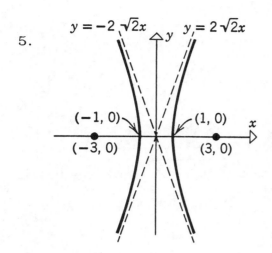

$x^2/1 - y^2/8 = 1$
$c^2 = 1 + 8 = 9$, $c = 3$

7.

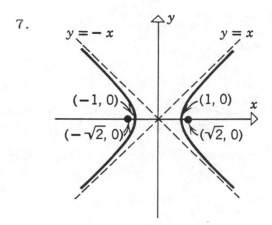

$c^2 = 1 + 1 = 2$, $c = \sqrt{2}$

9.

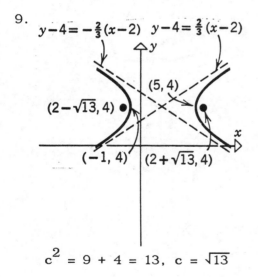

$c^2 = 9 + 4 = 13$, $c = \sqrt{13}$

11.

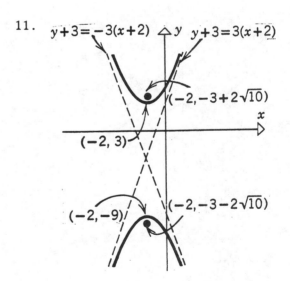

$(y + 3)^2/36 - (x + 2)^2/4 = 1$
$c^2 = 36 + 4 = 40$, $c = 2\sqrt{10}$

13.

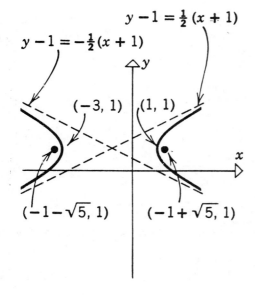

$y - 1 = \frac{1}{2}(x + 1)$

$y - 1 = -\frac{1}{2}(x + 1)$

$(-3, 1)$ $(1, 1)$

$(-1-\sqrt{5}, 1)$ $(-1+\sqrt{5}, 1)$

$(x + 1)^2/4 - (y - 1)^2/1 = 1$

$c^2 = 4 + 1 = 5, \ c = \sqrt{5}$

15.

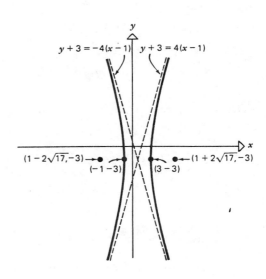

$y + 3 = -4(x - 1)$ $y + 3 = 4(x - 1)$

$(1 - 2\sqrt{17}, -3)$ $(-1 -3)$ $(3 -3)$ $(1 + 2\sqrt{17}, -3)$

$(x - 1)^2/4 - (y + 3)^2/64 = 1$

$c^2 = 4 + 64 = 68, \ c = \sqrt{17}$

17. $a = 2, \ c = 3, \ b^2 = 9 - 4 = 5; \ x^2/4 - y^2/5 = 1$

19. $a = 1, \ b/a = 2, \ b = 2; \ x^2 - y^2/4 = 1$

21. vertices along x-axis: $b/a = 3/2$ so $a = 8/3; \ x^2/(64/9) - y^2/16 = 1$
    vertices along y-axis: $a/b = 3/2$ so $a = 6; \ y^2/36 - x^2/16 = 1$

23. The form of the equation is $y^2/a^2 - x^2/b^2$ because $(5,9)$ is above the
    asymptote $y = x$, $a/b = 1$ and $81/a^2 - 25/b^2 = 1$, solve to get
    $a^2 = b^2 = 56; \ y^2/56 - x^2/56 = 1$

25. $a = 2$ so $x^2/4 - y^2/b^2 = 1$, $(4,2)$ is on the curve so
    $4 - 4/b^2 = 1, \ b^2 = 4/3; \ x^2/4 - y^2/(4/3) = 1$

27. the center is at $(2,-3)$, $a = 2$, $c = 3$, $b^2 = 9 - 4 = 5$;
    $(x - 2)^2/4 - (y + 3)^2/5 = 1$

29. the center is at $(6,4)$, $a = 4$, $c = 5$, $b^2 = 25 - 16 = 9$;
    $(x - 6)^2/16 - (y - 4)^2/9 = 1$

31. From the definition of a hyperbola,

$$\left| \sqrt{(x - 1)^2 + (y - 1)^2} - \sqrt{x^2 + y^2} \right| = 1,$$

$$\sqrt{(x - 1)^2 + (y - 1)^2} - \sqrt{x^2 + y^2} = \pm 1,$$

    transpose the second radical to the right hand side of the equation and

    square and simplify to get $\pm 2\sqrt{x^2 + y^2} = -2x - 2y + 1$, square and
    simplify again to get $8xy - 4x - 4y + 1 = 0$.

33. Substitute $x = 2y + 20$ into
    $x^2 - 4y^2 = 36$ to get $y = -91/20$,
    so $x = 2(-91/20) + 20 = 109/10$.
    The curves intersect at $(109/10, -91/20)$.

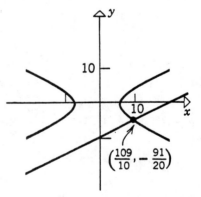

35. Eliminate x to get $y^2 = 1$, $y = \pm 1$.
    Use either equation to find that
    $x = \pm 2$ if $y = 1$ or if $y = -1$.
    The curves intersect at $(2,1)$,
    $(2,-1)$, $(-2,1)$, and $(-2,-1)$.

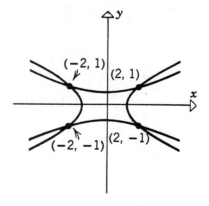

37. Let (x,y) be the coordinates of the point, use the formula for the
distance between a point and a line to get

$$(|mx - y|/\sqrt{m^2 + 1})(|mx + y|/\sqrt{m^2 + 1}) = k^2,$$
$$|m^2x^2 - y^2|/(m^2 + 1) = k^2 \text{ so } m^2x^2 - y^2 = \pm k^2(m^2 + 1) \text{ which are}$$
hyperbolas with $y = \pm mx$ as asymptotes.

39. Let (x_0, y_0) be one of the points then $\left. dy/dx \right|_{(x_0,y_0)} = 4x_0/y_0$, the

tangent line is $y = (4x_0/y_0)x + 4$, but (x_0, y_0) is on both the line and

the curve which leads to $4x_0^2 - y_0^2 + 4y_0 = 0$ and $4x_0^2 - y_0^2 = 36$, solve

to get $x_0 = \pm 3\sqrt{13}/2$, $y_0 = -9$.

41. Let (x_0, y_0) be such a point. The foci are at $(-\sqrt{5}, 0)$ and $(\sqrt{5}, 0)$, the
lines are perpendicular if the product of their slopes is -1 so

$$\frac{y_0}{x_0 + \sqrt{5}} \cdot \frac{y_0}{x_0 - \sqrt{5}} = -1, \quad y_0^2 = 5 - x_0^2 \text{ and } 4x_0^2 - y_0^2 = 4. \quad \text{Solve to get}$$

$x_0 = \pm 3/\sqrt{5}$, $y_0 = \pm 4/\sqrt{5}$. The coordinates are $(\pm 3/\sqrt{5}, 4/\sqrt{5})$,

$(\pm 3/\sqrt{5}, -4/\sqrt{5})$.

43. (a) $c > a$ so $c/a > 1$, $e > 1$.
     (b) Keeping a fixed, c approaches a if e approaches 1 so

     $b = \sqrt{c^2 - a^2}$ approaches 0; the two branches of the hyperbola flatten
     and approach the focal axis, excluding the portion between the

     vertices.   c gets larger if e gets larger so $b = \sqrt{c^2 - a^2}$ gets
     larger; the branches approach two lines that are perpendicular to
     the focal axis at the vertices.

45. Let d_1 and d_2 be the distances of the first and second observers,
respectively, from the point where the gun was fired, and let v be the
speed of sound.   Then t = (time for sound to reach the second observer)
– (time for sound to reach the first observer) = $d_2/v - d_1/v$ so

$d_2 - d_1 = vt$.   For constant v and t the difference of distances, d_2 and

d_1 is constant so the gun was fired somewhere on a branch of a hyperbola whose foci are where the observers are.

47.   Similar to the derivation in the text.

EXERCISE SET 12.5

1.   (a) $\sin\theta = \sqrt{3}/2$, $\cos\theta = 1/2$

$x' = (-2)(1/2) + (6)(\sqrt{3}/2) = -1 + 3\sqrt{3}$

$y' = -(-2)(\sqrt{3}/2) + 6(1/2) = 3 + \sqrt{3}$

(b) $x = \frac{1}{2} x' - \frac{\sqrt{3}}{2} y' = \frac{1}{2} (x' - \sqrt{3}\, y')$

$y = \frac{\sqrt{3}}{2} x' + \frac{1}{2} y' = \frac{1}{2} (\sqrt{3}\, x' + y')$

$\sqrt{3}\, [\frac{1}{2} (x' - \sqrt{3}\, y')][\frac{1}{2} (\sqrt{3}\, x' + y')] + [\frac{1}{2} (\sqrt{3}\, x' + y')]^2 = 6$

$\frac{\sqrt{3}}{4} (\sqrt{3}\, x'^2 - 2x'y' - \sqrt{3}\, y'^2) + \frac{1}{4} (3x'^2 + 2\sqrt{3}\, x'y' + y'^2) = 6$

$\frac{3}{2} x'^2 - \frac{1}{2} y'^2 = 6$, $\frac{x'^2}{4} - \frac{y'^2}{12} = 1$

(c)

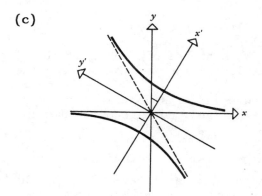

3. $\cot 2\theta = (0 - 0)/1 = 0, \quad 2\theta = 90^{\circ}, \quad \theta = 45^{\circ}$

    $x = (\sqrt{2}/2)(x' - y'), \quad y = (\sqrt{2}/2)(x' + y')$

    $y'^2/18 - x'^2/18 = 1, \quad \text{hyperbola}$

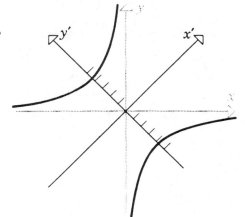

5. $\cot 2\theta = [1 - (-2)]/4 = 3/4$

    $\cos 2\theta = 3/5$

    $\sin \theta = \sqrt{(1 - 3/5)/2} = 1/\sqrt{5}$

    $\cos \theta = \sqrt{(1 + 3/5)/2} = 2/\sqrt{5}$

    $x = (1/\sqrt{5})(2x' - y')$

    $y = (1/\sqrt{5})(x' + 2y')$

    $x'^2/3 - y'^2 = 1, \quad \text{hyperbola}$

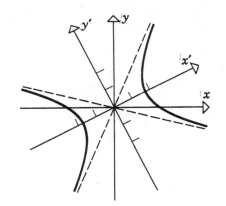

7. $\cot 2\theta = (1 - 3)/(2\sqrt{3}) = -1/\sqrt{3},$

    $2\theta = 120^{\circ}, \qquad \theta = 60^{\circ}$

    $x = (1/2)(x' - \sqrt{3}\, y')$

    $y = (1/2)(\sqrt{3}\, x' + y')$

    $y' = x'^2, \quad \text{parabola}$

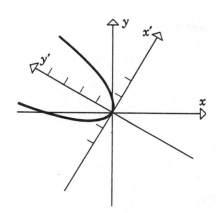

9. $\cot 2\theta = (9 - 16)/(-24) = 7/24$
    $\cos 2\theta = 7/25,$
    $\sin \theta = 3/5, \qquad \cos \theta = 4/5$
    $x = (1/5)(4x' - 3y'),$
    $y = (1/5)(3x' + 4y')$
    $y'^2 = 4(x' - 1),$ parabola

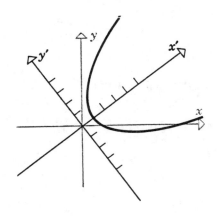

11. $\cot 2\theta = (52 - 73)/(-72) = 7/24$
    $\cos 2\theta = 7/25, \qquad \sin \theta = 3/5,$
    $\cos \theta = 4/5$
    $x = (1/5)(4x' - 3y'),$
    $y = (1/5)(3x' + 4y')$
    $(x' + 1)^2/4 + y'^2 = 1,$
    ellipse

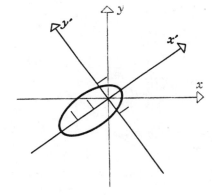

13. Let $x = x'\cos \theta - y'\sin \theta$, $y = x'\sin \theta + y'\cos \theta$ then $x^2 + y^2 = r^2$ becomes
    $(\sin^2\theta + \cos^2\theta)x'^2 + (\sin^2\theta + \cos^2\theta)y'^2 = r^2, \quad x'^2 + y'^2 = r^2.$
    Under a rotation transformation the center of the circle stays at the
    origin of both coordinate systems.

15. $x' = (\sqrt{2}/2)(x + y)$, $y' = (\sqrt{2}/2)(-x + y)$ which when substituted into
    $3x'^2 + y'^2 = 6$ yields $x^2 + xy + y^2 = 3.$

17. $\sqrt{x} + \sqrt{y} = 1$, $\sqrt{x} = 1 - \sqrt{y}$, $x = 1 - 2\sqrt{y} + y$, $2\sqrt{y} = 1 - x + y,$
    $4y = 1 + x^2 + y^2 - 2x + 2y - 2xy$, $x^2 - 2xy + y^2 - 2x - 2y + 1 = 0.$
    $\cot 2\theta = \dfrac{1 - 1}{2} = 0$, $2\theta = \pi/2$, $\theta = \pi/4.$ Let $x = x'/\sqrt{2} - y'/\sqrt{2},$
    $y = x'/\sqrt{2} + y'/\sqrt{2}$ to get $2y'^2 - 2\sqrt{2}x' + 1 = 0$, which is a parabola.

From $\sqrt{x} + \sqrt{y} = 1$ we see that $0 \le x \le 1$ and $0 \le y \le 1$, so the graph is just a portion of a parabola.

19. Use (9) to express $B' - 4A'C'$ in terms of A, B, C, and θ, then simplify.

21. $\cot 2\theta = (A - C)/B = 0$ if $A = C$ so $2\theta = 90°$, $\theta = 45°$.

23. $B^2 - 4AC = (-1)^2 - 4(1)(1) = -3 < 0$; ellipse, point, or no graph

25. $B^2 - 4AC = (2\sqrt{3})^2 - 4(1)(3) = 0$; parabola, line, pair of parallel lines, or no graph

27. $B^2 - 4AC = (-24)^2 - 4(34)(41) = -5000 < 0$; ellipse, point, or no graph

29. Part (b): from (18), $A'C' < 0$ so A' and C' have opposite signs. By multiplying (19) through by -1, if necessary, assume that $A' < 0$ and $C' > 0$ so (19) can be written as $(x' - h)^2/C' - (y' - k)^2/|A'| = K$. If $K \ne 0$ then the graph is a hyperbola (divide both sides by K), if $K = 0$ then we get the pair of intersecting lines

$(x' - h)/\sqrt{C'} = \pm(y' - k)/\sqrt{|A'|}$.

Part (c): from (18), $A'C' = 0$ so either $A' = 0$ or $C' = 0$ but not both (this would imply that $A = B = C = 0$ which results in (14) being linear). Suppose $A' \ne 0$ and $C' = 0$ then complete the square to get

$(x' - h)^2 = -\dfrac{E'}{A'} y' + K$. If $E' \ne 0$ the graph is a parabola, if $E' = 0$

and $K = 0$ the graph is the line $x' = h$, if $E' = 0$ and $K > 0$ the graph is the pair of parallel lines $x' = h \pm \sqrt{K}$, if $E' = 0$ and $K < 0$ there is no graph.

SUPPLEMENTARY EXERCISES CHAPTER 12

1. parabola, $(y - 3)^2 = -12(x + 2)$, $p = -3$; vertex $(-2,3)$, focus $(-5,3)$, directrix $x = 1$.

3. ellipse, $(x + 2)^2/4 + (y - 1)^2/9$, $a = 3$, $b = 2$, $c = \sqrt{5}$; center $(-2,1)$, foci $(-2,1\pm\sqrt{5})$, major axis 6, minor axis 4.

5. hyperbola, $(x - 2)^2/9 - (y - 1)^2/1 = 1$, $a = 3$, $b = 1$, $c = \sqrt{10}$, center $(2,1)$, foci $(2\pm\sqrt{10},1)$, vertices $(-1,1)$ and $(5,1)$, asymptotes $y - 1 = \pm(x - 2)/3$.

7. parabola, $(y - 1)^2 = (-3/2)(x - 3)$, $p = -3/8$; vertex $(3,1)$, focus $(21/8,1)$, directrix $x = 27/8$.

9. $p = 4$; $(y - 3)^2 = 16(x - 1)$.

11. center $(0,0)$, $c = 5$, $a = 6/2 = 3$, $b^2 = 16$, $y^2/9 - x^2/16 = 1$.

13. center $(0,0)$, $c = 3$, $a = 10/2 = 5$, $b^2 = 16$; $x^2/25 + y^2/16 = 1$.

15. The curve is a parabola with focus at $(3,4)$ and directrix $y = 2$ so the vertex is at $(3,3)$ and $p = 1$; $(x - 3)^2 = 4(y - 3)$.

17.

19.

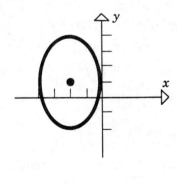

21. $\cot 2\theta = (3 - 3)/(-2) = 0$, $\theta = 45°$; use $x = (\sqrt{2}/2)(x' - y')$, $y = (\sqrt{2}/2)(x' + y')$ to get $x'^2/2 + y'^2/1 = 1$; ellipse.

23. $\cot 2\theta = (11 - 1)/(10\sqrt{3}) = 1/\sqrt{3}$, $\theta = 30°$; use $x = (1/2)(\sqrt{3}x' - y')$, $y = (1/2)(x' + \sqrt{3}y')$ to get $x'^2/(1/4) - y'^2/1 = 1$; hyperbola.

25. $\cot 2\theta = (16 - 9)/(-24) = -7/24$, $\cos 2\theta = -7/25$, $\sin \theta = \sqrt{(1 + 7/25)/2} = 4/5$,

$\cos \theta = \sqrt{(1 - 7/25)/2} = 3/5$, $\theta = \tan^{-1}(4/3)$; use $x = (1/5)(3x' - 4y')$, $y = (1/5)(4x' + 3y')$ to get $y'^2 = 4(x' - 1)$; parabola.

27.

29.

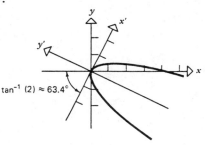

CHAPTER 13

EXERCISE SET 13.1

1.

3. (a) $(3\sqrt{3},3)$ (b) $(-7/2,7\sqrt{3}/2)$
    (c) $(4\sqrt{2},4\sqrt{2})$ (d) $(5,0)$
    (e) $(-7\sqrt{3}/2,7/2)$ (f) $(0,0)$

5. (a) $(5,\pi)$ (b) $(4,11\pi/6)$
    (c) $(2,3\pi/2)$ (d) $(8\sqrt{2},5\pi/4)$
    (e) $(6,2\pi/3)$ (f) $(\sqrt{2},\pi/4)$

7. (a) $(-5,0)$ (b) $(-4,5\pi/6)$
    (c) $(-2,\pi/2)$ (d) $(-8\sqrt{2},\pi/4)$
    (e) $(-6,5\pi/3)$ (f) $(-\sqrt{2},5\pi/4)$

9. $y = 4$; line

11. $2r - r\cos\theta = 6$, $2r = x + 6$, square to get $4r^2 = (x + 6)^2$,
    $4(x^2 + y^2) = x^2 + 12x + 36$, $3x^2 + 4y^2 - 12x = 36$; ellipse

13. $r^2 + 4r\cos\theta = 0$, $x^2 + y^2 + 4x = 0$; circle

15. $3r\cos\theta + 2r\sin\theta = 6$, $3x + 2y = 6$; line

17. $r\cos\theta = 7$

19. $r^4 = 16(r^2\cos^2\theta - r^2\sin^2\theta)$, $r^2 = 16\cos 2\theta$

21. $r^2\cos^2\theta = 9r\sin\theta$, $r = 9\sec\theta\tan\theta$

23. $r^2 - 6r\sin\theta = 0$, $r = 6\sin\theta$

25.   27.

29. $r^2 = ar\sin\theta + br\cos\theta$, $x^2 + y^2 = ay + bx$ which is a circle.

31. (r,θ) and $(-r,\theta+\pi)$ are polar coordinates of the same point so if $r < 0$ then $-r > 0$ and $x = (-r)\cos(\theta + \pi) = (-r)(-\cos\theta) = r\cos\theta$, $y = (-r)\sin(\theta + \pi) = (-r)(-\sin\theta) = r\sin\theta$.

EXERCISE SET 13.2

1.   3.   5.

Circle   Circle

7.

Circle

9.

Cardioid

11.

Cardioid

13.

Cardioid

15.

Limaçon

17.

Limaçon

19.

Limaçon

21.

Limaçon

23.

Limaçon

25.

Lemniscate

27.

Lemniscate

29.

Spiral

31.

Four-petal rose

33.

Three-petal rose

35.

Eight-petal rose

37.

39.

41.

43.

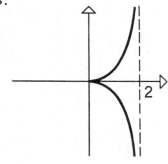

45. $\displaystyle\lim_{\theta\to0^+} y = \lim_{\theta\to0^+} \frac{\sin\theta}{\sqrt{\theta}} = 0$

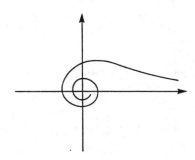

47. $y = r\sin\theta = (1 + \cos\theta)\sin\theta = \sin\theta + \sin\theta\cos\theta$,

$dy/d\theta = \cos\theta - \sin^2\theta + \cos^2\theta = 2\cos^2\theta + \cos\theta - 1$
$\qquad = (2\cos\theta - 1)(\cos\theta + 1)$
$dy/d\theta = 0$ if $\cos\theta = 1/2$ or if $\cos\theta = -1$; $\theta = \pi/3$ or π.

If $\theta = 0$, $\pi/3$, π, then $y = 0$, $3\sqrt{3}/4$, 0 so the maximum value of y is $3\sqrt{3}/4$.

49. $x = r\cos\theta = (1 + \cos\theta)\cos\theta = \cos\theta + \cos^2\theta$,
$dx/d\theta = -\sin\theta - 2\sin\theta\cos\theta = -\sin\theta(1 + 2\cos\theta)$,
$dx/d\theta = 0$ if $\sin\theta = 0$ or if $\cos\theta = -1/2$; $\theta = 0$, $2\pi/3$, or π. If $\theta = 0$, $2\pi/3$, π, then $x = 2$, $-1/4$, 0 so the minimum value of x is $-1/4$.

51. $r = r\cos\theta + k$,
$r = k/(1 - \cos\theta)$

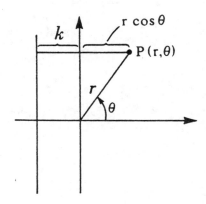

EXERCISE SET 13.3

1. $A = \int_{\pi/6}^{\pi/3} \frac{1}{2} \theta^2 d\theta = 7\pi^3/1296$ 3. $A = 2\int_0^{\pi} \frac{1}{2} (2 + 2\cos\theta)^2 d\theta = 6\pi$

5. $A = 2\int_{\pi/6}^{\pi/2} \frac{1}{2} [25\sin^2\theta - (2 + \sin\theta)^2] d\theta = 8\pi/3 + \sqrt{3}$

7. $A = 2\int_0^{\pi/3} \frac{1}{2} [(2 + 2\cos\theta)^2 - 9] d\theta = 9\sqrt{3}/2 - \pi$

9. $A = 2\int_0^{\pi/2} \frac{1}{2} \sin 2\theta \, d\theta = 1$ 11. $A = 6\int_0^{\pi/6} \frac{1}{2} (16\cos^2 3\theta) d\theta = 4\pi$

13. $A = 2\left[\int_0^{\pi/3} \frac{1}{2} (1 + \cos\theta)^2 d\theta + \int_{\pi/3}^{\pi/2} \frac{1}{2} (9\cos^2\theta) d\theta\right] = 5\pi/4$

15. $A = 2\int_0^{\cos^{-1}(3/5)} \frac{1}{2} (100 - 36\sec^2\theta) d\theta = 100\cos^{-1}(3/5) - 48$

17. $A = \int_0^{\pi/2} \frac{1}{2} a^2\sec^4(\theta/2) d\theta = 4a^2/3$

19. $A = \int_0^{\pi} \frac{9}{2} e^{-4\theta} d\theta = \frac{9}{8} (1 - e^{-4\pi})$ 21. $A = \int_{1/9}^{4} \frac{1}{2} \frac{1}{\theta} d\theta = \ln 6$

23. $A = 4\int_0^{\pi/6} \frac{1}{2} (4\cos 2\theta - 2) d\theta = 2\sqrt{3} - 2\pi/3$

EXERCISE SET 13.4

1. (a)

   (b) $y = x^2$

3. $\cos^2 t + \sin^2 t = 1$;
   $x^2 + y^2 = 1$

5. $t = (x + 4)/3$;
   $y = 2x + 10$

7. $\cos t = x/2$, $\sin t = y/5$;
   $x^2/4 + y^2/25 = 1$

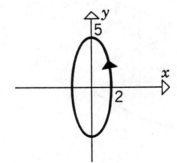

9. $\cos t = (x - 3)/2$,

$\sin t = (y - 2)/4$;

$(x - 3)^2/4 + (y - 2)^2/16 = 1$

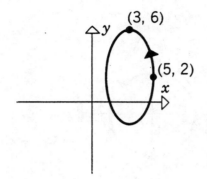

(3, 6)

(5, 2)

11. $\sin 2\pi t = x/4$, $\cos 2\pi t = y/4$;

$x^2/16 + y^2/16 = 1$

4

13. $\cos 2t = 1 - 2\sin^2 t$;

$x = 1 - 2y^2$, $-1 \leq y \leq 1$

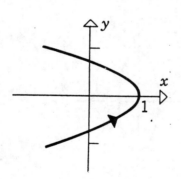

1

15. $y = \ln t^2 = \ln x$, $x \geq 1$

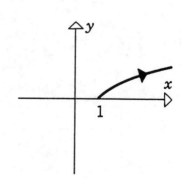

1

17. $x = 3y^2 - 1$,
      $-1 < x \leq 2$, $0 < y \leq 1$

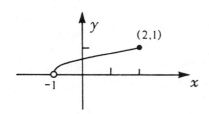

19. $x/2 + y/3 = 1$,
      $0 \leq x \leq 2$, $0 \leq y \leq 3$

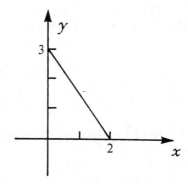

21. $x + (y - 1)^2 = 1$,
      $0 \leq x \leq 1$, $0 \leq y \leq 2$

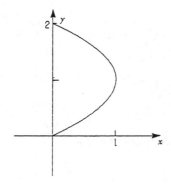

23. $x^2 + y^2 = 1$,
      $\cos 1 \leq x < 1$, $0 < y \leq \sin 1$

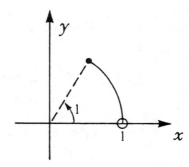

25. $(x_0, \sqrt{x_0})$ is on the curve for $x_0 \geq 0$. $y' = 1/(2\sqrt{x})$; the tangent line at

      $(x_0, \sqrt{x_0})$ is $y - \sqrt{x_0} = (x - x_0)/(2\sqrt{x_0})$, which crosses the x-axis at

      $x = -x_0 = t$ so $x_0 = -t$ hence $x = -t$, $y = \sqrt{-t}$ for $t \leq 0$.

27. $dy/dx = \dfrac{\cos\ t}{-\sin\ t} = -\cot t$, $dy/dx \big|_{t=3\pi/4} = 1$

29. $dy/dx = \dfrac{2}{1/(2\sqrt{t})} = 4\sqrt{t}$, $dy/dx \big|_{t=9} = 12$

31. $dy/dx = \dfrac{6\pi\cos 2\pi s}{-8\pi\sin 2\pi s} = -\dfrac{3}{4}\cot 2\pi s, \;\; dy/dx\big|_{s=-1/4} = 0$

33. $dy/dx = \dfrac{t^2}{t} = t, \;\; d^2y/dx^2 = \dfrac{1}{t}, \;\; d^2y/dx^2\big|_{t=2} = 1/2$

35. $dy/dx = \dfrac{2}{1/(2\sqrt{t})} = 4\sqrt{t}, \;\; d^2y/dx^2 = \dfrac{2/\sqrt{t}}{1/(2\sqrt{t})} = 4, \;\; d^2y/dx^2\big|_{t=1} = 4$

37. $(dx/dt)^2 + (dy/dt)^2 = (4)^2 + (3)^2 = 25, \;\; L = \displaystyle\int_0^2 5\,dt = 10$

39. $(dx/dt)^2 + (dy/dt)^2 = (t^2)^2 + (t)^2 = t^2(t^2 + 1),$

   $L = \displaystyle\int_0^1 t(t^2 + 1)^{1/2}dt = (2\sqrt{2} - 1)/3$

41. $(dx/dt)^2 + (dy/dt)^2 = (-2\sin 2t)^2 + (2\cos 2t)^2 = 4, \;\; L = \displaystyle\int_0^{\pi/2} 2\,dt = \pi$

43. $(dx/dt)^2 + (dy/dt)^2 = [2(1 + t)]^2 + [3(1 + t)^2]^2$
   $$= (1 + t)^2[4 + 9(1 + t)^2],$$

   $L = \displaystyle\int_0^1 (1 + t)[4 + 9(1 + t)^2]^{1/2}dt = (80\sqrt{10} - 13\sqrt{13})/27$

45. $(dx/dt)^2 + (dy/dt)^2 = [a(1 - \cos t)]^2 + [a\sin t]^2 = 2a^2(1 - \cos t)$
   $$= 4a^2\sin^2(t/2),$$

   $L = \displaystyle\int_0^{2\pi} 2a\sin(t/2)dt = 8a$

47. $x = (2 + 3\sin\theta)\cos\theta, \;\; y = (2 + 3\sin\theta)\sin\theta$

49. $dy/dx = \dfrac{-e^{-t}}{e^t} = -e^{-2t}$; for $t = 2$, $dy/dx = -e^{-4}$, $(x,y) = (e^2, e^{-2})$;

   $y - e^{-2} = -e^{-4}(x - e^2), \;\; y = -e^{-4}x + 2e^{-2}$

51. $dy/dx = \dfrac{2t + 1}{6t^2 - 30t + 24} = \dfrac{2t + 1}{6(t - 1)(t - 4)}$

  (a) $dy/dx = 0$ if $t = -1/2$

  (b) $dx/dy = \dfrac{6(t - 1)(t - 4)}{2t + 1} = 0$ if $t = 1,4$

53. If $x = 3$ then $t^2 - 3t + 5 = 3$, $t^2 - 3t + 2 = 0$, $(t - 1)(t - 2) = 0$, $t = 1$ or 2. If $t = 1$ or 2 then $y = 1$ so $(3,1)$ is reached when $t = 1$ or 2. $dy/dx = (3t^2 + 2t - 10)/(2t - 3)$. For $t = 1$, $dy/dx = 5$, the tangent line is $y - 1 = 5(x - 3)$, $y = 5x - 14$. For $t = 2$, $dy/dx = 6$, the tangent line is $y - 1 = 6(x - 3)$, $y = 6x - 17$.

55. Assuming that $a \neq 0$ and $b \neq 0$, eliminate the parameter to get $(x - h)^2/a^2 + (y - k)^2/b^2 = 1$. If $|a| = |b|$ the curve is a circle with center (h,k) and radius $|a|$; if $|a| \neq |b|$ the curve is an ellipse with center (h,k) and vertices $(h\pm|a|,k)$ when $|a| > |b|$ or vertices $(h,k\pm|b|)$ when $|a| < |b|$.

57. (a)

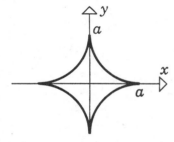

  (b) Use $b = a/4$ in the equations of Exercise 46 to get

  $x = \dfrac{3}{4} a \cos\phi + \dfrac{1}{4} a \cos 3\phi$, $y = \dfrac{3}{4} a \sin\phi - \dfrac{1}{4} a \sin 3\phi$;

  but trigonometric identities yield

  $\cos 3\phi = 4\cos^3\phi - 3 \cos\phi$, $\sin 3\phi = 3 \sin\phi - 4 \sin^3\phi$

  so $x = a \cos^3\phi$, $y = a \sin^3\phi$.

  (c) From the result in part (b), $\cos\phi = (x/a)^{1/3}$, $\sin\phi = (y/a)^{1/3}$ so $(x/a)^{2/3} + (y/a)^{2/3} = 1$, $x^{2/3} + y^{2/3} = a^{2/3}$

59. $x' = e^t(\cos t - \sin t)$, $y' = e^t(\cos t + \sin t)$, $(x')^2 + (y')^2 = 2e^{2t}$

  $S = 2\pi \displaystyle\int_0^{\pi/2} (e^t \sin t)\sqrt{2e^{2t}}\, dt = 2\sqrt{2}\, \pi \int_0^{\pi/2} e^{2t} \sin t\, dt$, use integration by parts to get

$$S = 2\sqrt{2}\ \pi \left[\frac{1}{5}\ e^{2t}(2\sin t - \cos t)\right]_0^{\pi/2} = \frac{2\sqrt{2}}{5}\ \pi(2e^{\pi} + 1)$$

61. $x' = 1,\ y' = 4t,\ (x')^2 + (y')^2 = 1 + 16t^2$

$$S = 2\pi \int_0^1 t\sqrt{1 + 16t^2}\ dt = \frac{\pi}{24}\ (17\sqrt{17} - 1)$$

63. $\frac{dx}{d\phi} = a(1 - \cos \phi),\ \frac{dy}{d\phi} = a\sin \phi,\ (\frac{dx}{d\phi})^2 + (\frac{dy}{d\phi})^2 = 2a^2(1 - \cos \phi)$

$$S = 2\pi \int_0^{2\pi} a(1 - \cos \phi)\ \sqrt{2a^2(1 - \cos \phi)}\ d\phi = 2\sqrt{2}\ \pi a^2 \int_0^{2\pi} (1 - \cos \phi)^{3/2} d\phi,$$

but $1 - \cos \phi = 2\sin^2 \frac{\phi}{2}$ so $(1 - \cos \phi)^{3/2} = 2\sqrt{2}\ \sin^3 \frac{\phi}{2}$ for $0 \leq \phi \leq \pi$
and, taking advantage of the symmetry of the cycloid,

$$S = 16\pi a^2 \int_0^{\pi} \sin^3 \frac{\phi}{2}\ d\phi = 64\pi a^2/3$$

EXERCISE SET 13.5

1. $\theta = \pi/3;\ dr/d\theta = -\sqrt{3},\ r = 1,\ \tan \theta = \sqrt{3},\ m = \dfrac{1 + (\sqrt{3})(-\sqrt{3})}{-\sqrt{3} + (-\sqrt{3})} = 1/\sqrt{3}$

3. $\theta = 2;\ dr/d\theta = -1/4,\ r = 1/2,\ \tan \theta = \tan 2,$
    $m = \dfrac{1/2 + (\tan 2)(-1/4)}{-(1/2)\tan 2 + (-1/4)} = \dfrac{\tan 2 - 2}{2\tan 2 + 1}$

5. $\theta = 3\pi/4;\ dr/d\theta = -3\sqrt{2}/2,\ r = \sqrt{2}/2,\ \tan \theta = -1,$
    $m = \dfrac{\sqrt{2}/2 + (-1)(-3\sqrt{2}/2)}{-(\sqrt{2}/2)(-1) + (-3\sqrt{2}/2)} = -2$

7. $\theta = \pi/2;\ dr/d\theta = 3,\ r = 3,\ \tan \psi = 1$

9. $\theta = \pi,\ dr/d\theta = -5,\ r = 0,\ \tan \psi = 0$

11. $\theta = 5\pi/6$; $dr/d\theta = -2\sqrt{3}$, $r = 1$, $\tan \psi = -1/(2\sqrt{3})$

13. $r^2 + (dr/d\theta)^2 = (e^{3\theta})^2 + (3e^{3\theta})^2 = 10e^{6\theta}$,

$$L = \int_0^2 \sqrt{10} \ e^{3\theta} d\theta = \sqrt{10} \ (e^6 - 1)/3$$

15. $r^2 + (dr/d\theta)^2 = (2a \cos \theta)^2 + (-2a \sin \theta)^2 = 4a^2$, $L = \int_0^\pi 2a \ d\theta = 2\pi a$

17. $r^2 + (dr/d\theta)^2 = (a\theta^2)^2 + (2 a \theta)^2 = a^2\theta^2(\theta^2 + 4)$,

$$L = \int_0^\pi a\theta(\theta^2 + 4)^{1/2} d\theta = \tfrac{a}{3}[(\pi^2 + 4)^{3/2} - 8]$$

19. $r^2 + (dr/d\theta)^2 = [a(1 - \cos \theta)]^2 + [a \sin \theta]^2 = 4a^2 \sin^2(\theta/2)$,

$$L = 2 \int_0^\pi 2a \sin(\theta/2)d\theta = 8a$$

21. $dx/d\theta = 4 \sin^2\theta - \sin \theta - 2$, $dy/d\theta = \cos \theta(1 - 4 \sin \theta)$. $dy/d\theta = 0$ when $\cos \theta = 0$ or $\sin \theta = 1/4$ so $\theta = \pi/2$, $3\pi/2$, $\sin^{-1}(1/4)$, or $\pi - \sin^{-1}(1/4)$; $dx/d\theta \neq 0$ at these points so there is a horizontal tangent at each one.

23. $(\sqrt{3}/2, \pi/6)$ satisfies both equations so it is a point of intersection. $\tan \psi_1 = \dfrac{r}{dr/d\theta} = \dfrac{\sin 2\theta}{2 \cos 2\theta} = \tfrac{1}{2} \tan 2\theta$, $\tan \psi_2 = \dfrac{r}{dr/d\theta} = \dfrac{\cos \theta}{-\sin \theta} = -\cot \theta$, at $\theta = \pi/6$, $\tan \psi_1 = \sqrt{3}/2$ and $\tan \psi_2 = -\sqrt{3}$ so

$$\tan \beta = \left| \frac{-\sqrt{3} - \sqrt{3}/2}{1 + (\sqrt{3}/2)(-\sqrt{3})} \right| = 3\sqrt{3}, \ \beta = \tan^{-1}3\sqrt{3}.$$

25. $\tan \psi = r/(dr/d\theta) = e^{a\theta}/(ae^{a\theta}) = 1/a$ so ψ is a constant.

SUPPLEMENTARY EXERCISES CHAPTER 13

1. (a) $(1,\sqrt{3})$ (b) $(0,-2)$
    (c) $(0,0)$ (d) $(-1,1)$
    (e) $(-3,0)$ (f) $(3/5,-4/5)$

3. (a)   (b)

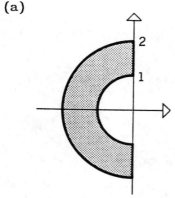

5. $r^2\sin 2\theta = 1$, $r^2(2\sin\theta\cos\theta) = 1$, $2(r\sin\theta)(r\cos\theta) = 1$, $2yx = 1$; hyperbola.

7. $r = -4\csc\theta$, $r = -4/\sin\theta$, $r\sin\theta = -4$, $y = -4$, line.

9. $\theta = \pi/3$, $\tan\theta = \sqrt{3}$, $y = \sqrt{3}x$; line.

11. $r = 0$, $x = 0$ and $y = 0$; point.   13. $x = -3$, $r\cos\theta = -3$

15. $y = 3x$, $\tan\theta = 3$, $\theta = \tan^{-1}3$

17.   19.

21.

23.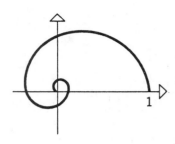

25. $a \cos 2\theta = a/2$, $\cos 2\theta = 1/2$;
one solution is $2\theta = \pi/3$, $\theta = \pi/6$
and from the symmetry of the graphs
the others are $\theta = -\pi/6$, $\pm\pi/3$, $\pm2\pi/3$,
$\pm5\pi/6$.   The points of intersection are
$(a/2, \pm\pi/6)$, $(a/2, \pm\pi/3)$, $(a/2, \pm2\pi/3)$,
$(a/2, \pm5\pi/6)$.

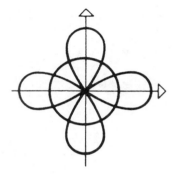

27. (a) $A = 2\left[\dfrac{1}{2}\displaystyle\int_0^{\pi/3}(1 + \cos\theta)^2 d\theta + \dfrac{1}{2}\int_{\pi/3}^{\pi/2}(3\cos\theta)^2 d\theta\right]$

$= \displaystyle\int_0^{\pi/3}(1 + \cos\theta)^2 d\theta + \int_{\pi/3}^{\pi/2} 9\cos^2\theta\, d\theta$

(b) $r^2 + (dr/d\theta)^2 = (1 + \cos\theta)^2 + (-\sin\theta)^2 = 2(1 + \cos\theta)$,

$L = 2\displaystyle\int_{\pi/3}^{\pi}\sqrt{2(1 + \cos\theta)}\, d\theta$

29. (a) $A = \displaystyle\int_{\pi/2}^{\pi}\dfrac{1}{2}[(2\sin\theta)^2 - (2 + 2\cos\theta)^2]d\theta$

$= 2\displaystyle\int_{\pi/2}^{\pi}[\sin^2\theta - (1 + \cos\theta)^2]d\theta$

(b) $r^2 + (dr/d\theta)^2 = (2\sin\theta)^2 + (2\cos\theta)^2 = 4$, $L = \displaystyle\int_0^{\pi/2} 2\, d\theta$

31. $A = 4\int_0^{\pi/6} \frac{1}{2}(2a^2\cos 2\theta - a^2)d\theta = 2a^2\int_0^{\pi/6}(2\cos 2\theta - 1)d\theta = a^2(\sqrt{3} - \pi/3)$

33. $A = 2\int_{\pi/2}^{\pi}[\sin^2\theta - (1 + \cos\theta)^2]d\theta = 2\int_{\pi/2}^{\pi}(-1 - 2\cos\theta - \cos 2\theta)d\theta$

     $= 4 - \pi$

35. (a) eliminate the parameter to get
     $(x - 1)^2/9 + (y + 1)^2/4 = 1$ for
     $-2 \leq x \leq 4$ and $-1 \leq y \leq 1$.

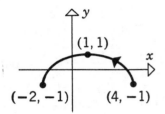

(1, 1)

(−2, −1)   (4, −1)

(b) $dy/dx = \dfrac{2\cos\theta}{-3\sin\theta} = -\dfrac{2}{3}\cot\theta$, $d^2y/dx^2 = \dfrac{(2/3)\csc^2\theta}{-3\sin\theta} = -\dfrac{2}{9}\csc^3\theta$;

     at $\theta_0 = \pi/2$, $dy/dx = 0$ and $d^2y/dx^2 = -2/9$, $x = 1$, $y = 1$ so the
     tangent line is $y = 1$.

37. (a) eliminate the parameter to get
     $y = \ln(1/x) = -\ln x$ for
     $1 \leq x \leq e^{-1}$ and $0 \leq y \leq 1$.

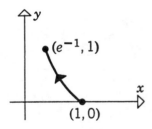

$(e^{-1}, 1)$

(1, 0)

(b) $dy/dx = \dfrac{1/t}{-1/t^2} = -t$, $d^2y/dx^2 = \dfrac{-1}{-1/t^2} = t^2$; at $t_0 = 2$, $dy/dx = -2$ and

     $d^2y/dx^2 = 4$, $x = 1/2$, $y = \ln 2$ so the tangent line is
     $y - \ln 2 = -2(x - 1/2)$, $y = -2x + 1 + \ln 2$.

39. $(dx/dt)^2 + (dy/dt)^2 = (-2\tan 2t)^2 + 2^2 = 4\sec^2 2t$,

     $L = \int_0^{\pi/6} 2\sec 2t\, dt = \ln(2 + \sqrt{3})$

41. $(dx/dt)^2 + (dy/dt)^2 = (6t)^2 + (3t^2 - 3)^2 = (3t^2 + 3)^2,$

   $L = \int_0^1 (3t^2 + 3)dt = 4$

43. $r^2 + (dr/d\theta)^2 = (e^\theta)^2 + (e^\theta)^2 = 2e^{2\theta},\ L = \int_0^{2\pi} \sqrt{2}e^\theta d\theta = \sqrt{2}(e^{2\pi} - 1)$

45. $dx/dt = -2\cos t,\ dy/dt = 1 - 2\sin t$
   (a) horizontal when $dy/dt = 0$ and $dx/dt \neq 0$; $1 - 2\sin t = 0$,

   $\sin t = 1/2$, $t = \pi/6,\ 5\pi/6$ so (x,y) is $(0,\pi/6+\sqrt{3})$ or $(0,5\pi/6-\sqrt{3})$.
   (b) vertical when $dx/dt = 0$ and $dy/dt \neq 0$; $-2\cos t = 0$, $\cos t = 0$,
   $t = \pi/2$ so (x,y) is $(-1,\pi/2)$.

47. $dy/dx = (2t - 4)/(1/t) = 2t^2 - 4t,\ d^2y/dx^2 = (4t - 4)/(1/t) = 4t(t - 1).$
   For $t > 0$, $d^2y/dx^2 = 0$ when $t = 1$ and d^2y/dx^2 changes sign there so an
   inflection point occurs at $t = 1$.

49. $r = 2(1 + \cos\theta),\ dr/d\theta = -2\sin\theta.$ When $\theta = \pi/2,\ \sin\theta = 1,\ \cos\theta = 0,$
   $r = 2,\ dr/d\theta = -2$ so $m = [(2)(0) + (1)(-2)]/[-(2)(1) + (0)(-2)] = 1$ and
   $\tan\psi = 2/(-2) = -1,\ \psi = 3\pi/4.$

CHAPTER 14

EXERCISE SET 14.1

1. (a) $d = \sqrt{(2 - 0)^2 + (1 - 0)^2 + (3 - 0)^2} = \sqrt{4 + 1 + 9} = \sqrt{14}$;
    midpoint $(1, 1/2, 3/2)$

    (b) $d = \sqrt{(4 - 5)^2 + (1 - 2)^2 + (6 - 3)^2} = \sqrt{1 + 1 + 9} = \sqrt{11}$;
    midpoint $(9/2, 3/2, 9/2)$

    (c) $d = \sqrt{(3 + 2)^2 + (0 + 1)^2 + (5 - 3)^2} = \sqrt{23 + 1 + 4} = \sqrt{30}$;
    midpoint $(1/2, -1/2, 4)$

    (d) $d = \sqrt{(4 + 1)^2 + (3 + 1)^2 + (-2 + 3)^2} = \sqrt{25 + 16 + 1} = \sqrt{42}$;
    midpoint $(3/2, 1, -5/2)$

3. vertices: $(4, 2, -2)$, $(4, 2, 1)$, $(4, 1, 1)$, $(4, 1, -2)$,
    $(-6, 1, 1)$, $(-6, 2, 1)$, $(-6, 2, -2)$, $(-6, 1, -2)$

5. (a) the sides have lengths 7, 14, and $7\sqrt{5}$; it is a right triangle
    because the sides satisfy the Pythagorean theorem,
    $(7\sqrt{5})^2 = 7^2 + 14^2$.

    (b) $(2, 1, 6)$ is the vertex of the $90°$ angle because it is opposite the
    longest side (the hypotenuse).

    (c) area $= (1/2)(\text{altitude})(\text{base}) = (1/2)(7)(14) = 49$

7. The distance to the z-axis is the distance between (x_0, y_0, z_0) and

    $(0, 0, z_0)$ which is $\sqrt{x_0^2 + y_0^2}$; similarly, the distance to the x-axis is

    $\sqrt{y_0^2 + z_0^2}$ and the distance to the y-axis is $\sqrt{x_0^2 + z_0^2}$.

9. $(x + 2)^2 + (y - 4)^2 + (z + 1)^2 = 36$

11. $x^2 + (y - 1)^2 + z^2 = 9$

13. $(x - 2)^2 + (y + 1)^2 + (z + 3)^2 = r^2$,

(a) $r^2 = 3^2 = 9$ (b) $r^2 = 1^2 = 1$ (c) $r^2 = 2^2 = 4$

15. $r = \frac{1}{2} \sqrt{(-1 - 0)^2 + (2 - 2)^2 + (1 - 3)^2} = \frac{1}{2} \sqrt{5}$, center $(-1/2, 2, 2)$,

$(x + 1/2)^2 + (y - 2)^2 + (z - 2)^2 = 5/4$

17. $r = \sqrt{(7 - 3)^2 + (2 + 2)^2 + (1 - 4)^2} = \sqrt{41}$,

$(x - 3)^2 + (y + 2)^2 + (z - 4)^2 = 41$

19. $r = |[\text{distance between } (0,0,0) \text{ and } (3,-2,4)] \pm 1| = \sqrt{29} \pm 1$,

$x^2 + y^2 + z^2 = r^2 = (\sqrt{29} \pm 1)^2 = 30 \pm 2\sqrt{29}$

21. $(x + 5)^2 + (y + 2)^2 + (z + 1)^2 = 49$; sphere, $C(-5,-2,-1)$, $r = 7$

23. $(x - 1/2)^2 + (y - 3/4)^2 + (z + 5/4)^2 = 54/16$; sphere, $C(1/2, 3/4, -5/4)$,

$r = 3\sqrt{6}/4$

25. $(x - 3/2)^2 + (y + 2)^2 + (z - 4)^2 = -11/4$; no graph

27. Complete the square to get $(x + 1)^2 + (y - 1)^2 + (z - 2)^2 = 9$; center $(-1,1,2)$, radius 3. The distance between the origin and the center is $\sqrt{6} < 3$ so the origin is inside the sphere. The largest distance is $3 + \sqrt{6}$, the smallest is $3 - \sqrt{6}$.

29. $(y + 3)^2 + (z - 2)^2 > 16$; all points outside the circular cylinder $(y + 3)^2 + (z - 2)^2 = 16$.

31. Let r be the radius of a styrofoam sphere. The distance from the origin to the center of the bowling ball is equal to the sum of the distance from the origin to the center of the styrofoam sphere nearest the origin and the distance between the center of this sphere and the center of the bowling ball so $\sqrt{3}R = \sqrt{3}r + r + R$, $(\sqrt{3} + 1)r = (\sqrt{3} - 1)R$,

$r = \dfrac{\sqrt{3} - 1}{\sqrt{3} + 1} R = (2 - \sqrt{3})R$.

33. (a)

(b)

(c)

35. (a)

$y = e^x$

(b)

$x = \ln z$

(c)

$yz = 1$

37. (a)

(b)

39. (a)

(b)

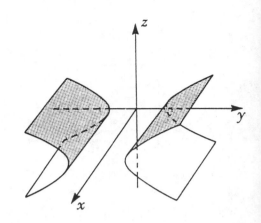

41. $(a \sin \phi \cos \theta)^2 + (a \sin \phi \sin \theta)^2 + (a \cos \phi)^2$

$= a^2 \sin^2 \phi \cos^2 \theta + a^2 \sin^2 \phi \sin^2 \theta + a^2 \cos^2 \phi$

$= a^2 \sin^2 \phi (\cos^2 \theta + \sin^2 \theta) + a^2 \cos^2 \phi$

$= a^2 \sin^2 \phi + a^2 \cos^2 \phi = a^2 (\sin^2 \phi + \cos^2 \phi) = a^2$

EXERCISE SET 14.2

1.

3.

5.

7.

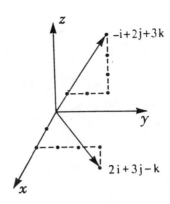

9. (a) $\langle 2-3,\ 8-5 \rangle = \langle -1,3 \rangle$ (b) $\langle 0-7,\ 0-(-2) \rangle = \langle -7,2 \rangle$
    (c) $\langle -4-(-6),\ -1-(-2) \rangle = \langle 2,1 \rangle$ (d) $\langle -8-0,\ 7-0 \rangle = \langle -8,7 \rangle$

11. (a) $\langle -3,6,1 \rangle$ (b) $\langle 1,-3,-5 \rangle$

13. Let (x,y) be the terminal point, then $x - 1 = 3$, $x = 4$ and
    $y - (-2) = -2$, $y = -4$. The terminal point is $(4,-4)$.

15. Let (x,y) be the initial point, then $2 - x = -2$, $x = 4$ and $0 - y = 4$,
    $y = -4$. The initial point is $(4,-4)$.

17. Let (x,y,z) be the initial point, then $5 - x = -3$, $-y = 1$, and
    $-1 - z = 2$ so $x = 8$, $y = -1$, and $z = -3$. The initial point is
    $(8,-1,-3)$.

19. (a) $-5i - 2j$ (b) $8i + 10j$
    (c) $-2i + 4j$ (d) $40i + 36j$
    (e) $-20i + 22j$ (f) $-8i - 8j$

21. (a) $-i + 4j - 2k$ (b) $18i + 12j - 6k$
    (c) $-i - 5j - 2k$ (d) $40i - 4j - 4k$
    (e) $-2i - 16j - 18k$ (f) $-i + 13j - 2k$

23. (a) $\|v\| = \sqrt{1 + 1} = \sqrt{2}$ (b) $\|v\| = \sqrt{4 + 0} = 2$
    (c) $\|v\| = \sqrt{2 + 7} = 3$

25. (a) $\|v\| = \sqrt{14}$ (b) $\|v\| = 3$

27. (a) $\|\langle 5,4 \rangle\| = \sqrt{41}$ (b) $2 + 5 = 7$
    (c) $3\|u\| + 4\|v\| = 3\sqrt{29} + 8$ (d) $\|\langle -3,-9 \rangle\| = 3\sqrt{10}$

(e) $\frac{1}{5}$ $\langle 3,4\rangle$ = $\langle 3/5, 4/5\rangle$ (f) 1

29.   (a) $\|\mathbf{u} + \mathbf{v}\| = \|2\mathbf{i} - 2\mathbf{j} + 2\mathbf{k}\| = 2\sqrt{3}$ (b) $\|\mathbf{u}\| + \|\mathbf{v}\| = \sqrt{14} + \sqrt{2}$

   (c) $\|-2\mathbf{u}\| + 2\|\mathbf{v}\| = 2\sqrt{14} + 2\sqrt{2}$

   (d) $\|3\mathbf{u} - 5\mathbf{v} + \mathbf{w}\| = \|-12\mathbf{j} + 2\mathbf{k}\| = 2\sqrt{37}$

   (e) $(1/\sqrt{6})\mathbf{i} + (1/\sqrt{6})\mathbf{j} - (2/\sqrt{6})\mathbf{k}$ (f) 1

31. $6\mathbf{x} = 2\mathbf{u} - \mathbf{v} - \mathbf{w} = \langle -4,6\rangle,\ \mathbf{x} = \langle -2/3, 1\rangle$

33. $\mathbf{u} = \frac{5}{7}\mathbf{i} + \frac{2}{7}\mathbf{j} + \frac{1}{7}\mathbf{k},\ \mathbf{v} = \frac{8}{7}\mathbf{i} - \frac{1}{7}\mathbf{j} - \frac{4}{7}\mathbf{k}$

35.   Take $\mathbf{z}$ as the diagonal of a
      parallelepiped with $\mathbf{u}$, $\mathbf{v}$, and $\mathbf{w}$ along
      its edges as shown. Then $\mathbf{z}$ can be
      written as the sum $c_1\mathbf{u} + c_2\mathbf{v} + c_3\mathbf{w}$.

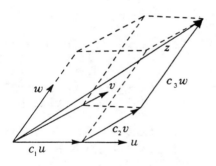

37. $c_1\mathbf{u} + c_2\mathbf{v} = \langle c_1 - 2c_2,\ -3c_1 + 6c_2\rangle = \langle 3,5\rangle$, so $c_1 - 2c_2 = 3$ and
      $-3c_1 + 6c_2 = 5$ which has no solution.

39.   Equate corresponding components to get the system of equations
$$c_1 + 3c_2 + 4c_3 = 2 \tag{1}$$
$$-c_1 \qquad\quad - c_3 = 1 \tag{2}$$
$$c_2 + c_3 = -1 \tag{3}$$
      From (2) and (3), $c_1 = -1 - c_3$ and $c_2 = -1 - c_3$; substitute these into
      (1) to get $-4 = 2$, which is nonsense so the system has no solution.

43. $\|3\mathbf{i} - 4\mathbf{j}\| = 5$ so the required vector is $-\frac{1}{5}(3\mathbf{i} - 4\mathbf{j}) = -\frac{3}{5}\mathbf{i} + \frac{4}{5}\mathbf{j}$

45. $\|6\mathbf{i} - 4\mathbf{j} + 2\mathbf{k}\| = 2\sqrt{14}$ so the required vector is $-(3\mathbf{i} - 2\mathbf{j} + \mathbf{k})/\sqrt{14}$

47. $\overrightarrow{AB} = 4i + j - k$, $\|\overrightarrow{AB}\| = 3\sqrt{2}$ so the required vector is $(4i + j - k)/(3\sqrt{2})$

49. $-\dfrac{1}{2} v = \langle -3/2, 2 \rangle$ 51. $-2v = 6i - 8j - 2k$

53. $\|r - r_0\| = \|\langle x - x_0, y - y_0 \rangle\| = \sqrt{(x - x_0)^2 + (y - y_0)^2} = 1$, circle of radius 1 and center at (x_0, y_0).

55. (a) $\|r\| = \sqrt{x^2 + y^2 + z^2} = 2$, sphere of radius 2 with center at $(0,0,0)$.

   (b) $\|r - r_0\| = \sqrt{(x - x_0)^2 + (y - y_0)^2 + (z - z_0)^2} = 3$, sphere of radius 3 with center at (x_0, y_0, z_0).

   (c) $\|r - r_0\| = \sqrt{(x - x_0)^2 + (y - y_0)^2 + (z - z_0)^2} \leq 1$, all points on and inside the sphere of radius 1 with center at (x_0, y_0, z_0).

57. (a) Choose two points on the line, for example $P_1(0,4)$ and $P_2(1,3)$ then $\overrightarrow{P_1 P_2} = \langle 1, -1 \rangle$ is parallel to the line, $\|\langle 1, -1 \rangle\| = \sqrt{2}$ so $\langle 1/\sqrt{2}, -1/\sqrt{2} \rangle$ and $\langle -1/\sqrt{2}, 1/\sqrt{2} \rangle$ are unit vectors parallel to the line.
   (b) Pick any line that is perpendicular to the line $x + y = 4$, for example $y = x$, and proceed as in part (a) to get $\langle 1/\sqrt{2}, 1/\sqrt{2} \rangle$ and $\langle -1/\sqrt{2}, -1/\sqrt{2} \rangle$.

59. Let $R(x,y)$ be the required point then $\overrightarrow{QR} = \dfrac{3}{4} \overrightarrow{QP}$,
   $\langle x-7, y+4 \rangle = \dfrac{3}{4} \langle -5, 7 \rangle = \langle -15/4, 21/4 \rangle$ so $x - 7 = -15/4$, $x = 13/4$ and $y + 4 = 21/4$, $y = 5/4$. The point is $(13/4, 21/4)$.

61. (a) $\langle \cos(\pi/3), \sin(\pi/3) \rangle = \langle 1/2, \sqrt{3}/2 \rangle$
   (b) $4\langle \cos(3\pi/4), \sin(3\pi/4) \rangle = \langle -2\sqrt{2}, 2\sqrt{2} \rangle$

63. Let A, B, C be the vertices $(0,0)$, $(1,3)$, $(2,4)$ and D the fourth vertex (x,y). For the parallelogram ABCD, $\overrightarrow{AD} = \overrightarrow{BC}$, $\langle x,y \rangle = \langle 1,1 \rangle$ so $x = 1$,

y = 1 and D is at (1,1). For the parallelogram ACBD, $\overrightarrow{AD} = \overrightarrow{CB}$,
⟨x,y⟩ = ⟨-1,-1⟩ so x = -1, y = -1 and D is at (-1,-1).

65.   Use an analytic approach as illustrated in the text.

67.   Draw the triangles with sides formed by the vectors **u, v, u + v** and k**u**,
k**v**, k**u** + k**v**. By similar triangles, k(**u** + **v**) = k**u** + k**v**.

69.   Let **a, b, c** be vectors along the sides
of the triangle and A,B the midpoints
of **a** and **b**, then

$$\mathbf{u} = \frac{1}{2}\,\mathbf{a} - \frac{1}{2}\,\mathbf{b} = \frac{1}{2}\,(\mathbf{a} - \mathbf{b}) = \frac{1}{2}\,\mathbf{c}$$

so **u** is parallel to **c** and half as long.

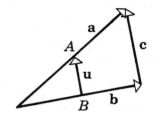

71. $\overrightarrow{AB} = \overrightarrow{AM} + \overrightarrow{MB}$,
$\overrightarrow{AC} = \overrightarrow{AM} + \overrightarrow{MC}$; add to get
$\overrightarrow{AB} + \overrightarrow{AC} = 2\overrightarrow{AM} + \overrightarrow{MB} + \overrightarrow{MC}$ but
$\overrightarrow{MB} = -\overrightarrow{MC}$ so $\overrightarrow{AB} + \overrightarrow{AC} = 2\overrightarrow{AM}$.

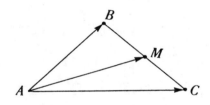

73. $\overrightarrow{AB} = \overrightarrow{AM} + \overrightarrow{MN} + \overrightarrow{NB}$,
$\overrightarrow{AD} = \overrightarrow{AM} + \overrightarrow{MN} + \overrightarrow{ND}$,
$\overrightarrow{CB} = \overrightarrow{CM} + \overrightarrow{MN} + \overrightarrow{NB}$,
$\overrightarrow{CD} = \overrightarrow{CM} + \overrightarrow{MN} + \overrightarrow{ND}$, where
$\overrightarrow{AM} = -\overrightarrow{CM}$ and $\overrightarrow{NB} = -\overrightarrow{ND}$;

add to get

$\overrightarrow{AB} + \overrightarrow{AD} + \overrightarrow{CB} + \overrightarrow{CD} = 4\overrightarrow{MN}$

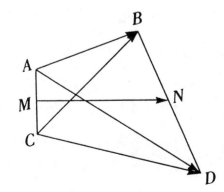

75. $\displaystyle\sum_{k=1}^{n} \overrightarrow{PP}_k = \sum_{k=1}^{n} (r_k - r) = \sum_{k=1}^{n} r_k - \sum_{k=1}^{n} r = \sum_{k=1}^{n} r_k - nr = 0, \; r = \frac{1}{n}\sum_{k=1}^{n} r_k.$

77. (a) $r_1 = -i + 2j$, $r_2 = 2i + 3j$, $r_3 = 5i - 2j$, $r_4 = -j$,

$r = \frac{1}{4}\displaystyle\sum_{k=1}^{4} r_k = \frac{3}{2}\,i + \frac{1}{2}\,j$; centroid is at $(3/2, 1/2)$.

EXERCISE SET 14.3

1. (a) $(1)(6) + (2)(-8) = -10$ (b) $(-7)(0) + (-3)(1) = -3$
    (c) $(1)(8) + (-3)(-2) + (7)(-2) = 0$
    (d) $(-3)(4) + (1)(2) + (2)(-5) = -20$

3. (a) $u \cdot v = -34 < 0$, obtuse   (b) $u \cdot v = 6 > 0$, acute
    (c) $u \cdot v = -1 < 0$, obtuse   (d) $u \cdot v = 0$, orthogonal

5. (a) $(14/13)i + (21/13)j$ (b) $\langle 2, 6 \rangle$
    (c) $-(11/13)i + j + (55/13)k$ (d) $\langle -32/89, \, -12/89, \, 73/89 \rangle$

9. (a) $\langle 1, 2 \rangle \cdot (\langle 28, -14 \rangle + \langle 6, 0 \rangle) = \langle 1, 2 \rangle \cdot \langle 34, -14 \rangle = 6$

    (b) $\|6w\| = 6\|w\| = 36$ (c) $24\sqrt{5}$ (d) $24\sqrt{5}$

11. Let A, B, and C be the vertices $(-1, 0)$, $(2, -1)$, and $(1, 4)$ with corresponding interior angles α, β, and γ, then

$$\cos\alpha = \frac{\overrightarrow{AB}\cdot\overrightarrow{AC}}{\|\overrightarrow{AB}\|\,\|\overrightarrow{AC}\|} = \frac{\langle 3, -1 \rangle \cdot \langle 2, 4 \rangle}{\sqrt{10}\,\sqrt{20}} = 1/(5\sqrt{2})$$

$$\cos\beta = \frac{\overrightarrow{BA}\cdot\overrightarrow{BC}}{\|\overrightarrow{BA}\|\,\|\overrightarrow{BC}\|} = \frac{\langle -3, 1 \rangle \cdot \langle -1, 5 \rangle}{\sqrt{10}\,\sqrt{26}} = 4/\sqrt{65}$$

$$\cos\gamma = \frac{\overrightarrow{CA}\cdot\overrightarrow{CB}}{\|\overrightarrow{CA}\|\,\|\overrightarrow{CB}\|} = \frac{\langle -2, -4 \rangle \cdot \langle 1, -5 \rangle}{\sqrt{20}\,\sqrt{26}} = 9/\sqrt{130}$$

13. $\overrightarrow{AB} = \langle 1, 3, -2 \rangle$, $\overrightarrow{BC} = \langle 4, -2, -1 \rangle$, $\overrightarrow{AB} \cdot \overrightarrow{BC} = 0$ so $\overrightarrow{AB}$ and $\overrightarrow{BC}$ are orthogonal; it is a right triangle with the right angle at vertex B.

15. (a) $\mathbf{r} \cdot \mathbf{r}_0 = x_0x + y_0y = 0$ where $\mathbf{r}$ is perpendicular to $\mathbf{r}_0$; the line through the origin and perpendicular to $\mathbf{r}_0$.

   (b) $(\mathbf{r} - \mathbf{r}_0) \cdot \mathbf{r}_0 = x_0(x - x_0) + y_0(y - y_0) = 0$ where the vector from (x_0,y_0) to (x,y) is perpendicular to $\mathbf{r}_0$; the line through (x_0,y_0) and perpendicular to $\mathbf{r}_0$.

   (c) $\mathbf{r} \cdot (\mathbf{r} - \mathbf{r}_0) = x(x - x_0) + y(y - y_0) = (x - x_0/2)^2 + (y - y_0/2)^2$
   $- x_0^2/4 - y_0^2/4 = 0$; circle with center at the midpoint of $\mathbf{r}_0$ and radius $\|\mathbf{r}_0\|/2$.

17. (a) $\mathbf{a} \cdot \mathbf{b} = 0$, $4k + 3 = 0$, $k = -3/4$

   (b) Use $\mathbf{a} \cdot \mathbf{b} = \|\mathbf{a}\| \|\mathbf{b}\| \cos\theta$ to get

   $4k + 3 = \sqrt{k^2 + 1} (5)\cos(\pi/4)$, $4k + 3 = 5\sqrt{k^2 + 1} /\sqrt{2}$
   Square both sides and rearrange to get
   $7k^2 + 48k - 7 = 0$, $(7k - 1)(k + 7) = 0$ so $k = -7$ (invalid)
   or $k = 1/7$.

   (c) Proceed as in (b) with $\theta = \pi/6$ to get $11k^2 - 96k + 39 = 0$ and use the quadratic formula to get $k = (48 \pm 25\sqrt{3})/11$.

   (d) If $\mathbf{a}$ and $\mathbf{b}$ are parallel then $\theta = 0$ or π so

   $\mathbf{a} \cdot \mathbf{b} = \pm \|\mathbf{a}\| \|\mathbf{b}\|$, $4k + 3 = \pm 5\sqrt{k^2 + 1}$,
   $9k^2 - 24k + 16 = 0$, $(3k - 4)^2 = 0$, $k = 4/3$.

19. $\cos^2\alpha + \cos^2\beta + \cos^2\gamma = \dfrac{u_1^2}{\|\mathbf{u}\|^2} + \dfrac{u_2^2}{\|\mathbf{u}\|^2} + \dfrac{u_3^2}{\|\mathbf{u}\|^2}$
   $= (u_1^2 + u_2^2 + u_3^2)/\|\mathbf{u}\|^2 = \|\mathbf{u}\|^2/\|\mathbf{u}\|^2 = 1$

21. (a) $D = |3 - 8 + 7|/\sqrt{9 + 16} = 2/5$

   (b) $D = |-6 + 5 - 1|/\sqrt{4 + 1} = 2/\sqrt{5}$

   (c) $D = |4 + 6 - 8|/\sqrt{4 + 1} = 2/\sqrt{5}$

23. (a) $\overrightarrow{AP} = 2\mathbf{i} + 2\mathbf{j} + 3\mathbf{k}$, $\overrightarrow{AB} = -2\mathbf{i} + \mathbf{j} + 2\mathbf{k}$,
   $\|\text{proj}_{\overrightarrow{AB}}\overrightarrow{AP}\| = \left\|\dfrac{4}{9} (-2\mathbf{i} + \mathbf{j} + 2\mathbf{k})\right\| = 4/3$

(b) $\|\overrightarrow{AP}\| = \sqrt{17}$, $\sqrt{17 - 16/9} = \sqrt{137}/3$

25. Let P and Q be the points $(1,3)$ and $(4,7)$ then $\overrightarrow{PQ} = 3\mathbf{i} + 4\mathbf{j}$ so
    $W = \mathbf{F} \cdot \overrightarrow{PQ} = -12$ foot-pounds.

27. The diagonals have lengths $\|\mathbf{u} + \mathbf{v}\|$ and $\|\mathbf{u} - \mathbf{v}\|$ but
    $\|\mathbf{u} + \mathbf{v}\|^2 = (\mathbf{u} + \mathbf{v}) \cdot (\mathbf{u} + \mathbf{v}) = \|\mathbf{u}\|^2 + 2\mathbf{u} \cdot \mathbf{v} + \|\mathbf{v}\|^2$, and
    $\|\mathbf{u} - \mathbf{v}\|^2 = (\mathbf{u} - \mathbf{v}) \cdot (\mathbf{u} + \mathbf{v}) = \|\mathbf{u}\|^2 - 2\mathbf{u} \cdot \mathbf{v} + \|\mathbf{v}\|^2$. If the
    parallelogram is a rectangle then $\mathbf{u} \cdot \mathbf{v} = 0$ so $\|\mathbf{u} + \mathbf{v}\|^2 = \|\mathbf{u} - \mathbf{v}\|^2$; the
    diagonals are equal. If the diagonals are equal, then $4\mathbf{u} \cdot \mathbf{v} = 0$,
    $\mathbf{u} \cdot \mathbf{v} = 0$ so $\mathbf{u}$ is perpendicular to $\mathbf{v}$ and hence the parallelogram is a
    rectangle.

29. $\|\mathbf{u} + \mathbf{v}\|^2 = (\mathbf{u} + \mathbf{v}) \cdot (\mathbf{u} + \mathbf{v}) = \|\mathbf{u}\|^2 + 2\mathbf{u} \cdot \mathbf{v} + \|\mathbf{v}\|^2$ and
    $\|\mathbf{u} - \mathbf{v}\|^2 = (\mathbf{u} - \mathbf{v}) \cdot (\mathbf{u} - \mathbf{v}) = \|\mathbf{u}\|^2 - 2\mathbf{u} \cdot \mathbf{v} + \|\mathbf{v}\|^2$,
    subtract to get $\|\mathbf{u} + \mathbf{v}\|^2 - \|\mathbf{u} - \mathbf{v}\|^2 = 4\mathbf{u} \cdot \mathbf{v}$, the result follows by
    dividing both sides by 4.

31. With the cube as shown in the diagram,
    and a the length of each edge,
    $\mathbf{d}_1 = a\mathbf{i} + a\mathbf{j} + a\mathbf{k}$, $\mathbf{d}_2 = a\mathbf{i} + a\mathbf{j} - a\mathbf{k}$,
    $\cos \theta = (\mathbf{d}_1 \cdot \mathbf{d}_2)/(\|\mathbf{d}_1\|\|\mathbf{d}_2\|) = 1/3$,

    $\theta \approx 71°$

33. $\mathbf{v} = c_1\mathbf{v}_1 + c_2\mathbf{v}_2 + c_3\mathbf{v}_3$ so
    $\mathbf{v} \cdot \mathbf{v}_i = c_i\mathbf{v}_i \cdot \mathbf{v}_i$ because $\mathbf{v}_i \cdot \mathbf{v}_j = 0$ if $i \neq j$,
    $= c_i\|\mathbf{v}_i\|^2$, $c_i = \mathbf{v} \cdot \mathbf{v}_i/\|\mathbf{v}_i\|^2$ for $i = 1,2,3$.

35. If **v** is orthogonal to $\mathbf{w}_1$ and $\mathbf{w}_2$ then $\mathbf{v} \cdot \mathbf{w}_1 = \mathbf{v} \cdot \mathbf{w}_2 = 0$,
$\mathbf{v} \cdot (k_1\mathbf{w}_1 + k_2\mathbf{w}_2) = k_1\mathbf{v} \cdot \mathbf{w}_1 + k_2\mathbf{v} \cdot \mathbf{w}_2 = k_1(0) + k_2(0) = 0$ so **v** is
orthogonal to $k_1\mathbf{w}_1 + k_2\mathbf{w}_2$ for all scalars k_1 and k_2.

EXERCISE SET 14.4

1. $\langle 7,10,9 \rangle$ 3. $\langle -4,-6,-3 \rangle$

5. (a) $\mathbf{v} \times \mathbf{w} = \langle -23,7,-1 \rangle$, $\mathbf{u} \times (\mathbf{v} \times \mathbf{w}) = \langle -20,-67,-9 \rangle$
    (b) $\mathbf{u} \times \mathbf{v} = \langle -10,-14,2 \rangle$, $(\mathbf{u} \times \mathbf{v}) \times \mathbf{w} = \langle -78,52,-26 \rangle$
    (c) $\mathbf{v} - 2\mathbf{w} = \langle -2,-7,-3 \rangle$, $\mathbf{u} \times (\mathbf{v} - 2\mathbf{w}) = \langle 24,0,-16 \rangle$
    (d) $(\mathbf{u} \times \mathbf{v}) - 2\mathbf{w} = \langle -10,-14,2 \rangle - \langle 2,8,10 \rangle = \langle -12,-22,-8 \rangle$
    (e) $(\mathbf{u} \times \mathbf{v}) \times (\mathbf{v} \times \mathbf{w}) = \langle -10,-14,2 \rangle \times \langle -23,7,-1 \rangle = \langle 0,-56,-392 \rangle$
    (f) $(\mathbf{v} \times \mathbf{w}) \times (\mathbf{u} \times \mathbf{v}) = \langle 0,56,392 \rangle$

9. A vector parallel to the yz-plane must be perpendicular to **i**;
$\mathbf{i} \times (3\mathbf{i} - \mathbf{j} + 2\mathbf{k}) = -2\mathbf{j} - \mathbf{k}$, $\|-2\mathbf{j} - \mathbf{k}\| = \sqrt{5}$, the unit vectors are
$\pm(2\mathbf{j} + \mathbf{k})/\sqrt{5}$.

11. $(\mathbf{u} + \mathbf{v}) \times (\mathbf{u} - \mathbf{v}) = \mathbf{u} \times \mathbf{u} - \mathbf{u} \times \mathbf{v} + \mathbf{v} \times \mathbf{u} - \mathbf{v} \times \mathbf{v} = 2\mathbf{v} \times \mathbf{u}$ because
$\mathbf{u} \times \mathbf{u} = \mathbf{v} \times \mathbf{v} = 0$ and $-\mathbf{u} \times \mathbf{v} = \mathbf{v} \times \mathbf{u}$.

13. (a) $A = \frac{1}{2} \|\overrightarrow{PQ} \times \overrightarrow{PR}\| = \frac{1}{2} \|\langle -1,-5,2 \rangle \times \langle 2,0,3 \rangle\| = \frac{1}{2} \|\langle -15,7,10 \rangle\| = \sqrt{374}/2$

      (b) $A = \frac{1}{2} \|\overrightarrow{PQ} \times \overrightarrow{PR}\| = \frac{1}{2} \|\langle -1,4,8 \rangle \times \langle 5,2,12 \rangle\| = \frac{1}{2} \|\langle 32,52,-22 \rangle\| = 9\sqrt{13}$

15. (a) $\overrightarrow{AB} = -\mathbf{i} + 2\mathbf{j} + 2\mathbf{k}$, $\overrightarrow{AC} = \mathbf{i} + \mathbf{j} - \mathbf{k}$,
      $\overrightarrow{AB} \times \overrightarrow{AC} = -4\mathbf{i} + \mathbf{j} - 3\mathbf{k}$, area $= \frac{1}{2} \|\overrightarrow{AB} \times \overrightarrow{AC}\| = \sqrt{26}/2$
      (b) area $= \frac{1}{2} h\|\overrightarrow{AB}\| = \frac{3}{2} h = \frac{1}{2} \sqrt{26}$, $h = \sqrt{26}/3$

17. (a) $\mathbf{u} = \overrightarrow{AP} = -4\mathbf{i} + 2\mathbf{k}$, $\mathbf{v} = \overrightarrow{AB} = -3\mathbf{i} + 2\mathbf{j} - 4\mathbf{k}$,
      $\mathbf{u} \times \mathbf{v} = -4\mathbf{i} - 22\mathbf{j} - 8\mathbf{k}$; distance $= \|\mathbf{u} \times \mathbf{v}\|/\|\mathbf{v}\| = 2\sqrt{141/29}$
      (b) $\mathbf{u} = \overrightarrow{AP} = 2\mathbf{i} + 2\mathbf{j} + 3\mathbf{k}$, $\mathbf{v} = \overrightarrow{AB} = -2\mathbf{i} + \mathbf{j} + 2\mathbf{k}$,
      $\mathbf{u} \times \mathbf{v} = \mathbf{i} - 10\mathbf{j} + 6\mathbf{k}$; distance $= \|\mathbf{u} \times \mathbf{v}\|/\|\mathbf{v}\| = \sqrt{137}/3$

19. ambiguous (needs parentheses)

21. 80   23.   1

25. (a) $V = |\mathbf{a} \cdot (\mathbf{b} \times \mathbf{c})| = |-16| = 16$ (b) $V = |\mathbf{a} \cdot (\mathbf{b} \times \mathbf{c})| = |45| = 45$

27. (a) $V = |\mathbf{a} \cdot (\mathbf{b} \times \mathbf{c})| = |-9| = 9$

(b) $A = \|\mathbf{a} \times \mathbf{c}\| = \|3\mathbf{i} - 8\mathbf{j} + 7\mathbf{k}\| = \sqrt{122}$

(c) $\mathbf{b} \times \mathbf{c} = -3\mathbf{i} - \mathbf{j} + 2\mathbf{k}$ is perpendicular to the plane determined by $\mathbf{b}$ and $\mathbf{c}$; let θ be the angle between $\mathbf{a}$ and $\mathbf{b} \times \mathbf{c}$ then

$$\cos \theta = \frac{\mathbf{a} \cdot (\mathbf{b} \times \mathbf{c})}{\|\mathbf{a}\| \, \|\mathbf{b} \times \mathbf{c}\|} = \frac{-9}{\sqrt{14}\sqrt{14}} = -9/14$$

so the acute angle ϕ that $\mathbf{a}$ makes with the plane determined by $\mathbf{b}$ and $\mathbf{c}$ is $\phi = \theta - \pi/2 = \sin^{-1}(9/14)$.

29. (a) $(\mathbf{u} + k\mathbf{v}) \times \mathbf{v} = \mathbf{u} \times \mathbf{v} + k\mathbf{v} \times \mathbf{v} = \mathbf{u} \times \mathbf{v} + k(\mathbf{0}) = \mathbf{u} \times \mathbf{v}$

(b) $\mathbf{u} \cdot \mathbf{v} \times \mathbf{z} = \mathbf{u} \cdot (\mathbf{v} \times \mathbf{z}) = \mathbf{z} \cdot (\mathbf{u} \times \mathbf{v})$ from formula (9)

$= (\mathbf{u} \times \mathbf{v}) \cdot \mathbf{z} = \mathbf{u} \times \mathbf{v} \cdot \mathbf{z}$

31. Part (b):   let $\mathbf{u} = \langle u_1, u_2, u_3 \rangle$, $\mathbf{v} = \langle v_1, v_2, v_3 \rangle$, and $\mathbf{w} = \langle w_1, w_2, w_3 \rangle$; show that $\mathbf{u} \times (\mathbf{v} + \mathbf{w})$ and $(\mathbf{u} \times \mathbf{v}) + (\mathbf{u} \times \mathbf{w})$ are the same.

Part (c): $(\mathbf{u} + \mathbf{v}) \times \mathbf{w} = -[\mathbf{w} \times (\mathbf{u} + \mathbf{v})]$ from part (a)

$= -[(\mathbf{w} \times \mathbf{u}) + (\mathbf{w} \times \mathbf{v})]$ from part (b)

$= (\mathbf{u} \times \mathbf{w}) + (\mathbf{v} \times \mathbf{w})$ from part (a)

33. Let $\mathbf{x} = x_1\mathbf{i} + x_2\mathbf{j} + x_3\mathbf{k}$ and $\mathbf{y} = y_1\mathbf{i} + y_2\mathbf{j} + y_3\mathbf{k}$. If $\mathbf{z} = \mathbf{i}$ then both $\mathbf{x} \times (\mathbf{y} \times \mathbf{i})$ and $(\mathbf{x} \cdot \mathbf{i})\mathbf{y} - (\mathbf{x} \cdot \mathbf{y})\mathbf{i}$ can be shown to be the same. The cases $\mathbf{z} = \mathbf{j}$ and $\mathbf{z} = \mathbf{k}$ are treated the same way. Finally, for $\mathbf{z} = z_1\mathbf{i} + z_2\mathbf{j} + z_3\mathbf{k}$,

$\mathbf{x} \times (\mathbf{y} \times \mathbf{z}) = \mathbf{x} \times (z_1\mathbf{y} \times \mathbf{i} + z_2\mathbf{y} \times \mathbf{j} + z_3\mathbf{y} \times \mathbf{k})$

$= z_1[\mathbf{x} \times (\mathbf{y} \times \mathbf{i})] + z_2[\mathbf{x} \times (\mathbf{y} \times \mathbf{j})] + z_3[\mathbf{x} \times (\mathbf{y} \times \mathbf{k})]$

$= z_1[(\mathbf{x} \cdot \mathbf{i})\mathbf{y} - (\mathbf{x} \cdot \mathbf{y})\mathbf{i}] + z_2[(\mathbf{x} \cdot \mathbf{j})\mathbf{y} - (\mathbf{x} \cdot \mathbf{y})\mathbf{j}]$

$+ z_3[(\mathbf{x} \cdot \mathbf{k})\mathbf{y} - (\mathbf{x} \cdot \mathbf{y})\mathbf{k}]$

$= [\mathbf{x} \cdot (z_1\mathbf{i} + z_2\mathbf{j} + z_3\mathbf{k})]\mathbf{y} - (\mathbf{x} \cdot \mathbf{y})(z_1\mathbf{i} + z_2\mathbf{j} + z_3\mathbf{k})$

$= (\mathbf{x} \cdot \mathbf{z})\mathbf{y} - (\mathbf{x} \cdot \mathbf{y})\mathbf{z}$

35. If $\mathbf{a}$, $\mathbf{b}$, $\mathbf{c}$, and $\mathbf{d}$ lie in the same plane then $\mathbf{a} \times \mathbf{b}$ and $\mathbf{c} \times \mathbf{d}$ are parallel so $(\mathbf{a} \times \mathbf{b}) \times (\mathbf{c} \times \mathbf{d}) = \mathbf{0}$

37. (a) $\overrightarrow{PQ} = \langle 3,-1,-3 \rangle$, $\overrightarrow{PR} = \langle 2,-2,1 \rangle$, $\overrightarrow{PS} = \langle 4,-4,3 \rangle$.

   $V = \frac{1}{6} \left| \overrightarrow{PQ} \cdot (\overrightarrow{PR} \times \overrightarrow{PS}) \right| = \frac{1}{6} |-4| = 2/3$.

   (b) $\overrightarrow{PQ} = \langle 1,2,-1 \rangle$, $\overrightarrow{PR} = \langle 3,4,0 \rangle$, $\overrightarrow{PS} = \langle -1,-3,4 \rangle$.

   $V = \frac{1}{6} \left| \overrightarrow{PQ} \cdot (\overrightarrow{PR} \times \overrightarrow{PS}) \right| = \frac{1}{6} |-3| = 1/2$.

EXERCISE SET 14.5

1. $\overrightarrow{P_1P_2} = \langle 2,3 \rangle$ so $x = 3 + 2t$, $y = -2 + 3t$

3. $\overrightarrow{P_1P_2} = \langle 0,2 \rangle$ so $x = 4$, $y = 1 + 2t$

5. $\overrightarrow{P_1P_2} = \langle -3,6,1 \rangle$ so $x = 5 - 3t$, $y = -2 + 6t$, $z = 1 + t$

7. $\overrightarrow{P_1P_2} = \langle -1,6,1 \rangle$ so $x = -t$, $y = 6t$, $z = t$

9. Exercise 1 with $0 \le t \le 1$ 11. Exercise 3 with $0 \le t \le 1$

13. Exercise 5 with $0 \le t \le 1$.   15. Exercise 7 with $0 \le t \le 1$.

17. $x = -5 + 2t$, $y = 2 - 3t$

19. $2x + 2yy' = 0$, $y' = -x/y = -(3)/(-4) = 3/4$, $\mathbf{v} = 4\mathbf{i} + 3\mathbf{j}$;
    $x = 3 + 4t$, $y = -4 + 3t$

21. $x = -1 + 3t$, $y = 2 - 4t$, $z = 4 + t$

23. The line is parallel to the vector $\langle 2,-1,2 \rangle$ so $x = -2 + 2t$, $y = -t$,
    $z = 5 + 2t$.

25. The line is parallel to the vector $\langle 1,0,0 \rangle$ so $x = 3 + t$, $y = 7$, $z = 0$.

27. $3 + 4t = 4t^2$, $4t^2 - 4t - 3 = 0$, $(2t + 1)(2t - 3) = 0$; $t = -1/2$, $3/2$. If
    $t = -1/2$, then $x = -1$, $y = 1$; if $t = 3/2$, then $x = 3$, $y = 9$. The points
    of intersection are $(-1,1)$ and $(3,9)$.

29. (a) z = 0 when t = 3 so the point is (-2,10,0)
     (b) y = 0 when t = -2 so the point is (-2,0,-5)
     (c) x is always -2 so the line does not intersect the yz-plane

31. $(1 + t)^2 + (3 - t)^2 = 16$, $t^2 - 2t - 3 = 0$, $(t + 1)(t - 3) = 0$;
     t = -1, 3. The points of intersection are (0,4,-2) and (4,0,6).

33. The line is parallel to the vector $\langle a,b,c \rangle$ so $x = x_1 + at$, $y = y_1 + bt$,
     $z = z_1 + ct$

35. The lines intersect if we can find values of t_1 and t_2 that satisfy the
     equations $2 + t_1 = 2 + t_2$, $2 + 3t_1 = 3 + 4t_2$, and $3 + t_1 = 4 + 2t_2$.
     Solutions of the first two of these equations are $t_1 = -1$, $t_2 = -1$ which
     also satisfy the third equation so the lines intersect at (1,-1,2).

37. The lines are parallel, respectively, to the vectors $\langle 7,1,-3 \rangle$ and
     $\langle -1,0,2 \rangle$. These vectors are not parallel so the lines are not parallel.
     The system of equations $1 + 7t_1 = 4 - t_2$, $3 + t_1 = 6$, and
     $5 - 3t_1 = 7 + 2t_2$ has no solution so the lines do not intersect.

39. The points lie on the same line if $\overrightarrow{P_1P_2}$ is parallel to $\overrightarrow{P_2P_3}$.

     (a) $\overrightarrow{P_1P_2} = \langle 3,-7,-7 \rangle$, $\overrightarrow{P_2P_3} = \langle -9,-7,-3 \rangle$; these vectors are not parallel
         so the points do not lie on the same line.
     (b) $\overrightarrow{P_1P_2} = \langle 2,-4,-4 \rangle$, $\overrightarrow{P_2P_3} = \langle 1,-2,-2 \rangle$; $\overrightarrow{P_1P_2} = 2\overrightarrow{P_2P_3}$ so the vectors are
         parallel and the points lie on the same line.

41. Let the desired point be $P(x_0,y_0,z_0)$, then $\overrightarrow{P_1P} = (2/3)\overrightarrow{P_1P_2}$,
     $\langle x_0 - 1, y_0 - 4, z_0 + 3 \rangle = (2/3)\langle 0,1,2 \rangle = \langle 0,2/3,4/3 \rangle$; equate
     corresponding components to get $x_0 = 1$, $y_0 = 14/3$, $z_0 = -5/3$.

43. Show that two different points on one line lie on the other line. For
     example, with t = 0 and 1 in the equations for the first line we find
     that (3,1) and (2,3) are on the line. These points are also on the
     second line for t = 4/3 and 1.

45. (a) $\langle x,y \rangle = \langle 2,-1 \rangle + t\langle -7,4 \rangle$ (b) $\langle x,y \rangle = \langle 0,3 \rangle + t\langle 4,0 \rangle$

47. The line segment joining the points $(1,0)$ and $(-3,6)$.

49. $(3,0,1)$ is on the line $(t = 0)$ so $u = -5i + j$, $v = -i + j + 2k$,
     $u \times v = 2i + 10j - 4k$; distance $= \|u \times v\|/\|v\| = 2\sqrt{5}$.

51. $(0,1,2)$ is on the given line $(t = 0)$ so $u = j - k$ is a vector from this
     point to the point $(0,2,1)$, $v = 2i - j + k$ is parallel to the given
     line.
     $u \times v = -2j - 2k$, and hence $w = j + k$, is perpendicular to both lines so
     $v \times w = -2i - 2j + 2k$, and hence $i + j - k$, is parallel to the line we
     seek. Thus $x = t$, $y = 2 + t$, $z = 1 - t$ are parametric equations of the
     line.

EXERCISE SET 14.6

1. $(x - 2) + 4(y - 6) + 2(z - 1) = 0$, $x + 4y + 2z = 28$

3. $z = 0$

5. Denote the given points by P_1, P_2, and P_3, respectively, then
    $\overrightarrow{P_1P_2} \times \overrightarrow{P_1P_3}$ is a normal to the plane.

    (a) $\overrightarrow{P_1P_2} \times \overrightarrow{P_1P_3} = \langle 2,1,2 \rangle \times \langle 3,-1,-2 \rangle = \langle 0,10,-5 \rangle$, for convenience
        choose $\langle 0,2,-1 \rangle$ which is also normal to the plane. Use any of the
        given points to get $2y - z = 1$
    (b) $\overrightarrow{P_1P_2} \times \overrightarrow{P_1P_3} = \langle -1,-1,-2 \rangle \times \langle -4,1,1 \rangle = \langle 1,9,-5 \rangle$, $x + 9y - 5z = 16$

7. (a) yes, because $\langle 2,-1,-4 \rangle$ and $\langle 3,2,1 \rangle$ are perpendicular
    (b) no, because $\langle 1,2,3 \rangle$ and $\langle 1,-1,2 \rangle$ are not perpendicular

9. (a) yes, because $\langle 2,1,-1 \rangle$ and $\langle 4,2,-2 \rangle$ are parallel
    (b) no, because $\langle -1,1,-3 \rangle$ and $\langle 2,2,0 \rangle$ are not parallel

11. (a) $n_1 = \langle 1,0,0 \rangle$, $n_2 = \langle 2,-1,1 \rangle$, $n_1 \cdot n_2 = 2$ so
    $$\cos \theta = \frac{n_1 \cdot n_2}{\|n_1\| \ \|n_2\|} = \frac{2}{\sqrt{1} \ \sqrt{6}} = 2/\sqrt{6}, \ \theta = \cos^{-1}(2/\sqrt{6}) \approx 35°$$

(b) $n_1 = \langle 1,2,-2 \rangle$, $n_2 = \langle 6,-3,2 \rangle$, $n_1 \cdot n_2 = -4$ so

$$\cos \theta = \frac{(-n_1) \cdot n_2}{\|-n_1\| \, \|n_2\|} = \frac{4}{(3)(7)} = 4/21, \; \theta = \cos^{-1}(4/21) \approx 79°$$

13. (a) $z = 0$ (b) $y = 0$ (c) $x = 0$

15. $\langle 4,-2,7 \rangle$ is normal to the desired plane and $(0,0,0)$ is a point on it;
$4x - 2y + 7z = 0$

17. Find two points P_1 and P_2 on the line of intersection of the given
planes and then find an equation of the plane that contains P_1, P_2, and
the given point $P_0(-1,4,2)$. Let (x_0,y_0,z_0) be on the line of
intersection of the given planes then $4x_0 - y_0 + z_0 - 2 = 0$ and
$2x_0 + y_0 - 2z_0 - 3 = 0$, eliminate y_0 by addition of the equations to get
$6x_0 - z_0 - 5 = 0$; if $x_0 = 0$ then $z_0 = -5$, if $x_0 = 1$ then $z_0 = 1$.
Substitution of these values of x_0 and z_0 into either of the equations
of the planes gives the corresponding values $y_0 = -7$ and $y_0 = 3$ so
$P_1(0,-7,-5)$ and $P_2(1,3,1)$ are on the line of intersection of the planes.
$\overrightarrow{P_0P_1} \times \overrightarrow{P_0P_2} = \langle 4,-13,21 \rangle$ is normal to the desired plane whose equation
is $4x - 13y + 21z = -14$.

19. The line is parallel to the line of intersection of the planes if it is
parallel to both planes. Normals to the given planes are $n_1 = \langle 1,-4,2 \rangle$
and $n_2 = \langle 2,3,-1 \rangle$ so $n_1 \times n_2 = \langle -2,5,11 \rangle$ is parallel to the line of
intersection of the planes and hence parallel to the desired line whose
equations are $x = 5 - 2t$, $y = 5t$, $z = -2 + 11t$.

21. $n_1 = \langle 2,1,1 \rangle$ and $n_2 = \langle 1,2,1 \rangle$ are normals to the given planes,
$n_1 \times n_2 = \langle -1,-1,3 \rangle$ so $\langle 1,1,-3 \rangle$ is normal to the desired plane whose
equation is $x + y - 3z = 6$.

23. $v_1 = \langle 1,2,-1 \rangle$ and $v_2 = \langle -1,-2,1 \rangle$ are parallel, respectively, to the
given lines and to each other so the lines are parallel. Let $t = 0$ to
find the points $P_1(-2,3,4)$ and $P_2(3,4,0)$ that lie, respectively, on the

given lines. $\mathbf{v}_1 \times \overrightarrow{P_1 P_2} = \langle -7,-1,-9 \rangle$ so $\langle 7,1,9 \rangle$ is normal to the desired plane whose equation is $7x + y + 9z = 25$.

25. The plane is the perpendicular bisector of the line segment that joins $P_1(2,-1,1)$ and $P_2(3,1,5)$. The midpoint of the line segment is $(5/2,0,3)$

and $\overrightarrow{P_1 P_2} = \langle 1,2,4 \rangle$ is normal to the plane so an equation is $x + 2y + 4z = 29/2$.

27. $\mathbf{v} = \langle 0,1,1 \rangle$ is parallel to the line.
   (a) $\mathbf{n} = \langle 6,4,-4 \rangle$ is normal to the plane, $\mathbf{v} \cdot \mathbf{n} = 0$ so the line is parallel to the plane because $\mathbf{v}$ and $\mathbf{n}$ are perpendicular. $(0,0,0)$ lies on the line and the plane so the entire line must lie in the plane.

   (b) $\mathbf{n} = \langle 5,-3,3 \rangle$ is normal to the plane, $\mathbf{v} \cdot \mathbf{n} = 0$ so the line is parallel to the plane. $(0,0,0)$ is on the line, $(0,0,1/3)$ is on the plane. The line is below the plane because $(0,0,0)$ is below $(0,0,1/3)$.

   (c) $\mathbf{n} = \langle 6,2,-2 \rangle$, $\mathbf{v} \cdot \mathbf{n} = 0$ so the line is parallel to the plane. $(0,0,0)$ is on the line, $(0,0,-3/2)$ is on the plane. The line is above the plane because $(0,0,0)$ is above $(0,0,-3/2)$.

29. (a) $\mathbf{n}_1 = \langle -2,3,7 \rangle$ and $\mathbf{n}_2 = \langle 1,2,-3 \rangle$ are normals to the planes, $\mathbf{n}_1 \times \mathbf{n}_2 = \langle -23,1,-7 \rangle$ is parallel to the line of intersection. Let $z = 0$ in both equations and solve for x and y to get $x = -11/7$, $y = -12/7$ so $(-11/7,-12/7,0)$ is on the line whose equations are $x = -11/7 - 23t$, $y = -12/7 + t$, $z = -7t$

   (b) Similar to part (a) with $\mathbf{n}_1 = \langle 3,-5,2 \rangle$, $\mathbf{n}_2 = \langle 0,0,1 \rangle$, $\mathbf{n}_1 \times \mathbf{n}_2 = \langle -5,-3,0 \rangle$. $z = 0$ so $3x - 5y = 0$, let $x = 0$ then $y = 0$ and $(0,0,0)$ is on the line whose equations are $x = -5t$, $y = -3t$, $z = 0$.

31. $D = |2(1) - 2(-2) + (3) - 4|/\sqrt{4 + 4 + 1} = 5/3$

33. $D = |20(7) - 4(2) - 5(-1)|/\sqrt{400 + 16 + 25} = 137/21$

35. $(0,0,0)$ is on the first plane so
   $D = |6(0) - 3(0) - 3(0) - 5|/\sqrt{36 + 9 + 9} = 5/\sqrt{54}$

37. $(1,3,5)$ and $(4,6,7)$ are on L_1 and L_2, respectively. $v_1 = \langle 7,1,-3 \rangle$ and $v_2 = \langle -1,0,2 \rangle$ are parallel to L_1 and L_2. $v_1 \times v_2 = \langle 2,-11,1 \rangle$ so the plane $2x - 11y + z + 51 = 0$ contains L_2.

$$D = |2(1) - 11(3) + (5) + 51|/\sqrt{4 + 121 + 1} = 25/\sqrt{126}$$

39. $(2,6,0)$ and $(3,5,6)$ are on L_1 and L_2, respectively. $v_1 = \langle 4,-4,5 \rangle$ and $v_2 = \langle 8,-3,1 \rangle$ are parallel to L_1 and L_2. $v_1 \times v_2 = \langle 11,36,20 \rangle$ so $11x + 36y + 20z - 333 = 0$ contains L_2.

$$D = |11(2) + 36(6) + 20(0) - 333|/\sqrt{121 + 1296 + 400} = 95/\sqrt{1817}.$$

41. $n_1 = \langle a_1,b_1,c_1 \rangle$ and $n_2 = \langle a_2,b_2,c_2 \rangle$ are normals to the planes, the planes are perpendicular if and only if their normals are perpendicular so $n_1 \cdot n_2 = 0$, $a_1 a_2 + b_1 b_2 + c_1 c_2 = 0$.

43. The distance between $(2,1,-3)$ and the plane is

$$|2 - 3(1) + 2(-3) - 4|/\sqrt{1 + 9 + 4} = 11/\sqrt{14}$$ which is the radius of the sphere; an equation is $(x - 2)^2 + (y - 1)^2 + (z + 3)^2 = 121/14$.

EXERCISE SET 14.7

1.

Ellipsoid

3.

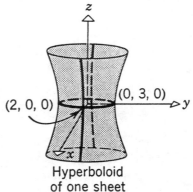

Hyperboloid
of one sheet

5.

Elliptic cone

7.

Elliptic
paraboloid

9.

(2, 0, 0)   (0, 4, 0)

Hyperboloid
of one sheet

11.

Hyperbolic
paraboloid

13.   (a)

**Hyperboloid
of one sheet**

(b)

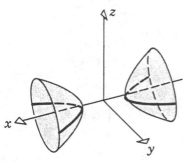

Hyperboloid
of two sheets

(c)

Paraboloid

(d)

Cone

(e)

Hyperbolic
paraboloid

(f)

Paraboloid

15.

17.

(0, 1, 0)

(1, 0, 0)

19.

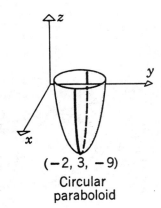

$(-2, 3, -9)$
Circular
paraboloid

21.

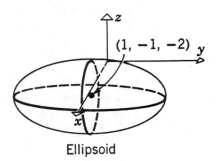

$(1, -1, -2)$

Ellipsoid

23. $(0, -1, 5)$

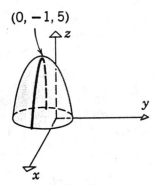

Circular
paraboloid

29.   (a) $z = k^2/9 + y^2/4$, $y^2 = 4(z - k^2/9)$ so $4p = 4$, $p = 1$; the focus is at
$(k,0,k^2/9 + 1)$ and the vertex is at $(k,0,k^2/9)$.

(b) $k = x^2/9 + y^2/4$, $x^2/(9k) + y^2/(4k) = 1$ so $a = 3\sqrt{k}$, $b = 2\sqrt{k}$, $c = \sqrt{5k}$;
the foci are at $(\pm\sqrt{5k},0,k)$, the endpoints of the major axis are
$(\pm3\sqrt{k},0,k)$, and the endpoints of the minor axis are $(0,\pm2\sqrt{k},k)$.

31. $|z - (-1)| = \sqrt{x^2 + y^2 + (z - 1)^2}$, $z^2 + 2z + 1 = x^2 + y^2 + z^2 - 2z + 1$,
$z = (x^2 + y^2)/4$; paraboloid.

33.   (a) $(3 + t)^2 - (2 + t)^2 = 5 + 2t$ and $(3 + t)^2 - (2 - t)^2 = 5 + 10t$ so
both lines lie completely on the surface $z = x^2 - y^2$.

(b) $(x_0 + t)^2 - (y_0 + at)^2 = x_0^2 - y_0^2 + 2(x_0 - ay_0)t + (1 - a^2)t^2$

$= z_0 + bt$ if $z_0 = x_0^2 - y_0^2$, $2(x_0 - ay_0) = b$, and $1 - a^2 = 0$. But $z_0 = x_0^2 - y_0^2$ because (x_0, y_0, z_0) is on the surface so it remains to find a and b so that $2(x_0 - ay_0) = b$ and $1 - a^2 = 0$. Solve the second equation to get $a = \pm 1$ so $b = 2(x_0 - y_0)$ and $b = 2(x_0 + y_0)$.

EXERCISE SET 14.8

1. (a) $(8, \pi/6, -4)$ (b) $(5\sqrt{2}, 3\pi/4, 6)$
    (c) $(2, \pi/2, 0)$ (d) $(8, 5\pi/3, 6)$
    (e) $(2, 7\pi/4, 1)$ (f) $(0, 0, 1)$

3. $\rho = \sqrt{x^2 + y^2 + z^2}$, $\tan\theta = y/x$, $\cos\phi = z/\rho$
    (a) $(2\sqrt{2}, \pi/3, 3\pi/4)$ (b) $(2, 7\pi/4, \pi/4)$
    (c) $(6, \pi/2, \pi/3)$ (d) $(10, 5\pi/6, \pi/2)$
    (e) $(8\sqrt{2}, \pi/4, \pi/6)$ (f) $(2\sqrt{2}, 5\pi/3, 3\pi/4)$

5. $\rho = \sqrt{r^2 + z^2}$, $\theta = \theta$, $\tan\phi = r/z$
    (a) $(2\sqrt{3}, \pi/6, \pi/6)$ (b) $(\sqrt{2}, \pi/4, 3\pi/4)$
    (c) $(2, 3\pi/4, \pi/2)$ (d) $(4\sqrt{3}, 1, 2\pi/3)$
    (e) $(4\sqrt{2}, 5\pi/6, \pi/4)$ (f) $(2\sqrt{2}, 0, 3\pi/4)$

7.

$(3, 0, 0)$

$x^2 + y^2 = 9$

9.

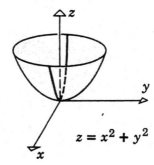

$z = x^2 + y^2$

11.

$(0, 4, 0)$

$x^2 + (y-2)^2 = 4$

13.

$(1, 0, 0)$

$x^2 + y^2 + z^2 = 1$

15.

$(3, 0, 0)$

$x^2 + y^2 + z^2 = 9$

17.

$z = \sqrt{x^2 + y^2}$

$r/z = \tan \phi = 1$
$z = r$ (cylindrical)

$z = \sqrt{x^2 + y^2}$

19.

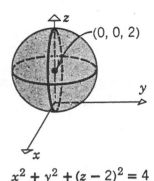

$$x^2 + y^2 + (z - 2)^2 = 4$$

$\rho = 4 \cos \phi$

$\rho^2 = 4\rho \cos \phi$

$x^2 + y^2 + z^2 = 4z$

$x^2 + y^2 + (z - 2)^2 = 4$

21.

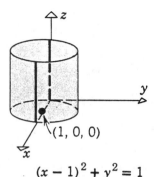

$$(x - 1)^2 + y^2 = 1$$

$\rho \sin \phi = 2 \cos \theta$

$r = 2 \cos \theta$ (cylindrical)

$r^2 = 2r \cos \theta$

$x^2 + y^2 = 2x$

$(x - 1)^2 + y^2 = 1$

23. (a) $r^2 + z^2 = 9$ (b) $\rho = 3$

25. (a) $2r \cos \theta + 3r \sin \theta + 4z = 1$
     (b) $2\rho \sin \phi \cos \theta + 3\rho \sin \phi \sin \theta + 4\rho \cos \phi = 1$

27. (a) $r^2 \cos^2 \theta = 16 - z^2$

     (b) $x^2 = 16 - z^2$, $x^2 + y^2 + z^2 = 16 + y^2$, $\rho^2 = 16 + \rho^2 \sin^2 \phi \sin^2 \theta$,
     $\rho^2(1 - \sin^2 \phi \sin^2 \theta) = 16$

29. $\theta = \pi/6$, $\phi = \pi/6$, spherical $(4000, \pi/6, \pi/6)$,
     rectangular $(1000\sqrt{3}, \ 1000, \ 2000\sqrt{3})$

31.

SUPPLEMENTARY EXERCISES CHAPTER 14

1. (a) $\overrightarrow{P_1P_2} = (5 - 2)\mathbf{i} + (-1 - 3)\mathbf{j} = 3\mathbf{i} - 4\mathbf{j}$, $\|\overrightarrow{P_1P_2}\| = 5$

    (b) $\overrightarrow{P_1P_2} = (1 - 2)\mathbf{i} + (3 + 1)\mathbf{j} = -\mathbf{i} + 4\mathbf{j}$, $\|\overrightarrow{P_1P_2}\| = \sqrt{17}$

3. $-(3\mathbf{i} - 4\mathbf{j}) = -3\mathbf{i} + 4\mathbf{j}$

5. $4\mathbf{i} + 3\mathbf{j}$ is the vector from $(1,2)$ to $(5,5)$ so the desired vector is $(3/5)(4\mathbf{i} + 3\mathbf{j})$

7. $12 \cos 120°\mathbf{i} + 12 \sin 120°\mathbf{j} = -6\mathbf{i} + 6\sqrt{3}\mathbf{j}$

9. $3\mathbf{u} - (\mathbf{i} + \mathbf{j}) = \mathbf{i} + \mathbf{u}$, $2\mathbf{u} = 2\mathbf{i} + \mathbf{j}$, $\mathbf{u} = \mathbf{i} + (1/2)\mathbf{j}$

11. The effect of F_1 and F_2 is the same as the effect of $F_1 + F_2 = -\mathbf{i} - 5\mathbf{j}$ acting at the point, to cancel the effect a force $F_3 = -(F_1 + F_2) = \mathbf{i} + 5\mathbf{j}$ must be applied at the point.

13. (a) $\sqrt{6}$ (b) -3 (c) $\langle 5,-5,-5 \rangle$

    (d) $\langle -5,5,5 \rangle$ (e) $5\sqrt{3}/2$ (f) $\langle -1,8,-9 \rangle$

15. (a) $2/3$ (b) $2/5$

    (c) $\cos^{-1}(-2/15)$ (d) $3/5,-4/5,0$

17. Both sides reduce to $2\mathbf{i} - 2\mathbf{j} + \mathbf{k}$

19. $\langle -3/\sqrt{2},0,3/\sqrt{2} \rangle$

21. (a) $\text{proj}_{\mathbf{v}}\mathbf{u} = \dfrac{\mathbf{u} \cdot \mathbf{v}}{\|\mathbf{v}\|^2}\mathbf{v} = \dfrac{(-3)}{6}(\mathbf{i} + \mathbf{j} + 2\mathbf{k}) = -\dfrac{1}{2}\mathbf{i} - \dfrac{1}{2}\mathbf{j} - \mathbf{k}$

    (b) $\mathbf{u} - \text{proj}_{\mathbf{v}}\mathbf{u} = \dfrac{3}{2}\mathbf{i} + \dfrac{5}{2}\mathbf{j} - 2\mathbf{k}$

23. $\cos^2\alpha + \cos^2\beta + \cos^2\gamma = 1$, let $\alpha = 50°$, $\beta = 70°$,

    $\cos^2\gamma = 1 - \cos^2(50°) - \cos^2(70°) \approx 0.46985$, $\gamma \approx 62°$.

25. (a) $(\mathbf{u} + \mathbf{v}) \cdot (\mathbf{u} - \mathbf{v}) = 0$, $\mathbf{u} \cdot \mathbf{u} - \mathbf{u} \cdot \mathbf{v} + \mathbf{v} \cdot \mathbf{u} - \mathbf{v} \cdot \mathbf{v} = 0$,

   $\|\mathbf{u}\|^2 - \|\mathbf{v}\|^2 = 0$, $\|\mathbf{u}\| = \|\mathbf{v}\|$

   (b) $(\mathbf{a} \cdot \mathbf{b})^2 + \|\mathbf{a} \times \mathbf{b}\|^2 = (\|\mathbf{a}\| \, \|\mathbf{b}\| \cos \theta)^2 + (\|\mathbf{a}\| \, \|\mathbf{b}\| \sin \theta)^2$

   $\qquad\qquad = \|\mathbf{a}\|^2 \|\mathbf{b}\|^2 (\cos^2 \theta + \sin^2 \theta) = \|\mathbf{a}\|^2 \|\mathbf{b}\|^2$

27. $\mathbf{a} \times \mathbf{b} = \langle 5, 7, -1 \rangle$ is orthogonal to both $\mathbf{a}$ and $\mathbf{b}$, $\|\mathbf{a} \times \mathbf{b}\| = 5\sqrt{3}$ so $\pm\langle 1/\sqrt{3}, 7/(5\sqrt{3}), -1/(5\sqrt{3}) \rangle$ are unit vectors orthogonal to both $\mathbf{a}$ and $\mathbf{b}$.

29. The plane contains $\overrightarrow{AB}$ and is parallel to $\mathbf{v}$ thus $\mathbf{v} \times \overrightarrow{AB}$ is normal to the plane, $\mathbf{v} \times \overrightarrow{AB} = \langle 5, -5, -5 \rangle$ so $\langle 1, -1, -1 \rangle$ is also a normal to the plane whose equation is $x - y - z = -4$.

31. $\overrightarrow{PQ} \times \overrightarrow{PR} = \langle 1, 2, -1 \rangle \times \langle 1, 0, 1 \rangle = \langle 2, -2, -2 \rangle$ is normal to the plane and hence so is $\langle 1, -1, -1 \rangle$, an equation of the plane is $x - y - z = -1$.

33. (a) Parametric equations of L are $x = 1 + 3t$, $y = 2 - t$, $z = 8 - 4t$. If Q is on L then for some t_0, $k = 1 + 3t_0$, $3 = 2 - t_0$, $\ell = 8 - 4t_0$. The second of these equations yields $t_0 = -1$ so $k = -2$, $\ell = 12$.

   (b) Use parametric equations in part (a) for L, solve the system $1 + 3t_1 = -8 - 3t_2$, $2 - t_1 = 5 + t_2$, $8 - 4t_1 = 0$ to get $t_1 = 2$, $t_2 = -5$ so L' intersects L at $(7, 0, 0)$.

   (c) An equation of the plane is $3x - 2y + 6z = 6$, use the parametric equations in part (a) to get $3(1 + 3t) - 2(2 - t) + 6(8 - 4t) = 6$, $t = 41/13$ so L intersects the plane at $(136/13, -15/13, -60/13)$.

35. (a) $\overrightarrow{P_1 P_2} = \langle 2, 3, -3 \rangle$, use P_1 to get $x = 1 + 2t$, $y = -1 + 3t$, $z = 2 - 3t$.

   (b) $\overrightarrow{P_1 P_2} = \langle 0, 5, -7 \rangle$, use P_1 to get $x = 1$, $y = -3 + 5t$, $z = 4 - 7t$.

37. (a) $\mathbf{n}_1 = \langle 2, 1, -1 \rangle$ and $\mathbf{n}_2 = \langle 1, 2, 1 \rangle$ are normals to the planes so $\mathbf{n}_1 \times \mathbf{n}_2 = \langle 3, -3, 3 \rangle$ is parallel to the line of intersection and hence so is $\langle 1, -1, 1 \rangle$. To find a point on the line of intersection, let $x = 0$ in the equations of the planes to get $y - z = 3$ and $2y + z = 3$ which yield $y = 2$, $z = -1$ so $(0, 2, -1)$ is on the line whose equations are $x = t$, $y = 2 - t$, $z = -1 + t$.

(b) $n_1 \cdot n_2 = 3 > 0$ so $\cos \theta = \dfrac{n_1 \cdot n_2}{\|n_1\| \ \|n_2\|} = \dfrac{3}{\sqrt{6}\sqrt{6}} = 1/2$, $\theta = 60°$.

39.   (a) the region above the elliptic paraboloid $z = 4x^2 + 9y^2$
      (b) the point $(0,0,0)$

41. $z^2 = \dfrac{x^2}{(36/100)} + \dfrac{y^2}{(36/225)}$, elliptic cone

43. $x^2 + y^2/16 + z^2/25 = 1$, ellipsoid

45. $x^2/25 + y^2/4 - z^2/16 = -1$, hyperboloid of two sheets

47. $W = F \cdot \overrightarrow{PQ} = (3i - 4j + k) \cdot (i - j + 6k) = 13$ ft-lbs

49. $\rho = \sqrt{r^2 + z^2}$, $\theta = \theta$, $\phi = \cos^{-1} \dfrac{z}{\sqrt{r^2 + z^2}}$

51.   (a) $(1,1,1)$ (b) $(\sqrt{3}, \pi/4, \tan^{-1}\sqrt{2})$

53.   (a) $z = r^2(\cos^2\theta - \sin^2\theta)$, $z = x^2 - y^2$
      (b) $(\rho \sin\phi \cos\theta)(\rho \cos\phi) = 1$, $xz = 1$

CHAPTER 15

EXERCISE SET 15.1

1. $(-\infty,+\infty)$; $f(\pi) = -i - 3\pi j$.   3. $[2,+\infty)$; $f(3) = -i - \ln 3j + k$.

5. $r = 3\cos ti + (t + \sin t)j$ 7. $r = 2ti + 2\sin 3tj + 5\cos 3tk$.

9. $x = 3t^2$, $y = -2$, $z = 0$.   11. $x = 2t - 1$, $y = -3\sqrt{t}$, $z = \sin 3t$.

13. The line is 2-space through the point $(2,0)$ and parallel to the vector $-3i - 4j$.

15. The line in 3-space through the point $(0,-3,1)$ and parallel to the vector $2i + 3k$.

17. The line is parallel to the vector $-2i + 3j$; the slope is $-3/2$.

19. $y = 0$ in the xz-plane so $1 - 2t = 0$, $t = 1/2$ thus $x = 2 + 1/2 = 5/2$ and $z = 3(1/2) = 3/2$; the coordinates are $(5/2,0,3/2)$.

21.

$x = 2$

23.

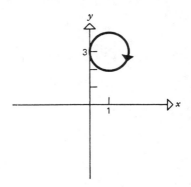

$(x - 1)^2 + (y - 3)^2 = 1$

25.

$$x^2 - y^2 = 1, \ x \geq 1$$

27.

29.

31.

33.   The helix makes one turn as t varies from 0 to 2π so $z = c(2\pi) = 3$, $c = 3/(2\pi)$.

35. $x^2 + y^2 = t^2\cos^2 t + t^2\sin^2 t = t^2$, $\sqrt{x^2 + y^2} = t = z$; a conical helix.

EXERCISE SET 15.2

1. $9\mathbf{i} + 6\mathbf{j}$

3. $\mathbf{j}$

5. $2\mathbf{i} - 3\mathbf{j} + 4\mathbf{k}$

7. $\frac{\pi}{2}\mathbf{i} + \mathbf{k}$

9. $\lim\limits_{t \to \pi/2} \mathbf{f}(t) = 3\mathbf{i} - \pi\mathbf{j} = \mathbf{f}(\pi/2)$

11. $\mathbf{f}'(t) = 5\mathbf{i} + (1 - 2t)\mathbf{j}$

13. $-\dfrac{1}{t^2}\mathbf{i} + \sec^2 t\,\mathbf{j} + 2e^{2t}\mathbf{k}$

15. $\mathbf{r}'(t) = \langle 1, 2t \rangle$,
      $\mathbf{r}'(2) = \langle 1, 4 \rangle$,
       $\mathbf{r}(2) = \langle 2, 4 \rangle$

17. $\mathbf{r}'(t) = \langle -e^{-t}, 2e^{2t} \rangle$,
      $\mathbf{r}'(\ln 2) = \langle -1/2, 8 \rangle$,
       $\mathbf{r}(\ln 2) = \langle 1/2, 4 \rangle$

19. $\mathbf{r}'(t) = 2 \cos ti - 2 \sin tk$,
      $\mathbf{r}'(\pi/2) = -2k$,
       $\mathbf{r}(\pi/2) = 2i + j$

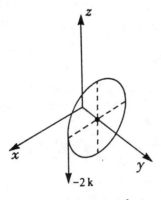

21. $\mathbf{r}'(t) = j - 2tk$
      $\mathbf{r}'(1) = j - 2k$,
       $\mathbf{r}(1) = 3i + j + k$

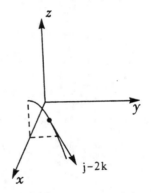

23. $\mathbf{r}'(t) = 2ti - \frac{1}{t} j$, $\mathbf{r}'(1) = 2i - j$, $\mathbf{r}(1) = i + 2j$; $x = 1 + 2t$, $y = 2 - t$, $z = 0$.

25. $\mathbf{r}'(t) = -2\pi \sin \pi ti + 2\pi \cos \pi tj + 3k$, $\mathbf{r}'(1/3) = -\sqrt{3}\pi i + \pi j + 3k$,
      $\mathbf{r}(1/3) = i + \sqrt{3}j + k$; $x = 1 - \sqrt{3}\pi t$, $y = \sqrt{3} + \pi t$, $z = 1 + 3t$.

27. $\mathbf{r}'(t) = 2i + \dfrac{3}{2\sqrt{3t + 4}} j$, $t = 0$ at P_0 so $\mathbf{r}'(0) = 2i + \dfrac{3}{4} j$,

      $\mathbf{r}(0) = -i + 2j$; $\mathbf{r} = (-i + 2j) + t(2i + \dfrac{3}{4} j)$.

29. $\mathbf{r}'(t) = 2t\mathbf{i} + \dfrac{1}{(t + 1)^2}\mathbf{j} - 2t\mathbf{k}$, $t = -2$ at P_0 so $\mathbf{r}'(-2) = -4\mathbf{i} + \mathbf{j} + 4\mathbf{k}$,

$\mathbf{r}(0) = 4\mathbf{i} + \mathbf{j}$; $\mathbf{r} = (4\mathbf{i} + \mathbf{j}) + t(-4\mathbf{i} + \mathbf{j} + 4\mathbf{k})$.

31. $\mathbf{r}' = 3\cos t\,\mathbf{i} + 2\sin t\,\mathbf{j} + \mathbf{k}$, $\mathbf{r}'(\pi/2) = 2\mathbf{j} + \mathbf{k}$, $\mathbf{r}(\pi/2) = 3\mathbf{i} + \dfrac{\pi}{2}\mathbf{k}$;

$2\mathbf{j} + \mathbf{k}$ is normal to the plane, and $(3,0,\pi/2)$ is on the plane so an equation of the plane is $2y + z = \pi/2$.

33. (a) $2t - t^2 - 3t = -2$, $t^2 + t - 2 = 0$, $(t + 2)(t - 1) = 0$ so $t = -2, 1$.
The points of intersection are $(-2,4,6)$ and $(1,1,-3)$.
(b) $\mathbf{r}' = \mathbf{i} + 2t\mathbf{j} - 3\mathbf{k}$; $\mathbf{r}'(-2) = \mathbf{i} - 4\mathbf{j} - 3\mathbf{k}$, $\mathbf{r}'(1) = \mathbf{i} + 2\mathbf{j} - 3\mathbf{k}$, and $\mathbf{n} = 2\mathbf{i} - \mathbf{j} + \mathbf{k}$ is normal to the plane. Let θ be the acute angle, then

for $t = -2$: $\cos(90^\circ - \theta) = |\mathbf{n} \cdot \mathbf{r}|/(\|\mathbf{n}\|\,\|\mathbf{r}\|) = 3/\sqrt{156}$, $\theta \approx 14^\circ$;

for $t = 1$: $\cos(90^\circ - \theta) = |\mathbf{n} \cdot \mathbf{r}|/(\|\mathbf{n}\|\,\|\mathbf{r}\|) = 3/\sqrt{84}$, $\theta \approx 19^\circ$.

35. Eliminate the parameter to get $y = x^2$ for both graphs. $\mathbf{r}_1'(t) = \mathbf{i} + 2t\mathbf{j}$ which is never 0, but $\mathbf{r}_2'(t) = 3t^2\mathbf{i} + 6t^5\mathbf{j} = 0$ when $t = 0$.

37. (a) $7t^6$
(b) $12(t\tan t + 1)\sec t - (\sin t)/t - (\cos t)\ln t$

39. $(d\mathbf{u}/dt)(dt/dw) = (\mathbf{i} + 2t\mathbf{j})(4) = 4\mathbf{i} + 8t\mathbf{j} = 4\mathbf{i} + 8(4w + 1)\mathbf{j}$

41. $(d\mathbf{u}/dt)(dt/dw) = (e^t\mathbf{i} - 4e^{-t}\mathbf{j})(2w) = 2we^{w^2}\mathbf{i} - 8we^{-w^2}\mathbf{j}$

43. $\dfrac{d}{dt}[\mathbf{r}(t) \times \mathbf{r}'(t)] = \mathbf{r}(t) \times \mathbf{r}''(t) + \mathbf{r}'(t) \times \mathbf{r}'(t)$

$= \mathbf{r}(t) \times \mathbf{r}''(t) + 0 = \mathbf{r}(t) \times \mathbf{r}''(t)$

45. In Exercise 44, write each triple scalar product as a determinant.

47. $\dfrac{d}{dt}\left[\dfrac{1}{\|\mathbf{r}\|}\mathbf{r}\right] = \dfrac{1}{\|\mathbf{r}\|}\mathbf{r}' + \mathbf{r}\dfrac{d}{dt}\left[\|\mathbf{r}\|^{-1}\right]$

$= \dfrac{1}{\|\mathbf{r}\|}\mathbf{r}' + \mathbf{r}(-\|\mathbf{r}\|^{-2})\dfrac{d}{dt}\left[\|\mathbf{r}\|\right] = \dfrac{1}{\|\mathbf{r}\|}\mathbf{r}' - \dfrac{\mathbf{r} \cdot \mathbf{r}'}{\|\mathbf{r}\|^3}\mathbf{r}$

49. Let $r_1(t) = x_1(t)i + y_1(t)j + z_1(t)k$ and
    $r_2(t) = x_2(t)i + y_2(t)j + z_2(t)k$, in both (5) and (6); show that the
    left and right members of the equalities are the same.

51. Let $r(t) = x(t)i + y(t)j$ and use Theorem 15.2.3 and the chain rule.

53. Let $f(t) = x(t)i + y(t)j$ and use Definition 15.2.1, part (a), and the
    conditions for f to be continuous at t_0.

EXERCISE SET 15.3

1. $3ti + 2t^2j + C$

3. $\langle \frac{1}{3} \sin 3t, \ \frac{1}{3} \cos 3t \rangle \Big]_0^{\pi/3} = \langle 0, \ -2/3 \rangle$

5. $(\frac{2}{3} t^{3/2}i + 2t^{1/2}j) \Big]_1^9 = \frac{52}{3} i + 4j$

7. $\langle (t - 1)e^t, \ t(\ln t - 1) \rangle + C$

9. $(t^3/3)i - t^2j + \ln|t|k + C$ 11. $\frac{1}{2}(e^2 - 1)i + (1 - e^{-1})j + \frac{1}{2}k$

13. (a) $F = 3ti - 2j - 3t^2k$

    (b) $\int_0^2 F \cdot (dr/dt)dt = \int_0^2 (3t + 12t)dt = \int_0^2 15t \, dt = 30$

15. $r(t) = \int r'(t)dt = (\sin t)i - (\cos t)j + C$,
    $r(0) = -j + C = i - j$ so $C = i$ and $r(t) = (1 + \sin t)i - (\cos t)j$.

17. $\mathbf{r}'(t) = \int \mathbf{r}''(t)dt = 4t^3\mathbf{i} - 2t\mathbf{j} + C_1$, $\mathbf{r}'(0) = C_1 = 0$, $\mathbf{r}'(t) = 4t^3\mathbf{i} - 2t\mathbf{j}$

   $\mathbf{r}(t) = \int \mathbf{r}'(t)dt = t^4\mathbf{i} - t^2\mathbf{j} + C_2$,

   $\mathbf{r}(0) = C_2 = 2\mathbf{i} - 4\mathbf{j}$, $\mathbf{r}(t) = (t^4 + 2)\mathbf{i} - (t^2 + 4)\mathbf{j}$

19. $\mathbf{r}(t) = \int \mathbf{r}'(t)dt = 2t\mathbf{i} + \frac{1}{2}\ln(t^2 + 1)\mathbf{j} + \frac{1}{2}t^2\mathbf{k} + C$,

   $\mathbf{r}(1) = 2\mathbf{i} + \frac{1}{2}\ln 2\mathbf{j} + \frac{1}{2}\mathbf{k} + C = 0$ so $C = -2\mathbf{i} - \frac{1}{2}\ln 2\mathbf{j} - \frac{1}{2}\mathbf{k}$ and

   $\mathbf{r}(t) = 2(t - 1)\mathbf{i} + \frac{1}{2}\ln\frac{t^2 + 1}{2}\mathbf{j} + \frac{1}{2}(t^2 - 1)\mathbf{k}$.

21. $\mathbf{r}'(t) = 3\mathbf{i} - 2\mathbf{j} + \mathbf{k}$, $\|\mathbf{r}'(t)\| = \sqrt{14}$, $L = \int_3^4 \sqrt{14}\,dt = \sqrt{14}$

23. $\mathbf{r}'(t) = 3t^2\mathbf{i} + \mathbf{j} + \sqrt{6}t\mathbf{k}$, $\|\mathbf{r}'(t)\| = 3t^2 + 1$, $L = \int_1^3 (3t^2 + 1)dt = 28$

25. $\mathbf{r}'(t) = \langle e^t, -e^{-t}, \sqrt{2}\rangle$, $\|\mathbf{r}'(t)\| = e^t + e^{-t}$, $L = \int_0^1 (e^t + e^{-t})dt = e - e^{-1}$

27. $dx/dt = -a\sin t$, $dy/dt = a\cos t$, $dz/dt = c$,

   $L = \int_0^{t_0} \sqrt{a^2\sin^2 t + a^2\cos^2 t + c^2}\,dt = \int_0^{t_0} \sqrt{a^2 + c^2}\,dt = t_0\sqrt{a^2 + c^2}$

29. $x = 3u - 2$, $y = 4u + 3$, $(dx/du)^2 + (dy/du)^2 = 25$,

   $s = \int_0^t 5\,du = 5t$ so $t = s/5$, $x = (3/5)s - 2$, $y = (4/5)s + 3$.

31. $x = 3 + \cos u$, $y = 2 + \sin u$, $(dx/du)^2 + (dy/du)^2 = 1$,

   $s = \int_0^t du = t$ so $t = s$, $x = 3 + \cos s$, $y = 2 + \sin s$ for $0 \le s \le 2\pi$.

33. $x = u^3/3$, $y = u^2/2$, $(dx/du)^2 + (dy/du)^2 = u^2(u^2 + 1)$,

$s = \int_0^t u(u^2 + 1)^{1/2}du = \frac{1}{3}[(t^2 + 1)^{3/2} - 1]$ so $t = [(3s + 1)^{2/3} - 1]^{1/2}$,

$x = \frac{1}{3}[(3s + 1)^{2/3} - 1]^{3/2}$, $y = \frac{1}{2}[(3s + 1)^{2/3} - 1]$ for $s \geq 0$.

35. $x = e^u\cos u$, $y = e^u\sin u$, $(dx/du)^2 + (dy/du)^2 = 2e^{2u}$,

$s = \int_0^t \sqrt{2}\, e^u du = \sqrt{2}(e^t - 1)$ so $t = \ln(s/\sqrt{2} + 1)$,

$x = (s/\sqrt{2} + 1)\cos[\ln(s/\sqrt{2} + 1)]$, $y = (s/\sqrt{2} + 1)\sin[\ln(s/\sqrt{2} + 1)]$
for $0 \leq s \leq \sqrt{2}(e^{\pi/2} - 1)$.

37. $x = 1 - 3u$, $y = 5u$, $z = 2 + u$, $(dx/du)^2 + (dy/du)^2 + (dz/du)^2 = 35$,

$s = \int_0^t \sqrt{35}du = \sqrt{35}t$ so $t = s/\sqrt{35}$,

$x = 1 - 3s/\sqrt{35}$, $y = 5s/\sqrt{35}$, $z = 2 + s/\sqrt{35}$.

39. $x = u\cos u$, $y = u\sin u$, $z = \frac{2}{3}\sqrt{2}\, u^{3/2}$,

$(dx/du)^2 + (dy/du)^2 + (dz/du)^2 = u^2 + 2u + 1 = (u + 1)^2$

$s = \int_0^t (u + 1)du = \frac{1}{2}t^2 + t$ so $t = \sqrt{2s + 1} - 1$,

$x = (\sqrt{2s + 1} - 1)\cos(\sqrt{2s + 1} - 1)$, $y = (\sqrt{2s + 1} - 1)\sin(\sqrt{2s + 1} - 1)$,
$z = \frac{2}{3}\sqrt{2}[\sqrt{2s + 1} - 1]^{3/2}$ for $s \geq 0$.

41. (a) $x = OB + CP = a\cos\theta + a\theta\sin\theta$,
        $y = AB - AC = a\sin\theta - a\theta\cos\theta$.
     (b) $x = a(\cos u + u\sin u)$,
        $y = a(\sin u - u\cos u)$,
        $dx/du = au\cos u$, $dy/du = au\sin u$,
        $(dx/dy)^2 + (dy/du)^2 = a^2u^2$,
        $$s = \int_0^\theta au\,du = \frac{1}{2}a\theta^2 \quad\text{so}\quad \theta = \sqrt{2s/a},$$
        $x = a(\cos\sqrt{2s/a} + \sqrt{2s/a}\sin\sqrt{2s/a})$,
        $y = a(\sin\sqrt{2s/a} - \sqrt{2s/a}\cos\sqrt{2s/a})$ for $s \geq 0$.

43. (a) $(dr/dt)^2 + r^2(d\theta/dt)^2 + (dz/dt)^2 = 9e^{4t}$,
        $$L = \int_0^{\ln 2} 3e^{2t}dt = \frac{3}{2}e^{2t}\Big]_0^{\ln 2} = 9/2.$$
     (b) $(dr/dt)^2 + r^2(d\theta/dt)^2 + (dz/dt)^2 = 5t^2 + t^4 = t^2(5 + t^2)$,
        $$L = \int_1^2 t(5 + t^2)^{1/2}dt = 9 - 2\sqrt{6}.$$

45. (a) $(d\rho/dt)^2 + \rho^2\sin^2\phi(d\theta/dt)^2 + \rho^2(d\phi/dt)^2 = 3e^{-2t}$,
        $$L = \int_0^2 \sqrt{3}e^{-t}dt = \sqrt{3}(1 - e^{-2}).$$
     (b) $(d\rho/dt)^2 + \rho^2\sin^2\phi(d\theta/dt)^2 + \rho^2(d\phi/dt)^2 = 5$,
        $$L = \int_1^5 \sqrt{5}dt = 4\sqrt{5}.$$

47. $\mathbf{r}'(t) = (1/t)\mathbf{i} + 2\mathbf{j} + 2t\mathbf{k}$.
     (a) $2t + 1/t$ (b) $2t + 1/t$ (c) $8 + \ln 3$

49. Let $\mathbf{F}(t) = X(t)\mathbf{i} + Y(t)\mathbf{j}$ and $\mathbf{f}(t) = x(t)\mathbf{i} + y(t)\mathbf{j}$ where $X'(t) = x(t)$;
     $Y'(t) = y(t)$ and use Definition 15.3.1 and properties of indefinite and
     definite integrals.

51. Similar to proof of part (a).

EXERCISE SET 15.4

1. $\mathbf{r}'(t) = -5\sin t\mathbf{i} + 5\cos t\mathbf{j}$, $\|\mathbf{r}'(t)\| = 5$
    $\mathbf{T}(t) = -\sin t\mathbf{i} + \cos t\mathbf{j}$, $\mathbf{T}'(t) = -\cos t\mathbf{i} - \sin t\mathbf{j}$;

$$\mathbf{T}(\pi/3) = -\frac{\sqrt{3}}{2}\mathbf{i} + \frac{1}{2}\mathbf{j}, \quad \mathbf{T}'(\pi/3) = -\frac{1}{2}\mathbf{i} - \frac{\sqrt{3}}{2}\mathbf{j}, \quad \mathbf{N}(\pi/3) = -\frac{1}{2}\mathbf{i} - \frac{\sqrt{3}}{2}\mathbf{j}.$$

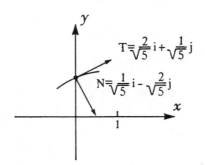

3. $\mathbf{r}'(t) = 2t\mathbf{i} + \mathbf{j}$, $\|\mathbf{r}'(t)\| = \sqrt{4t^2 + 1}$, $\mathbf{T}(t) = (4t^2 + 1)^{-1/2}(2t\mathbf{i} + \mathbf{j})$,
    $\mathbf{T}'(t) = (4t^2 + 1)^{-1/2}(2\mathbf{i}) - 4t(4t^2 + 1)^{-3/2}(2t\mathbf{i} + \mathbf{j})$;
    $\mathbf{T}(1) = \frac{2}{\sqrt{5}}\mathbf{i} + \frac{1}{\sqrt{5}}\mathbf{j}$, $\mathbf{T}'(1) = \frac{2}{5\sqrt{5}}(\mathbf{i} - 2\mathbf{j})$, $\mathbf{N}(1) = \frac{1}{\sqrt{5}}\mathbf{i} - \frac{2}{\sqrt{5}}\mathbf{j}$.

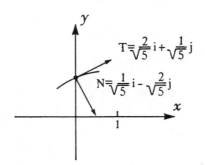

5. $\mathbf{r}'(t) = t\mathbf{i} + t^2\mathbf{j}$, $T(t) = (t^2 + t^4)^{-1/2}(t\mathbf{i} + t^2\mathbf{j})$,

$T'(t) = (t^2 + t^4)^{-1/2}(\mathbf{i} + 2t\mathbf{j}) - (t + 2t^3)(t^2 + t^4)^{-3/2}(t\mathbf{i} + t^2\mathbf{j})$;

$T(1) = \dfrac{1}{\sqrt{2}}\,\mathbf{i} + \dfrac{1}{\sqrt{2}}\,\mathbf{j}$, $T'(1) = \dfrac{1}{2\sqrt{2}}\,(-\mathbf{i} + \mathbf{j})$, $N(1) = -\,\dfrac{1}{\sqrt{2}}\,\mathbf{i} + \dfrac{1}{\sqrt{2}}\,\mathbf{j}$.

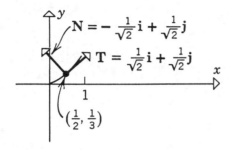

7. $\mathbf{r}'(t) = -4\sin t\,\mathbf{i} + 9\cos t\,\mathbf{j}$,

$T(t) = (16\sin^2 t + 81\cos^2 t)^{-1/2}(-4\sin t\,\mathbf{i} + 9\cos t\,\mathbf{j})$,

$T'(t) = (16\sin^2 t + 81\cos^2 t)^{-1/2}(-4\cos t\,\mathbf{i} - 9\sin t\,\mathbf{j})$

$\qquad\qquad + 65\sin t\cos t(16\sin^2 t + 81\cos^2 t)^{-3/2}(-4\sin t\,\mathbf{i} + 9\cos t\,\mathbf{j})$;

$T(\pi/4) = -\,\dfrac{4}{\sqrt{97}}\,\mathbf{i} + \dfrac{9}{\sqrt{97}}\,\mathbf{j}$, $T'(\pi/4) = \dfrac{72}{97\sqrt{97}}\,(-9\mathbf{i} - 4\mathbf{j})$,

$N(\pi/4) = -\,\dfrac{9}{\sqrt{97}}\,\mathbf{i} - \dfrac{4}{\sqrt{97}}\,\mathbf{j}$.

9. $\mathbf{r}'(t) = -4\sin t\,\mathbf{i} + 4\cos t\,\mathbf{j} + \mathbf{k}$,

$T(t) = \dfrac{1}{\sqrt{17}}\,(-4\sin t\,\mathbf{i} + 4\cos t\,\mathbf{j} + \mathbf{k})$,

$$T'(t) = \frac{1}{\sqrt{17}}(-4\cos ti - 4\sin tj), \quad T(\pi/2) = -\frac{4}{\sqrt{17}}i + \frac{1}{\sqrt{17}}k$$

$$T'(\pi/2) = -\frac{4}{\sqrt{17}}j, \quad N(\pi/2) = -j.$$

11. $r'(t) = i + tj + t^2k, \quad T(t) = (1 + t^2 + t^4)^{-1/2}(i + tj + t^2k),$
     $T'(t) = (1 + t^2 + t^4)^{-1/2}(j + 2tk)$
     $\qquad - (t + 2t^3)(1 + t^2 + t^4)^{-3/2}(i + tj + t^2k),$
     $T(0) = i, \quad T'(0) = j = N(0).$

13. $r'(t) = -3\sin ti + 4\cos tj + k,$
     $T(t) = (9\sin^2 t + 16\cos^2 t + 1)^{-1/2}(-3\sin ti + 4\cos tj + k),$
     $T'(t) = (9\sin^2 t + 16\cos^2 t + 1)^{-1/2}(-3\cos ti - 4\sin tj)$
     $\qquad + 7\sin t\cos t(9\sin^2 t + 16\cos^2 t + 1)^{-3/2}(-3\sin ti + 4\cos tj + k),$
     $T(\pi/2) = -\frac{3}{\sqrt{10}}i + \frac{1}{\sqrt{10}}k, \quad T'(\pi/2) = -\frac{4}{\sqrt{10}}j, \quad N(\pi/2) = -j.$

15. $r'(t) = j + 2tk, \quad T(t) = (1 + 4t^2)^{-1/2}(j + 2tk),$
     $T'(t) = (1 + 4t^2)^{-1/2}(2k) - 4t(1 + 4t^2)^{-3/2}(j + 2tk),$
     $T(1) = \frac{1}{\sqrt{5}}j + \frac{2}{\sqrt{5}}k, \quad T'(1) = \frac{2}{5\sqrt{5}}(-2j + k), \quad N(1) = -\frac{2}{\sqrt{5}}j + \frac{1}{\sqrt{5}}k.$

17. (a) $r(s) = (3s/5 + 1)i + (4s/5 - 2)j,$
     $\qquad T = dr/ds = (3/5)i + (4/5)j$
     (b) $T(5) = (3/5)i + (4/5)j$
     $\qquad r(5) = 4i + 2j$

19. $r(t) = a\cos ti + a\sin tj + ctk$ where $a\cos ti + a\sin tj$ points from the z-axis to a point on the curve, but N is oppositely directed so N points directly toward the z-axis.

21. $\mathbf{r}'(t) = -5 \sin t\mathbf{i} + 5 \cos t\mathbf{j}$, $\mathbf{r}''(t) = -5 \cos t\mathbf{i} - 5 \sin t\mathbf{j}$,

$\mathbf{r}'(\pi/3) = -\dfrac{5\sqrt{3}}{2}\,\mathbf{i} + \dfrac{5}{2}\,\mathbf{j}$, $\mathbf{r}''(\pi/3) = -\dfrac{5}{2}\,\mathbf{i} - \dfrac{5\sqrt{3}}{2}\,\mathbf{j}$,

$\mathbf{n} = 25(-5/2\mathbf{i} - 5\sqrt{3}/2\mathbf{j}) - (0)\mathbf{r}' = \dfrac{125}{2}\,(-\mathbf{i} - \sqrt{3}\mathbf{j})$; $\mathbf{N} = -1/2\mathbf{i} - \sqrt{3}/2\mathbf{j}$.

23. $\mathbf{r}'(t) = 2t\mathbf{i} + \mathbf{j}$, $\mathbf{r}''(t) = 2\mathbf{i}$, $\mathbf{r}'(1) = 2\mathbf{i} + \mathbf{j}$, $\mathbf{r}''(1) = 2\mathbf{i}$,

$\mathbf{n} = 5(2\mathbf{i}) - (4)(2\mathbf{i} + \mathbf{j}) = 2(\mathbf{i} - 2\mathbf{j})$; $\mathbf{N} = 1/\sqrt{5}\mathbf{i} - 2/\sqrt{5}\mathbf{j}$.

25. $\mathbf{r}'(t) = -4 \sin t\mathbf{i} + 4 \cos t\mathbf{j} + \mathbf{k}$, $\mathbf{r}''(t) = -4 \cos t\mathbf{i} - 4 \sin t\mathbf{j}$,
$\mathbf{r}'(\pi/2) = -4\mathbf{i} + \mathbf{k}$, $\mathbf{r}''(\pi/2) = -4\mathbf{j}$, $\mathbf{n} = 17(-4\mathbf{j}) - (0)\mathbf{r}' = -68\mathbf{j}$; $\mathbf{N} = -\mathbf{j}$.

27. $\mathbf{r}'(t) = \mathbf{i} + t\mathbf{j} + t^2\mathbf{k}$, $\mathbf{r}''(t) = \mathbf{j} + 2t\mathbf{k}$, $\mathbf{r}'(0) = \mathbf{i}$, $\mathbf{r}''(0) = \mathbf{j}$,
$\mathbf{n} = (1)(\mathbf{j}) - (0)\mathbf{r}' = \mathbf{j} = \mathbf{N}$.

29. $\mathbf{N}(s) = \mathbf{T}'(s)/\|\mathbf{T}'(s)\|$, but $\mathbf{T}'(s) = \mathbf{r}''(s)$ so $\mathbf{N}(s) = \mathbf{r}''(s)/\|\mathbf{r}''(s)\|$.

EXERCISE SET 15.5

1. $\mathbf{r}'(t) = 2t\mathbf{i} + 3t^2\mathbf{j}$, $\mathbf{r}''(t) = 2\mathbf{i} + 6t\mathbf{j}$, $\mathbf{r}'(1/2) = \mathbf{i} + 3/4\mathbf{j}$,
$\mathbf{r}''(1/2) = 2\mathbf{i} + 3\mathbf{j}$; $\kappa = \|3/2\mathbf{k}\|/\|\mathbf{i} + 3/4\mathbf{j}\|^3 = 96/125$.

3. $\mathbf{r}'(t) = 3e^{3t}\mathbf{i} - e^{-t}\mathbf{j}$, $\mathbf{r}''(t) = 9e^{3t}\mathbf{i} + e^{-t}\mathbf{j}$,
$\mathbf{r}'(0) = 3\mathbf{i} - \mathbf{j}$, $\mathbf{r}''(0) = 9\mathbf{i} + \mathbf{j}$; $\kappa = \|12\mathbf{k}\|/\|3\mathbf{i} - \mathbf{j}\|^3 = 6/(5\sqrt{10})$.

5. $\mathbf{r}'(t) = (\cos t - t \sin t)\mathbf{i} + (\sin t + t \cos t)\mathbf{j}$,
$\mathbf{r}''(t) = -(2 \sin t + t \cos t)\mathbf{i} + (2 \cos t - t \sin t)\mathbf{j}$,
$\|\mathbf{r}'(t) \times \mathbf{r}''(t)\| = 2 + t^2$, $\|\mathbf{r}'(t)\| = (1 + t^2)^{1/2}$; $\kappa(1) = 3/(2\sqrt{2})$.

7. $\mathbf{r}'(t) = -4 \sin t\mathbf{i} + 4 \cos t\mathbf{j} + \mathbf{k}$, $\mathbf{r}''(t) = -4 \cos t\mathbf{i} - 4 \sin t\mathbf{j}$,
$\mathbf{r}'(\pi/2) = -4\mathbf{i} + \mathbf{k}$, $\mathbf{r}''(\pi/2) = -4\mathbf{j}$; $\kappa = \|4\mathbf{i} + 16\mathbf{k}\|/\|-4\mathbf{i} + \mathbf{k}\|^3 = 4/17$.

9. $\mathbf{r}'(t) = \mathbf{i} + t\mathbf{j} + t^2\mathbf{k}$, $\mathbf{r}''(t) = \mathbf{j} + 2t\mathbf{k}$, $\mathbf{r}'(0) = \mathbf{i}$, $\mathbf{r}''(0) = \mathbf{j}$;
$\kappa = \|\mathbf{k}\|/\|\mathbf{i}\|^3 = 1$.

11. $\mathbf{r}'(t) = -3 \sin t\mathbf{i} + 4 \cos t\mathbf{j} + \mathbf{k}$, $\mathbf{r}''(t) = -3 \cos t\mathbf{i} - 4 \sin t\mathbf{j}$,
$\mathbf{r}'(\pi/2) = -3\mathbf{i} + \mathbf{k}$, $\mathbf{r}''(\pi/2) = -4\mathbf{j}$; $\kappa = \|4\mathbf{i} + 12\mathbf{k}\|/\|-3\mathbf{i} + \mathbf{k}\|^3 = 2/5$.

13. $\mathbf{r}'(t) = \mathbf{j} + 2t\mathbf{k}$, $\mathbf{r}''(t) = 2\mathbf{k}$, $\mathbf{r}'(1) = \mathbf{j} + 2\mathbf{k}$, $\mathbf{r}''(1) = 2\mathbf{k}$;

$\kappa = \|2\mathbf{i}\|/\|\mathbf{j} + 2\mathbf{k}\|^3 = 2/(5\sqrt{5})$.

15. $\mathbf{r}'(x) = \mathbf{i} + (dy/dx)\mathbf{j}$, $\mathbf{r}''(x) = (d^2y/dx^2)\mathbf{j}$;

$\kappa(x) = \|(d^2y/dx^2)\mathbf{k}\|/\|\mathbf{i} + (dy/dx)\mathbf{j}\|^3 = |d^2y/dx^2|/[1 + (dy/dx)^2]^{3/2}$.

17. $\kappa(x) = \dfrac{|\sin x|}{(1 + \cos^2 x)^{3/2}}$, $\kappa(\pi/2) = 1$

19. $\kappa(x) = \dfrac{2|x|^3}{(x^4 + 1)^{3/2}}$, $\kappa(1) = 1/\sqrt{2}$

21. $\kappa(x) = \dfrac{2\sec^2 x |\tan x|}{(1 + \sec^4 x)^{3/2}}$, $\kappa(\pi/4) = 4/(5\sqrt{5})$

23. $\kappa(x) = \dfrac{\sec^2 x}{(1 + \tan^2 x)^{3/2}} = \cos x$; $\kappa(x)$ is maximum for $x = 0$.

25. $x'(t) = 2t$, $y'(t) = 3t^2$, $x''(t) = 2$, $y''(t) = 6t$,

$x'(1/2) = 1$, $y'(1/2) = 3/4$, $x''(1/2) = 2$, $y''(1/2) = 3$; $\kappa = 96/125$.

27. $x'(t) = 3e^{3t}$, $y'(t) = -e^{-t}$, $x''(t) = 9e^{3t}$, $y''(t) = e^{-t}$,

$x'(0) = 3$, $y'(0) = -1$, $x''(0) = 9$, $y''(0) = 1$; $\kappa = 6/(5\sqrt{10})$.

29. $x'(t) = \cos t - t\sin t$, $y'(t) = \sin t + t\cos t$,

$x''(t) = -2\sin t - t\cos t$, $y''(t) = 2\cos t - t\sin t$,

$x'y'' - y'x'' = 2 + t^2$, $x'^2 + y'^2 = 1 + t^2$; $\kappa(1) = 3/(2\sqrt{2})$.

31. $x'(t) = -a\sin t$, $y'(t) = b\cos t$, $x''(t) = -a\cos t$, $y''(t) = -b\sin t$;

$\kappa(0) = a/b^2$, $\kappa(\pi/2) = b/a^2$.

33. $\kappa(\theta) = 1$, $\kappa(\pi/6) = 1$

35. $\kappa(\theta) = \dfrac{3}{2\sqrt{2}a(1 + \cos\theta)^{1/2}}$, $\kappa(\pi/2) = \dfrac{3}{2\sqrt{2}a}$

37. $x'(t) = a(1 - \cos t)$, $y'(t) = a \sin t$,
     $x''(t) = a \sin t$, $y''(t) = a \cos t$;

$$\kappa(t) = \frac{a^2(1 - \cos t)}{[2a^2(1 - \cos t)]^{3/2}}$$

$$= \frac{1}{2\sqrt{2}a(1 - \cos t)^{1/2}},$$

but $1 - \cos t = 2 \sin^2(t/2)$ so
$$\kappa(t) = \frac{1}{4a \sin(t/2)} = \frac{1}{4a} \csc(t/2).$$

39. $\kappa(x) = \dfrac{|\sin x|}{(1 + \cos^2 x)^{3/2}}$,

$\kappa(\pi/2) = 1$, $\rho = 1/\kappa = 1$

$\rho = 1$

41. $\kappa(x) = (1 + x^2)^{-3/2}$,

$\kappa(-1) = 2^{-3/2}$, $\rho = 2^{3/2} = 2\sqrt{2}$

$\rho = 2\sqrt{2}$

43. $\kappa(x) = \dfrac{|x|}{(x^2 + 1)^{3/2}}$,

$\kappa(1) = 2^{-3/2}$, $\rho = 2^{3/2} = 2\sqrt{2}$

$\rho = 2\sqrt{2}$

45. $\kappa(t) = \dfrac{|\cos t - 1|}{(2 - 2 \cos t)^{3/2}}$,

$\kappa(\pi) = 1/4$, $\rho = 4$

$\rho = 4$

47. $\kappa(t) = \dfrac{2}{(4\sin^2 t + \cos^2 t)^{3/2}}$,

$\rho(t) = \dfrac{1}{2}(4\sin^2 t + \cos^2 t)^{3/2}$,

$\rho(0) = 1/2$, $\rho(\pi/2) = 4$

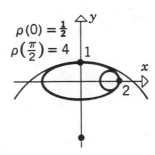

$\rho(0) = \frac{1}{2}$

$\rho(\frac{\pi}{2}) = 4$

49. Let $y = t$, then $x = \dfrac{t^2}{4p}$ and $\kappa(t) = \dfrac{1/|2p|}{[t^2/(4p^2) + 1]^{3/2}}$;

$t = 0$ when $(x,y) = (0,0)$ so $\kappa(0) = 1/|2p|$, $\rho = 2|p|$.

51. Let $x = 3\cos t$, $y = 2\sin t$ for $0 \le t < 2\pi$, then

$\kappa(t) = \dfrac{6}{(9\sin^2 t + 4\cos^2 t)^{3/2}}$

so $\rho(t) = \dfrac{1}{6}(9\sin^2 t + 4\cos^2 t)^{3/2} = \dfrac{1}{6}(5\sin^2 t + 4)^{3/2}$ which, by

inspection, is minimum when $t = 0$ or π. The radius of curvature is

minimum at $(3,0)$ and $(-3,0)$.

53. $\mathbf{r}'(t) = -\sin t\,\mathbf{i} + \cos t\,\mathbf{j} - \sin t\,\mathbf{k}$, $\mathbf{r}''(t) = -\cos t\,\mathbf{i} - \sin t\,\mathbf{j} - \cos t\,\mathbf{k}$,

$\|\mathbf{r}'(t) \times \mathbf{r}''(t)\| = \|-\mathbf{i} + \mathbf{k}\| = \sqrt{2}$, $\|\mathbf{r}'(t)\| = (1 + \sin^2 t)^{1/2}$;

$\kappa(t) = \sqrt{2}/(1 + \sin^2 t)^{3/2}$, $\rho(t) = (1 + \sin^2 t)^{3/2}/\sqrt{2}$. The minimum value

of ρ is $1/\sqrt{2}$; the maximum value is 2.

55. Use the formula in Exercise 32:

$dr/d\theta = ae^{a\theta} = ar$, $d^2r/d\theta^2 = a^2 e^{a\theta} = a^2 r$; $\kappa = \dfrac{1}{(1 + a^2)^{1/2} r}$.

57. $|d\phi/ds| = \kappa = \|dT/ds\| = 0.05$ radians/cm $= (9/\pi)^\circ$/cm $\approx 2.86^\circ$/cm.

59. $\kappa = 0$ along $y = 0$; along $y = x^2$, $\kappa(x) = 2/(1 + 4x^2)^{3/2}$, $\kappa(0) = 2$.

Along $y = x^3$, $\kappa(x) = 6|x|/(1 + 9x^4)^{3/2}$, $\kappa(0) = 0$.

61. $\kappa = 1/r$ along the circle; along $y = ax^2$, $\kappa(x) = 2a/(1 + 4a^2 x)^{3/2}$,

$\kappa(0) = 2a$ so $2a = 1/r$, $a = 1/(2r)$.

63. $\kappa = \|dT/ds\|$, but $T = dr/ds$ so

$$\kappa = \|d^2r/ds^2\| = \sqrt{(d^2x/ds^2)^2 + (d^2y/ds^2)^2 + (d^2z/ds^2)^2}.$$

EXERCISE SET 15.6

1. $v(t) = -3\sin t\, i + 3\cos t\, j$
    $a(t) = -3\cos t\, i - 3\sin t\, j$

    $\|v(t)\| = \sqrt{9\sin^2 t + 9\cos^2 t} = 3$

    $r(\pi/3) = (3/2)i + (3\sqrt{3}/2)j$

    $v(\pi/3) = -(3\sqrt{3}/2)i + (3/2)j$

    $a(\pi/3) = -(3/2)i - (3\sqrt{3}/2)j$

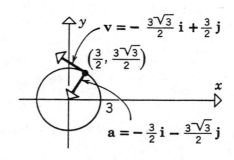

3. $v(t) = e^t i - e^{-t} j$
    $a(t) = e^t i + e^{-t} j$

    $\|v(t)\| = \sqrt{e^{2t} + e^{-2t}}$
    $r(0) = i + j$
    $v(0) = i - j$
    $a(0) = i + j$

5. $v(t) = \sinh t\, i + \cosh t\, j$
    $a(t) = \cosh t\, i + \sinh t\, j$

    $\|v(t)\| = \sqrt{\sinh^2 t + \cosh^2 t}$
    $r(\ln 2) = (5/4)i + (3/4)j$
    $v(\ln 2) = (3/4)i + (5/4)j$
    $a(\ln 2) = (5/4)i + (3/4)j$

7. $v = i + tj + t^2 k$, $a = j + 2tk$;

    at $t = 1$, $v = i + j + k$, $\|v\| = \sqrt{3}$, $a = j + 2k$

9. $\mathbf{v} = -2\sin t\,\mathbf{i} + 2\cos t\,\mathbf{j} + \mathbf{k}$, $\mathbf{a} = -2\cos t\,\mathbf{i} - 2\sin t\,\mathbf{j}$;
    at $t = \pi/4$, $\mathbf{v} = -\sqrt{2}\mathbf{i} + \sqrt{2}\mathbf{j} + \mathbf{k}$, $\|\mathbf{v}\| = \sqrt{5}$, $\mathbf{a} = -\sqrt{2}\mathbf{i} - \sqrt{2}\mathbf{j}$

11. $\mathbf{v} = e^t(\cos t + \sin t)\mathbf{i} + e^t(\cos t - \sin t)\mathbf{j} + \mathbf{k}$,
    $\mathbf{a} = 2e^t\cos t\,\mathbf{i} - 2e^t\sin t\,\mathbf{j}$; at $t = \pi/2$,
    $\mathbf{v} = e^{\pi/2}\mathbf{i} - e^{\pi/2}\mathbf{j} + \mathbf{k}$, $\|\mathbf{v}\| = (1 + 2e^\pi)^{1/2}$, $\mathbf{a} = -2e^{\pi/2}\mathbf{j}$

13. $\mathbf{v}(t) = \displaystyle\int(-32\mathbf{j})dt = -32t\,\mathbf{j} + \mathbf{C}_1$, $\mathbf{v}(0) = \mathbf{C}_1 = 0$ so $\mathbf{v}(t) = -32t\,\mathbf{j}$;
    $\mathbf{r}(t) = \displaystyle\int(-32t\,\mathbf{j})dt = -16t^2\mathbf{j} + \mathbf{C}_2$, $\mathbf{r}(0) = \mathbf{C}_2 = 0$ so $\mathbf{r}(t) = -16t^2\mathbf{j}$

15. $\mathbf{v}(t) = -\sin t\,\mathbf{i} + \cos t\,\mathbf{j} + \mathbf{C}_1$, $\mathbf{v}(0) = \mathbf{j} + \mathbf{C}_1 = \mathbf{i}$,
    $\mathbf{C}_1 = \mathbf{i} - \mathbf{j}$ so $\mathbf{v}(t) = (1 - \sin t)\mathbf{i} + (\cos t - 1)\mathbf{j}$;
    $\mathbf{r}(t) = (t + \cos t)\mathbf{i} + (\sin t - t)\mathbf{j} + \mathbf{C}_2$, $\mathbf{r}(0) = \mathbf{i} + \mathbf{C}_2 = \mathbf{j}$,
    $\mathbf{C}_2 = -\mathbf{i} + \mathbf{j}$ so $\mathbf{r}(t) = (t + \cos t - 1)\mathbf{i} + (\sin t - t + 1)\mathbf{j}$

17. $\mathbf{v}(t) = \displaystyle\int(\mathbf{i} + t\mathbf{k})dt = t\mathbf{i} + \frac{1}{2}t^2\mathbf{k} + \mathbf{C}_1$, $\mathbf{v}(0) = \mathbf{C}_1 = 0$ so
    $\mathbf{v}(t) = t\mathbf{i} + \frac{1}{2}t^2\mathbf{k}$; $\mathbf{r}(t) = \displaystyle\int(t\mathbf{i} + \frac{1}{2}t^2\mathbf{k})dt = \frac{1}{2}t^2\mathbf{i} + \frac{1}{6}t^3\mathbf{k} + \mathbf{C}_2$,
    $\mathbf{r}(0) = \mathbf{C}_2 = \mathbf{j}$ so $\mathbf{r}(t) = \frac{1}{2}t^2\mathbf{i} + \mathbf{j} + \frac{1}{6}t^3\mathbf{k}$.

19. $\mathbf{v}(t) = -\cos t\,\mathbf{i} + \sin t\,\mathbf{j} + e^t\mathbf{k} + \mathbf{C}_1$, $\mathbf{v}(0) = -\mathbf{i} + \mathbf{k} + \mathbf{C}_1 = \mathbf{k}$ so
    $\mathbf{C}_1 = \mathbf{i}$, $\mathbf{v}(t) = (1 - \cos t)\mathbf{i} + \sin t\,\mathbf{j} + e^t\mathbf{k}$;
    $\mathbf{r}(t) = (t - \sin t)\mathbf{i} - \cos t\,\mathbf{j} + e^t\mathbf{k} + \mathbf{C}_2$,
    $\mathbf{r}(0) = -\mathbf{j} + \mathbf{k} + \mathbf{C}_2 = -\mathbf{i} + \mathbf{k}$ so $\mathbf{C}_2 = -\mathbf{i} + \mathbf{j}$,
    $\mathbf{r}(t) = (t - \sin t - 1)\mathbf{i} + (1 - \cos t)\mathbf{j} + e^t\mathbf{k}$.

21. $\mathbf{v} = -2\sin t\,\mathbf{i} + 2\cos t\,\mathbf{j}$, $\mathbf{a} = -2\cos t\,\mathbf{i} - 2\sin t\,\mathbf{j}$,
    $ds/dt = \|\mathbf{v}\| = 2$, $d^2s/dt^2 = 0$. When $t = \pi/3$, $d^2s/dt^2 = 0$ and
    $\mathbf{a} = -\mathbf{i} - \sqrt{3}\mathbf{j}$ so $a_T = 0$, $a_N^2 = \|\mathbf{a}\|^2 - a_T^2 = 4 - 0 = 4$, $a_N = 2$

23. $\mathbf{v} = -e^{-t}\mathbf{i} + e^{t}\mathbf{j}$, $\mathbf{a} = e^{-t}\mathbf{i} + e^{t}\mathbf{j}$, $ds/dt = \sqrt{e^{-2t} + e^{2t}}$,

$d^2s/dt^2 = (-e^{-2t} + e^{2t})/\sqrt{e^{-2t} + e^{2t}}$. When $t = 0$, $d^2s/dt^2 = 0$ and

$\mathbf{a} = \mathbf{i} + \mathbf{j}$ so $a_T = 0$, $a_N^2 = 2 - 0 = 2$, $a_N = \sqrt{2}$.

25. $\mathbf{v} = (3t^2 - 2)\mathbf{i} + 2t\mathbf{j}$, $\mathbf{a} = 6t\mathbf{i} + 2\mathbf{j}$,

$ds/dt = \sqrt{9t^4 - 8t^2 + 4}$, $d^2s/dt^2 = (18t^3 - 8t)/\sqrt{9t^4 - 8t^2 + 4}$. When

$t = 1$, $d^2s/dt^2 = 10/\sqrt{5} = 2\sqrt{5}$ and $\mathbf{a} = 6\mathbf{i} + 2\mathbf{j}$ so $a_T = 2\sqrt{5}$,

$a_N^2 = 40 - 20 = 20$, $a_N = \sqrt{20} = 2\sqrt{5}$.

27. $\mathbf{v} = \mathbf{i} + 2t\mathbf{j} + 3t^2\mathbf{k}$, $\mathbf{a} = 2\mathbf{j} + 6t\mathbf{k}$,

$ds/dt = \sqrt{1 + 4t^2 + 9t^4}$, $d^2s/dt^2 = (4t + 18t^3)/\sqrt{1 + 4t^2 + 9t^4}$.

when $t = 1$, $d^2s/dt^2 = 22/\sqrt{14}$ and $\mathbf{a} = 2\mathbf{j} + 6\mathbf{k}$ so $a_T = 22/\sqrt{14}$,

$a_N^2 = 40 - 242/7 = 38/7$, $a_N = \sqrt{38/7}$.

29. $\mathbf{v} = 3\cos t\,\mathbf{i} - 2\sin t\,\mathbf{j} - 2\cos 2t\,\mathbf{k}$, $\mathbf{a} = -3\sin t\,\mathbf{i} - 2\cos t\,\mathbf{j} + 4\sin 2t\,\mathbf{k}$,

$ds/dt = \sqrt{9\cos^2 t + 4\sin^2 t + 4\cos^2 2t}$,

$d^2s/dt^2 = -(5\sin t\cos t + 8\sin 2t\cos 2t)/\sqrt{9\cos^2 t + 4\sin^2 t + 4\cos^2 2t}$.

When $t = \pi/2$, $d^2s/dt^2 = 0$ and $\mathbf{a} = -3\mathbf{i}$ so $a_T = 0$, $a_N^2 = 9 - 0 = 9$, $a_N = 3$.

31. $\mathbf{T} = \mathbf{v}/\|\mathbf{v}\| = -\mathbf{j}$, $a_T = \mathbf{a} \cdot \mathbf{T} = -3$, $a_N^2 = 13 - 9 = 4$, $a_N = 2$;

$\mathbf{a} = a_T\mathbf{T} + a_N\mathbf{N}$ so $\mathbf{N} = (\mathbf{a} - a_T\mathbf{T})/a_N = \mathbf{i}$.

33. $\mathbf{T} = \mathbf{v}/\|\mathbf{v}\| = 2/3\mathbf{i} + 2/3\mathbf{j} + 1/3\mathbf{k}$, $a_T = \mathbf{a} \cdot \mathbf{T} = 4/3$, $a_N^2 = 5 - 16/9 = 29/9$,

$a_N = \sqrt{29}/3$; $\mathbf{a} = a_T\mathbf{T} + a_N\mathbf{N}$ so $\mathbf{N} = (\mathbf{a} - a_T\mathbf{T})/a_N = (\mathbf{i} - 8\mathbf{j} + 14\mathbf{k})/(3\sqrt{29})$.

35. $a = (d^2s/dt^2)T + \kappa(ds/dt)^2 N$ where $\kappa = \dfrac{|d^2y/dx^2|}{[1 + (dy/dx)^2]^{3/2}}$.   Assuming that

$d^2y/dx^2 = 0$ at a point of inflection then $\kappa = 0$ there and $a = (d^2s/dt^2)T$ so a is tangent to the curve.

37. $a_N = \kappa(ds/dt)^2 = (1/\rho)(ds/dt)^2 = (1/1)(3 \times 10^5)^2$

$$= 9 \times 10^{10} \text{ kilometers/sec}^2$$

39. $a = r''(t) = 0$, $r'(t) = \langle b_1, b_2, b_3 \rangle$, $r(t) = \langle b_1 t + c_1, \ b_2 t + c_2, \ b_3 t + c_3 \rangle$ where b_1, b_2, b_3, c_1, c_2, c_3 are constants so $x = b_1 t + c_1$, $y = b_2 t + c_2$, $z = b_3 t + c_3$ which is a line.

41.   (a) $r' = -3t^2 i - 3t^2 j + 3t^2 k$, $a = r'' = -6ti - 6tj + 6tk$;
      $a = 0$ when $t = 0$.

   (b) Let $t' = t^3$, then $r(t') = (2 - t')i - t'j + (t' - 5)k$, which is a line with parameter t'. This does not contradict the result in Exercise 39; a particle can have a nonzero acceleration along a line.

43. $v = 3 \cos 3ti + 6 \sin 3tj$, $\|v\| = \sqrt{9 \cos^2 3t + 36 \sin^2 3t} = 3\sqrt{1 + 3 \sin^2 3t}$; the maximum speed is 6, the minimum speed is 3.

45. $v = 3t^2 i + 2tj$, $a = 6ti + 2j$; $v = 3i + 2j$ and $a = 6i + 2j$ when $t = 1$ so $\cos \theta = (v \cdot a)/(\|v\| \ \|a\|) = 11/\sqrt{130}$, $\theta \approx 15°$.

47. $v = (2t - 5)i + 2j + 6tk$, $a = 2i + 6k$;
      $v \cdot a = 40t - 10 = 0$ if $t = 1/4$, $r = -19/16i + 3/2j + 3/16k$.

49.   displacement $= r(3) - r(1) = 8i + 26/3j$; $v = 2ti + t^2 j$,

$$L = \int_1^3 t\sqrt{4 + t^2}\,dt = (13\sqrt{13} - 5\sqrt{5})/3.$$

51.   displacement $= r(2\pi) - r(0) = 0$; $v = 6 \cos 3ti - 6 \sin 3tj$,

$$L = \int_0^{2\pi} 6\,dt = 12\pi.$$

53. displacement $= \mathbf{r}(\ln 3) - \mathbf{r}(0) = 2\mathbf{i} - 2/3\mathbf{j} + \sqrt{2}\ln 3\mathbf{k}$;
     $\mathbf{v} = e^t\mathbf{i} - e^{-t}\mathbf{j} + \sqrt{2}\mathbf{k}$, $L = \int_0^{\ln 3}(e^t + e^{-t})dt = 8/3$.

55. (a) $\mathbf{v}_0 = 320\cos 60°\mathbf{i} + 320\sin 60°\mathbf{j} = 160\mathbf{i} + 160\sqrt{3}\mathbf{j}$, $\mathbf{s}_0 = 0$,
         $\mathbf{r}(t) = -16t^2\mathbf{j} + (160\mathbf{i} + 160\sqrt{3}\mathbf{j})t = 160t\mathbf{i} + (160\sqrt{3}t - 16t^2)\mathbf{j}$
         so $x = 160t$, $y = 160\sqrt{3}t - 16t^2$.
     (b) $dy/dt = 160\sqrt{3} - 32t$, $dy/dt = 0$ when $t = 5\sqrt{3}$ so
         $y_{max} = 160\sqrt{3}(5\sqrt{3}) - 16(5\sqrt{3})^2 = 1200$ ft.
     (c) $y = 16t(10\sqrt{3} - t)$, $y = 0$ when $t = 0$ or $10\sqrt{3}$ so
         $x_{max} = 160(10\sqrt{3}) = 1600\sqrt{3}$ ft.
     (d) $\mathbf{v}(t) = 160\mathbf{i} + (160\sqrt{3} - 32t)\mathbf{j}$, $\mathbf{v}(10\sqrt{3}) = 160(\mathbf{i} - \sqrt{3}\mathbf{j})$,
         $\|\mathbf{v}(10\sqrt{3})\| = 320$ ft/sec.

57. $\mathbf{v}_0 = 80\cos(-60°)\mathbf{i} + 80\sin(-60°)\mathbf{j} = 40(\mathbf{i} - \sqrt{3}\mathbf{j})$, so $\mathbf{s}_0 = 168$,
     $\mathbf{r}(t) = (-16t^2 + 168)\mathbf{j} + 40(\mathbf{i} - \sqrt{3}\mathbf{j})t = 40t\mathbf{i} + (168 - 40\sqrt{3}t - 16t^2)\mathbf{j}$
     so $x = 40t$, $y = 168 - 40\sqrt{3}t - 16t^2$; $y = 0$ when
     $t = -7\sqrt{3}/2$ (invalid) or $t = \sqrt{3}$ so $x(\sqrt{3}) = 40\sqrt{3}$ ft.

59. Let $v_0 = \|\mathbf{v}_0\|$ then $\mathbf{v}_0 = (\sqrt{3}v_0/2)\mathbf{i} + (v_0/2)\mathbf{j}$,
     $\mathbf{s}_0 = 0$, $\mathbf{r}(t) = (\sqrt{3}v_0t/2)\mathbf{i} + (v_0t/2 - 16t^2)\mathbf{j}$ so
     $x = \sqrt{3}v_0t/2$, $y = v_0t/2 - 16t^2$. $dy/dt = v_0/2 - 32t$, $dy/dt = 0$ when
     $t = v_0/64$ so $y_{max} = v_0^2/256 = 2500$, $v_0 = 800$ ft/sec.

61. $\mathbf{v}_0 = 800\cos\theta\,\mathbf{i} + 800\sin\theta\,\mathbf{j}$, $\mathbf{s}_0 = 0$,
     $\mathbf{r}(t) = 800t\cos\theta\,\mathbf{i} + (800t\sin\theta - 16t^2)\mathbf{j}$ so
     $x = 800t\cos\theta$, $y = 800t\sin\theta - 16t^2 = 16t(50\sin\theta - t)$.
     $y = 0$ when $t = 0$ or $50\sin\theta$ so

$x_{max} = 40,000 \sin\theta\cos\theta = 20,000 \sin 2\theta = 10,000,$

$2\theta = 30^\circ$ or 150°, $\theta = 15^\circ$ or 75°.

63. (a) $\mathbf{v}_0 = v_0\cos\alpha\mathbf{i} + v_0\sin\alpha\mathbf{j}$, $\mathbf{s}_0 = 0$,

   $\mathbf{r}(t) = (v_0\cos\alpha)t\mathbf{i} + [(v_0\sin\alpha)t - gt^2/2]\mathbf{j}$ so

   $x = (v_0\cos\alpha)t$, $y = (v_0\sin\alpha)t - gt^2/2$

   (b) $dy/dt = v_0\sin\alpha - gt$ so $dy/dt = 0$ when $t = (v_0\sin\alpha)/g$ and

   $y_{max} = (v_0\sin\alpha)^2/(2g)$

   (c) $y = 0$ when $t = 0$ or $(2v_0\sin\alpha)/g$,

   $x = (2v_0^2\sin\alpha\cos\alpha)/g = (v_0^2\sin 2\alpha)/g$ at

   $t = (2v_0\sin\alpha)/g$ so x is maximum when $2\alpha = 90^\circ$, $\alpha = 45^\circ$.

65. (a) $\mathbf{r}(t) = -16t^2\mathbf{k}$, $\mathbf{v}(t) = -32t\mathbf{k}$, $\|\mathbf{v}(t)\| = 32t$

   (b) $\mathbf{r}(t) = -16t^2\mathbf{k} + (\mathbf{i} + 2\mathbf{j} + \mathbf{k}) = \mathbf{i} + 2\mathbf{j} + (1 - 16t^2)\mathbf{k}$,
   $\mathbf{v}(t) = -32t\mathbf{k}$, $\|\mathbf{v}(t)\| = 32t$

   (c) $\mathbf{r}(t) = -16t^2\mathbf{k} + t(\mathbf{i} + \mathbf{k}) + (3\mathbf{i} - \mathbf{j} + 4\mathbf{k})$

   $= (3 + t)\mathbf{i} - \mathbf{j} + (4 + t - 16t^2)\mathbf{k}$,

   $\mathbf{v}(t) = \mathbf{i} + (1 - 32t)\mathbf{k}$, $\|\mathbf{v}(t)\| = \sqrt{1 + (1 - 32t)^2}$.

SUPPLEMENTARY EXERCISES CHAPTER 15

1. (a) $\mathbf{v} = \frac{1}{2}(t + 4)^{-1/2}\mathbf{i} + 2\mathbf{j}$, $\mathbf{a} = -\frac{1}{4}(t + 4)^{-3/2}\mathbf{i}$

    (b) $\mathbf{r}'(-3) = (1/2)\mathbf{i} + 2\mathbf{j}$, $\mathbf{r}''(-3) = -(1/4)\mathbf{i}$
    $\mathbf{r}'(0) = (1/4)\mathbf{i} + 2\mathbf{j}$, $\mathbf{r}''(0) = -(1/32)\mathbf{i}$

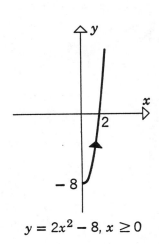

$y = 2x^2 - 8,\ x \geq 0$

3. (a) $\mathbf{v} = \langle 6t^2, 3t^2 \rangle$, $\mathbf{a} = \langle 12t, 6t \rangle$

    (b) $\mathbf{r}'(0) = \langle 0,0 \rangle$, $\mathbf{r}''(0) = \langle 0,0 \rangle$
    $\mathbf{r}'(-1/2) = \langle 3/2,\ 3/4 \rangle$, $\mathbf{r}''(-1/2) = \langle -6,-3 \rangle$

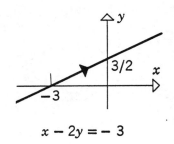

$x - 2y = -3$

5. (a) $(k\mathbf{i} + m\mathbf{j})t + \mathbf{C}$

    (b) $\langle e^{2t}/2,\ 2e^t \rangle \Big]_0^{\ln 3} = \langle 9/2,6 \rangle - \langle 1/2,2 \rangle = \langle 4,4 \rangle$

    (c) $\displaystyle\int_0^2 dt = 2$ (d) $\sqrt{t^2 + 3}\ \mathbf{i} + \ln(\sin t)\mathbf{j} + \mathbf{C}$

7. (a) $ds/dt = \|r'(t)\| = \|\langle(t^2 - 1)/t^2, 2/t\rangle\| = 1 + 1/t^2$

(b) $s = \int_1^t (1 + 1/u^2)du = t - 1/t, \quad t^2 - st - 1 = 0$

$t = (s \pm \sqrt{s^2 + 4})/2$, but $t \geq 0$ so $t = (s + \sqrt{s^2 + 4})/2$

and $x = \sqrt{s^2 + 4}$, $y = 2 \ln[(s + \sqrt{s^2 + 4})/2]$.

9.

11. $r(t) = a \sin t\, i + a \cos t\, j + a \ln(\cos t)k$

$v = a \cos t\, i - a \sin t\, j - a \tan t\, k$

$\|v\| = a(\cos^2 t + \sin^2 t + \tan^2 t)^{1/2}$

$\quad = a(1 + \tan^2 t)^{1/2} = a \sec t$ (for t near 0)

$a = -a \sin t\, i - a \cos t\, j - a \sec^2 t\, k$

$T = v/\|v\| = \cos^2 t\, i - \sin t \cos t\, j - \sin t\, k$

$dT/dt = -2 \sin t \cos t\, i - (\cos^2 t - \sin^2 t)j - \cos t\, k;$

at $t = 0$, $v = a i$, $\|v\| = a$, $a = -a(j + k)$,

$T = i$, $N = (dT/dt)/\|dT/dt\| = (-j - k)/\sqrt{2}$,

$\kappa = \|v \times a\|/\|v\|^3 = \|a^2 j - a^2 k\|/a^3 = \sqrt{2}/a$

13. $(dx/dt)^2 + (dy/dt)^2 + (dz/dt)^2 = 4 + 144 \cos^2 3t + 144 \sin^2 3t = 148,$

$L = \int_0^{2\pi} \sqrt{148}dt = 2\pi\sqrt{148} = 4\pi\sqrt{37}.$

15. $r'(t) = \langle -e^{-t}, 2e^{2t}, 3t^2\rangle$, $r'(0) = \langle -1,2,0\rangle$ is parallel to the tangent line to the curve at the tip of $r(0) = \langle 1,1,1\rangle$ so parametric equations of the tangent line are $x = 1 - t$, $y = 1 + 2t$, $z = 1$.

17. (a) $\int_0^3 \langle 2t,3,-t^2\rangle dt = \langle t^2,3t,-t^3/3\rangle\Big]_0^3 = \langle 9,9,-9\rangle$

(b) $\mathbf{u} \times \mathbf{v} = \langle 3t+t^4, -2t^2, 2t^3\rangle$, $d(\mathbf{u} \times \mathbf{v})/dt = \langle 3+4t^3, -4t, 6t^2\rangle$.

19. (a) $\mathbf{r}'(t) = 2t\mathbf{i} - (1/t^2)\mathbf{j}$, $t = 1$ at P_0 so

$\mathbf{T} = \mathbf{r}'(1)/\|\mathbf{r}'(1)\| = (2\mathbf{i} - \mathbf{j})/\sqrt{5}$

(b) $\mathbf{N} = (\mathbf{i} + 2\mathbf{j})/\sqrt{5}$

(c) $\kappa(t) = \dfrac{6t^4}{(4t^6 + 1)^{3/2}}$ so $\kappa(1) = \dfrac{6}{5^{3/2}}$

21. (a) Let $y = t$, then $\mathbf{r}(t) = (t - 1)^2\mathbf{i} + t\mathbf{j}$,
$\mathbf{r}'(t) = 2(t - 1)\mathbf{i} + \mathbf{j}$, $t = 1$ at P_0 so $\mathbf{T} = \mathbf{r}'(1)/\|\mathbf{r}'(1)\| = \mathbf{j}$

(b) $\mathbf{N} = \mathbf{i}$

(c) $\kappa(t) = 2/[4(t - 1)^2 + 1]^{3/2}$, $\kappa(1) = 2$

23. $x = t + t^3$, $y = t + t^2$, $\kappa(t) = \dfrac{|2 - 6t - 6t^2|}{[(1 + 3t^2)^2 + (1 + 2t)^2]^{3/2}}$, $t = 1$ at P_0,

$\kappa(1) = 2/25$

25. Let $y = t$, then $x = \ln(\sec t)$, $\kappa(t) = |\cos t|$, $t = 0$ at P_0, $\kappa(0) = 1$

27. $\kappa(x) = \dfrac{2}{[1 + 4(x - 1)^2]^{3/2}}$, $\kappa(1) = 2$, $\rho = 1/2$. The parabola opens upward
and has its vertex at $(1,0)$ so the center of curvature is at $(1,1/2)$ and
the oscillating circle is $(x - 1)^2 + (y - 1/2)^2 = 1/4$. $y' = 0$ and
$y'' = 2$ at $(1,0)$ for both the parabola and the circle.

29. $du/dw = (du/dt)(dt/dw) = (1/w)\langle e^t, 4e^{2t}\rangle = \langle 1,4w\rangle$

31. $\mathbf{v} = \mathbf{r}'(t) = \langle t\sin t, t\cos t\rangle$,
$\mathbf{a} = \mathbf{r}''(t) = \langle \sin t + t\cos t, \cos t - t\sin t\rangle$
$ds/dt = \|\mathbf{v}\| = t$, $d^2s/dt^2 = 1 = a_T$,

$a_N^2 = \|\mathbf{a}\|^2 - a_T^2 = (1 + t^2) - 1 = t^2$, $a_N = t$.

33. (a) $v = \langle -e^{-t}, e^{t} \rangle$, $a = \langle e^{-t}, e^{t} \rangle$, $ds/dt = \|v\| = (e^{-2t} + e^{2t})^{1/2}$, $t = 0$
     at P_0 so $v = \langle -1,1 \rangle$, $a = \langle 1,1 \rangle$, $ds/dt = \sqrt{2}$

  (b) $T = v/\|v\| = \langle -1,1 \rangle/\sqrt{2}$, $N = \langle 1,1 \rangle/\sqrt{2}$ $\kappa = 1/\sqrt{2}$

  (c) $d^2s/dt^2 = (-e^{-2t} + e^{2t})^{-1/2}$, $a_T = 0$, $a_N = \kappa(ds/dt)^2 = \sqrt{2}$

  (d) The trajectory is the branch of the hyperbola $xy = 1$ in the first
      quadrant, traced so that y increases with t.

  (e) If $C(h,k)$ is the center of the circle then $\overrightarrow{P_0C} = (1/\kappa)N$,

      $\langle h - 1, k - 1 \rangle = \sqrt{2}\langle 1,1 \rangle/\sqrt{2} = \langle 1,1 \rangle$ so $h - 1 = 1$, $h = 2$ and
      $k - 1 = 1$, $k = 2$. The center is at $(2,2)$.

35. $r(0) = 0$, $v(0) = i + 2j$. Use $F = ma$ with $m = 1$ to get
     $a = \sin t\,i + 4e^{2t}j$, $v(t) = \int a\,dt = -\cos t\,i + 2e^{2t}j + C_1$,
     $v(0) = -i + 2j + C_1 = i + 2j$, $C_1 = 2i$ so, $v(t) = (2 - \cos t)i + 2e^{2t}j$,
     $r(t) = \int v\,dt = (2t - \sin t)i + e^{2t}j + C_2$, $r(0) = j + C_2 = 0$ so $C_2 = -j$,
     $r(t) = (2t - \sin t)i + (e^{2t} - 1)j$

37. $dx/dt = 4$, by the chain rule $dy/dt = (2 - 2x)(dx/dt) = 8(1 - x)$,
     $ds/dt = [(dx/dt)^2 + (dy/dt)^2]^{1/2} = 4[1 + 4(1 - x)^2]^{1/2}$,
     $d^2s/dt^2 = 2[1 + 4(1 - x)^2]^{-1/2}[8(1 - x)](-dx/dt)$
     $\qquad\quad = -64(1 - x)[1 + 4(1 - x)^2]^{-1/2}$
     so $a_T = -64(1 - x)/\sqrt{1 + 4(1 - x)^2}$; $d^2x/dt^2 = 0$,
     $d^2y/dt^2 = -8dx/dt = -32$,
     $\|a\|^2 = (d^2x/dt^2)^2 + (d^2y/dt^2)^2 = 0 + (-32)^2 = 1024$.

  (a) $a_T = 0$, $a_N^2 = \|a\|^2 - a_T^2 = 1024$, $a_N = 32$

  (b) $a_T = -64/\sqrt{5}$, $a_N^2 = 1024 - (64/\sqrt{5})^2 = 1024/5$, $a_N = 32/\sqrt{5}$

39. In one revolution the weight travels a distance that is equal to the
     circumference of a circle of radius 2 m so $ds/dt = 2\pi(2) = 4\pi$ m/sec,
     $a_T = d^2s/dt^2 = 0$, $a_N = \kappa(ds/dt)^2 = (1/2)(4\pi)^2 = 8\pi^2$ m/sec^2.

41. (a) $\mathbf{r}'(t) = e^t \langle 2 \cos 2t, \ -2 \sin 2t \rangle + e^t \langle \sin 2t, \ \cos 2t \rangle,$

$= e^t \langle 2 \cos 2t + \sin 2t, \ \cos 2t - 2 \sin 2t \rangle,$

$ds/dt = \|\mathbf{r}'(t)\|$

$= e^t [(2 \cos 2t + \sin 2t)^2 + (\cos 2t - 2 \sin 2t)^2]^{1/2} = \sqrt{5} e^t.$

(b) $L = \displaystyle\int_0^{\ln 3} (ds/dt)dt = \int_0^{\ln 3} \sqrt{5} e^t dt = 2\sqrt{5}.$

CHAPTER 16

EXERCISE SET 16.1

1.   (a) $f(2,1) = (2)^2(1) + 1 = 5$ (b) $f(1,2) = (1)^2(2) + 1 = 3$
     (c) $f(0,0) = (0)^2(0) + 1 = 1$ (d) $f(1,-3) = (1)^2(-3) + 1 = -2$
     (e) $f(3a,a) = (3a)^2(a) + 1 = 9a^3 + 1$
     (f) $f(ab, a - b) = (ab)^2(a - b) + 1 = a^3b^2 - a^2b^3 + 1$

3.   (a) $f(x+y, x-y) = (x + y)(x - y) + 3 = x^2 - y^2 + 3$
     (b) $f(xy, 3x^2y^3) = (xy)(3x^2y^3) + 3 = 3x^3y^4 + 3$

5. $F(g(x),h(y)) = F(x^3, 3y + 1) = x^3e^{x^3(3y+1)}$

7.   (a) $t^2 + 3t^{10}$ (b) 0   (c) 3076

9.   (a) 19   (b) −9   (c) 3
     (d) $a^6 + 3$ (d) $-t^8 + 3$ (e) $(a + b)(a - b)^2b^3 + 3$

11. $F(x^2,y+1,z^2) = (y + 1)e^{x^2(y+1)z^2}$

13. (a) t^{14} (b) 0   (c) 16,384

15.   17.   19.

            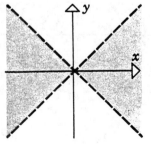

21. all points above or on the line $y = -2$

23.   all points above the line y = 2x

25.   all points not on the plane x + y + z = 0

27.   all points inside the cylinder $x^2 + y^2 = 1$

29.

31

33.

35.

37.

39.

41.

43.

45.

47.

49.

51.

53. concentric spheres, common center at $(2,0,0)$

55. concentric cylinders, common axis the y-axis

57. (a) $f(-1,1) = 0$; $x^2 - 2x^3 + 3xy = 0$ (b) $f(0,0) = 0$; $x^2 - 2x^3 + 3xy = 0$
    (c) $f(2,-1) = -18$; $x^2 - 2x^3 + 3xy = -18$

59. (a) $f(1,-2,0) = 5$; $x^2 + y^2 - z = 5$
    (b) $f(1,0,3) = -2$; $x^2 + y^2 - z = -2$ (c) $f(0,0,0) = 0$; $x^2 + y^2 - z = 0$

61. $V = 8/\sqrt{16 + x^2 + y^2}$

$\sqrt{16 + x^2 + y^2} = 8/V$

$x^2 + y^2 = 64/V^2 - 16$,

the equipotential curves are circles.

63. (a) A   (b) B

65. (a) decrease   (b) increase   (c) increase   (d) decrease

67. (a) open   (b) neither   (c) closed   (d) closed

69. (a) bounded   (b) unbounded   (c) unbounded   (d) unbounded

EXERCISE SET 16.2

1.

3.

5.

7.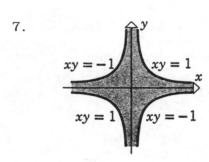

9.   all of 3-space

11.   all points not on the cylinder $x^2 + z^2 = 1$

13.   35   15.   -8   17.   0

19.   along $y = 0$: $\lim\limits_{x\to 0} \dfrac{x}{x^2} = \lim\limits_{x\to 0} \dfrac{1}{x}$ does not exist because $\left|\dfrac{1}{x}\right| \to +\infty$ as $x \to 0$ so
       the original limit does not exist.

21.   Let $z = x^2 + y^2$, then $\lim\limits_{(x,y)\to(0,0)} \dfrac{\sin(x^2 + y^2)}{x^2 + y^2} = \lim\limits_{z\to 0^+} \dfrac{\sin z}{z} = 1$

23. $\lim\limits_{(x,y)\to(0,0)} \dfrac{(x^2 + y^2)(x^2 - y^2)}{x^2 + y^2} = \lim\limits_{(x,y)\to(0,0)} (x^2 - y^2) = 0$

25. along $y = 0$: $\lim\limits_{x \to 0} \dfrac{0}{3x^2} = \lim\limits_{x \to 0} 0 = 0$;

    along $y = x$: $\lim\limits_{x \to 0} \dfrac{x^2}{5x^2} = \lim\limits_{x \to 0} 1/5 = 1/5$

    so the limit does not exist.

27. Let $z = x^2 + y^2$, then $\lim\limits_{(x,y) \to (0,0)} e^{-1/(x^2+y^2)} = \lim\limits_{z \to 0^+} e^{-1/z} = 0$

29. Use polar coordinates: $y = r \sin \theta$ and $x^2 + y^2 = r^2$ so

    $y \ln(x^2 + y^2) = r \sin \theta \ln r^2 = 2r \sin \theta \ln r$. But $|\sin \theta| \leq 1$ so

    $|y \ln(x^2 + y^2)| \leq |2r \ln r|$; $\lim\limits_{r \to 0^+} 2r \ln r = 0$ thus

    $\lim\limits_{(x,y) \to (0,0)} y \ln(x^2 + y^2) = 0$

31. 8/3

33. Let $t = \sqrt{x^2 + y^2 + z^2}$, then

    $\lim\limits_{(x,y,z) \to (0,0,0)} \dfrac{\sin(x^2 + y^2 + z^2)}{\sqrt{x^2 + y^2 + z^2}} = \lim\limits_{t \to 0^+} \dfrac{\sin(t^2)}{t} = 0$

35. along the z-axis: $\lim\limits_{z \to 0} (0/z^2) = \lim\limits_{z \to 0} 0 = 0$;

    along the line $x = t$, $y = t$, $z = t$: $\lim\limits_{t \to 0} \dfrac{t^2}{3t^2} = \lim\limits_{t \to 0} 1/3 = 1/3$

    so the limit does not exist.

37. (a) $\lim\limits_{x \to 0} \dfrac{mx^3}{x^4 + m^2 x^2} = \lim\limits_{x \to 0} \dfrac{mx}{x^2 + m^2} = 0$ (b) $\lim\limits_{x \to 0} \dfrac{x^4}{2x^4} = \lim\limits_{x \to 0} 1/2 = 1/2$

39. (a) $\lim\limits_{t \to 0} \dfrac{abct^3}{a^2 t^2 + b^4 t^4 + c^4 t^4} = \lim\limits_{t \to 0} \dfrac{abct}{a^2 + b^4 t^2 + c^4 t^2} = 0$

(b) $\displaystyle\lim_{t\to 0} \frac{t^4}{t^4 + t^4 + t^4} = \lim_{t\to 0} 1/3 = 1/3$

41. $-\pi/2$ because $\displaystyle\frac{x^2 - 1}{x^2 + (y - 1)^2} \to -\infty$ as $(x,y) \to (0,1)$

43. No, because $\displaystyle\lim_{(x,y)\to(0,0)} \frac{x^2}{x^2 + y^2}$ does not exist.

   Along $x = 0$: $\displaystyle\lim_{y\to 0} (0/y^2) = \lim_{y\to 0} 0 = 0$;

   along $y = 0$: $\displaystyle\lim_{x\to 0} (x^2/x^2) = \lim_{x\to 0} 1 = 1$.

45. $x^2 + y^2 = r^2$, $|x^2 + y^2 - 0| = |r^2| = r^2 < \epsilon$ if $r < \sqrt{\epsilon}$; choose $\delta = \sqrt{\epsilon}$.

47. $x^2 + y^2 + z^2 = \rho^2$, $|x^2 + y^2 + z^2 - 0| = |\rho^2| = \rho^2 < \epsilon$ if $\rho < \sqrt{\epsilon}$; choose $\delta = \sqrt{\epsilon}$.

EXERCISE SET 16.3

1. $\partial z/\partial x = 9x^2 y^2$, $\partial z/\partial y = 6x^3 y$

3. $\partial z/\partial x = 8xy^3 e^{x^2 y^3}$, $\partial z/\partial y = 12x^2 y^2 e^{x^2 y^3}$

5. $\partial z/\partial x = x^3/(y^{3/5} + x) + 3x^2 \ln(1 + xy^{-3/5})$, $\partial z/\partial y = -(3/5)x^4/(y^{8/5} + xy)$

7. $f_x(x,y) = (3/2)x^2 y(5x^2 - 7)(3x^5 y - 7x^3 y)^{-1/2}$

   $f_y(x,y) = (1/2)x^3(3x^2 - 7)(3x^5 y - 7x^3 y)^{-1/2}$

9. $f_x(x,y) = \dfrac{y^{-1/2}}{y^2 + x^2}$, $f_y(x,y) = -\dfrac{xy^{-3/2}}{y^2 + x^2} - \dfrac{3}{2} y^{-5/2}\tan^{-1}(x/y)$

11. $f_x(x,y) = -(4/3)y^2\sec^2 x(y^2\tan x)^{-7/3}$

$f_y(x,y) = -(8/3)y\tan x(y^2\tan x)^{-7/3}$

13. (a) $f_x(x,y) = -2x$, $f_x(3,1) = -6$ (b) $f_y(x,y) = -21y^2$, $f_y(3,1) = -21$

15. (a) $\partial z/\partial x = x(x^2 + 4y^2)^{-1/2}$, $\partial z/\partial x\big|_{(1,2)} = 1/\sqrt{17}$

(b) $\partial z/\partial y = 4y(x^2 + 4y^2)^{-1/2}$, $\partial z/\partial y\big|_{(1,2)} = 8/\sqrt{17}$

17. $\frac{3}{2}(x^2 + y^2 + z^2)^{1/2}(2x + 2z\frac{\partial z}{\partial x}) = 0$, $\partial z/\partial x = -x/z$; similarly,

$\partial z/\partial y = -y/z$

19. $2x + z(xy\frac{\partial z}{\partial x} + yz)\cos xyz + \frac{\partial z}{\partial x}\sin xyz = 0$,

$\frac{\partial z}{\partial x} = -\dfrac{2x + yz^2\cos xyz}{xyz\cos xyz + \sin xyz}$; $z(xy\frac{\partial z}{\partial y} + xz)\cos xyz + \frac{\partial z}{\partial y}\sin xyz = 0$,

$\frac{\partial z}{\partial y} = -\dfrac{xz^2\cos xyz}{xyz\cos xyz + \sin xyz}$

21. $f_{xx} = 8$, $f_{yy} = -96xy^2 + 140y^3$, $f_{xy} = f_{yx} = -32y^3$

23. $f_{xx} = e^x\cos y$, $f_{yy} = -e^x\cos y$, $f_{xy} = f_{yx} = -e^x\sin y$

25. $f_{xx} = -16/(4x - 5y)^2$, $f_{yy} = -25/(4x - 5y)^2$, $f_{xy} = f_{yx} = 20/(4x - 5y)^2$

27. (a) $30xy^4 - 4$ (b) $60x^2y^3$ (c) $60x^3y^2$

29. (a) $f_{xyy} = -30ye^{-5x}$, $f_{xyy}(0,1) = -30$

(b) $f_{xxx} = -125y^3e^{-5x}$, $f_{xxx}(0,1) = -125$

(c) $f_{yyxx} = 150ye^{-5x}$, $f_{yyxx}(0,1) = 150$

31. (a) $\dfrac{\partial^3 f}{\partial x^3}$ (b) $\dfrac{\partial^3 f}{\partial y^2 \partial x}$ (c) $\dfrac{\partial^4 f}{\partial x^2 \partial y^2}$ (d) $\dfrac{\partial^4 f}{\partial y^3 \partial x}$

33. $\partial w/\partial x = 2xy^4 z^3 + y$, $\partial w/\partial y = 4x^2 y^3 z^3 + x$, $\partial w/\partial z = 3x^2 y^4 z^2 + 2z$

35. $\partial w/\partial x = 2x/(y^2 + z^2)$, $\partial w/\partial y = -2y(x^2 + z^2)/(y^2 + z^2)^2$,
     $\partial w/\partial z = 2z(y^2 - x^2)/(y^2 + z^2)^2$

37. $\partial w/\partial x = x/\sqrt{x^2 + y^2 + z^2}$, $\partial w/\partial y = y/\sqrt{x^2 + y^2 + z^2}$,
     $\partial w/\partial z = z/\sqrt{x^2 + y^2 + z^2}$

39. $f_x = -y^2 z^3/(1 + x^2 y^4 z^6)$, $f_y = -2xyz^3/(1 + x^2 y^4 z^6)$,
     $f_z = -3xy^2 z^2/(1 + x^2 y^4 z^6)$

41. $f_x = 4xyz \cosh\sqrt{z} \sinh(x^2 yz)\cosh(x^2 yz)$
     $f_y = 2x^2 z \cosh\sqrt{z} \sinh(x^2 yz)\cosh(x^2 yz)$
     $f_z = 2x^2 y \cosh\sqrt{z} \sinh(x^2 yz)\cosh(x^2 yz) + (1/2)z^{-1/2}\sinh\sqrt{z} \sinh^2(x^2 yz)$

43. (a) -80 (b) 40 (c) -60

45. (a) $2/\sqrt{7}$ (b) $4/\sqrt{7}$ (c) $1/\sqrt{7}$

47. $(3/2)(x^2 + y^2 + z^2 + w^2)^{1/2}(2x + 2w \frac{\partial w}{\partial x}) = 0$, $\partial w/\partial x = -x/w$; similarly,
     $\partial w/\partial y = -y/w$ and $\partial w/\partial z = -z/w$

49. $\dfrac{\partial w}{\partial x} = -\dfrac{yzw \cos xyz}{2w + \sin xyz}$, $\dfrac{\partial w}{\partial y} = -\dfrac{xzw \cos xyz}{2w + \sin xyz}$, $\dfrac{\partial w}{\partial z} = -\dfrac{xyw \cos xyz}{2w + \sin xyz}$

51. (a) $f_{xy} = 15x^2 y^4 z^7 + 2y$ (b) $f_{yz} = 35x^3 y^4 z^6 + 3y^2$
     (c) $f_{xz} = 21x^2 y^5 z^6$ (d) $f_{zz} = 42x^3 y^5 z^5$

(e) $f_{zyy} = 140x^3y^3z^6 + 6y$ (f) $f_{xxy} = 30xy^4z^7$

(g) $f_{zyx} = 105x^2y^4z^6$ (h) $f_{xxyz} = 210xy^4z^6$

53.   (a) $\partial^2 f/\partial x^2 = e^x\sin y - e^y\cos x = -\partial^2 f/\partial y^2$

       (b) $\partial^2 f/\partial x^2 = 2(y^2 - x^2)/(x^2 + y^2)^2 = -\partial^2 f/\partial y^2$

       (c) $\partial^2 f/\partial x^2 = 4xy/(x^2 + y^2)^2 = -\partial^2 f/\partial y^2$

55. $\partial z/\partial y = 6y$, $\partial z/\partial y\big|_{(2,1)} = 6$

57.   (a) $\partial z/\partial y = 8y$, $\partial z/\partial y\big|_{(-1,1)} = 8$ (b) $\partial z/\partial x = 2x$, $\partial z/\partial x\big|_{(-1,1)} = -2$

59.   (a) $\partial V/\partial r = 2\pi rh$ (b) $\partial V/\partial h = \pi r^2$

       (c) $\partial V/\partial r\big|_{r=6,\ h=4} = 48\pi$ (d) $\partial V/\partial h\big|_{r=8,\ h=10} = 64\pi$

61.   (a) $P = 10T/V$, $\partial P/\partial T = 10/V$, $\partial P/\partial T\big|_{T=80,\ V=50} = 1/5$

       (b) $V = 10T/P$, $\partial V/\partial P = -10T/P^2$, if $V = 50$ and $T = 80$ then

           $P = 10(80)/(50) = 16$, $\partial V/\partial P\big|_{T=80,\ P=16} = -25/8$

63. $(1 + \frac{\partial z}{\partial x})\cos(x + z) + \cos(x - y) = 0$, $\frac{\partial z}{\partial x} = -1 - \frac{\cos(x - y)}{\cos(x + z)}$;

       $\frac{\partial z}{\partial y}\cos(x + z) - \cos(x - y) = 0$, $\frac{\partial z}{\partial y} = \frac{\cos(x - y)}{\cos(x + z)}$;

       $\dfrac{\partial^2 z}{\partial x\partial y} = \dfrac{-\cos(x + z)\sin(x - y) + \cos(x - y)\sin(x + z)(\partial z/\partial x)}{\cos^2(x + z)}$,

       substitute for $\partial z/\partial x$ and simplify to get

       $\dfrac{\partial^2 z}{\partial x\partial y} = -\dfrac{\cos^2(x+z)\sin(x-y)+\cos^2(x-y)\sin(x+z)+\cos(x-y)\cos(x+z)\sin(x+z)}{\cos^3(x + z)}$

65.   (a) $\partial T/\partial x = 3x^2 + 1$, $\partial T/\partial x\big|_{(1,2)} = 4$ (b) $\partial T/\partial y = 4y$, $\partial T/\partial y\big|_{(1,2)} = 8$

67. $\partial u/\partial x = \partial v/\partial y$ and $\partial u/\partial y = -\partial v/\partial x$ so $\partial^2 u/\partial x^2 = \partial^2 v/\partial x\partial y$, and

       $\partial^2 u/\partial y^2 = -\partial^2 v/\partial y\partial x$, $\partial^2 u/\partial x^2 + \partial^2 u/\partial y^2 = \partial^2 v/\partial x\partial y - \partial^2 v/\partial y\partial x$, if

       $\partial^2 v/\partial x\partial y = \partial^2 v/\partial y\partial x$ then $\partial^2 u/\partial x^2 + \partial^2 u/\partial y^2 = 0$

thus u satisfies Laplace's equation. The proof that v satisfies Laplace's equation is similar.

69.   (a) Both are positive; $\partial T/\partial x$ has the largest absolute value.
(b) All are negative.

EXERCISE SET 16.4

1. $\Delta f = f(1.1, 3.2) - f(1,3) = (1.1)^2(3.2) - (1)^2(3) = 0.872$

3. $\Delta f = f(3,1) - f(-1,2) = 3/1 - (-1)/2 = 7/2$

5. $f_x(x,y) = y$, $f_y(x,y) = x$,
$\Delta f = (x + \Delta x)(y + \Delta y) - xy = y\Delta x + x\Delta y + \Delta x\Delta y$
$= f_x(x,y)\Delta x + f_y(x,y)\Delta y + (0)\Delta x + (\Delta x)\Delta y$
where $\epsilon_1 = 0$ and $\epsilon_2 = \Delta x$.

7. $f_x(x,y) = 2xy$, $f_y(x,y) = x^2$,
$\Delta f = (x + \Delta x)^2(y + \Delta y) - x^2 y$
$= 2xy\Delta x + x^2\Delta y + y(\Delta x)^2 + (\Delta x)^2\Delta y + 2x\Delta x\Delta y$
$= f_x(x,y)\Delta x + f_y(x,y)\Delta y + (y\Delta x + \Delta x\Delta y)\Delta x + (2x\Delta x)\Delta y$
where $\epsilon_1 = y\Delta x + \Delta x\Delta y$ and $\epsilon_2 = 2x\Delta x$.

9.   (a) $\lim\limits_{(x,y)\to(0,0)} f(x,y) = 0 = f(0,0)$

(b) $\lim\limits_{h\to 0} \dfrac{f(0+h,0) - f(0,0)}{h} = \lim\limits_{h\to 0} \dfrac{f(h,0)}{h} = \lim\limits_{h\to 0} \dfrac{|h|}{h}$, which does not exist

because $\lim\limits_{h\to 0^+} |h|/h = 1$ and $\lim\limits_{h\to 0^-} |h|/h = -1$.

11. $f_x(0,0) = \lim\limits_{h\to 0} \dfrac{f(h,0) - f(0,0)}{h} = \lim\limits_{h\to 0} \dfrac{0 - 0}{h} = \lim\limits_{h\to 0} 0 = 0$

$f_y(0,0) = \lim\limits_{h\to 0} \dfrac{f(0,h) - f(0,0)}{h} = \lim\limits_{h\to 0} \dfrac{0 - 0}{h} = 0$;

along $y = 0$, $\lim\limits_{x \to 0} \dfrac{0}{x^2} = \lim\limits_{x \to 0} 0 = 0$

along $y = x$, $\lim\limits_{x \to 0} \dfrac{x^2}{2x^2} = \lim\limits_{x \to 0} 1/2 = 1/2$ so $\lim\limits_{(x,y) \to (0,0)} f(x,y)$ does not exist

13. $f_{xy} = f_{yx} = 12x^2 + 6x$ 15. $f_{xy} = f_{yx} = -6xy^2\sin(x^2 + y^3)$

17. (a) 4: f_{xxx}, $f_{xxy} = f_{xyx} = f_{yxx}$, $f_{xyy} = f_{yxy} = f_{yyx}$, f_{yyy}

    (b) 5: f_{xxxx}, $f_{xxxy} = f_{xxyx} = f_{xyxx} = f_{yxxx}$,

              $f_{xxyy} = f_{xyxy} = f_{xyyx} = f_{yxyx} = f_{yyxx} = f_{yxxy}$,

              $f_{xyyy} = f_{yxyy} = f_{yyxy} = f_{yyyx}$, f_{yyyy}

19. $42t^{13}$ 21. $3t^{-2}\sin(1/t)$ 23. $-\dfrac{10}{3}t^{7/3}e^{1-t^{10/3}}$

25. $\partial z/\partial u = 24u^2v^2 - 16uv^3 - 2v + 3$, $\partial z/\partial v = 16u^3v - 24u^2v^2 - 2u - 3$

27. $\partial z/\partial u = -\dfrac{2}{3}\dfrac{\sin u}{\sin v}$, $\partial z/\partial v = -\dfrac{2\cos u \cos v}{3\sin^2 v}$

29. $\partial z/\partial u = e^u$, $\partial z/\partial v = 0$ 31. $\partial z/\partial u = 2e^{2u}/(1 + e^{4u})$, $\partial z/\partial v = 0$

33. $\partial T/\partial r = 3r^2\sin\theta\cos^2\theta - 4r^3\sin^3\theta\cos\theta$

    $\partial T/\partial\theta = -2r^3\sin^2\theta\cos\theta + r^4\sin^4\theta + r^3\cos^3\theta - 3r^4\sin^2\theta\cos^2\theta$

35. $\partial t/\partial x = (x^2 + y^2)/(4x^2y^3)$, $\partial t/\partial y = (y^2 - 3x^2)/(4xy^4)$

37. $-\pi$ 39. $\sqrt{3}\,e^{\sqrt{3}}$, $(2 - 4\sqrt{3})e^{\sqrt{3}}$

41. $F(x,y) = x^3 - 3xy^2 + y^3 - 5$, $\dfrac{dy}{dx} = -\dfrac{3x^2 - 3y^2}{-6xy + 3y^2} = \dfrac{x^2 - y^2}{2xy - y^2}$

43. $F(x,y) = x - (xy)^{1/2} + 3y - 4$, $\dfrac{dy}{dx} = -\dfrac{1 - (1/2)(xy)^{-1/2}y}{-(1/2)(xy)^{-1/2}x + 3} = \dfrac{2\sqrt{xy} - y}{x - 6\sqrt{xy}}$

45. $D = (x^2 + y^2)^{1/2}$ where x and y are the distances of cars A and B, respectively, from the intersection and D is the distance between them.
$dD/dt = [x/(x^2 + y^2)^{1/2}](dx/dt) + [y/(x^2 + y^2)^{1/2}](dy/dt)$,
$dx/dt = -25$ and $dy/dt = -30$ when $x = 0.3$ and $y = 0.4$ so
$dD/dt = (0.3/0.5)(-25) + (0.4/0.5)(-30) = -39$ mph.

47. $A = \frac{1}{2} ab \sin\theta$ but $\theta = \pi/6$ when $a = 4$ and $b = 3$ so

$A = \frac{1}{2}(4)(3)\sin(\pi/6) = 3$.   Solve $\frac{1}{2} ab \sin\theta = 3$ for θ to get

$\theta = \sin^{-1}(\frac{6}{ab})$, $0 \le \theta \le \pi/2$.

$\dfrac{d\theta}{dt} = \dfrac{\partial\theta}{\partial a}\dfrac{da}{dt} + \dfrac{\partial\theta}{\partial b}\dfrac{db}{dt}$

$= \dfrac{1}{\sqrt{1 - \dfrac{36}{a^2 b^2}}}(-\dfrac{6}{a^2 b})\dfrac{da}{dt} + \dfrac{1}{\sqrt{1 - \dfrac{36}{a^2 b^2}}}(-\dfrac{6}{ab^2})\dfrac{db}{dt}$

$= -\dfrac{6}{\sqrt{a^2 b^2 - 36}}(\dfrac{1}{a}\dfrac{da}{dt} + \dfrac{1}{b}\dfrac{db}{dt})$,

$da/dt = 1$ and $db/dt = 1$ when $a = 4$ and $b = 3$ so
$\dfrac{d\theta}{dt} = -\dfrac{6}{\sqrt{144 - 36}}(\dfrac{1}{4} + \dfrac{1}{3}) = -\dfrac{7}{12\sqrt{3}} = -\dfrac{7}{36}\sqrt{3}$ radians/sec

49. $\dfrac{dT}{dt} = \dfrac{\partial T}{\partial x}\dfrac{dx}{dt} + \dfrac{\partial T}{\partial y}\dfrac{dy}{dt} = \dfrac{y^2}{x}\dfrac{dx}{dt} + 2y \ln x \dfrac{dy}{dt}$,
$dx/dt = 1$ and $dy/dt = -4$ at $(3,2)$ so
$dT/dt = (4/3)(1) + (4 \ln 3)(-4) = 4/3 - 16 \ln 3 \ ^\circ$C/sec.

51. $z = f(u)$, $u = x^2 - y^2$; $\partial z/\partial x = (dz/du)(\partial u/\partial x) = 2x\, dz/du$,
$\partial z/\partial y = (dz/du)(\partial u/\partial y) = -2y\, dz/du$,
$y\, \partial z/\partial x + x\, \partial z/\partial y = 2xy\, dz/du - 2xy\, dz/du = 0$.

53. (a) $\dfrac{\partial z}{\partial x} = \dfrac{dz}{du}\dfrac{\partial u}{\partial x}$, $\dfrac{\partial^2 z}{\partial x^2} = \dfrac{dz}{du}\dfrac{\partial^2 u}{\partial x^2} + \dfrac{\partial}{\partial x}(\dfrac{dz}{du})\dfrac{\partial u}{\partial x} = \dfrac{dz}{du}\dfrac{\partial^2 u}{\partial x^2} + \dfrac{d^2 z}{du^2}(\dfrac{\partial u}{\partial x})^2$;

proceed in a similar fashion for $\partial^2 z/\partial y^2$.

(b) $\dfrac{\partial^2 z}{\partial y \partial x} = \dfrac{dz}{du}\dfrac{\partial^2 u}{\partial y \partial x} + \dfrac{\partial}{\partial y}(\dfrac{dz}{du})\dfrac{\partial u}{\partial x} = \dfrac{dz}{du}\dfrac{\partial^2 u}{\partial y \partial x} + \dfrac{d^2 z}{du^2}\dfrac{\partial u}{\partial x}\dfrac{\partial u}{\partial y}$.

55. (a) $\dfrac{\partial^2 z}{\partial x^2} + \dfrac{\partial^2 z}{\partial y^2} = \dfrac{dz}{dr}\dfrac{\partial^2 r}{\partial x^2} + \dfrac{d^2 z}{dr^2}\left(\dfrac{\partial r}{\partial x}\right)^2 + \dfrac{dz}{dr}\dfrac{\partial^2 r}{\partial y^2} + \dfrac{d^2 z}{dr^2}\left(\dfrac{\partial r}{\partial y}\right)^2$

$$= \dfrac{dz}{dr}\dfrac{y^2}{r^3} + \dfrac{d^2 z}{dr^2}\dfrac{x^2}{r^2} + \dfrac{dz}{dr}\dfrac{x^2}{r^3} + \dfrac{d^2 z}{dr^2}\dfrac{y^2}{r^2}$$

$$= \dfrac{d^2 z}{dr^2} + \dfrac{1}{r}\dfrac{dz}{dr} = 0 \ \text{ so } \ r\dfrac{d^2 z}{dr^2} + \dfrac{dz}{dr} = 0.$$

(b) $r\dfrac{d^2 z}{dr^2} + \dfrac{dz}{dr} = \dfrac{d}{dr}\left(r\dfrac{dz}{dr}\right) = 0$, $r\dfrac{dz}{dr} = C_1$, $\dfrac{dz}{dr} = C_1/r$, $z = C_1 \ln r + C_2$.

57. $z = f(u) + g(v)$ where $u = y + cx$ and $v = y - cx$,

$\dfrac{\partial z}{\partial x} = \dfrac{\partial z}{\partial u}\dfrac{\partial u}{\partial x} + \dfrac{\partial z}{\partial v}\dfrac{\partial v}{\partial x} = cf'(u) - cg'(v)$ and

$\dfrac{\partial^2 z}{\partial x^2} = \dfrac{\partial}{\partial u}\left(\dfrac{\partial z}{\partial x}\right)\dfrac{\partial u}{\partial x} + \dfrac{\partial}{\partial v}\left(\dfrac{\partial z}{\partial x}\right)\dfrac{\partial v}{\partial x} = c^2 f''(u) + c^2 g''(v)$,

similarly we find that $\dfrac{\partial^2 z}{\partial y^2} = f''(u) + g''(v)$ so $\dfrac{\partial^2 z}{\partial x^2} = c^2\dfrac{\partial^2 z}{\partial y^2}$.

59. The result follows by a direct application of the chain rule.

61. (a) $1 = -r\sin\theta\,\dfrac{\partial\theta}{\partial x} + \cos\theta\,\dfrac{\partial r}{\partial x}$ and $0 = r\cos\theta\,\dfrac{\partial\theta}{\partial x} + \sin\theta\,\dfrac{\partial r}{\partial x}$;
solve for $\partial r/\partial x$ and $\partial\theta/\partial x$.

(b) $0 = -r\sin\theta\,\dfrac{\partial\theta}{\partial y} + \cos\theta\,\dfrac{\partial r}{\partial y}$ and $1 = r\cos\theta\,\dfrac{\partial\theta}{\partial y} + \sin\theta\,\dfrac{\partial r}{\partial y}$;
solve for $\partial r/\partial y$ and $\partial\theta/\partial y$.

63. Square and add the results of parts (a) and (b).

65. (a) $f(tx,ty) = 3t^2 x^2 + t^2 y^2 = t^2 f(x,y)$; $n = 2$.

(b) $f(tx,ty) = \sqrt{t^2 x^2 + t^2 y^2} = tf(x,y)$; $n = 1$.

(c) $f(tx,ty) = t^3 x^2 y - 2t^3 y^3 = t^3 f(x,y)$; $n = 3$.

(d) $f(tx,ty) = 5/(t^2 x^2 + 2t^2 y^2)^2 = t^{-4}f(x,y)$; $n = -4$.

67. (a) $\dfrac{\partial w}{\partial x} = \dfrac{\partial f}{\partial x} + \dfrac{\partial f}{\partial y}\dfrac{\partial y}{\partial x}$ (b) $\dfrac{\partial w}{\partial z} = \dfrac{\partial f}{\partial y}\dfrac{\partial y}{\partial z}$

69. Represent the line segment C that joins A and B by $x = x_0 + (x_1 - x_0)t$, $y = y_0 + (y_1 - y_0)t$ for $0 \leq t \leq 1$. $f(x,y) = F(t)$ for (x,y) on C, moreover $f(x_1,y_1) - f(x_0,y_0) = F(1) - F(0)$. Apply the mean value theorem to $F(t)$ on the interval $[0,1]$ to get
$[F(1) - F(0)]/(1 - 0) = F'(t^*)$, $F(1) - F(0) = F'(t^*)$ for some t^* in $(0,1)$ so $f(x_1,y_1) - f(x_0,y_0) = F'(t^*)$. By the chain rule,
$F'(t) = f_x(x,y)(dx/dt) + f_y(x,y)(dy/dt)$
$\qquad = f_x(x,y)(x_1 - x_0) + f_y(x,y)(y_1 - y_0)$.
Let (x^*,y^*) be the point on C for $t = t^*$ then
$f(x_1,y_1) - f(x_0,y_0) = F'(t^*) = f_x(x^*,y^*)(x_1 - x_0) + f_y(x^*,y^*)(y_1 - y_0)$.

EXERCISE SET 16.5

1. At P, $\partial z/\partial x = 48$ and $\partial z/\partial y = -14$, tangent plane $48x - 14y - z = 64$, normal line $x = 1 + 48t$, $y = -2 - 14t$, $z = 12 - t$.

3. At P, $\partial z/\partial x = 1$ and $\partial z/\partial y = -1$, tangent plane $x - y - z = 0$, normal line $x = 1 + t$, $y = -t$, $z = 1 - t$.

5. At P, $\partial z/\partial x = 0$ and $\partial z/\partial y = 3$, tangent plane $3y - z = -1$, normal line $x = \pi/6$, $y = 3t$, $z = 1 - t$.

7. By implicit differentiation $\partial z/\partial x = -x/z$, $\partial z/\partial y = -y/z$ so at P, $\partial z/\partial x = 3/4$ and $\partial z/\partial y = 0$, tangent plane $3x - 4z = -25$, normal line $x = -3 + 3t/4$, $y = 0$, $z = 4 - t$.

9. $dz = 7dx - 2dy$

11. $dz = [y/(1 + x^2y^2)]dx + [x/(1 + x^2y^2)]dy$

13. $df = (2x + 2y - 4)dx + 2xdy$; $x = 1$, $y = 2$, $dx = 0.01$, $dy = 0.04$ so $df = 0.10$

15. $df = -x^{-2}dx - y^{-2}dy$; $x = -1$, $y = -2$, $dx = -0.02$, $dy = -0.04$ so $df = 0.03$

17. The tangent plane is horizontal if the normal $\partial z/\partial x\ \mathbf{i} + \partial z/\partial y\ \mathbf{j} - \mathbf{k}$ is parallel to $\mathbf{k}$ which occurs when $\partial z/\partial x = \partial z/\partial y = 0$.

(a) $\partial z/\partial x = 3x^2 y^2$, $\partial z/\partial y = 2x^3 y$; $3x^2 y^2 = 0$ and $2x^3 y = 0$ for all (x,y) on the x-axis or y-axis, and $z = 0$ for these points, the tangent plane is horizontal at all points on the x-axis or y-axis.

(b) $\partial z/\partial x = 2x - y - 2$, $\partial z/\partial y = -x + 2y + 4$; solve the system $2x - y - 2 = 0$, $-x + 2y + 4 = 0$, to get $x = 0$, $y = -2$. $z = -4$ at $(0,-2)$, the tangent plane is horizontal at $(0,-2,-4)$.

19. $\partial z/\partial x = -6x$, $\partial z/\partial y = -4y$ so $-6x_0 \mathbf{i} - 4y_0 \mathbf{j} - \mathbf{k}$ is normal to the surface at a point (x_0, y_0, z_0) on the surface. This normal must be parallel to the given line and hence to the vector $-3\mathbf{i} + 8\mathbf{j} - \mathbf{k}$ which is parallel to the line so $-6x_0 = -3$, $x_0 = 1/2$ and $-4y_0 = 8$, $y_0 = -2$. $z = -3/4$ at $(1/2,-2)$. The point on the surface is $(1/2,-2,-3/4)$.

21. $(2,2,2\sqrt{2})$ satisfies both equations. $\mathbf{n}_1 = -(1/\sqrt{2})\mathbf{i} - (1/\sqrt{2})\mathbf{j} - \mathbf{k}$ and $\mathbf{n}_2 = (1/\sqrt{2})\mathbf{i} + (1/\sqrt{2})\mathbf{j} - \mathbf{k}$ are normal, respectively, to each of the surfaces at $(2,2,2\sqrt{2})$. $\mathbf{n}_1 \cdot \mathbf{n}_2 = 0$ so the normals are perpendicular and hence so are the tangent planes.

23. $z = \sqrt{x^2 + y^2}$, $dz = x(x^2 + y^2)^{-1/2}dx + y(x^2 + y^2)^{-1/2}dy$; $x = 3$, $y = 4$, $dx = 0.2$, $dy = -0.04$ so $dz = 0.088$ cm.

25. $A = xy$, $dA = ydx + xdy$, $dA/A = dx/x + dy/y$, $|dx/x| \le 0.03$ and $|dy/y| \le 0.05$, $|dA/A| \le |dx/x| + |dy/y| \le 0.08 = 8\%$

27. $z = \sqrt{x^2 + y^2}$, $dz = \dfrac{x}{\sqrt{x^2 + y^2}}\ dx + \dfrac{y}{\sqrt{x^2 + y^2}}\ dy$,

$\dfrac{dz}{z} = \dfrac{x}{x^2 + y^2}\ dx + \dfrac{y}{x^2 + y^2}\ dy = \dfrac{x^2}{x^2 + y^2}(\dfrac{dx}{x}) + \dfrac{y^2}{x^2 + y^2}(\dfrac{dy}{y})$,

$\left|\dfrac{dz}{z}\right| \le \dfrac{x^2}{x^2 + y^2}\left|\dfrac{dx}{x}\right| + \dfrac{y^2}{x^2 + y^2}\left|\dfrac{dy}{y}\right|$, if $\left|\dfrac{dx}{x}\right| \le r/100$ and $\left|\dfrac{dy}{y}\right| \le r/100$

then $\left|\dfrac{dz}{z}\right| \le \dfrac{x^2}{x^2 + y^2}(r/100) + \dfrac{y^2}{x^2 + y^2}(r/100) = r/100$ so the percentage error in z is at most r%.

29. $dR = \dfrac{R_2^2}{(R_1 + R_2)^2} dR_1 + \dfrac{R_1^2}{(R_1 + R_2)^2} dR_2$,

$\dfrac{dR}{R} = \dfrac{R_2}{R_1 + R_2} \left(\dfrac{dR_1}{R_1}\right) + \dfrac{R_1}{R_1 + R_2} \left(\dfrac{dR_2}{R_2}\right)$;

$\left|\dfrac{dR}{R}\right| \leq \dfrac{R_2}{R_1 + R_2} \left|\dfrac{dR_1}{R_1}\right| + \dfrac{R_1}{R_1 + R_2} \left|\dfrac{dR_2}{R_2}\right|$; if $R_1 = 200$, $R_2 = 400$,

$\left|dR_1/R_1\right| \leq 0.02$, and $\left|dR_2/R_2\right| \leq 0.02$ then

$\left|dR/R\right| \leq (400/600)(0.02) + (200/600)(0.02) = 0.02 = 2\%$.

31. $d\theta = \dfrac{1}{\sqrt{c^2 - a^2}} da - \dfrac{a}{c\sqrt{c^2 - a^2}} dc$; if $a = 3$, $c = 5$, $\left|da\right| \leq 0.01$, and

$\left|dc\right| \leq 0.01$ then $\left|d\theta\right| \leq (1/4)(0.01) + (3/20)(0.01) = 0.004$ radians.

33. $dT = \dfrac{\pi}{g\sqrt{L/g}} dL - \dfrac{\pi L}{g^2 \sqrt{L/g}} dg$, $\dfrac{dT}{T} = \dfrac{1}{2}\dfrac{dL}{L} - \dfrac{1}{2}\dfrac{dg}{g}$;

$\left|dL/L\right| \leq 0.005$ and $\left|dg/g\right| \leq 0.001$ so

$\left|dT/T\right| \leq (1/2)(0.005) + (1/2)(0.001) = 0.003 = 0.3\%$

35. (a) $z = xy$, $dz = y\,dx + x\,dy$, $dz/z = dx/x + dy/y$; $(r + s)\%$.

(b) $z = x/y$, $dz = dx/y - x\,dy/y^2$, $dz/z = dx/x - dy/y$; $(r + s)\%$.

(c) $z = x^2 y^3$, $dz = 2xy^3 dx + 3x^2 y^2 dy$, $dz/z = 2dx/x + 3dy/y$; $(2r + 3s)\%$.

(d) $z = x^3 y^{1/2}$, $dz = 3x^2 y^{1/2} dx + x^3 dy/(2y^{1/2})$, $dz/z = 3dx/x + (1/2)dy/y$; $(3r + s/2)\%$.

37. (a) $2t + 7 = (-1 + t)^2 + (2 + t)^2$, $t^2 = 1$, $t = \pm 1$ so the points of intersection are $(-2,1,5)$ and $(0,3,9)$.

(b) $\partial z/\partial x = 2x$, $\partial z/\partial y = 2y$ so at $(-2,1,5)$ the vector $\mathbf{n} = -4\mathbf{i} + 2\mathbf{j} - \mathbf{k}$ is normal to the surface. $\mathbf{v} = \mathbf{i} + \mathbf{j} + 2\mathbf{k}$ is parallel to the line. $\mathbf{n} \cdot \mathbf{v} = -4$ so the cosine of the acute angle is

$[\mathbf{n} \cdot (-\mathbf{v})]/(\|\mathbf{n}\| \|-\mathbf{v}\|) = 4/(\sqrt{21}\sqrt{6}) = 4/(3\sqrt{14})$. Similarly, at $(0,3,9)$ the vector $\mathbf{n} = 6\mathbf{j} - \mathbf{k}$ is normal to the surface,

$\mathbf{n} \cdot \mathbf{v} = 4$ so the cosine of the acute angle is $4/(\sqrt{37}\sqrt{6}) = 4/\sqrt{222}$.

39. Use implicit differentiation to get $\partial z/\partial x = -c^2 x/(a^2 z)$, $\partial z/\partial y = -c^2 y/(b^2 z)$. At (x_0, y_0, z_0), $z_0 \neq 0$, a normal to the surface is $-[c^2 x_0/(a^2 z_0)]\mathbf{i} - [c^2 y_0/(b^2 z_0)]\mathbf{j} - \mathbf{k}$ so the tangent plane is

$$-\frac{c^2 x_0}{a^2 z_0} x - \frac{c^2 y_0}{b^2 z_0} y - z = -\frac{c^2 x_0^2}{a^2 z_0} - \frac{c^2 y_0^2}{b^2 z_0} - z_0,$$

$$\frac{x_0 x}{a^2} + \frac{y_0 y}{b^2} - \frac{z_0 z}{c^2} = \frac{x_0^2}{a^2} + \frac{y_0^2}{b^2} + \frac{z_0^2}{c^2} = 1$$

41. $\mathbf{n}_1 = f_x(x_0, y_0)\mathbf{i} + f_y(x_0, y_0)\mathbf{j} - \mathbf{k}$ and $\mathbf{n}_2 = g_x(x_0, y_0)\mathbf{i} + g_y(x_0, y_0)\mathbf{j} - \mathbf{k}$ are normal, respectively, to $z = f(x,y)$ and $z = g(x,y)$ at P. $\mathbf{n}_1$ and $\mathbf{n}_2$ are perpendicular if and only if $\mathbf{n}_1 \cdot \mathbf{n}_2 = 0$,

$f_x(x_0, y_0)g_x(x_0, y_0) + f_y(x_0, y_0)g_y(x_0, y_0) + 1 = 0$,
$f_x(x_0, y_0)g_x(x_0, y_0) + f_y(x_0, y_0)g_y(x_0, y_0) = -1$.

EXERCISE SET 16.6

1. $\nabla z = 4\mathbf{i} - 8\mathbf{j}$

3. $\nabla z = \frac{x}{x^2 + y^2}\mathbf{i} + \frac{y}{x^2 + y^2}\mathbf{j}$

5. $\nabla f(x,y) = 3(2x + y)(x^2 + xy)^2\mathbf{i} + 3x(x^2 + xy)^2\mathbf{j}$
   $\nabla f(-1,-1) = -36\mathbf{i} - 12\mathbf{j}$

7. $\nabla f(x,y) = [y/(x + y)]\mathbf{i} + [y/(x + y) + \ln(x + y)]\mathbf{j}$,
   $\nabla f(-3,4) = 4\mathbf{i} + 4\mathbf{j}$

9. $\nabla f(x,y) = (3y/2)(1 + xy)^{1/2}\mathbf{i} + (3x/2)(1 + xy)^{1/2}\mathbf{j}$,
   $\nabla f(3,1) = 3\mathbf{i} + 9\mathbf{j}$, $D_{\mathbf{u}}f = \nabla f \cdot \mathbf{u} = 12/\sqrt{2} = 6\sqrt{2}$

11. $\nabla f(x,y) = [2x/(1 + x^2 + y)]\mathbf{i} + [1/(1 + x^2 + y)]\mathbf{j}$, $\nabla f(0,0) = \mathbf{j}$,
    $D_{\mathbf{u}}f = -3/\sqrt{10}$

13. $\nabla f(x,y) = 12x^2y^2i + 8x^3yj$, $\nabla f(2,1) = 48i + 64j$,
     $u = (4/5)i - (3/5)j$, $D_uf = \nabla f \cdot u = 0$

15. $\nabla f(x,y) = (y^2/x)i + 2y \ln x \; j$, $\nabla f(1,4) = 16i$,
     $u = (-i + j)/\sqrt{2}$, $D_uf = -8\sqrt{2}$

17. $\nabla f(x,y) = -[y/(x^2 + y^2)]i + [x/(x^2 + y^2)]j$,
     $\nabla f(-2,2) = -(i + j)/4$, $u = -(i + j)/\sqrt{2}$, $D_uf = \sqrt{2}/4$

19. $\nabla f(x,y) = (y/2)(xy)^{-1/2}i + (x/2)(xy)^{-1/2}j$, $\nabla f(1,4) = i + (1/4)j$,
     $u = \cos\theta \; i + \sin\theta \; j = (1/2)i + (\sqrt{3}/2)j$, $D_uf = 1/2 + \sqrt{3}/8$

21. $\nabla f(x,y) = 2 \sec^2(2x + y)i + \sec^2(2x + y)j$, $\nabla f(\pi/6,\pi/3) = 8i + 4j$,
     $u = (i - j)/\sqrt{2}$, $D_uf = 2\sqrt{2}$

23. $f(1,2) = 3$, level curve
     $4x - 2y + 3 = 3$,
     $4x - 2y = 0$.
     $\nabla f(x,y) = 4i - 2j$

25. $f(-2,0) = 4$, level curve
     $x^2 + 4y^2 = 4$, $x^2/4 + y^2 = 1$.
     $\nabla f(x,y) = 2xi + 8yj$
     $\nabla f(-2,0) = -4i$

27. $\nabla f(x,y) = 12x^2y^2i + 8x^3yj$, $\nabla f(-1,1) = 12i - 8j$,
     $u = (3i - 2j)/\sqrt{13}$, $\|\nabla f(-1,1)\| = 4\sqrt{13}$

29. $\nabla f(x,y) = x(x^2 + y^2)^{-1/2}i + y(x^2 + y^2)^{-1/2}j$, $\nabla f(4,-3) = (4i - 3j)/5$,
    $u = (4i - 3j)/5$, $\|\nabla f(4,-3)\| = 1$

31. $\nabla f(x,y) = -2xi - 2yj$, $\nabla f(-1,-3) = 2i + 6j$, $u = -(i + 3j)/\sqrt{10}$,
    $-\|\nabla f(-1,-3)\| = -2\sqrt{10}$

33. $\nabla f(x,y) = -3 \sin(3x - y)i + \sin(3x - y)j$,
    $\nabla f(\pi/6, \pi/4) = (-3i + j)/\sqrt{2}$, $u = (3i - j)/\sqrt{10}$, $-\|\nabla f(\pi/6, \pi/4)\| = -\sqrt{5}$

35. $\nabla f(x,y) = y(x + y)^{-2}i - x(x + y)^{-2}j$, $\nabla f(1,0) = -j$, $\overrightarrow{PQ} = -2i - j$,
    $u = (-2i - j)/\sqrt{5}$, $D_u f = 1/\sqrt{5}$

37. $\nabla f(x,y) = \dfrac{ye^y}{2\sqrt{xy}} i + (\sqrt{xy} \, e^y + \dfrac{xe^y}{2\sqrt{xy}})j$, $\nabla f(1,1) = (e/2)(i + 3j)$,
    $u = -j$, $D_u f = -3e/2$

39. $\nabla f(x,y) = 8xyi + 4x^2 j$, $\nabla f(1,-2) = -16i + 4j$ is perpendicular to the
    level curve through P so $u = \pm(-4i + j)/\sqrt{17}$.

41. Solve the system $(3/5)f_x(1,2) - (4/5)f_y(1,2) = -5$,
    $(4/5)f_x(1,2) + (3/5)f_y(1,2) = 10$ for $f_x(1,2)$ and $f_y(1,2)$ to get
    $f_x(1,2) = 5$, $f_y(1,2) = 10$.   For (c), $\nabla f(1,2) = 5i + 10j$,
    $u = (-i - 2j)/\sqrt{5}$, $D_u f = -5\sqrt{5}$.

43. $\nabla f(4,-5) = 2i - j$, $u = (5i + 2j)/\sqrt{29}$, $D_u f = 8/\sqrt{29}$

45. $\nabla z = 6xi - 2yj$, $\|\nabla z\| = \sqrt{36x^2 + 4y^2} = 6$ if $36x^2 + 4y^2 = 36$; all points on
    the ellipse $9x^2 + y^2 = 9$.

47. $r = ti - t^2 j$, $dr/dt = i - 2tj = i - 4j$ at the point $(2,-4)$,
    $u = (i - 4j)/\sqrt{17}$; $dz/ds = D_u z = \nabla z \cdot u = 36/\sqrt{17}$.

49.   (a) $\nabla V(x,y) = -2e^{-2x}\cos 2y\ i - 2e^{-2x}\sin 2y\ j$,

$E = -\nabla V(\pi/4,0) = 2e^{-\pi/2}i$

(b) $V(x,y)$ decreases most rapidly in the direction of $-\nabla V(x,y)$ which is E.

51.   (a) $\nabla r = \dfrac{x}{\sqrt{x^2 + y^2}}\ i + \dfrac{y}{\sqrt{x^2 + y^2}}\ j = r/r$

(b) $\nabla f(r) = \dfrac{\partial f(r)}{\partial x}\ i + \dfrac{\partial f(r)}{\partial y}\ j = f'(r)\dfrac{\partial r}{\partial x}\ i + f'(r)\dfrac{\partial r}{\partial y}\ j = f'(r)\nabla r$

53. $u_r = \cos\theta\ i + \sin\theta\ j$, $u_\theta = -\sin\theta\ i + \cos\theta\ j$,

$\nabla z = \dfrac{\partial z}{\partial x}\ i + \dfrac{\partial z}{\partial y}\ j$

$= (\dfrac{\partial z}{\partial r}\cos\theta - \dfrac{1}{r}\dfrac{\partial z}{\partial\theta}\sin\theta)i + (\dfrac{\partial z}{\partial r}\sin\theta + \dfrac{1}{r}\dfrac{\partial z}{\partial x}\cos\theta)j$

$= \dfrac{\partial z}{\partial r}(\cos\theta\ i + \sin\theta\ j) + \dfrac{1}{r}\dfrac{\partial z}{\partial\theta}(-\sin\theta\ i + \cos\theta\ j)$

$= \dfrac{\partial z}{\partial r}\ u_r + \dfrac{1}{r}\dfrac{\partial z}{\partial\theta}\ u_\theta$

55. $dz/dt = (\partial z/\partial x)(dx/dt) + (\partial z/\partial y)(dy/dt)$
$= (\partial z/\partial x\ i + \partial z/\partial y\ j)\cdot(dx/dt\ i + dy/dt\ j) = \nabla z\cdot r'(t)$

57.   Let u_1 and u_2 be nonparallel unit vectors for which the directional derivative is zero. Let u be any other unit vector, then $u = c_1 u_1 + c_2 u_2$ for some choice of scalars c_1 and c_2,

$D_u f(x,y) = \nabla f(x,y)\cdot u = c_1\nabla f(x,y)\cdot u_1 + c_2\nabla f(x,y)\cdot u_2$

$= c_1 D_{u_1}f(x,y) + c_2 D_{u_2}f(x,y) = 0.$

EXERCISE SET 16.7

1. $165t^{32}$ 3. $-2t\cos(t^2)$ 5.   3264

7. $\nabla f(x,y,z) = 20x^4 y^2 z^3 i + 8x^5 yz^3 j + 12x^5 y^2 z^2 k$,

$\nabla f(2,-1,1) = 320i - 256j + 384k$, $D_u f = -320$

9. $\nabla f(x,y,z) = \dfrac{2x}{x^2 + 2y^2 + 3z^2}\, i + \dfrac{4y}{x^2 + 2y^2 + 3z^2}\, j + \dfrac{6z}{x^2 + 2y^2 + 3z^2}\, k$,

$\nabla f(-1,2,4) = (-2/57)i + (8/57)j + (24/57)k$, $D_u f = -314/741$

11. $\nabla f(x,y,z) = (3x^2 z - 2xy)i - x^2 j + (x^3 + 2z)k$,

$\nabla f(2,-1,1) = 16i - 4j + 10k$, $u = (3i - j + 2k)/\sqrt{14}$, $D_u f = 72/\sqrt{14}$

13. $\nabla f(x,y,z) = -\dfrac{1}{z + y}\, i - \dfrac{z - x}{(z + y)^2}\, j + \dfrac{y + x}{(z + y)^2}\, k$,

$\nabla f(1,0,-3) = (1/3)i + (4/9)j + (1/9)k$, $u = (-6i + 3j - 2k)/7$,
$D_u f = -8/63$

15. $\nabla f(1,1,-1) = 3i - 3j$, $u = (i - j)/\sqrt{2}$, $\|\nabla f(1,1,-1)\| = 3\sqrt{2}$

17. $\nabla f(1,2,-2) = (-i + j)/2$, $u = (-i + j)/\sqrt{2}$, $\|\nabla f(1,2,-2)\| = 1/\sqrt{2}$

19. $\nabla f(5,7,6) = -i + 11j - 12k$, $u = (i - 11j + 12k)/\sqrt{266}$,
$-\|\nabla f(5,7,6)\| = -\sqrt{266}$

21. $\nabla f(2,1,-1) = -i + j - k$, $\overrightarrow{PQ} = -3i + j + k$, $u = (-3i + j + k)/\sqrt{11}$,
$D_u f = 3/\sqrt{11}$

23. Let u be the unit vector in the direction of a, then
$D_u f(3,-2,1) = \nabla f(3,-2,1) \cdot u = \|\nabla f(3,-2,1)\|\cos\theta = 5\cos\theta = -5$,
$\cos\theta = -1$, $\theta = \pi$ so $\nabla f(3,-2,1)$ is oppositely directed to u;
$\nabla f(3,-2,1) = -5u = -10/3i + 5/3j + 10/3k$.

25. $f(x,y,z) = x^2 + y^2 + z^2$, $\nabla f(-3,2,-6) = -2(3i - 2j + 6k)$; tangent plane
$3x - 2y + 6z = -49$; normal line $x = -3 + 3t$, $y = 2 - 2t$, $z = -6 + 6t$

27. $f(x,y,z) = \sqrt{\dfrac{z + x}{y - 1}} - z^2$, $\nabla f(3,5,1) = (i - j - 15k)/8$; tangent plane
$x - y - 15z = -17$; normal line $x = 3 + t$, $y = 5 - t$, $z = 1 - 15t$

29. $f(x,y,z) = x^2 + y^2 + z^2$, if (x_0,y_0,z_0) is on the sphere then
$\nabla f(x_0,y_0,z_0) = 2(x_0 i + y_0 j + z_0 k)$ is normal to the sphere at (x_0,y_0,z_0),

the normal line is $x = x_0 + x_0 t$, $y = y_0 + y_0 t$, $z = z_0 + z_0 t$ which passes through the origin when $t = -1$.

31. $f(x,y,z) = x^2 + y^2 - z^2$, if (x_0, y_0, z_0) is on the surface then

$\nabla f(x_0, y_0, z_0) = 2(x_0 i + y_0 j - z_0 k)$ is normal there and hence so is

$n_1 = x_0 i + y_0 j - z_0 k$. n_1 must be parallel to $\vec{PQ} = 3i + 2j - 2k$ so

$n_1 = c\vec{PQ}$ for some constant c. Equate components to get $x_0 = 3c$, $y_0 = 2c$

and $z_0 = 2c$ which when substituted into the equation of the surface

yields $9c^2 + 4c^2 - 4c^2 = 1$, $c^2 = 1/9$, $c = \pm 1/3$ so the points are
(1,2/3,2/3) and (-1,-2/3,-2/3).

33. $dw = 8dx - 3dy + 4dz$

35. $dw = \dfrac{yz}{1 + x^2 y^2 z^2} dx + \dfrac{xz}{1 + x^2 y^2 z^2} dy + \dfrac{xy}{1 + x^2 y^2 z^2} dz$

37. $df = 2y^2 z^3 dx + 4xyz^3 dy + 6xy^2 z^2 dz$
$= 2(-1)^2 (2)^3 (-0.01) + 4(1)(-1)(2)^3 (-0.02) + 6(1)(-1)^2 (2)^2 (0.02)$
$= 0.96$

39. $V = \ell wh$, $dV = whd\ell + \ell hdw + \ell wdh$,
$|dV| \le wh|d\ell| + \ell h|dw| + \ell w|dh|$
$\le (4)(5)(0.05) + (3)(5)(0.05) + (3)(4)(0.05) = 2.35 \text{ cm}^3$

41. $dA = \dfrac{1}{2} b \sin\theta \, da + \dfrac{1}{2} a \sin\theta \, db + \dfrac{1}{2} ab \cos\theta \, d\theta$,

$|dA| \le \dfrac{1}{2} b \sin\theta |da| + \dfrac{1}{2} a \sin\theta |db| + \dfrac{1}{2} ab |\cos\theta| |d\theta|$

$\le \dfrac{1}{2} (50)(1/2)(1/2) + \dfrac{1}{2} (40)(1/2)(1/4) + \dfrac{1}{2} (40)(50)(\sqrt{3}/2)(\pi/90)$

$= 35/4 + 50\pi\sqrt{3}/9 \approx 39 \text{ ft}^2$

43. $dw = y^2 z^3 dx + 2xyz^3 dy + 3xy^2 z^2 dz$,
$|dw/w| \le |dx/x| + 2|dy/y| + 3|dz/z|$
$\le (0.01) + 2(0.02) + 3(0.03) = 0.14 = 14\%$

45. $f(x,y,z) = x^2 + y^2 + z^2 - a^2$, $g(x,y,z) = z^2 - x^2 - y^2$. If (x_0,y_0,z_0) is
a point of intersection then
$$f_x g_x + f_y g_y + f_z g_z = (2x_0)(-2x_0) + (2y_0)(-2y_0) + (2z_0)(2z_0)$$
$$= 4(z_0^2 - x_0^2 - y_0^2),$$
but $z_0^2 - x_0^2 - y_0^2 = 0$ because (x_0,y_0,z_0) is on $z^2 = x^2 + y^2$.

EXERCISE SET 16.8

1. $\partial f/\partial v = 8vw^3 x^4 y^5$, $\partial f/\partial w = 12v^2 w^2 x^4 y^5$, $\partial f/\partial x = 16v^2 w^3 x^3 y^5$,
$\partial f/\partial y = 20v^2 w^3 x^4 y^4$

3. $\partial f/\partial v_1 = 2v_1/(v_3^2 + v_4^2)$, $\partial f/\partial v_2 = -2v_2/(v_3^2 + v_4^2)$,
$\partial f/\partial v_3 = -2v_3(v_1^2 - v_2^2)/(v_3^2 + v_4^2)^2$, $\partial f/\partial v_4 = -2v_4(v_1^2 - v_2^2)/(v_3^2 + v_4^2)^2$

5. 128, –512, 32, 64/3   7. $210t^{29}$

9. $\partial z/\partial r = (dz/dx)(\partial x/\partial r) = 2r\cos^2\theta/(r^2\cos^2\theta + 1)$,
$\partial z/\partial\theta = (dz/dx)(\partial x/\partial\theta) = -2r^2\sin\theta\cos\theta/(r^2\cos^2\theta + 1)$

11. $\partial w/\partial\rho = 2\rho(4\sin^2\phi + \cos^2\phi)$, $\partial w/\partial\phi = 6\rho^2\sin\phi\cos\phi$, $\partial w/\partial\theta = 0$

13. $\dfrac{dw}{dy} = \dfrac{\partial w}{\partial x}\dfrac{dx}{dy} + \dfrac{\partial w}{\partial y} + \dfrac{\partial w}{\partial z}\dfrac{dz}{dy} = (-2\sin 2y\cos 2y + y + 1/2)/\sqrt{\cos^2 2y + y^2 + y}$

15. Let a, b, and c be the lengths of the sides opposite angles A, B, and C,
respectively. By the law of cosines
$$a = (b^2 + c^2 - 2bc\cos A)^{1/2}$$
$$\frac{da}{dt} = \frac{b - c\cos A}{a}\frac{db}{dt} + \frac{c - b\cos A}{a}\frac{dc}{dt} + \frac{bc\sin A}{a}\frac{dA}{dt}$$
$$= \frac{10 - 10}{10\sqrt{3}}(4) + \frac{20 - 5}{10\sqrt{3}}(2) + \frac{100\sqrt{3}}{10\sqrt{3}}(\pi/60)$$
$$= \sqrt{3} + \pi/6 \text{ cm/sec, increasing}$$

17. $2x + 4x\,\partial z/\partial x + 4z + 2z\,\partial z/\partial x - 3y\,\partial z/\partial x = 0,$
$\partial z/\partial x = -(2x + 4z)/(4x + 2z - 3y);$
$4x\,\partial z/\partial y + 2z\,\partial z/\partial y - 3y\,\partial z/\partial y - 3z = 0, \quad \partial z/\partial y = 3z/(4x + 2z - 3y)$

19. $f_{ww} = 0,\ f_{xx} = -2wxyz/(x^2 + y^2)^2,\ f_{yy} = 2wxyz/(x^2 + y^2)^2,\ f_{zz} = 0$

21. Let $z = f(u)$ where $u = x^2 + y^2$ then $\partial z/\partial x = (dz/du)(\partial u/\partial x) = 2x\,dz/du$,
$\partial z/\partial y = (dz/du)(\partial u/\partial y) = 2y\,dz/du$ so
$y\,\partial z/\partial x - x\,\partial z/\partial y = 2xy\,dz/du - 2xy\,dz/du = 0$

23. Let $w = f(r,s,t)$ where $r = x - y,\ s = y - z,\ t = z - x;$
$\partial w/\partial x = (\partial w/\partial r)(\partial r/\partial x) + (\partial w/\partial t)(\partial t/\partial x) = \partial w/\partial r - \partial w/\partial t,$
similarly $\partial w/\partial y = -\partial w/\partial r + \partial w/\partial s$ and
$\partial w/\partial z = -\partial w/\partial s + \partial w/\partial t$ so $\partial w/\partial x + \partial w/\partial y + \partial w/\partial z = 0$

25. $\dfrac{\partial F}{\partial x} + \dfrac{\partial F}{\partial z}\dfrac{\partial z}{\partial x} = 0$ so $\dfrac{\partial z}{\partial x} = -\dfrac{\partial F/\partial x}{\partial F/\partial z},\ \dfrac{\partial F}{\partial y} + \dfrac{\partial F}{\partial z}\dfrac{\partial z}{\partial y} = 0$ so $\dfrac{\partial z}{\partial y} = -\dfrac{\partial F/\partial y}{\partial F/\partial z}.$

27. $ye^x - 5\sin 3z - 3z = 0;\ \dfrac{\partial z}{\partial x} = -\dfrac{ye^x}{-15\cos 3z - 3} = \dfrac{ye^x}{15\cos 3z + 3},$
$\dfrac{\partial z}{\partial y} = \dfrac{e^x}{15\cos 3z + 3}.$

29. $\nabla f(u,v,w) = \dfrac{\partial f}{\partial x}\mathbf{i} + \dfrac{\partial f}{\partial y}\mathbf{j} + \dfrac{\partial f}{\partial z}\mathbf{k}$
$= \left(\dfrac{\partial f}{\partial u}\dfrac{\partial u}{\partial x} + \dfrac{\partial f}{\partial v}\dfrac{\partial v}{\partial x} + \dfrac{\partial f}{\partial w}\dfrac{\partial w}{\partial x}\right)\mathbf{i}$
$+ \left(\dfrac{\partial f}{\partial u}\dfrac{\partial u}{\partial y} + \dfrac{\partial f}{\partial v}\dfrac{\partial v}{\partial y} + \dfrac{\partial f}{\partial w}\dfrac{\partial w}{\partial y}\right)\mathbf{j} + \left(\dfrac{\partial f}{\partial u}\dfrac{\partial u}{\partial z} + \dfrac{\partial f}{\partial v}\dfrac{\partial v}{\partial z} + \dfrac{\partial f}{\partial w}\dfrac{\partial w}{\partial z}\right)\mathbf{k}$
$= \dfrac{\partial f}{\partial u}\nabla u + \dfrac{\partial f}{\partial v}\nabla v + \dfrac{\partial f}{\partial w}\nabla w$

31. $w_r = e^r/(e^r + e^s + e^t + e^u),\ w_{rs} = -e^r e^s/(e^r + e^s + e^t + e^u)^2$
$w_{rst} = 2e^r e^s e^t/(e^r + e^s + e^t + e^u)^3$
$w_{rstu} = -6e^r e^s e^t e^u/(e^r + e^s + e^t + e^u)^4$
$= -6e^{r+s+t+u}/e^{4w} = -6e^{r+s+t+u-4w}$

33.   (a) $dw/dt = \sum\limits_{i=1}^{4} (\partial w/\partial x_i)(dx_i/dt)$

   (b) $\partial w/\partial v_j = \sum\limits_{i=1}^{4} (\partial w/\partial x_i)(\partial x_i/\partial v_j)$ for $j = 1,2,3$

35. $dF/dx = (\partial F/\partial u)(du/dx) + (\partial F/\partial v)(dv/dx)$
      $= -f(u)a'(x) + f(v)b'(x) = f(b(x))b'(x) - f(a(x))a'(x)$

EXERCISE SET 16.9

1. $f_x = 6x + 2y = 0$, $f_y = 2x + 2y = 0$; critical point $(0,0)$; $D > 0$ and $f_{xx} > 0$ at $(0,0)$, relative minimum.

3. $f_x = y + 2 = 0$, $f_y = 2y + x + 3 = 0$; critical point $(1,-2)$; $D < 0$ at $(1,-2)$, saddle point.

5. $f_x = 2x + y - 3 = 0$, $f_y = x + 2y = 0$; critical point $(2,-1)$; $D > 0$ and $f_{xx} > 0$ at $(2,-1)$, relative minimum.

7. $f_x = 2x - 2xy = 0$, $f_y = 4y - x^2 = 0$; critical points $(0,0)$ and $(\pm 2,1)$; $D > 0$ and $f_{xx} > 0$ at $(0,0)$, relative minimum; $D < 0$ at $(\pm 2,1)$, saddle points.

9. $f_x = 2x - 2/(x^2 y) = 0$, $f_y = 2y - 2/(xy^2) = 0$; critical points $(-1,-1)$ and $(1,1)$; $D > 0$ and $f_{xx} > 0$ at $(-1,-1)$ and $(1,1)$, relative minima.

11. $f_x = 2x = 0$, $f_y = 1 - e^y = 0$; critical point $(0,0)$; $D < 0$ at $(0,0)$, saddle point.

13. $f_x = e^x \sin y = 0$, $f_y = e^x \cos y = 0$, $\sin y = \cos y = 0$ is impossible, no critical points.

15. $f_x = 2y^2 - 2xy + 4y = 0$, $f_y = 4xy - x^2 + 4x = 0$; $2y(y - x + 2) = 0$ and $x(4y - x + 4) = 0$, critical points $(0,0)$, $(0,-2)$, $(4,0)$, and $(4/3,-2/3)$; $D < 0$ at $(0,0)$, $(0,-2)$, and $(4,0)$, saddle points; $D > 0$ and $f_{xx} > 0$ at $(4/3,-2/3)$, relative minimum.

17. $f_x = -2(x + 1)e^{-(x^2+y^2+2x)} = 0$, $f_y = -2ye^{-(x^2+y^2+2x)} = 0$; critical point $(-1,0)$; $D > 0$ and $f_{xx} < 0$ at $(-1,0)$, relative maximum.

19. $f_x = \cos x = 0$, $f_y = \cos y = 0$; critical point $(\pi/2,\pi/2)$; $D > 0$ and $f_{xx} < 0$ at $(\pi/2,\pi/2)$, relative maximum.

21. (a) critical point $(0,0)$; $D = 0$
    (b) $f(0,0) = 0$, $x^4 + y^4 \geq 0$ so $f(x,y) \geq f(0,0)$, relative minimum.

23. (a) $f_x = 3e^y - 3x^2 = 3(e^y - x^2) = 0$,

    $f_y = 3xe^y - 3e^{3y} = 3e^y(x - e^{2y}) = 0$, $e^y = x^2$ and $e^{2y} = x$, $x^4 = x$,

    $x(x^3 - 1) = 0$ so $x = 0,1$; critical point $(1,0)$; $D > 0$ and $f_{xx} < 0$ at $(1,0)$, relative maximum.

    (b) $\lim_{x \to \infty} f(x,0) = \lim_{x \to \infty} (3x - x^3 - 1) = +\infty$ so no absolute maximum.

25. $f_x = y - 1 = 0$, $f_y = x - 3 = 0$; critical point $(3,1)$.
    Along $y = 0$: $u(x) = -x$; no critical points,
    along $x = 0$: $v(x) = -3y$; no critical points,
    along $y = -\dfrac{4}{5}x + 4$: $w(x) = -\dfrac{4}{5}x^2 + \dfrac{27}{5}x - 12$; critical point $(27/8,13/10)$.

(x,y)	$(3,1)$	$(0,0)$	$(5,0)$	$(0,4)$	$(27/8,13/10)$
$f(x,y)$	-3	0	-5	-12	$-231/80$

Absolute maxium value is 0, absolute minimum value is -12.

27. $f_x = 2x - 2 = 0$, $f_y = -6y + 6 = 0$; critical point $(1,1)$.

Along $y = 0$: $u_1(x) = x^2 - 2x$; critical point $(1,0)$,

along $y = 2$: $u_2(x) = x^2 - 2x$; critical point $(1,2)$

along $x = 0$: $v_1(x) = -3y^2 + 6y$; critical point $(0,1)$,

along $x = 2$: $v_2(x) = -3y^2 + 6y$; critical point $(2,1)$

(x,y)	$(1,1)$	$(1,0)$	$(1,2)$	$(0,1)$	$(2,1)$	$(0,0)$	$(0,2)$	$(2,0)$	$(2,2)$
$f(x,y)$	2	-1	-1	3	3	0	0	0	0

Absolute maxium value is 3, absolute minimum value is -1.

29. $f_x = 2x - 1 = 0$, $f_y = 4y = 0$; critical point $(1/2,0)$.

Along $x^2 + y^2 = 4$: $y^2 = 4 - x^2$, $u(x) = 8 - x - x^2$ for $-2 \leq x \leq 2$;
critical points $(-1/2, \pm\sqrt{15}/2)$.

(x,y)	$(1/2,0)$	$(-1/2,\sqrt{15}/2)$	$(-1/2,-\sqrt{15}/2)$	$(-2,0)$	$(2,0)$
$f(x,y)$	-1/4	33/4	33/4	6	2

Absolute maximum value is 33/4, absolute minimum value is -1/4.

31. $f(x,y) = (y - x)^2$; if (x_0,y_0) is on the line $y = x$ then $f(x_0,y_0) = 0$,
but $(y - x)^2 \geq 0$ so $f(x,y) \geq f(x_0,y_0)$ for all points (x,y) thus f has an
absolute minimum at (x_0,y_0).

33. Minimize $S = x^2 + y^2 + z^2$ subject to $x + y + z = 27$, $x > 0$, $y > 0$, $z > 0$.

$z = 27 - x - y$ so $S = x^2 + y^2 + (27 - x - y)^2$, $S_x = 4x + 2y - 54 = 0$,

$S_y = 2x + 4y - 54 = 0$; critical point $(9,9)$; $S_{xx}S_{yy} - S_{xy}^2 > 0$ and
$S_{xx} > 0$ at $(9,9)$, relative minimum. $z = 9$ when $x = y = 9$, the sum of
the squares is minimum for the numbers $9,9,9$.

35. Minimize $w = D^2 = x^2 + y^2 + z^2$ subject to $x^2 - yz = 5$. $x^2 = 5 + yz$ so $w = 5 + yz + y^2 + z^2$, $w_y = z + 2y = 0$, $w_z = y + 2z = 0$; critical point when $y = z = 0$; $w_{yy}w_{zz} - w_{yz}^2 > 0$ and $w_{yy} > 0$ when $y = z = 0$, relative minimum. $x^2 = 5$, $x = \pm\sqrt{5}$ when $y = z = 0$. The points $(\pm\sqrt{5},0,0)$ are closest to the origin.

37. Maximize $V = xyz$ subject to $x + y + z = 1$, $x > 0$, $y > 0$, $z > 0$. $z = 1 - x - y$ so $V = xy - x^2y - xy^2$, $V_x = y(1 - 2x - y) = 0$, $V_y = x(1 - x - 2y) = 0$, $1 - 2x - y = 0$ and $1 - x - 2y = 0$; critical point $(1/3,1/3)$; $V_{xx}V_{yy} - V_{xy}^2 > 0$ and $V_{xx} < 0$ at $(1/3,1/3)$, relative maximum. The maximum volume is $V = (1/3)(1/3)(1/3) = 1/27$.

39. Let x, y, and z be, respectively, the length, width, and height of the box. Minimize $C = 10(2xy) + 5(2xz + 2yz) = 10(2xy + xz + yz)$ subject to $xyz = 16$. $z = 16/(xy)$ so $C = 20(xy + 8/y + 8/x)$, $C_x = 20(y - 8/x^2) = 0$, $C_y = 20(x - 8/y^2) = 0$; critical point $(2,2)$; $C_{xx}C_{yy} - C_{xy}^2 > 0$ and $C_{xx} > 0$ at $(2,2)$, relative minimum. $z = 4$ when $x = y = 2$. The cost of materials is minimum if the length and width are 2 ft and the height is 4 ft.

41. Minimize $w = D^2 = (x + 1)^2 + (y - 3)^2 + (z - 2)^2$ subject to $x - 2y + z = 4$.
$z = 4 - x + 2y$ so $w = (x + 1)^2 + (y - 3)^2 + (2 - x + 2y)^2$,
$w_x = 2(2x - 2y - 1) = 0$, $w_y = 2(-2x + 5y + 1) = 0$; critical point $(1/2,0)$; $w_{xx}w_{yy} - w_{xy}^2 > 0$ and $w_{xx} > 0$ at $(1/2,0)$, relative minimum.
$w = 27/2$, $D = \sqrt{w} = 3\sqrt{6}/2$ is the minimum distance.

43. Minimize $S = xy + 2xz + 2yz$ subject to $xyz = V$, $x > 0$, $y > 0$, $z > 0$ where x, y, and z are, respectively, the length, width, and height of the box. $z = V/(xy)$ so $S = xy + 2V/y + 2V/x$, $S_x = y - 2V/x^2 = 0$, $S_y = x - 2V/y^2 = 0$; critical point $(\sqrt[3]{2V}, \sqrt[3]{2V})$; $S_{xx}S_{yy} - S_{xy}^2 > 0$ and

$S_{xx} > 0$ at this point so there is a relative minimum there. The length and width are each $\sqrt[3]{2V}$, the height is $z = \sqrt[3]{2V}/2$.

45. (a) $\dfrac{\partial f}{\partial a} = \Sigma 2(ax_k + b - y_k)x_k = 2(a\Sigma x_k^2 + b\Sigma x_k - \Sigma x_k y_k) = 0$

   if $(\Sigma x_k^2)a + (\Sigma x_k)b = \Sigma x_k y_k$,

   $\dfrac{\partial f}{\partial b} = \Sigma 2(ax_k + b - y_k) = 2(a\Sigma x_k + bn - \Sigma y_k) = 0$

   if $(\Sigma x_k)a + nb = \Sigma y_k$.

47. $\Sigma x_k = 10$, $\Sigma y_k = 8.2$, $\Sigma x_k^2 = 30$, $\Sigma x_k y_k = 23$, $n = 4$;
   $a = 0.5$, $b = 0.8$, $y = 0.5x + 0.8$.

EXERCISE SET 16.10

1. $y = 8x\lambda$, $x = 16y\lambda$; $y/(8x) = x/(16y)$, $x^2 = 2y^2$ so $4(2y^2) + 8y^2 = 16$,
   $y^2 = 1$, $y = \pm 1$. Test $(\pm\sqrt{2}, -1)$ and $(\pm\sqrt{2}, 1)$. $f(-\sqrt{2}, -1) = f(\sqrt{2}, 1) = \sqrt{2}$,
   $f(-\sqrt{2}, 1) = f(\sqrt{2}, -1) = -\sqrt{2}$. Maximum $\sqrt{2}$ at $(-\sqrt{2}, -1)$ and $(\sqrt{2}, 1)$, minimum
   $-\sqrt{2}$ at $(-\sqrt{2}, 1)$ and $(\sqrt{2}, -1)$.

3. $12x^2 = 4x\lambda$, $2y = 2y\lambda$. If $y \neq 0$ then $\lambda = 1$ and $12x^2 = 4x$,
   $12x(x - 1/3) = 0$, $x = 0$ or $x = 1/3$ so from $2x^2 + y^2 = 1$ we find that
   $y = \pm 1$ when $x = 0$, $y = \pm\sqrt{7}/3$ when $x = 1/3$. If $y = 0$ then
   $2x^2 + (0)^2 = 1$, $x = \pm 1/\sqrt{2}$. Test $(0, \pm 1)$, $(1/3, \pm\sqrt{7}/3)$, and $(\pm 1/\sqrt{2}, 0)$.
   $f(0, \pm 1) = 1$, $f(1/3, \pm\sqrt{7}/3) = 25/27$, $f(1/\sqrt{2}, 0) = \sqrt{2}$, $f(-1/\sqrt{2}, 0) = -\sqrt{2}$.
   Maximum $\sqrt{2}$ at $(1/\sqrt{2}, 0)$, minimum $-\sqrt{2}$ at $(-1/\sqrt{2}, 0)$.

5. $2 = 2x\lambda$, $1 = 2y\lambda$, $-2 = 2z\lambda$; $1/x = 1/(2y) = -1/z$ thus $x = 2y$, $z = -2y$ so
   $(2y)^2 + y^2 + (-2y)^2 = 4$, $y^2 = 4/9$, $y = \pm 2/3$. Test $(-4/3, -2/3, 4/3)$ and
   $(4/3, 2/3, -4/3)$. $f(-4/3, -2/3, 4/3) = -6$, $f(4/3, 2/3, -4/3) = 6$. Maximum 6
   at $(4/3, 2/3, -4/3)$, minimum -6 at $(-4/3, -2/3, 4/3)$.

7. $yz = 2x\lambda$, $xz = 2y\lambda$, $xy = 2z\lambda$; $yz/(2x) = xz/(2y) = xy/(2z)$ thus $y^2 = x^2$, $z^2 = x^2$ so $x^2 + x^2 + x^2 = 1$, $x = \pm 1/\sqrt{3}$. Test the eight possibilities with $x = \pm 1/\sqrt{3}$, $y =$ $\pm 1/\sqrt{3}$, and $z = \pm 1/\sqrt{3}$ to find the maximum is $1/(3\sqrt{3})$ at $(1/\sqrt{3}, 1/\sqrt{3}, 1/\sqrt{3})$, $(1/\sqrt{3}, -1/\sqrt{3}, -1/\sqrt{3})$, $(-1/\sqrt{3}, 1/\sqrt{3}, -1/\sqrt{3})$, and $(-1/\sqrt{3}, -1/\sqrt{3}, 1/\sqrt{3})$; the minimum is $-1/(3\sqrt{3})$ at $(1/\sqrt{3}, 1/\sqrt{3}, -1/\sqrt{3})$, $(1/\sqrt{3}, -1/\sqrt{3}, 1/\sqrt{3})$, $(-1/\sqrt{3}, 1/\sqrt{3}, 1/\sqrt{3})$, and $(-1/\sqrt{3}, -1/\sqrt{3}, -1/\sqrt{3})$.

9. $f(x,y) = (x - 4)^2 + (y - 2)^2$, $g(x,y) = y - 2x$; $2(x - 4) = -2\lambda$, $2(y - 2) = \lambda$; $x - 4 = -2(y - 2)$, $x = -2y + 8$ so $y = 2(-2y + 8) + 3$, $y = 19/5$. The point is $(2/5, 19/5)$.

11. $f(x,y,z) = (x - 1)^2 + (y + 1)^2 + (z - 1)^2$; $2(x - 1) = 4\lambda$, $2(y + 1) = 3\lambda$, $2(z - 1) = \lambda$; $x = 4z - 3$, $y = 3z - 4$ so $4(4z - 3) + 3(3z - 4) + z = 2$, $z = 1$. The point is $(1, -1, 1)$.

13. $f(x,y) = \sin x \sin y$; $x + y = \pi/2$; $\cos x \sin y = \lambda$, $\sin x \cos y = \lambda$; $\cos x \sin y = \sin x \cos y$ but $\cos x \neq 0$ and $\cos y \neq 0$ thus $\tan y = \tan x$, $y = x$ so $x + x = \pi/2$, $x = \pi/4$. The maximum value of $f(x,y)$ is $f(\pi/4, \pi/4) = 1/2$.

15. $f(x,y) = (x - 1)^2 + (y - 2)^2$; $2(x - 1) = 2x\lambda$, $2(y - 2) = 2y\lambda$; $(x - 1)/x = (y - 2)/y$, $y = 2x$ so $x^2 + (2x)^2 = 45$, $x = \pm 3$. $f(-3, -6) = 80$ and $f(3, 6) = 20$ so $(3, 6)$ is closest and $(-3, -6)$ is farthest.

17. $f(x,y,z) = x^2 + y^2 + z^2$, $x + y + z = 27$; $2x = \lambda$, $2y = \lambda$, $2z = \lambda$, $y = x$, $z = x$ so $x + x + x = 27$, $x = 9$. The numbers are $9, 9, 9$.

19. $f(x,y,z) = x^2 + y^2 + z^2$; $2x = 2x\lambda$, $2y = -z\lambda$, $2z = -y\lambda$. If $x \neq 0$ then $\lambda = 1$ thus $2y = -z$ and $2z = -y$ so $y = z = 0$; use $x^2 - yz = 5$ to get $x = \pm\sqrt{5}$. If $x = 0$ then $(0)^2 - yz = 5$, $y = -5/z$ and also (from $2y = -z\lambda$, $2z = -y\lambda$) $y^2 = z^2$ so $(-5/z)^2 = z^2$, $z^4 = 25$, $z = \pm\sqrt{5}$. $f(\pm\sqrt{5}, 0, 0) = 5$ and $f(0, \sqrt{5}, -\sqrt{5}) = f(0, -\sqrt{5}, \sqrt{5}) = 10$ so $(\pm\sqrt{5}, 0, 0)$ are closest to the origin.

21. $f(x,y,z) = 20xy + 10xz + 10yz$, $xyz = 16$; $20y + 10z = yz\lambda$, $20x + 10z = xz\lambda$, $10x + 10y = xy\lambda$; $(20y + 10z)/(yz) = (20x + 10z)/(xz) = (10x + 10y)/(xy)$; $y = x$, $z = 2x$ so

x(x)(2x) = 16, x = 2.   The length and width are each 2 ft and the height
is 4 ft.

23. $f(a,b,\alpha) = ab\sin\alpha$, $2a + 2b = \ell$; $b\sin\alpha = 2\lambda$, $a\sin\alpha = 2\lambda$,
      $ab\cos\alpha = 0$; $\cos\alpha = 0$, $\alpha = \pi/2$ and $a = b$ so $2b + 2b = \ell$, $b = \ell/4$.

SUPPLEMENTARY EXERCISES CHAPTER 16

1.   (a) 1   (b) xy   (c) $e^{r+s}\ln rs$

3.   (a) upper half of the elliptic cone $z^2 = x^2 + 4y^2$
     (b) the plane with x, y, and z intercepts of 1, a, and b.

5. $1/(x\sin yz) - 3y(\csc yz \cot yz)\ln xy$

7. $\pi/2$, 0, 1, $-\pi^2/4$ 9.   2

11. $(2x - 2y + 4r)/(x^2 + y^2 + 2z) = \dfrac{2}{r + s}$

13. $\partial^2 w/\partial x^2 = \partial^2 w/\partial y^2 = -(x - y)^{-2} - \cos(x + y)$

15. $f_{xyzx} = f_{zxxy} = 0$

17.   (a) $dP/dt = (\partial P/\partial T)(dT/dt) = (10/V)(dT/dt)$
                      $= (10/2.5)(3) = 12$ newtons/m^2/min
      (b) $dP/dt = (\partial P/\partial V)(dV/dt) = -(10T/V^2)(dV/dt)$
                      $= -(500/6.25)(-3) = 240$ newtons/m^2/min

19.   (a) $\displaystyle\lim_{(x,y)\to(0,0)} \frac{(x^2 - y^2)(x^2 + y^2)}{x^2 + y^2} = \lim_{(x,y)\to(0,0)} (x^2 - y^2) = 0$
      (b) continuous at (0,0) because $\displaystyle\lim_{(x,y)\to(0,0)} f(x,y) = f(0,0)$

21.   (a) $-(6x - 5y + y\sec^2 xy)/(-5x + x\sec^2 xy)$
      (b) $-[\ln y + \cos(x - y)]/[x/y - \cos(x - y)]$

23. $dV/dt = (\partial V/\partial E)(dE/dt) + (\partial V/\partial r)(dr/dt) = \dfrac{R}{r + R}\dfrac{dE}{dt} - \dfrac{RE}{(r + R)^2}\dfrac{dr}{dt}$

25.   (a) $\nabla f(3,1) = \langle 6,45\rangle$

(b) $\overrightarrow{P_0P_1} = \langle 1,-4\rangle$, $\mathbf{u} = \langle 1,-4\rangle/\sqrt{17}$, $D_{\mathbf{u}}f = -174/\sqrt{17}$

27.   (a) $\nabla f(3,2,6) = \langle 1/3,1/2,1/6\rangle$ (b) $D_{\mathbf{u}}f = \sqrt{3}/9$

29.   (a) $\nabla f(1,-1,2) = \langle 1,3,0\rangle$

(b) $\overrightarrow{P_0P_1} = \langle 10,11,-2\rangle$, $\mathbf{u} = \langle 10,11,-2\rangle/15$, $D_{\mathbf{u}}f = 43/15$

31.   (a) $\nabla f(1,-1) = 2(-\mathbf{i} + \mathbf{j})$, $D_{\mathbf{u}}f = 0$ if $\mathbf{u}$ is normal to ∇f so

$\mathbf{u} = \pm(\mathbf{i} + \mathbf{j})/\sqrt{2}$

(b) $\nabla f(-2,0) = \mathbf{i} - 2\mathbf{j}$, $\mathbf{u} = \pm(2\mathbf{i} + \mathbf{j})/\sqrt{5}$

33. $\nabla f(1,2) = a\mathbf{i} + b\mathbf{j}$. $D_{\mathbf{u}}f = 2\sqrt{2}$ when $\mathbf{u} = (\mathbf{i} + \mathbf{j})/\sqrt{2}$ and $D_{\mathbf{u}}f = -3$ when

$\mathbf{u} = -\mathbf{j}$ so $(a + b)/\sqrt{2} = 2\sqrt{2}$ and $-b = -3$, $a = 1$, $b = 3$.   If

$\mathbf{u} = -(\mathbf{i} + 2\mathbf{j})/\sqrt{5}$ then $D_{\mathbf{u}}f = -7/\sqrt{5}$.

35.   (a) $\langle f_x(4,-3),f_y(4,-3),-1\rangle = \langle 8/5,-6/5,-1\rangle$, let $\mathbf{N} = \langle 8,-6,-5\rangle$

(b) $8x - 6y - 5z = 0$

37.   Let $f(x,y,z) = z + xy$. $\nabla f(x_0,y_0,z_0) = \langle y_0,x_0,1\rangle$ is normal to the surface at $P_0(x_0,y_0,z_0)$. The normal line passes through the origin when $\langle x_0,y_0,z_0\rangle$ and $\langle y_0,x_0,1\rangle$ are parallel so $\langle x_0,y_0,z_0\rangle = k\langle y_0,x_0,1\rangle = \langle ky_0,kx_0,k\rangle$ for some value of k. Equate the third component of these vectors to find that $k = z_0$ so $x_0 = y_0z_0$ and $y_0 = x_0z_0$, eliminate y_0 to get $x_0 = x_0z_0^2$, $x_0(1 - z_0^2) = 0$, $x_0 = 0$ or $z_0 = \pm 1$. If $x_0 = 0$ then $y_0 = (0)z_0 = 0$ and, from the equation of the surface, $z_0 = 2 - (0)(0) = 2$ so $(0,0,2)$ is one of the points. If $z_0 = 1$ then $y_0 = x_0$ so $1 = 2 - x_0^2$, $x_0^2 = 1$, $x_0 = \pm 1$ so $(1,1,1)$ and $(-1,-1,1)$ are

also points where the normal line passes through the origin. If $z_0 = -1$ then $y_0 = -x_0$ so $-1 = 2 + x_0^2$, $x_0^2 = -3$ which has no real solution.

39. $\langle 18x_0, 8y_0, -1 \rangle$ is normal to the surface at a point (x_0, y_0, z_0).

    $\overrightarrow{PQ} = \langle -6, -4, -1 \rangle$ so the normal line is parallel to $\overrightarrow{PQ}$ if $\langle 18x_0, 8y_0, -1 \rangle = k\langle -6, -4, -1 \rangle$ for some value of k. By inspection k = 1 so $18x_0 = -6$ and $8y_0 = -4$, $x_0 = -1/3$ and $y_0 = -1/2$ thus $z_0 = 2$. The only point is $(-1/3, -1/2, 2)$.

41. $dV = (2/3)xh\,dx + (1/3)x^2 dh$
    $= (2/3)(1)(2)(-0.1) + (1/3)(1)^2(0.2) = -0.2/3 \approx -0.067\ m^3$
    $\Delta V = (1/3)(0.9)^2(2.2) - (1/3)(1)^2(2) = -0.218/3 \approx -0.073\ m^3$

43. $f_x = 2x + 3y - 6 = 0$, $f_y = 3x + 6y + 3 = 0$; critical point $(15, -8)$;
    $f_{xx}f_{yy} - f_{xy}^2 > 0$ and $f_{xx} > 0$ at $(15, -8)$, relative minimum.

45. $f_x = 3x^2 - 3y = 0$, $f_y = -3x + y = 0$; critical points $(0,0)$ and $(3,9)$;
    $D < 0$ at $(0,0)$, saddle point; $D > 0$ and $f_{xx} > 0$ at $(3,9)$, relative minimum.

47. (a) Let (x,y,z) be a point on the portion of the ellipsoid that is in the first octant then $V = (2x)(2y)(2z) = 8xyz$. For convenience introduce the new variables $u = x/a$, $v = y/b$, and $w = z/c$ so $V = (8abc)uvw$ where $u^2 + v^2 + w^2 = 1$. Also for convenience we will maximize $S = u^2v^2w^2$ instead of V. $w^2 = 1 - u^2 - v^2$ so $S = u^2v^2 - u^4v^2 - u^2v^4$, $S_u = 2uv^2(1 - 2u^2 - v^2) = 0$,

    $S_v = 2vu^2(1 - u^2 - 2v^2) = 0$; critical point $(1/\sqrt{3}, 1/\sqrt{3})$;

    $S_{uu}S_{vv} - S_{uv}^2 > 0$ d $S_{uu} < 0$ at this point so a relative maximum occurs

    there. If $u = v = 1/\sqrt{3}$ then $w = 1/\sqrt{3}$ so $x = a/\sqrt{3}$, $y = b/\sqrt{3}$, and $z = c/\sqrt{3}$. The dimensions of the box are $2a/\sqrt{3}$, $2b/\sqrt{3}$, and $2c/\sqrt{3}$.

(b) $f(x,y,z) = 8xyz$, $(x/a)^2 + (y/b)^2 + (z/c)^2 = 1$; $8yz = (2x/a^2)\lambda$, $8xz = (2y/b^2)\lambda$, $8xy = (2z/c^2)\lambda$; $4a^2yz/x = 4b^2xz/y = 4c^2xy/z$, $y^2/b^2 = x^2/a^2$ and $z^2/c^2 = x^2/a^2$ so $3(x^2/a^2) = 1$, $x = a/\sqrt{3}$ and therefore $y = b/\sqrt{3}$ and $z = c/\sqrt{3}$. The dimensions agree with those in part (a).

49. $f(I_1, I_2, I_3) = I_1^2 R_1 + I_2^2 R_2 + I_3^2 R_3$, $I_1 + I_2 + I_3 = I$. $2I_1 R_1 = \lambda$, $2I_2 R_2 = \lambda$, $2I_3 R_3 = \lambda$; $2I_1 R_1 = 2I_2 R_2 = 2I_3 R_3$, $I_1/I_2 = R_2/R_1 = R_1^{-1}/R_2^{-1}$ and $I_2/I_3 = R_2^{-1}/R_3^{-1}$ so $I_1 : I_2 : I_3 = R_1^{-1} : R_2^{-1} : R_3^{-1}$.

CHAPTER 17

EXERCISE SET 17.1

1. $\int_0^1 \int_0^2 (x + 3)dy\, dx = \int_0^1 (2x + 6)dx = 7$

3. $\int_2^4 \int_0^1 x^2 y\, dx\, dy = \int_2^4 \frac{1}{3}\, y\, dy = 2$

5. $\int_0^{\ln 3} \int_0^{\ln 2} e^{x+y}dy\, dx = \int_0^{\ln 3} e^x dx = 2$

7. $\int_0^3 \int_0^1 x(x^2 + y)^{1/2}dx\, dy = \int_0^3 \frac{1}{3}\, [(1 + y)^{3/2} - y^{3/2}]dy = 2(31 - 9\sqrt{3})/15$

9. $\int_{-1}^0 \int_2^5 dx\, dy = \int_{-1}^0 3dy = 3$

11. $\int_0^1 \int_0^1 \frac{x}{(xy + 1)^2}\, dy\, dx = \int_0^1 (1 - \frac{1}{x + 1})dx = 1 - \ln 2$

13. $\int_0^{\ln 2} \int_0^1 xy\, e^{y^2 x}dy\, dx = \int_0^{\ln 2} \frac{1}{2}\, (e^x - 1)dx = (1 - \ln 2)/2$

15. $\int_{-1}^1 \int_{-2}^2 4xy^3 dy\, dx = \int_{-1}^1 0\, dx = 0$

17. $\int_0^1 \int_2^3 x\sqrt{1 - x^2}\, dy\, dx = \int_0^1 x(1 - x^2)^{1/2}dx = 1/3$

19. $\displaystyle \int_{-\pi/4}^{\pi/4} \int_{0}^{\pi/4} \cos(x + y)dy\,dx \; = \; \int_{-\pi/4}^{\pi/4} [\sin(x + \pi/4) - \sin x]dx \; = \; 1$

21.

23.

25.

27. $\displaystyle V = \int_{1}^{3} \int_{0}^{2} (3x^3 + 3x^2 y)dy\,dx \; = \; \int_{1}^{3} (6x^3 + 6x^2)dx \; = \; 172$

29. $\displaystyle V = \int_{0}^{2} \int_{0}^{3} x^2 dy\,dx \; = \; \int_{0}^{2} 3x^2 dx \; = \; 8$

31. $\displaystyle \iint_{R} f(x,y)dA = \int_{a}^{b} \left[\int_{c}^{d} g(x)h(y)dy \right] dx = \int_{a}^{b} g(x) \left[\int_{c}^{d} h(y)dy \right] dx$

$\displaystyle = \left[\int_{a}^{b} g(x)dx \right] \left[\int_{c}^{d} h(y)dy \right]$

EXERCISE SET 17.2

1. $\displaystyle\int_0^1 \int_{x^2}^x xy^2 dy\, dx = \int_0^1 \frac{1}{3}(x^4 - x^7)dx = 1/40$

3. $\displaystyle\int_0^3 \int_0^{\sqrt{9-y^2}} y\, dx\, dy = \int_0^3 y\sqrt{9 - y^2}\, dy = 9$

5. $\displaystyle\int_{\sqrt{\pi}}^{\sqrt{2\pi}} \int_0^{x^3} \sin(y/x)dy\, dx = \int_{\sqrt{\pi}}^{\sqrt{2\pi}} [-x\cos(x^2) + x]dx = \pi/2$

7. $\displaystyle\int_{\pi/2}^{\pi} \int_0^{x^2} \frac{1}{x}\cos(y/x)dy\, dx = \int_{\pi/2}^{\pi} \sin x\, dx = 1$

9. $\displaystyle\int_0^a \int_0^{\sqrt{a^2-x^2}} (x + y)dy\, dx = \int_0^a [x\sqrt{a^2 - x^2} + (a^2 - x^2)/2]dx = 2a^3/3$

11. $\displaystyle\int_0^1 \int_0^x y\sqrt{x^2 - y^2}\, dy\, dx = \int_0^1 \frac{1}{3}x^3 dx = 1/12$

13. $\displaystyle\int_0^2 \int_0^{x^2} 6xy\, dy\, dx = \int_0^2 3x^5 dx = 32$

15. $\displaystyle\int_1^2 \int_{\pi/2}^{2\pi/x} x\cos xy\, dy\, dx = -\int_1^2 \sin(\pi x/2)dx = -2/\pi$

17. $\displaystyle\int_4^8 \int_{16/x}^x x^2 dy\, dx = \int_4^8 (x^3 - 16x)dx = 576$

19. $\displaystyle\int_0^4 \int_0^{\sqrt{y}} x(1 + y^2)^{-1/2}dx\, dy = \int_0^4 \frac{1}{2} y(1 + y^2)^{-1/2}dy = (\sqrt{17} - 1)/2$

21. $\displaystyle\int_{-1}^{1}\int_{-\sqrt{1-x^2}}^{\sqrt{1-x^2}} (3x - 2y)dy\,dx = \int_{-1}^{1} 6x\sqrt{1 - x^2}\,dx = 0$

23. $\displaystyle\int_{0}^{1}\int_{x}^{1} \frac{1}{1 + x^2}\,dy\,dx = \int_{0}^{1}\left[\frac{1}{1 + x^2} - \frac{x}{1 + x^2}\right]dx = \frac{\pi}{4} - \frac{1}{2}\ln 2$

25. $\displaystyle\int_{0}^{2}\int_{y^2}^{6-y} xy\,dx\,dy = \int_{0}^{2}\frac{1}{2}(36y - 12y^2 + y^3 - y^5)dy = 50/3$

27. $\displaystyle\int_{-1}^{0}\int_{x}^{x^3} (x - 1)dy\,dx + \int_{0}^{1}\int_{x^3}^{x} (x - 1)dy\,dx$

$\displaystyle = \int_{-1}^{0}(x^4 - x^3 - x^2 + x)dx + \int_{0}^{1}(-x^4 + x^3 + x^2 - x)dx = -1/2$

29. $A = \displaystyle\int_{0}^{5}\int_{0}^{5-x} dy\,dx = \int_{0}^{5} (5 - x)dx = 25/2$

31. $A = \displaystyle\int_{0}^{\pi/4}\int_{\sin x}^{\cos x} dy\,dx = \int_{0}^{\pi/4} (\cos x - \sin x)dx = \sqrt{2} - 1$

33. $A = \displaystyle\int_{-3}^{3}\int_{1-y^2/9}^{9-y^2} dx\,dy = \int_{-3}^{3} 8(1 - y^2/9)dy = 32$

35. $V = \displaystyle\int_{0}^{5/2}\int_{0}^{5-2x} (5 - 2x - y)dy\,dx = \int_{0}^{5/2}\frac{1}{2}(5 - 2x)^2dx = 125/12$

37. $V = \displaystyle\int_{-1}^{1}\int_{0}^{1-x^2} (x + 2y + 2)dy\,dx = \int_{-1}^{1}(x^4 - x^3 - 4x^2 + x + 3)dx = 56/15$

39. $V = \displaystyle\int_{0}^{3}\int_{0}^{2} (9x^2 + y^2)dy\,dx = \int_{0}^{3}(18x^2 + 8/3)dx = 170$

41. $V = \int_{-3/2}^{3/2} \int_{-\sqrt{9-4x^2}}^{\sqrt{9-4x^2}} (y + 3)\,dy\,dx = \int_{-3/2}^{3/2} 6\sqrt{9 - 4x^2}\,dx = 27\pi/2$

43. $V = \int_{0}^{4} \int_{0}^{2-x/2} (x/4 + 2y)\,dy\,dx = \int_{0}^{4} (x^2/8 - 3x/2 + 4)\,dx = 20/3$

45. $V = 4\int_{0}^{1} \int_{0}^{\sqrt{1-x^2}} (1 - x^2 - y^2)\,dy\,dx = \frac{8}{3}\int_{0}^{1} (1 - x^2)^{3/2}\,dx = \pi/2$

47. $V = 8\int_{0}^{5} \int_{0}^{\sqrt{25-x^2}} \sqrt{25 - x^2}\,dy\,dx = 8\int_{0}^{5} (25 - x^2)\,dx = 2000/3$

49. $\int_{0}^{\sqrt{2}} \int_{y^2}^{2} f(x,y)\,dx\,dy$

51. $\int_{1}^{e^2} \int_{\ln x}^{2} f(x,y)\,dy\,dx$

53. $\int_{-1}^{1} \int_{-2\sqrt{1-y^2}}^{2\sqrt{1-y^2}} f(x,y)\,dx\,dy$

55. $\int_{0}^{\pi/2} \int_{0}^{\sin x} f(x,y)\,dy\,dx$

57. $\int_{0}^{4} \int_{0}^{y/4} e^{-y^2}\,dx\,dy = \int_{0}^{4} \frac{1}{4} y e^{-y^2}\,dy = (1 - e^{-16})/8$

59. $\int_{0}^{2} \int_{0}^{x^2} e^{x^3}\,dy\,dx = \int_{0}^{2} x^2 e^{x^3}\,dx = (e^8 - 1)/3$

61. $\int_{0}^{\pi/2} \int_{0}^{\cos y} x\,dx\,dy = \frac{1}{2}\int_{0}^{\pi/2} \cos^2 y\,dy = \pi/8$

63. $\int_{0}^{2} \int_{0}^{y^2} \sin(y^3)\,dx\,dy = \int_{0}^{2} y^2 \sin(y^3)\,dy = (1 - \cos 8)/3$

65. (a) $\int_{-2}^{-1} \int_{0}^{2} xy^2 dy\ dx + \int_{-1}^{1} \int_{1}^{2} xy^2 dy\ dx + \int_{1}^{2} \int_{0}^{2} xy^2 dy\ dx$

$= \int_{-2}^{-1} \frac{8}{3} x\ dx + \int_{-1}^{1} \frac{7}{3} x\ dx + \int_{1}^{2} \frac{8}{3} x\ dx = 0$

(b) $\int_{1}^{4} \int_{-\sqrt{y}}^{2-y} xy^2 dx\ dy + \int_{1/2}^{1} \int_{-\sqrt{y}}^{\sqrt{y}} xy^2 dx\ dy$

$= \int_{1}^{4} \frac{1}{2} (4y^2 - 5y^3 + y^4) dy + \int_{1/2}^{1} (0) dy = -603/40$

67. $2\int_{-1}^{1} \int_{0}^{\sqrt{1-x^2}} dy\ dx + \int_{-1}^{1} \int_{0}^{\sqrt{1-x^2}} x\sqrt{9 - y^2}\ dy\ dx = 2 \cdot \frac{1}{2} \pi(1)^2 + 0 = \pi$ because

the first integral gives the area of a semicircle of radius 1, and with

$f(x,y) = x\sqrt{9 - y^2}$, $f(-x,y) = -f(x,y)$ so the second integral evaluates
to 0.

69. $f(x,y) = x^3y$, $f(-x,y) = -f(x,y)$ so $\iint\limits_{R} x^3y\ dA = 0$

EXERCISE SET 17.3

1. $\int_{0}^{\pi/2} \int_{0}^{\sin \theta} r \cos \theta\ dr\ d\theta = \int_{0}^{\pi/2} \frac{1}{2} \sin^2\theta \cos \theta\ d\theta = 1/6$

3. $\int_{-\pi/2}^{\pi/2} \int_{0}^{a \sin \theta} r^2 dr\ d\theta = \int_{-\pi/2}^{\pi/2} \frac{a^3}{3} \sin^3\theta\ d\theta = 0$

5. $\int_{0}^{\pi} \int_{0}^{1-\sin \theta} r^2 \cos\theta\ dr\ d\theta = \int_{0}^{\pi} \frac{1}{3} (1 - \sin \theta)^3 \cos \theta\ d\theta = 0$

7. $A = \int_0^{2\pi} \int_0^{1-\cos\theta} r\,dr\,d\theta = \int_0^{2\pi} \frac{1}{2}(1 - \cos\theta)^2 d\theta = 3\pi/2$

9. $A = \int_{\pi/4}^{\pi/2} \int_{\sin 2\theta}^{1} r\,dr\,d\theta = \int_{\pi/4}^{\pi/2} \frac{1}{2}(1 - \sin^2 2\theta)d\theta = \pi/16$

11. $A = 2\int_{\pi/6}^{\pi/2} \int_{2}^{4\sin\theta} r\,dr\,d\theta = \int_{\pi/6}^{\pi/2}(16\sin^2\theta - 4)d\theta = 4\pi/3 + 2\sqrt{3}$

13. $V = 8\int_0^{\pi/2} \int_0^1 r\sqrt{9 - r^2}\,dr\,d\theta = \frac{8}{3}(27 - 16\sqrt{2})\int_0^{\pi/2} d\theta = 4(27 - 16\sqrt{2})\pi/3$

15. $V = 2\int_0^{\pi/2} \int_0^{2\sin\theta} r^2 dr\,d\theta = \frac{16}{3}\int_0^{\pi/2} \sin^3\theta\,d\theta = 32/9$

17. $V = 2\int_0^{\pi/2} \int_0^{\cos\theta}(1 - r^2)r\,dr\,d\theta = \frac{1}{2}\int_0^{\pi/2}(1 - \sin^4\theta)d\theta = 5\pi/32$

19. $\int_0^{2\pi} \int_0^1 e^{-r^2} r\,dr\,d\theta = \frac{1}{2}(1 - e^{-1})\int_0^{2\pi} d\theta = (1 - e^{-1})\pi$

21. $\int_0^{\pi/4} \int_0^2 \frac{1}{1 + r^2} r\,dr\,d\theta = \frac{1}{2}\ln 5 \int_0^{\pi/4} d\theta = \frac{\pi}{8}\ln 5$

23. $\int_0^{\pi/2} \int_0^1 r^3 dr\,d\theta = \frac{1}{4}\int_0^{\pi/2} d\theta = \pi/8$

25. $\int_0^{\pi/2} \int_0^{2\cos\theta} r^2 dr\,d\theta = \frac{8}{3}\int_0^{\pi/2} \cos^3\theta\,d\theta = 16/9$

27. $\int_0^{\pi/2} \int_0^a \frac{r}{(1 + r^2)^{3/2}} dr\,d\theta = \frac{\pi}{2}(1 - 1/\sqrt{1 + a^2})$

29. $\displaystyle\int_0^{\pi/4}\int_0^2 \frac{r}{\sqrt{1+r^2}}\,dr\,d\theta = \frac{\pi}{4}(\sqrt{5}-1)$

31. $\displaystyle V = 2\int_0^{\pi/2}\int_0^{a\sin\theta} \frac{c}{a}(a^2-r^2)^{1/2}r\,dr\,d\theta = \frac{2}{3}a^2c\int_0^{\pi/2}(1-\cos^3\theta)\,d\theta$

$$= (3\pi-4)a^2c/9$$

33. $\displaystyle A = \int_{\pi/6}^{\pi/4}\int_{\sqrt{8\cos 2\theta}}^{4\sin\theta} r\,dr\,d\theta + \int_{\pi/4}^{\pi/2}\int_0^{4\sin\theta} r\,dr\,d\theta$

$$= \int_{\pi/6}^{\pi/4}(8\sin^2\theta - 4\cos 2\theta)\,d\theta + \int_{\pi/4}^{\pi/2} 8\sin^2\theta\,d\theta = 4\pi/3 + 2\sqrt{3} - 2$$

35. (a) $\displaystyle I^2 = \left[\int_0^{+\infty} e^{-x^2}dx\right]\left[\int_0^{+\infty} e^{-y^2}dy\right] = \int_0^{+\infty}\left[\int_0^{+\infty} e^{-x^2}dx\right]e^{-y^2}dy$

$$= \int_0^{+\infty}\int_0^{+\infty} e^{-x^2}e^{-y^2}dx\,dy = \int_0^{+\infty}\int_0^{+\infty} e^{-(x^2+y^2)}dx\,dy$$

(b) $\displaystyle I^2 = \int_0^{\pi/2}\int_0^{+\infty} e^{-r^2}r\,dr\,d\theta = \frac{1}{2}\int_0^{\pi/2}d\theta = \pi/4$

(c) $I = \sqrt{\pi}/2$

EXERCISE SET 17.4

1. $z = \sqrt{9-y^2}$, $z_x = 0$, $z_y = -y/\sqrt{9-y^2}$, $z_x^2 + z_y^2 + 1 = 9/(9-y^2)$,

$\displaystyle S = \int_0^2\int_{-3}^3 \frac{3}{\sqrt{9-y^2}}\,dy\,dx = \int_0^2 3\pi\,dx = 6\pi$

3. $z^2 = 4x^2 + 4y^2$, $2zz_x = 8x$ so $z_x = 4x/z$, similarly $z_y = 4y/z$ thus

$z_x^2 + z_y^2 + 1 = (16x^2 + 16y^2)/z^2 + 1 = 5$,

$S = \int_0^1 \int_{x^2}^x \sqrt{5}\, dy\, dx = \sqrt{5} \int_0^1 (x - x^2)dx = \sqrt{5}/6$

5. $z_x = -2x$, $z_y = -2y$, $z_x^2 + z_y^2 + 1 = 4x^2 + 4y^2 + 1$,

$S = \iint_R \sqrt{4x^2 + 4y^2 + 1}\, dA = \int_0^{2\pi} \int_0^1 r\sqrt{4r^2 + 1}\, dr\, d\theta$

$= \frac{1}{12}(5\sqrt{5} - 1)\int_0^{2\pi} d\theta = (5\sqrt{5} - 1)\pi/6$

7. $z_x = y$, $z_y = x$, $z_x^2 + z_y^2 + 1 = x^2 + y^2 + 1$,

$S = \iint_R \sqrt{x^2 + y^2 + 1}\, dA = \int_0^{\pi/6} \int_0^3 r\sqrt{r^2 + 1}\, dr\, d\theta$

$= \frac{1}{3}(10\sqrt{10} - 1)\int_0^{\pi/6} d\theta = (10\sqrt{10} - 1)\pi/18$

9. On the sphere, $z_x = -x/z$ and $z_y = -y/z$ so

$z_x^2 + z_y^2 + 1 = (x^2 + y^2 + z^2)/z^2 = 16/(16 - x^2 - y^2)$; the planes $z = 1$

and $z = 2$ intersect the sphere along the circles $x^2 + y^2 = 15$ and
$x^2 + y^2 = 12$;

$S = \iint_R \frac{4}{\sqrt{16 - x^2 - y^2}}\, dA = \int_0^{2\pi} \int_{\sqrt{12}}^{\sqrt{15}} \frac{4r}{\sqrt{16 - r^2}}\, dr\, d\theta = 4\int_0^{2\pi} d\theta = 8\pi$

11. On both upper and lower halves of the sphere, $z_x = -x/z$ and $z_y = -y/z$ so

$z_x^2 + z_y^2 + 1 = (x^2 + y^2 + z^2)/z^2 = a^2/(a^2 - x^2 - y^2)$,

$$S = (2)(2)\int_0^{\pi/2}\int_0^{a\sin\theta} \frac{ar}{\sqrt{a^2 - r^2}} \, dr \, d\theta$$

$$= 4a^2\int_0^{\pi/2} (1 - \cos\theta)d\theta = 2(\pi - 2)a^2$$

13. $z_x = -x/z$ and $z_y = 0$ on $x^2 + z^2 = 16$ so

$$z_x^2 + z_y^2 + 1 = (x^2 + z^2)/z^2 = 16/(16 - x^2),$$

$$S = 8\int_0^4 \int_0^{\sqrt{16-x^2}} \frac{4}{\sqrt{16 - x^2}} \, dy \, dx = 32\int_0^4 dx = 128$$

15. $z_x = \dfrac{h}{a}\dfrac{x}{\sqrt{x^2 + y^2}}$, $z_y = \dfrac{h}{a}\dfrac{y}{\sqrt{x^2 + y^2}}$,

$$z_x^2 + z_y^2 + 1 = \frac{h^2x^2 + h^2y^2}{a^2(x^2 + y^2)} + 1 = (a^2 + h^2)/a^2,$$

$$S = \int_0^{2\pi}\int_0^a \frac{\sqrt{a^2 + h^2}}{a} r \, dr \, d\theta = \frac{1}{2} a \sqrt{a^2 + h^2}\int_0^{2\pi} d\theta = \pi a\sqrt{a^2 + h^2}$$

EXERCISE SET 17.5

1. $\displaystyle\int_{-1}^1\int_0^2\int_0^1 (x^2 + y^2 + z^2)dx \, dy \, dz = \int_{-1}^1\int_0^2 (1/3 + y^2 + z^2)dy \, dz$

$$= \int_{-1}^1 (10/3 + 2z^2)dz = 8$$

3. $\displaystyle\int_0^2\int_{-1}^{y^2}\int_1^z yz \, dx \, dz \, dy = \int_0^2\int_{-1}^{y^2} (yz^2 - yz)dz \, dy$

$$= \int_0^2 (\frac{1}{3} y^7 - \frac{1}{2} y^5 + \frac{5}{6} y)dy = 7$$

5. $\displaystyle\int_0^3 \int_0^{\sqrt{9-z^2}} \int_0^x xy\,dy\,dx\,dz \;=\; \int_0^3 \int_0^{\sqrt{9-z^2}} \tfrac{1}{2}\,x^3 dx\,dz$

$\displaystyle = \int_0^3 \tfrac{1}{8}\,(81 - 18z^2 + z^4)dz = 81/5$

7. $\displaystyle\int_0^2 \int_0^{\sqrt{4-x^2}} \int_{-5+x^2+y^2}^{3-x^2-y^2} x\,dz\,dy\,dx \;=\; \int_0^2 \int_0^{\sqrt{4-x^2}} [2x(4 - x^2) - 2xy^2]dy\,dx$

$\displaystyle = \int_0^2 \tfrac{4}{3}\,x(4 - x^2)^{3/2}dx = 128/15$

9. $\displaystyle\int_0^\pi \int_0^1 \int_0^{\pi/6} xy\sin yz\,dz\,dy\,dx \;=\; \int_0^\pi \int_0^1 x[1 - \cos(\pi y/6)]dy\,dx$

$\displaystyle = \int_0^\pi (1 - 3/\pi)x\,dx = \pi(\pi - 3)/2$

11. $\displaystyle\int_0^{\sqrt{2}} \int_0^x \int_0^{2-x^2} xyz\,dz\,dy\,dx \;=\; \int_0^{\sqrt{2}} \int_0^x \tfrac{1}{2}\,xy(2 - x^2)^2 dy\,dx$

$\displaystyle = \int_0^{\sqrt{2}} \tfrac{1}{4}\,x^3(2 - x^2)^2 dx = 1/6$

13. $\displaystyle V = \int_0^4 \int_0^{(4-x)/2} \int_0^{(12-3x-6y)/4} dz\,dy\,dx$

$\displaystyle = \int_0^4 \int_0^{(4-x)/2} \tfrac{1}{4}\,(12 - 3x - 6y)dy\,dx \;=\; \int_0^4 \tfrac{3}{16}\,(4 - x)^2 dx = 4$

15. $\displaystyle V = 2\int_0^2 \int_{x^2}^4 \int_0^{4-y} dz\,dy\,dx = 2\int_0^2 \int_{x^2}^4 (4 - y)dy\,dx$

$\displaystyle = 2\int_0^2 (8 - 4x^2 + \tfrac{1}{2}\,x^4)dx = 256/15$

17. $V = 2\int_{-3}^{3}\int_{0}^{\sqrt{9-x^2}/3}\int_{0}^{x+3} dz\,dy\,dx = 2\int_{-3}^{3}\int_{0}^{\sqrt{9-x^2}/3} (x+3)dy\,dx$

$\qquad\qquad\qquad = 2\int_{-3}^{3}\frac{1}{3}(x+3)\sqrt{9-x^2}\,dx = 9\pi$

19. The projection of the curve of intersection onto the xy-plane is $x^2 + y^2 = 1$,

$V = 4\int_{0}^{1}\int_{0}^{\sqrt{1-x^2}}\int_{4x^2+y^2}^{4-3y^2} dz\,dy\,dx = 16\int_{0}^{1}\int_{0}^{\sqrt{1-x^2}}(1-x^2-y^2)dy\,dx$

$\qquad\qquad\qquad = \frac{32}{3}\int_{0}^{1}(1-x^2)^{3/2}dx = 2\pi$

21. The projection of the curve of intersection onto the xy-plane is $x^2 + y^2 = a^2$,

$V = 4\int_{0}^{a}\int_{0}^{\sqrt{a^2-x^2}}\int_{(x^2+y^2)/a}^{\sqrt{2a^2-x^2-y^2}} dz\,dy\,dx$

$\quad = 4\int_{0}^{a}\int_{0}^{\sqrt{a^2-x^2}}[\sqrt{2a^2-x^2-y^2}-\frac{1}{a}(x^2+y^2)]dy\,dx$

$\quad = 4\int_{0}^{\pi/2}\int_{0}^{a}(r\sqrt{2a^2-r^2}-\frac{1}{a}r^3)dr\,d\theta$

$\quad = 4\int_{0}^{\pi/2}\frac{1}{12}(8\sqrt{2}-7)a^3 d\theta = (8\sqrt{2}-7)\pi a^3/6$

23. (a)   (b)

25. (a) $\displaystyle\int_0^a \int_0^{b(1-x/a)} \int_0^{c(1-x/a-y/b)} dz\, dy\, dx,$

$\displaystyle\int_0^b \int_0^{a(1-y/b)} \int_0^{c(1-x/a-y/b)} dz\, dx\, dy,$

$\displaystyle\int_0^c \int_0^{a(1-z/c)} \int_0^{b(1-x/a-z/c)} dy\, dx\, dz,$

$\displaystyle\int_0^a \int_0^{c(1-x/a)} \int_0^{b(1-x/a-z/c)} dy\, dz\, dx,$

$\displaystyle\int_0^c \int_0^{b(1-z/c)} \int_0^{a(1-y/b-z/c)} dx\, dy\, dz,$

$\displaystyle\int_0^b \int_0^{c(1-y/b)} \int_0^{a(1-y/b-z/c)} dx\, dz\, dy$

(b) Use the first integral in part (a) to get

$$\int_0^a \int_0^{b(1-x/a)} c(1 - \frac{x}{a} - \frac{y}{b})dy\, dx \;=\; \int_0^a \frac{1}{2}\, bc(1 - \frac{x}{a})^2 dx \;=\; \frac{1}{6}\, abc$$

27. $\displaystyle V = \int_0^a \int_0^x \int_0^{x-y} dz\, dy\, dx = \int_0^a \int_0^x (x - y)dy\, dx = \int_0^a \frac{1}{2}\, x^2 dx = a^3/6$

29. (a) $\displaystyle\left[\int_{-1}^1 x\, dx\right]\left[\int_0^1 y^2 dy\right]\left[\int_0^{\pi/2} \sin z\, dz\right] = (0)(1/3)(1) = 0$

(b) $\displaystyle\left[\int_0^1 e^{2x} dx\right]\left[\int_0^{\ln 3} e^y dy\right]\left[\int_0^{\ln 2} e^{-z} dz\right] = [(e^2 - 1)/2](2)(1/2)$

$$= (e^2 - 1)/2$$

31. (a) 10   (b) 0

33. The region is symmetric with respect to the xz-plane, and for
$f(x,y,z) = x^2 y^3 e^z$, $f(x,-y,z) = -f(x,y,z)$ so the integral evaluates to 0.

EXERCISE SET 17.6

1. Introduce an x-axis with origin at m_1, let a be the unknown coordinate of the fulcrum then the total moment about the fulcrum is $5(0 - a) + 10(5 - a) + 20(15 - a) = 350 - 35a$ so for equilibrium $350 - 35a = 0$, $a = 10$. The fulcrum should be placed 10 ft to the right of m_1.

3. $M = \int_0^1 \int_0^{\sqrt{x}} (x + y)dy\,dx = 13/20$, $M_x = \int_0^1 \int_0^{\sqrt{x}} (x + y)y\,dy\,dx = 3/10$,

    $M_y = \int_0^1 \int_0^{\sqrt{x}} (x + y)x\,dy\,dx = 19/42$, $\bar{x} = M_y/M = 190/273$, $\bar{y} = M_x/M = 6/13$; the mass is 13/30 and the center of gravity is at (190/273, 6/13).

5. $M = \int_0^{\pi/2} \int_0^a r^3 \sin\theta \cos\theta\,dr\,d\theta = a^4/8$, $\bar{x} = \bar{y}$ from the symmetry of the density and the region, $M_y = \int_0^{\pi/2} \int_0^a r^4 \sin\theta \cos^2\theta\,dr\,d\theta = a^5/15$,

    $\bar{x} = 8a/15$; mass $a^4/8$, center of gravity (8a/15, 8a/15).

7. $A = 1/2$, $\iint_R x\,dA = \int_0^1 \int_0^x x\,dy\,dx = 1/3$, $\iint_R y\,dA = \int_0^1 \int_0^x y\,dy\,dx = 1/6$; centroid (2/3, 1/3)

9. $A = \int_{-2}^1 \int_x^{2-x^2} dy\,dx = 9/2$, $\iint_R x\,dA = \int_{-2}^1 \int_x^{2-x^2} x\,dy\,dx = -9/4$,

    $\iint_R y\,dA = \int_{-2}^1 \int_x^{2-x^2} y\,dy\,dx = 9/5$; centroid (-1/2, 2/5)

11. $\bar{x} = 0$ from the symmetry of the region,

$A = \frac{1}{2} \pi(b^2 - a^2)$, $\iint\limits_{R} y \, dA = \int_0^{\pi} \int_a^b r^2 \sin \theta \, dr \, d\theta = \frac{2}{3}(b^3 - a^3)$;

centroid $\bar{x} = 0$, $\bar{y} = \dfrac{4(b^3 - a^3)}{3\pi(b^2 - a^2)}$.

13. $\bar{x} = 0$ from the symmetry of the region, $A = 16$,

$\iint\limits_{R} y \, dA = \int_{-4}^{4} \int_{|x|}^{4} y \, dy \, dx = 128/3$; centroid $(0, 8/3)$

15. $M = \int_0^a \int_0^a \int_0^a (a - x) dz \, dy \, dx = a^4/2$, $\bar{y} = \bar{z} = a/2$ from the symmetry of

density and region,

$\bar{x} = \frac{1}{M} \int_0^a \int_0^a \int_0^a x(a - x) dz \, dy \, dx = (2/a^4)(a^5/6) = a/3$; mass $a^4/2$,

center of gravity $(a/3, a/2, a/2)$

17. $M = \int_{-1}^{1} \int_0^1 \int_0^{1-y^2} yz \, dz \, dy \, dx = 1/6$, $\bar{x} = 0$ by the symmetry of density and

region, $\bar{y} = \frac{1}{M} \iiint\limits_{G} y^2 z \, dV = (6)(8/105) = 16/35$,

$\bar{z} = \frac{1}{M} \iiint\limits_{G} yz^2 dV = (6)(1/12) = 1/2$; mass $1/6$, center of gravity

$(0, 16/35, 1/2)$

19. $\bar{x} = \bar{y} = \bar{z}$ from the symmetry of the region,

$V = 1/6$, $\bar{x} = \frac{1}{V} \int_0^1 \int_0^{1-x} \int_0^{1-x-y} x \, dz \, dy \, dx = (6)(1/24) = 1/4$;

centroid $(1/4, 1/4, 1/4)$

21. $\bar{x} = 1/2$ and $\bar{y} = 0$ from the symmetry of the region,

$$V = \int_0^1 \int_{-1}^1 \int_{y^2}^1 dz\,dy\,dx = 4/3, \quad \bar{z} = \frac{1}{V} \iiint_G z\,dV = (3/4)(4/5) = 3/5;$$

centroid $(1/2,0,3/5)$

23. $\bar{x} = \bar{y} = \bar{z}$ from the symmetry of the region, $V = \pi a^3/6$,

$$\bar{x} = \frac{1}{V} \int_0^a \int_0^{\sqrt{a^2-x^2}} \int_0^{\sqrt{a^2-x^2-y^2}} x\,dz\,dy\,dx$$

$$= \frac{1}{V} \int_0^a \int_0^{\sqrt{a^2-x^2}} x\sqrt{a^2 - x^2 - y^2}\,dy\,dx$$

$$= \frac{1}{V} \int_0^{\pi/2} \int_0^a r^2\sqrt{a^2 - r^2}\,\cos\theta\,dr\,d\theta = \frac{6}{\pi a^3}(\pi a^4/16) = 3a/8;$$

centroid $(3a/8, 3a/8, 3a/8)$

25. $$M = \iint_R k\sqrt{x^2 + y^2}\,dA = \int_0^{2\pi} \int_0^a kr^2\,dr\,d\theta = \frac{2}{3}\pi ka^3$$

27. Let $x = r\cos\theta$, $y = r\sin\theta$, and $dA = r\,dr\,d\theta$ in formulas (15a) and (15b).

29. $\bar{x} = \bar{y}$ from the symmetry of the region, $A = \int_0^{\pi/2} \int_0^{\sin 2\theta} r\,dr\,d\theta = \pi/8$,

$$\bar{x} = \frac{1}{A} \int_0^{\pi/2} \int_0^{\sin 2\theta} r^2\cos\theta\,dr\,d\theta = (8/\pi)(16/105) = \frac{128}{105\pi};$$

centroid $(\frac{128}{105\pi}, \frac{128}{105\pi})$

31. $\bar{x} = 0$ from the symmetry of the region, $\pi a^2/2$ is the area of the semicircle, $2\pi\bar{y}$ is the distance traveled by the centroid to generate the sphere so $4\pi a^3/3 = (\pi a^2/2)(2\pi\bar{y})$, $\bar{y} = 4a/(3\pi)$

33. $\bar{x} = k$ so $V = (\pi ab)(2\pi k) = 2\pi^2 abk$

35.   The region generates a cone of volume $\frac{1}{3}\pi ab^2$ when it is revolved about the x-axis, the area of the region is $\frac{1}{2}$ ab so $\frac{1}{3}\pi ab^2 = (\frac{1}{2} ab)(2\pi\bar{y})$, $\bar{y} = b/3$. A cone of volume $\frac{1}{3}\pi a^2 b$ is generated when the region is revolved about the y-axis so $\frac{1}{3}\pi a^2 b = (\frac{1}{2} ab)(2\pi\bar{x})$, $\bar{x} = a/3$. The centroid is (a/3, b/3).

37. $I_z = \int_{-a/2}^{a/2}\int_{-b/2}^{b/2}(x^2 + y^2)\delta\,dy\,dx = 4\delta\int_0^{a/2}\int_0^{b/2}(x^2 + y^2)dy\,dx$

   $= \delta ab(a^2 + b^2)/12$; $M = \delta ab$ so $I_z = M(a^2 + b^2)/12$.

39. $I_y = \int_0^{2\pi}\int_0^R r^3\cos^2\theta\,\delta\,dr\,d\theta = \delta\pi R^4/4$; $M = \delta\pi R^2$ so $I_y = MR^2/4$.

41. $I_z = \int_0^\pi\int_0^{2R\sin\theta} r^3\delta\,dr\,d\theta = 3\delta\pi R^4/2$; $M = \delta\pi R^2$ so $I_z = 3MR^2/2$.

43. $I_z = \int_0^a\int_0^a\int_0^a (x^2 + y^2)\,\delta\,dz\,dy\,dx = 2\delta a^5/3$; $M = \delta a^3$ so $I_z = 2Ma^2/3$.

EXERCISE SET 17.7

1. $\int_0^{2\pi}\int_0^1\int_0^{\sqrt{1-r^2}} zr\,dz\,dr\,d\theta = \int_0^{2\pi}\int_0^1 \frac{1}{2}(1 - r^2)r\,dr\,d\theta = \int_0^{2\pi}\frac{1}{8}\,d\theta = \pi/4$

3. $\int_0^{\pi/2}\int_0^{\pi/2}\int_0^1 \rho^3\sin\phi\cos\phi\,d\rho\,d\phi\,d\theta = \int_0^{\pi/2}\int_0^{\pi/2}\frac{1}{4}\sin\phi\cos\phi\,d\phi\,d\theta$

   $= \int_0^{\pi/2}\frac{1}{8}\,d\theta = \pi/16$

5. $V = \int_0^{2\pi} \int_0^3 \int_{r^2}^9 r\,dz\,dr\,d\theta = \int_0^{2\pi} \int_0^3 r(9 - r^2)dr\,d\theta = \int_0^{2\pi} \frac{81}{4}\,d\theta = 81\pi/2$

7. $r^2 + z^2 = 20$ intersects $z = r^2$ in a circle of radius 2,

$V = \int_0^{2\pi} \int_0^2 \int_{r^2}^{\sqrt{20-r^2}} r\,dz\,dr\,d\theta = \int_0^{2\pi} \int_0^2 (r\sqrt{20 - r^2} - r^3)dr\,d\theta$

$= \frac{4}{3}(10\sqrt{5} - 19)\int_0^{2\pi} d\theta = 8(10\sqrt{5} - 19)\pi/3$

9. $x^2 + y^2 = 4x$ becomes $r = 4\cos\theta$ in cylindrical coordinates,

$V = \int_0^{\pi/2} \int_0^{4\cos\theta} \int_0^{\sqrt{16-r^2}} r\,dz\,dr\,d\theta = \int_0^{\pi/2} \int_0^{4\cos\theta} r\sqrt{16 - r^2}\,dr\,d\theta$

$= \int_0^{\pi/2} \frac{64}{3}(1 - \sin^3\theta)d\theta$

$= 32(3\pi - 4)/9$

11. $V = \int_0^{\pi/2} \int_{\pi/6}^{\pi/3} \int_0^2 \rho^2\sin\phi\,d\rho\,d\phi\,d\theta = \int_0^{\pi/2} \int_{\pi/6}^{\pi/3} \frac{8}{3}\sin\phi\,d\phi\,d\theta$

$= \frac{4}{3}(\sqrt{3} - 1)\int_0^{\pi/2} d\theta = 2(\sqrt{3} - 1)\pi/3$

13. $V = \int_0^{2\pi} \int_{\pi/4}^{\pi/2} \int_0^3 \rho^2\sin\phi\,d\rho\,d\phi\,d\theta = \int_0^{2\pi} \int_{\pi/4}^{\pi/2} 9\sin\phi\,d\phi\,d\theta$

$= \frac{9\sqrt{2}}{2}\int_0^{2\pi} d\theta = 9\sqrt{2}\pi$

15.   (a) $V = 2\int_0^{2\pi} \int_0^a \int_0^{\sqrt{a^2-r^2}} r\,dz\,dr\,d\theta = 4\pi a^3/3$

(b) $V = \int_0^{2\pi} \int_0^\pi \int_0^a \rho^2\sin\phi\,d\rho\,d\phi\,d\theta = 4\pi a^3/3$

17. $M = \int_0^{\pi/2} \int_0^{2\cos\theta} \int_0^{4-r^2} zr\,dz\,dr\,d\theta = \int_0^{\pi/2} \int_0^{2\cos\theta} \frac{1}{2}r(4-r^2)^2 dr\,d\theta$

$$= \frac{16}{3} \int_0^{\pi/2} (1 - \sin^6\theta)d\theta$$

$$= (16/3)(11\pi/32) = 11\pi/6$$

19. $M = \int_0^{2\pi} \int_0^{\pi/4} \int_0^1 \rho^3 \sin\phi\,d\rho\,d\phi\,d\theta = \int_0^{2\pi} \int_0^{\pi/4} \frac{1}{4}\sin\phi\,d\phi\,d\theta$

$$= \frac{1}{8}(2-\sqrt{2}) \int_0^{2\pi} d\theta = (2-\sqrt{2})\pi/4$$

21. $M = \int_0^{2\pi} \int_0^{\pi} \int_0^a k\rho^3 \sin\phi\,d\rho\,d\phi\,d\theta = \int_0^{2\pi} \int_0^{\pi} \frac{1}{4}ka^4\sin\phi\,d\phi\,d\theta$

$$= \frac{1}{2}ka^4 \int_0^{2\pi} d\theta = \pi k a^4$$

23. $\overline{x} = \overline{y} = 0$ from the symmetry of the region,

$$V = \int_0^{2\pi} \int_0^1 \int_{r^2}^{\sqrt{2-r^2}} r\,dz\,dr\,d\theta = \int_0^{2\pi} \int_0^1 (r\sqrt{2-r^2} - r^3)dr\,d\theta$$

$$= (8\sqrt{2} - 7)\pi/6,$$

$\overline{z} = \frac{1}{V} \int_0^{2\pi} \int_0^1 \int_{r^2}^{\sqrt{2-r^2}} zr\,dz\,dr\,d\theta = \dfrac{6}{(8\sqrt{2}-7)\pi}(7\pi/12) = 7/(16\sqrt{2}-14)$;

centroid $\left(0,0,\dfrac{7}{16\sqrt{2}-14}\right)$

25. $\overline{x} = \overline{y} = \overline{z}$ from the symmetry of the region, $V = \pi a^3/6$,

$\overline{z} = \frac{1}{V} \int_0^{\pi/2} \int_0^{\pi/2} \int_0^a \rho^3 \cos\phi\sin\phi\,d\rho\,d\phi\,d\theta = \dfrac{6}{\pi a^3}(\pi a^4/16) = 3a/8$;

centroid $(3a/8, 3a/8, 3a/8)$

27. $\overline{x} = \overline{z} = 0$ from the symmetry of the region,
$V = 54\pi/3 - 16\pi/3 = 38\pi/3,$

$$\overline{y} = \frac{1}{V} \int_0^{\pi} \int_0^{\pi} \int_2^3 \rho^3 \sin^2\phi \sin\theta \, d\rho \, d\phi \, d\theta = \frac{1}{V} \int_0^{\pi} \int_0^{\pi} \frac{65}{4} \sin^2\phi \sin\theta \, d\phi \, d\theta$$

$$= \frac{1}{V} \int_0^{\pi} \frac{65\pi}{8} \sin\theta \, d\theta$$

$$= \frac{3}{38\pi} (65\pi/4) = 195/152;$$

centroid $(0, 195/152, 0)$

29. $\displaystyle\int_0^{\pi} \int_0^{\pi/2} \int_0^1 e^{-\rho^3} \rho^2 \sin\phi \, d\rho \, d\phi \, d\theta = \frac{1}{3}(1 - e^{-1}) \int_0^{\pi} \int_0^{\pi/2} \sin\phi \, d\phi \, d\theta$

$$= (1 - e^{-1})\pi/3$$

31. $\displaystyle\int_0^{2\pi} \int_0^{\pi} \int_0^3 \rho^3 \sin\phi \, d\rho \, d\phi \, d\theta = 81\pi$

33. $\overline{x} = \overline{y} = 0$ from the symmetry of density and region,

$$M = \int_0^{2\pi} \int_0^{\pi/2} \int_0^a k\rho^3 \sin\phi \, d\rho \, d\phi \, d\theta = \pi k a^4/2,$$

$$\overline{z} = \frac{1}{M} \int_0^{2\pi} \int_0^{\pi/2} \int_0^a k\rho^4 \sin\phi \cos\phi \, d\rho \, d\phi \, d\theta = \frac{2}{\pi k a^4} (\pi k a^5/5) = 2a/5;$$

center of gravity $(0, 0, 2a/5)$

35. $\overline{x} = \overline{y} = 0$ from the symmetry of density and region,

$$M = \int_0^{2\pi} \int_0^1 \int_0^{1-r^2} (r^2 + z^2)r \, dz \, dr \, d\theta = \pi/4,$$

$$\overline{z} = \frac{1}{M} \int_0^{2\pi} \int_0^1 \int_0^{1-r^2} z(r^2 + z^2)r \, dz \, dr \, d\theta = (4/\pi)(11\pi/120) = 11/30;$$

center of gravity $(0, 0, 11/30)$

37. In spherical coordinates the spheres are $\rho = 3$ and $\rho = 4\cos\phi$, respectively. They intersect when $\phi = \cos^{-1}(3/4)$ so

$$V = \int_0^{2\pi} \int_0^{\cos^{-1}(3/4)} \int_0^3 \rho^2 \sin\phi \, d\rho \, d\phi \, d\theta$$

$$+ \int_0^{2\pi} \int_{\cos^{-1}(3/4)}^{\pi/2} \int_0^{4\cos\phi} \rho^2 \sin\phi \, d\rho \, d\phi \, d\theta$$

$$= \int_0^{2\pi} \int_0^{\cos^{-1}(3/4)} 9 \sin\phi \, d\phi \, d\theta + \int_0^{2\pi} \int_{\cos^{-1}(3/4)}^{\pi/2} \frac{64}{3} \sin\phi \cos^3\phi \, d\phi \, d\theta$$

$$= \frac{9}{4} \int_0^{2\pi} d\theta + \frac{27}{16} \int_0^{2\pi} d\theta = 63\pi/8$$

39. (a) The sphere and cone intersect in a circle of radius $\rho_0 \sin\phi_0$.

$$V = \int_{\theta_1}^{\theta_2} \int_0^{\rho_0 \sin\phi_0} \int_{r\cot\phi_0}^{\sqrt{\rho_0^2 - r^2}} r \, dz \, dr \, d\theta$$

$$= \int_{\theta_1}^{\theta_2} \int_0^{\rho_0 \sin\phi_0} (r\sqrt{\rho_0^2 - r^2} - r^2\cot\phi_0) dr \, d\theta$$

$$= \int_{\theta_1}^{\theta_2} \frac{1}{3} \rho_0^3 (1 - \cos^3\phi_0 - \sin^3\phi_0 \cot\phi_0) d\theta$$

$$= \frac{1}{3} \rho_0^3 (1 - \cos^3\phi_0 - \sin^2\phi_0 \cos\phi_0)(\theta_2 - \theta_1)$$

$$= \frac{1}{3} \rho_0^3 (1 - \cos\phi_0)(\theta_2 - \theta_1).$$

(b) From part (a), the volume of the solid bounded by $\theta = \theta_1$, $\theta = \theta_2$, $\phi = \phi_1$, $\phi = \phi_2$, and $\rho = \rho_0$ is

$$\frac{1}{3} \rho_0^3 (1 - \cos\phi_2)(\theta_2 - \theta_1) - \frac{1}{3} \rho_0^3 (1 - \cos\phi_1)(\theta_2 - \theta_1)$$

$$= \frac{1}{3} \rho_0^3 (\cos\phi_1 - \cos\phi_2)(\theta_2 - \theta_1)$$

so the volume of the spherical wedge between $\rho = \rho_1$ and $\rho = \rho_2$ is

$$\Delta V = \frac{1}{3} \rho_2^3 (\cos\phi_1 - \cos\phi_2)(\theta_2 - \theta_1) - \frac{1}{3} \rho_1^3 (\cos\phi_1 - \cos\phi_2)(\theta_2 - \theta_1)$$

$$= \frac{1}{3} (\rho_2^3 - \rho_1^3)(\cos\phi_1 - \cos\phi_2)(\theta_2 - \theta_1)$$

(c) $\frac{d}{d\phi} \cos\phi = -\sin\phi$ so from the mean value theorem

$$\cos\phi_2 - \cos\phi_1 = -\sin\phi^*(\phi_2 - \phi_1) \text{ where } \phi^* \text{ is between } \phi_1 \text{ and } \phi_2.$$

Similarly $\frac{d}{d\rho} \rho^3 = 3\rho^2$ so $\rho_2^3 - \rho_1^3 = 3\rho^{*2}(\rho_2 - \rho_1)$ where ρ^* is between ρ_1 and ρ_2. Thus $\cos\phi_1 - \cos\phi_2 = \sin\phi^* \Delta\phi$ and $\rho_2^3 - \rho_1^3 = 3\rho^{*2}\Delta\rho$ so $\Delta V = \rho^{*2}\sin\phi^* \Delta\rho \, \Delta\phi \, \Delta\theta$.

41. $I_y = \int_0^{2\pi} \int_0^R \int_0^h (r^2\cos^2\theta + z^2)\delta r \, dz \, dr \, d\theta$

$= \delta\int_0^{2\pi} \int_0^R (hr^3\cos^2\theta + \frac{1}{3}h^3 r)dr \, d\theta$

$= \delta\int_0^{2\pi} (\frac{1}{4}R^4 h\cos^2\theta + \frac{1}{6}R^2 h^3)d\theta = \delta(\frac{\pi}{4}R^4 h + \frac{\pi}{3}R^2 h^3)$;

$M = \delta\pi R^2 h$ so $I_y = M(R^2/4 + h^2/3)$.

43. $I_z = \int_0^{2\pi} \int_0^{\pi} \int_0^R (\rho^2\sin^2\phi)\delta \, \rho^2\sin\phi \, d\rho \, d\phi \, d\theta$

$= \delta\int_0^{2\pi} \int_0^{\pi} \int_0^R \rho^4\sin^3\phi \, d\rho \, d\phi \, d\theta = \frac{8}{15}\delta\pi R^5$; $M = \frac{4}{3}\delta\pi R^3$ so $I_z = \frac{2}{5}MR^2$.

45. (a) $\rho = \sqrt{r^2 + z^2}$, $\cos\phi = z/\rho = z/\sqrt{r^2 + z^2}$,

$F_z = \int_0^{2\pi} \int_0^R \int_a^{a+h} \frac{k\delta zr}{(r^2 + z^2)^{3/2}} \, dz \, dr \, d\theta$

$= k\delta\int_0^{2\pi} \int_0^R \left[\frac{r}{\sqrt{r^2 + a^2}} - \frac{r}{\sqrt{r^2 + (a + h)^2}}\right]dr \, d\theta$

$= 2\pi k\delta(\sqrt{R^2 + a^2} - \sqrt{R^2 + (a + h)^2} + h)$.

(b) The components of the force in the x and y directions are zero because of the symmetry of the solid with respect to the yz and xz planes.

47. In spherical coordinates the plane $z = h$ is $\rho \cos \phi = h$ so $\rho = h/\cos \phi$,

$$F_z = \int_0^{2\pi} \int_0^{\tan^{-1}(R/h)} \int_0^{h/\cos \phi} \frac{k\delta \cos \phi}{\rho^2} \rho^2 \sin \phi \, d\rho \, d\phi \, d\theta$$

$$= k\delta \int_0^{2\pi} \int_0^{\tan^{-1}(R/h)} \int_0^{h/\cos \phi} \sin \phi \cos \phi \, d\rho \, d\phi \, d\theta$$

$$= 2\pi k\delta h[1 - \cos(\tan^{-1} \frac{R}{h})] = 2\pi k\delta h(1 - h/\sqrt{R^2 + h^2}).$$

SUPPLEMENTARY EXERCISES CHAPTER 17

1. $\displaystyle\int_{1/2}^1 \int_0^{2x} \cos(\pi x^2) dy \, dx = \int_{1/2}^1 2x \cos(\pi x^2) dx = -1/(\sqrt{2}\pi)$

3. $\displaystyle\int_{-1}^0 \int_0^{y^2} \int_{xy}^1 2y \, dz \, dx \, dy = \int_{-1}^0 \int_0^{y^2} (2y - 2xy^2) dx \, dy$

$$= \int_{-1}^0 (2y^3 - y^6) dy = -9/14$$

5. $\displaystyle\int_0^1 \int_{2y}^2 e^x e^y dx \, dy$

7. $A = \displaystyle\int_0^1 \int_0^{2x^3} dy \, dx + \int_1^2 \int_0^{4-2x} dy \, dx = 1/2 + 1 = 3/2$

9. (a)   (b)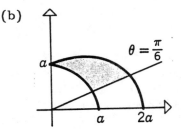

11. $\displaystyle\int_0^8 \int_{3\sqrt{-y}}^{3\sqrt{y}} x^2 \sin(y^2)\,dx\,dy \; = \; \int_0^8 \frac{2}{3}\, y \sin(y^2)\,dy \; = \; (1 - \cos 64)/3$

13. $V = \displaystyle\int_0^2 \int_0^{(6-3x)/2} (6 - 3x - 2y)\,dy\,dx \; = \; \int_0^2 \frac{1}{4}\,(6 - 3x)^2 dx \; = \; 6$

15. $V = \displaystyle\int_0^{2\pi} \int_{\sqrt{2}}^2 \frac{1}{r^3}\,dr\,d\theta \; = \; \int_0^{2\pi} \frac{1}{8}\,d\theta \; = \; \pi/4$

17. $\displaystyle\int_0^{2a} \int_0^{\sqrt{2ay-y^2}} \frac{2xy}{x^2 + y^2}\,dx\,dy \; = \; \int_0^{2a} (y\ln 2a - y\ln y)\,dy \; = \; a^2$

19. $A = 6\displaystyle\int_0^{\pi/6} \int_0^{\cos 3\theta} r\,dr\,d\theta \; = \; 3\int_0^{\pi/6} \cos^2 3\theta\,d\theta \; = \; \pi/4$

21. $z_x = 6x,\; z_y = 6y,\; z_x^2 + z_y^2 + 1 = 36(x^2 + y^2) + 1;$

   $S = \displaystyle\int_0^{2\pi} \int_0^1 r\sqrt{36r^2 + 1}\,dr\,d\theta \; = \; (37\sqrt{37} - 1)\pi/54$

23. $z_x = x/z,\; z_y = y/z,\; z_x^2 + z_y^2 + 1 = 2;\; S = \displaystyle\int_0^{2\pi} \int_1^4 \sqrt{2}\,r\,dr\,d\theta \; = \; 15\sqrt{2}\pi$

25. $\displaystyle\int_0^{2\pi} \int_0^4 \int_0^{4-r\sin\theta} r^2\,dz\,dr\,d\theta \; = \; \int_0^{2\pi} \int_0^4 (4r^2 - r^3\sin\theta)\,dr\,d\theta$

   $= \displaystyle\int_0^{2\pi} \frac{64}{3}\,(4 - 3\sin\theta)\,d\theta \; = \; 512\pi/3$

27. (a) G is the region in the first octant bounded by the coordinate planes
   and the plane $z = 1 - x - 2y$; the integral is

   $\displaystyle\int_0^{1/2} \int_0^{1-2y} \int_0^{1-2y-z} z\,dx\,dz\,dy$

(b) G is the region in the first octant bounded by the planes x = 0,
z = 0, z = 4 − y, and the parabolic cylinder y = x^2; the integral is

$$\int_0^4 \int_0^{4-y} \int_0^{\sqrt{y}} 3 \, dx \, dz \, dy$$

29. (a) $\displaystyle\int_0^{2\pi} \int_0^{\pi/3} \int_0^a \rho^4 \sin^3\phi \, d\rho \, d\phi \, d\theta$ (b) $\displaystyle\int_0^{2\pi} \int_0^{\sqrt{3}a/2} \int_{r/\sqrt{3}}^{\sqrt{a^2-r^2}} r^3 dz \, dr \, d\theta$

(c) $\displaystyle\int_{-\sqrt{3}a/2}^{\sqrt{3}a/2} \int_{-\sqrt{3a^2/4-x^2}}^{\sqrt{3a^2/4-x^2}} \int_{\sqrt{x^2+y^2}/\sqrt{3}}^{\sqrt{a^2-x^2-y^2}} (x^2 + y^2) dz \, dy \, dx$

31. $V = \displaystyle\int_0^{2\pi} \int_0^{\pi} \int_0^{a(1+\cos\phi)} \rho^2 \sin\phi \, d\rho \, d\phi \, d\theta = \int_0^{2\pi} \int_0^{\pi} \frac{1}{3} a^3 (1 + \cos\phi)^3 \sin\phi \, d\phi \, d\theta$

$$= \frac{4}{3} a^3 \int_0^{2\pi} d\theta = 8\pi a^3 / 3$$

33. $V = \displaystyle\int_0^{2\pi} \int_0^{a/\sqrt{3}} \int_{\sqrt{3}r}^a r \, dz \, dr \, d\theta = \int_0^{2\pi} \int_0^{a/\sqrt{3}} (ar - \sqrt{3}r^2) dr \, d\theta$

$$= \frac{1}{18} a^3 \int_0^{2\pi} d\theta = \pi a^3 / 9$$

35. $\bar{x} = 0$ from the symmetry of the region,

$A = \displaystyle\int_0^{2\pi} \int_0^{a(1+\sin\theta)} r \, dr \, d\theta = 3\pi a^2 / 2,$

$\bar{y} = \frac{1}{A} \displaystyle\int_0^{2\pi} \int_0^{a(1+\sin\theta)} r^2 \sin\theta \, dr \, d\theta = \frac{1}{A} \int_0^{2\pi} \frac{1}{3} a^3 (1 + \sin\theta)^3 \sin\theta \, d\theta$ but

$(1 + \sin\theta)^3 \sin\theta = \sin\theta + 3\sin^2\theta + 3\sin^3\theta + \sin^4\theta$ and

$\displaystyle\int_0^{2\pi} \sin\theta \, d\theta = \int_0^{2\pi} 3\sin^3\theta \, d\theta = 0$ so

$$\bar{y} = \frac{1}{A} \int_0^{2\pi} \frac{1}{3} a^3(3\sin^2\theta + \sin^4\theta)d\theta = \frac{2}{3\pi a^2}(5\pi a^3/4) = 5a/6;$$

centroid $(0,5a/6)$

37. $\bar{x} = 0$ from the symmetry of density and region,

$$M = 2\int_0^a \int_0^{b(1-x/a)} kx\,dy\,dx = ka^2b/3,$$

$$\bar{y} = \frac{2}{M} \int_0^a \int_0^{b(1-x/a)} k\,xy\,dy\,dx = \frac{6}{ka^2b}(ka^2b^2/24) = b/4;$$

center of gravity $(0,b/4)$

39. $M = \int_0^a \int_0^{b(1-x/a)} \int_0^{c(1-x/a-y/b)} kz\,dz\,dy\,dx$

$$= \frac{1}{2} kc^2 \int_0^a \int_0^{b(1-x/a)} (1 - x/a - y/b)^2 dy\,dx$$

$$= \frac{1}{6} kbc^2 \int_0^a (1 - x/a)^3 dx = \frac{1}{24} kabc^2$$

41. $\bar{x} = 0$ from the symmetry of the region,

$$V = 2\int_0^2 \int_{x^2}^4 \int_0^{4-y} dz\,dy\,dx = 256/15,$$

$$\bar{y} = \frac{2}{V} \int_0^2 \int_{x^2}^4 \int_0^{4-y} y\,dz\,dy\,dx = (15/128)(512/35) = 12/7,$$

$$\bar{z} = \frac{2}{V} \int_0^2 \int_{x^2}^4 \int_0^{4-y} z\,dz\,dy\,dx = (15/128)(1024/105) = 8/7;$$

centroid $(0,12/7,8/7)$

43. $\bar{x} = \bar{y} = 0$ from the symmetry of the region, $V = \pi R^2 h/3$,

$$\bar{z} = \frac{1}{V} \int_0^{2\pi} \int_0^R \int_0^{h(1-r/R)} zr\,dz\,dr\,d\theta = \frac{3}{\pi R^2 h}(\pi R^2 h^2/12) = h/4;$$

centroid $(0,0,h/4)$

CHAPTER 18

EXERCISE SET 18.1

1. (a) $\int_0^1 (2t + t^2)dt = 4/3$ (b) $\int_0^1 (t^2 - t^2)2t \, dt = 0$

    (c) $\int_C (2x + y)dx + \int_C (x^2 - y)dy = 4/3 + 0 = 4/3$

3. $\int_0^2 (t^2/2 - t^3)dt = -8/3$

5. $\int_0^{\pi/4} (8\cos^2 t - 16\sin^2 t - 20\sin t \cos t)dt = 1 - \pi$

7. C: $x = (3 - t)^2/3$, $y = 3 - t$, $0 \le t \le 3$; $\int_0^3 \frac{1}{3}(3 - t)^2 dt = 3$

9. C: $x = \cos t$, $y = \sin t$, $0 \le t \le \pi/2$;
    $\int_0^{\pi/2} (-\sin t - \cos^2 t)dt = -1 - \pi/4$

11. $\int_0^\pi (0)dt = 0$ 13. $\int_0^1 e^{-t}dt = 1 - e^{-1}$

15. $\int_0^1 (-3)e^{3t}dt = 1 - e^3$

17. $\int_0^{\pi/2} (7\sin^2 t \cos t + 3\sin t \cos t)dt = 23/6$

19.   C: $x = t^2$, $y = t$, $0 \leq t \leq 1$; $W = \int_0^1 3t^4 dt = 3/5$

21.   (a) C_1:   (1,1) to (2,2); $x = 1 + t$, $y = 1 + t$, $0 \leq t \leq 1$
        C_2:   (2,2) to (4,2); $x = 2 + 2t$, $y = 2$, $0 \leq t \leq 1$

   $\int_0^1 \frac{5/2}{(1 + t)^2} dt + \int_0^1 \frac{1/2}{1 + (1 + t)^2} dt = 5/4 - \pi/8 + \frac{1}{2} \tan^{-1} 2$

      (b) C_1:   (0,3) to (6,3); $x = 6t$, $y = 3$, $0 \leq t \leq 1$
        C_2:   (6,3) to (6,0); $x = 6$, $y = 3 - 3t$, $0 \leq t \leq 1$

   $\int_0^1 \frac{6}{36t^2 + 9} dt + \int_0^1 \frac{-12}{36 + 9(1 - t)^2} dt = \frac{1}{3} \tan^{-1} 2 - \frac{2}{3} \tan^{-1}(1/2)$

      (c) C: $x = 4\cos t$, $y = 4 \sin t$, $0 \leq t \leq \pi/2$
        $\int_0^{\pi/2} (-\frac{1}{4} \sin t + \cos t) dt = 3/4$

23. C_1:   (0,0,0) to (1,3,1); $x = t$, $y = 3t$, $z = t$, $0 \leq t \leq 1$
     C_2:   (1,3,1) to (2,-1,4); $x = 1 + t$, $y = 3 - 4t$, $z = 1 + 3t$, $0 \leq t \leq 1$

   $W = \int_0^1 (4t + 8t^2) dt + \int_0^1 (-11 - 17t - 11t^2) dt = -37/2$

25. $W = \int_0^1 \lambda[(1 - \lambda)t + (3\lambda - 1)t^2 - (1 + 2\lambda)t^3] dt = -\lambda/12$, $W = 1$
     when $\lambda = -12$

27.   C: $x = a\cos t$, $y = a \sin t$, $0 \leq t \leq \pi$;
     $F = \frac{k(x\mathbf{i} + y\mathbf{j})}{(x^2 + y^2)^{3/2}} = \frac{k}{a^2} (\cos t\, \mathbf{i} + \sin t\, \mathbf{j})$,

     $d\mathbf{r}/dt = a(-\sin t\, \mathbf{i} + \cos t\, \mathbf{j})$, $W = \int_0^\pi (0) dt = 0$

EXERCISE SET 18.2

1. $\partial x/\partial y = 0 = \partial y/\partial x$, conservative so $\partial\phi/\partial x = x$ and $\partial\phi/\partial y = y$,
    $\phi = x^2/2 + k(y)$, $k'(y) = y$, $k(y) = y^2/2 + K$, $\phi = x^2/2 + y^2/2 + K$

3. $\partial(x^2 y)/\partial y = x^2$ and $\partial(5xy^2)/\partial x = 5y^2$, not conservative

5. $\partial(\cos y + y\cos x)/\partial y = -\sin y + \cos x = \partial(\sin x - x\sin y)/\partial x$, conservative
    so $\partial\phi/\partial x = \cos y + y\cos x$ and $\partial\phi/\partial y = \sin x - x\sin y$,
    $\phi = x\cos y + y\sin x + k(y)$, $-x\sin y + \sin x + k'(y) = \sin x - x\sin y$,
    $k'(y) = 0$, $k(y) = K$, $\phi = x\cos y + y\sin x + K$

7. $\partial(y^2)/\partial y = 2y = \partial(2xy)/\partial x$, independent of path
    (a) $\partial\phi/\partial x = y^2$ and $\partial\phi/\partial y = 2xy$, $\phi = xy^2 + k(y)$, $2xy + k'(y) = 2xy$,
        $k'(y) = 0$, $k(y) = K$, $\phi = xy^2 + K$. Let $K = 0$ to get
        $\phi(1,3) - \phi(-1,2) = 9 - (-4) = 13$
    (b) C: $x = -1 + 2t$, $y = 2 + t$, $0 \leq t \leq 1$; $\displaystyle\int_0^1 (4 + 14t + 6t^2)dt = 13$

9. $\partial(3y)/\partial y = 3 = \partial(3x)/\partial x$, $\phi = 3xy$, $\phi(4,0) - \phi(1,2) = -6$

11. $\partial(2xe^y)/\partial y = 2xe^y = \partial(x^2 e^y)/\partial x$, $\phi = x^2 e^y$, $\phi(3,2) - \phi(0,0) = 9e^2$

13. $\partial(2xy^3)/\partial y = 6xy^2 = \partial(3x^2 y^2)/\partial x$, $\phi = x^2 y^3$, $\phi(-1,0) - \phi(2,-2) = 32$

15. $\phi = x^2 y^2/2$, $W = \phi(0,0) - \phi(1,1) = -1/2$

17. $\phi = \tan^{-1}(x/y)$, $W = \phi(3,3) - \phi(0,2) = \pi/4$

19. $\partial(e^y + ye^x)/\partial y = e^y + e^x = \partial(xe^y + e^x)/\partial x$ so $\mathbf{F}$ is conservative,
    $\phi = xe^y + ye^x$ so the work done in parts (a), (b), and (c) is
    $\phi(-a,0) - \phi(a,0) = -2a$. From Theorem 18.2.2 the work done in part
    (d) is 0.

21. $F = \dfrac{kx}{(x^2 + y^2)^{3/2}} i + \dfrac{ky}{(x^2 + y^2)^{3/2}} j$,

$\dfrac{\partial}{\partial y} \dfrac{kx}{(x^2 + y^2)^{3/2}} = - \dfrac{3k\,xy}{(x^2 + y^2)^{5/2}} = \dfrac{\partial}{\partial x} \dfrac{ky}{(x^2 + y^2)^{3/2}}$

so **F** is conservative

23. (a) If **F** is conservative then $F = (\partial\phi/\partial x)i + (\partial\phi/\partial y)j$ for some function
ϕ so $f(x,y) = \partial\phi/\partial x$ and $g(x,y) = \partial\phi/\partial y$. But
$d\phi = (\partial\phi/\partial x)dx + (\partial\phi/\partial y)dy$ so $d\phi = f(x,y)dx + g(x,y)dy$. If ϕ is
such that $d\phi = f(x,y)dx + g(x,y)dy$ then $f(x,y) = \partial\phi/\partial x$ and
$g(x,y) = \partial\phi/\partial y$ so $F = \nabla\phi$ and hence **F** is conservative.

(b) $\displaystyle\int_{(x_0,y_0)}^{(x_1,y_1)} d\phi = \int_{(x_0,y_0)}^{(x_1,y_1)} f(x,y)dx + g(x,y)dy = \phi(x_1,y_1) - \phi(x_0,y_0)$

EXERCISE SET 18.3

1. $\displaystyle\iint_R (2x - 2y)dA = \int_0^1 \int_0^1 (2x - 2y)dy\,dx = 0$

3. $\displaystyle\int_{-2}^4 \int_1^2 (2y - 3x)dy\,dx = 0$

5. $\displaystyle\int_0^{\pi/2} \int_0^{\pi/2} (-y\cos x + x\sin y)dy\,dx = 0$

7. $\displaystyle\iint_R [1 - (-1)]dA = 2\iint_R dA = 8\pi$

9. $\displaystyle\iint_R (- \frac{y}{1 + y} - \frac{1}{1 + y})dA = -\iint_R dA = -4$

11. $\displaystyle\iint_R (- \frac{y^2}{1 + y^2} - \frac{1}{1 + y^2})dA = -\iint_R dA = -1$

13. $\int_0^1 \int_{x^2}^{\sqrt{x}} (y^2 - x^2) dy\, dx = 0$

15. (a) $\int_C x\, dy = \int_0^{2\pi} ab\cos^2 t\, dt = \pi ab$

(b) $\int_C -y\, dx = \int_0^{2\pi} ab\sin^2 t\, dt = \pi ab$

17. C_1: (0,0) to (a,0); $x = at$, $y = 0$, $0 \le t \le 1$
    C_2: (a,0) to (0,b); $x = a - at$, $y = bt$, $0 \le t \le 1$
    C_3: (0,b) to (0,0); $x = 0$, $y = b - bt$, $0 \le t \le 1$

$$A = \int_C x\, dy = \int_0^1 (0)dt + \int_0^1 ab(1 - t)dt + \int_0^1 (0)dt = \tfrac{1}{2} ab$$

19. $W = \iint_R y\, dA = \int_0^\pi \int_0^5 r^2\sin\theta\, dr\, d\theta = 250/3$

21. (a) $\bar{x} = \frac{1}{A}\iint_R x\, dA$, but $\int_C \frac{1}{2} x^2 dy = \iint_R x\, dA$ from Green's Theorem so

$\bar{x} = \frac{1}{A}\int_C \frac{1}{2} x^2 dy = \frac{1}{2A}\int_C x^2 dy$. Similarly, $\bar{y} = -\frac{1}{2A}\int_C y^2 dx$.

(b) $\bar{x} = 0$ from the symmetry of the region,

C_1: (a,0) to (-a,0) along $y = \sqrt{a^2 - x^2}$; $x = a\cos t$, $y = a\sin t$, $0 \le t \le \pi$

C_2: (-a,0) to (a,0); $x = -a + 2at$, $y = 0$, $0 \le t \le 1$

$$A = \pi a^2/2, \quad \bar{y} = -\frac{1}{2A}\left[\int_0^\pi -a^3\sin^3 t\, dt + \int_0^1 (0)dt\right]$$

$$= -\frac{1}{\pi a^2}\left(-\frac{4a^3}{3}\right) = \frac{4a}{3\pi}; \text{ centroid } (0, \frac{4a}{3\pi})$$

23.   (a) C: $x = a + (c - a)t, \; y = b + (d - b)t, \; 0 \le t \le 1$

$$\int_C x\,dy - y\,dx = \int_0^1 (ad - bc)dt = ad - bc$$

     (b) Let C_1, C_2, and C_3 be the line segments from (x_1, y_1) to (x_2, y_2), (x_2, y_2) to (x_3, y_3), and (x_3, y_3) to (x_1, y_1), then from (4c)

$$A = \sum_{i=1}^{3} \frac{1}{2} \int_{C_i} x\,dy - y\,dx \text{ and from the result of part (a)}$$

$$A = \frac{1}{2} [(x_1 y_2 - x_2 y_1) + (x_2 y_3 - x_3 y_2) + (x_3 y_1 - x_1 y_3)]$$

     (c) $A = \frac{1}{2} [(x_1 y_2 - x_2 y_1) + (x_2 y_3 - x_3 y_2) + \cdots + (x_n y_1 - x_1 y_n)]$

     (d) $A = \frac{1}{2} [(0 - 0) + (6 + 8) + (0 + 2) + (0 - 0)] = 8$

25.   (a) $\displaystyle\int_0^{2\pi} (\sin^2 t + \cos^2 t)dt = \int_0^{2\pi} dt = 2\pi$

     (b) $\partial g / \partial x = \dfrac{y^2 - x^2}{(x^2 + y^2)^2} = \partial f / \partial y$

     (c) f and g do not have continuous first partial derivatives at the point (0,0) in R so Green's Theorem is not applicable.

EXERCISE SET 18.4

1.   R is the circular region enclosed by $x^2 + y^2 = 1$;

$$\iint_\sigma \delta_0\,dS = \delta_0 \iint_R \sqrt{4x^2 + 4y^2 + 1}\;dA = \delta_0 \int_0^{2\pi} \int_0^1 \sqrt{4r^2 + 1}\; r\,dr\,d\theta$$

$$= \frac{1}{12}(5\sqrt{5} - 1)\delta_0 \int_0^{2\pi} d\theta = \frac{\pi}{6}(5\sqrt{5} - 1)\delta_0 .$$

3. $z = \sqrt{4 - x^2}$, $\dfrac{\partial z}{\partial x} = -\dfrac{x}{\sqrt{4 - x^2}}$, $\dfrac{\partial z}{\partial y} = 0$;

$$\iint_\sigma \delta_0 dS = \delta_0 \iint_R \sqrt{\frac{x^2}{4 - x^2} + 1}\, dA = 2\delta_0 \int_0^4 \int_0^1 \frac{1}{\sqrt{4 - x^2}}\, dx\, dy = \frac{4}{3}\, \pi\delta_0.$$

5. R is the annular region between $x^2 + y^2 = 1$ and $x^2 + y^2 = 4$;

$$\iint_\sigma z^2 dS = \iint_R (x^2 + y^2)\sqrt{\frac{x^2}{x^2 + y^2} + \frac{y^2}{x^2 + y^2} + 1}\, dA$$

$$= \sqrt{2} \iint_R (x^2 + y^2)\, dA = \sqrt{2} \int_0^{2\pi} \int_1^2 r^3 dr\, d\theta = \frac{15}{2}\, \pi\sqrt{2}.$$

7. $z = \sqrt{1 - x^2}$, R is the rectangular region enclosed by $x = -1$, $x = 1$,

y = 0 and y = 1; $\partial z/\partial x = -x/\sqrt{1 - x^2}$ which does not exist along the boundaries $x = \pm 1$. We avoid them by using $x = \pm x_0$ as boundaries where x_0 is slightly smaller than 1 and then let x_0 approach 1. By the symmetry of the surface and the integrand we can integrate over the region R' enclosed by $x = 0$, $x = x_0$, y = 0 and y = 1 and double the result so

$$\iint_\sigma x^2 y\, dS = \lim_{x_0 \to 1^-} 2 \iint_{R'} x^2 y \sqrt{\frac{x^2}{1 - x^2} + 1}\, dA = \lim_{x_0 \to 1^-} 2 \int_0^{x_0} \int_0^1 \frac{x^2 y}{\sqrt{1 - x^2}}\, dy\, dx$$

$$= \lim_{x_0 \to 1^-} \int_0^{x_0} \frac{x^2}{\sqrt{1 - x^2}}\, dx; \text{ use } x = \sin\theta \text{ to get}$$

$$\iint_\sigma x^2 y\, dS = \lim_{\theta_0 \to \pi/2^-} \int_0^{\theta_0} \sin^2\theta\, d\theta \quad (\theta_0 = \sin^{-1} x_0)$$

$$= \lim_{\theta_0 \to \pi/2^-} \left(\frac{1}{2}\, \theta_0 - \frac{1}{4} \sin 2\theta_0\right) = \frac{\pi}{4}.$$

9. If we use the projection of σ onto the xz-plane then $y = 1 - x$ and R is the rectangular region in the xz-plane enclosed by $x = 0$, $x = 1$, $z = 0$

and $z = 1$;

$$\iint_{\sigma}(x + y + z)dS = \iint_{R}(1 + z)\sqrt{2}dA = \sqrt{2}\int_{0}^{1}\int_{0}^{1}(1 + z)dz\,dx = \frac{3\sqrt{2}}{2}.$$

11. There are six surfaces:
σ_1: $z = 0$; $0 \leq x \leq 1$, $0 \leq y \leq 1$ (project onto xy-plane),
σ_2: $x = 0$; $0 \leq y \leq 1$, $0 \leq z \leq 1$ (project onto yz-plane),
σ_3: $y = 0$; $0 \leq x \leq 1$, $0 \leq z \leq 1$ (project onto xz-plane),
σ_4: $z = 1$; $0 \leq x \leq 1$, $0 \leq y \leq 1$ (project onto xy-plane),
σ_5: $x = 1$; $0 \leq y \leq 1$, $0 \leq z \leq 1$ (project onto yz-plane),
σ_6: $y = 1$; $0 \leq x \leq 1$, $0 \leq z \leq 1$ (project onto xz-plane), so

$$\iint_{\sigma_1}(x + y + z)dS = \int_{0}^{1}\int_{0}^{1}(x + y)dx\,dy = 1,$$

$$\iint_{\sigma_2}(x + y + z)dS = \int_{0}^{1}\int_{0}^{1}(y + z)dy\,dz = 1,$$

$$\iint_{\sigma_3}(x + y + z)dS = \int_{0}^{1}\int_{0}^{1}(x + z)dx\,dz = 1,$$

$$\iint_{\sigma_4}(x + y + z)dS = \int_{0}^{1}\int_{0}^{1}(x + y + 1)dx\,dy = 2,$$

$$\iint_{\sigma_5}(x + y + z)dS = \int_{0}^{1}\int_{0}^{1}(1 + y + z)dy\,dz = 2,$$

$$\iint_{\sigma_6}(x + y + z)dS = \int_{0}^{1}\int_{0}^{1}(x + 1 + z)dx\,dz = 2,$$

thus, $\iint_{\sigma}(x + y + z)dS = 1 + 1 + 1 + 2 + 2 + 2 = 9.$

13. R is the circular region enclosed by $x^2 + y^2 = 1$;

$$\iint_\sigma \sqrt{x^2 + y^2 + z^2}\, dS = \iint_R \sqrt{2(x^2 + y^2)} \sqrt{\frac{x^2}{x^2 + y^2} + \frac{y^2}{x^2 + y^2} + 1}\, dA$$

$$= \lim_{r_0 \to 0^+} 2 \iint_{R'} \sqrt{x^2 + y^2}\, dA$$

where R' is the annular region enclosed by $x^2 + y^2 = 1$ and $x^2 + y^2 = r_0^2$

with r_0 slightly larger than 0 because $\sqrt{\frac{x^2}{x^2 + y^2} + \frac{y^2}{x^2 + y^2} + 1}$ is not

defined for $x^2 + y^2 = 0$, so

$$\iint_\sigma \sqrt{x^2 + y^2 + z^2}\, dS = \lim_{r_0 \to 0^+} 2 \int_0^{2\pi} \int_{r_0}^1 r^2 dr\, d\theta = \lim_{r_0 \to 0^+} \frac{4\pi}{3}(1 - r_0^3) = \frac{4\pi}{3}.$$

15. Use $z = \sqrt{a^2 - x^2 - y^2}$ and let R be the circular region enclosed by
$x^2 + y^2 = a^2$, $x = 0$, and $y = 0$ then by symmetry of the surface and the
integrand

$$\iint_\sigma (x^2 + y^2) dS = 2 \iint_R (x^2 + y^2) \sqrt{\frac{x^2}{a^2 - x^2 - y^2} + \frac{y^2}{a^2 - x^2 - y^2} + 1}\, dA$$

$$= 2a \iint_R \frac{x^2 + y^2}{\sqrt{a^2 - x^2 - y^2}}\, dA,$$

but the integrand is not defined along the boundary $x^2 + y^2 = a^2$ so

$$\iint_\sigma (x^2 + y^2) dS = \lim_{r_0 \to a^-} 2a \int_0^{2\pi} \int_0^{r_0} \frac{r^3}{\sqrt{a^2 - r^2}}\, dr\, d\theta$$

$$= \lim_{r_0 \to a^-} 4\pi a \left[\frac{2}{3} a^3 + \frac{1}{3}(a^2 - r_0^2)^{3/2} - a^2\sqrt{a^2 - r_0^2}\right] = \frac{8\pi}{3} a^4.$$

17. $z = 4 - y^2$, R is the rectangular region enclosed by $x = 0$, $x = 3$, $y = 0$ and $y = 3$;

$$\iint_{\sigma} y \, dS = \iint_{R} y\sqrt{4y^2 + 1} \, dA = \int_0^3 \int_0^3 y\sqrt{4y^2 + 1} \, dy \, dx = \frac{1}{4}(37\sqrt{37} - 1).$$

19. (a) $\dfrac{\sqrt{29}}{16} \displaystyle\int_0^6 \int_0^{(12-2x)/3} xy(12 - 2x - 3y) dy \, dx$

   (b) $\dfrac{\sqrt{29}}{4} \displaystyle\int_0^3 \int_0^{(12-4z)/3} yz(12 - 3y - 4z) dy \, dz$

   (c) $\dfrac{\sqrt{29}}{9} \displaystyle\int_0^3 \int_0^{6-2z} xz(12 - 2x - 4z) dx \, dz$

21. $\displaystyle\int_0^4 \int_1^2 y^3 z\sqrt{4y^2 + 1} \, dy \, dz$; $\dfrac{1}{2} \displaystyle\int_0^4 \int_1^4 xz\sqrt{1 + 4x} \, dx \, dz$

23. $M = \displaystyle\iint_{\sigma} \delta(x,y,z) dS = \iint_{\sigma} \delta_0 dS = \delta_0 \iint_{\sigma} dS = \sigma_0 S$

EXERCISE SET 18.5

1. From the orientation of the plane surface we see that upward, right, and forward unit normals should be used; in each case the result is

$$\frac{2}{\sqrt{29}} i + \frac{3}{\sqrt{29}} j + \frac{4}{\sqrt{29}} k.$$

3. (a) Use an upward unit normal to get $n = -\dfrac{2}{\sqrt{21}} i - \dfrac{4}{\sqrt{21}} j + \dfrac{1}{\sqrt{21}} k.$

   (b) Use an upward unit normal to get $n = \dfrac{3}{5\sqrt{2}} i - \dfrac{4}{5\sqrt{2}} j + \dfrac{1}{\sqrt{2}} k.$

   (c) Use a downward unit normal with $z = -\sqrt{25 - x^2}$ to get $n = \dfrac{3}{5} i - \dfrac{4}{5} k.$

5. R is the circular region enclosed by $x^2 + y^2 = 1$;

$$\iint_\sigma \mathbf{F} \cdot \mathbf{n} \, dS = \iint_R (2x^2 + 2y^2 + 2z) dA = 2\iint_R dA = (2)(\text{area of } R) = 2\pi.$$

7. R is the circular reigon enclosed by $x^2 + y^2 = 1$;

$$\iint_\sigma \mathbf{F} \cdot \mathbf{n} \, dS \lim_{r_0 \to 1^-} \iint_{R_{r_0}} z^2 dA = \lim_{r_0 \to 1^-} \iint_{R_{r_0}} (1 - x^2 - y^2) dA$$

where R_{r_0} is the region enclosed by $x^2 + y^2 = r_0^2$ with r_0 slightly less

than 1 because $-\dfrac{\partial z}{\partial x} \mathbf{i} - \dfrac{\partial z}{\partial y} \mathbf{j} + \mathbf{k}$ is not defined along the boundary

$x^2 + y^2 = 1$, so

$$\iint_\sigma \mathbf{F} \cdot \mathbf{n} \, dS \lim_{r_0 \to 1^-} \int_0^{2\pi} \int_0^{r_0} (1 - r^2) r \, dr \, d\theta = \lim_{r_0 \to 1^-} 2\pi(\tfrac{1}{2} r_0^2 - \tfrac{1}{4} r_0^4) = \tfrac{\pi}{2}.$$

9. R is the circular region enclosed by $x^2 + y^2 = 9$;

$$\iint_\sigma \mathbf{F} \cdot \mathbf{n} \, dS = \iint_R \left(\frac{x^2}{\sqrt{9 - x^2 - y^2}} + \frac{y^2}{\sqrt{9 - x^2 - y^2}} + z \right) dA$$

$$= 9\iint_R \frac{1}{\sqrt{9 - x^2 - y^2}} \, dA = \lim_{r_0 \to 3^-} 9 \int_0^{2\pi} \int_0^{r_0} \frac{r}{\sqrt{9 - r^2}} \, dr \, d\theta$$

$$= \lim_{r_0 \to 3^-} 18\pi(3 - \sqrt{9 - r_0^2}) = 54\pi.$$

11. R is the annular region enclosed by $x^2 + y^2 = 1$ and $x^2 + y^2 = 4$;

$$\iint_\sigma \mathbf{F} \cdot \mathbf{n} \, dS = \iint_R \left(-\frac{x^2}{\sqrt{x^2 + y^2}} - \frac{y^2}{\sqrt{x^2 + y^2}} + 2z \right) dA$$

$$= \iint_R \sqrt{x^2 + y^2} \, dA = \int_0^{2\pi} \int_1^2 r^2 dr \, d\theta = \frac{14\pi}{3}.$$

13. R is the circular region enclosed by $x^2 + y^2 - y = 0$;
$\iint_\sigma \mathbf{F} \cdot \mathbf{n}\, dS = \iint_R (-x)\, dA$, but in polar coordinates the boundary
$x^2 + y^2 - y = 0$ is $r = \sin\theta$, so
$$\iint_\sigma \mathbf{F} \cdot \mathbf{n}\, dS = -\int_0^\pi \int_0^{\sin\theta} r^2 \cos\theta\, dr\, d\theta = 0.$$

15. Divide the surface into two parts σ_1 and σ_2 corresponding to
$z = \sqrt{a^2 - x^2 - y^2}$ and $z = -\sqrt{a^2 - x^2 - y^2}$, respectively. For each part R
is the circular region enclosed by $x^2 + y^2 = a^2$. σ_1 is oriented by
upward normals and σ_2 by downward normals so
$$\iint_{\sigma_1} \mathbf{F} \cdot \mathbf{n}\, dS = \iint_R \left(\frac{x^2}{\sqrt{a^2 - x^2 - y^2}} + \frac{y^2}{\sqrt{a^2 - x^2 - y^2}} + z \right) dA$$
$$= a^2 \iint_R \frac{1}{\sqrt{a^2 - x^2 - y^2}}\, dA = \lim_{r_0 \to a^-} a^2 \int_0^{2\pi} \int_0^{r_0} \frac{r}{\sqrt{a^2 - r^2}}\, dr\, d\theta$$
$$= \lim_{r_0 \to a^-} 2\pi a^2 (a - \sqrt{a^2 - r_0^2}) = 2\pi a^3;$$
similarly $\iint_{\sigma_2} \mathbf{F} \cdot \mathbf{n}\, dS = a^2 \iint_R \dfrac{1}{\sqrt{a^2 - x^2 - y^2}}\, dA = 2\pi a^3$ so
$$\iint_\sigma \mathbf{F} \cdot \mathbf{n}\, dS = 2\pi a^3 + 2\pi a^3 = 4\pi a^3.$$

17. Replace $\mathbf{n}$ by $-\mathbf{n}$ to reverse the orientation of σ, so
$$\iint_\sigma \mathbf{F} \cdot (-\mathbf{n})\, dS = -\iint_\sigma \mathbf{F} \cdot \mathbf{n}\, dS.$$

19. (a) $\iint\limits_{R} F \cdot (i - \frac{\partial x}{\partial y} j - \frac{\partial x}{\partial z} k)dA$, σ oriented by forward normals, and

$\iint\limits_{R} F \cdot (-i + \frac{\partial x}{\partial y} j + \frac{\partial x}{\partial z} k)dA$, σ oriented by backward normals where R

is the projection of σ onto the yz-plane.

(b) $\iint\limits_{R} F \cdot (- \frac{\partial y}{\partial x} i + j - \frac{\partial y}{\partial z} k)dA$, σ oriented by right normals, and

$\iint\limits_{R} F \cdot (\frac{\partial y}{\partial x} i - j + \frac{\partial y}{\partial z} k)dA$, σ oriented by left normals, where R is

the projection of σ onto the xz-plane.

21. R is the circular region in the xz-plane enclosed by $x^2 + z^2 = 1$;

$$\iint\limits_{\sigma} F \cdot n \, dS = \iint\limits_{R} (\frac{x^2}{\sqrt{1 - x^2 - z^2}} + y + \frac{z^2}{\sqrt{1 - x^2 - z^2}})dA = \iint\limits_{R} \frac{1}{\sqrt{1 - x^2 - z^2}} dA,$$

use polar coordinates in the xz-plane with $x = r \cos \theta$ and $z = r \sin \theta$ to

get

$$\iint\limits_{\sigma} F \cdot n \, dS = \lim_{r_0 \to 1^-} \int_0^{2\pi} \int_0^{r_0} \frac{r}{\sqrt{1 - r^2}} \, dr \, d\theta = \lim_{r_0 \to 1^-} 2\pi(1 - \sqrt{1 - r_0^2}) = 2\pi$$

EXERCISE SET 18.6

1. div $F = z^3 + 8y^3 x^2 + 10zy$ 3. div $F = ye^{xy} + \sin y + 2 \sin z \cos z$

5. div $F = \frac{1}{x} + xze^{xyz} + \frac{x}{x^2 + z^2}$

7. G is the cube;

$$\iiint\limits_{G} div \, F \, dV = 8 \iiint\limits_{G} dV = (8)(\text{volume of cube}) = (8)(1) = 8.$$

9. G is the cylindrical solid;

$$\iiint_G \text{div}\, \mathbf{F}\, dV = 3\iiint_G dV = (3)(\text{volume of cylinder}) = (3)[\pi a^2(1)] = 3\pi a^2.$$

11. G is the cylindrical solid;

$$\iiint_G \text{div}\, \mathbf{F}\, dV = 3\iiint_G (x^2 + y^2 + z^2)dV = 3\int_0^{2\pi}\int_0^2\int_0^3 (r^2 + z^2)r\, dz\, dr\, d\theta$$

$$= 180\pi.$$

13. G is the tetrahedron;

$$\iiint_G \text{div}\, \mathbf{F}\, dV = \iiint_G x\, dV = \int_0^1\int_0^{1-x}\int_0^{1-x-y} x\, dz\, dy\, dx = \frac{1}{24}.$$

15. G is the conical solid;

$$\iiint_G \text{div}\, \mathbf{F}\, dV = 2\iiint_G (x + y + z)dV$$

$$= 2\int_0^{2\pi}\int_0^1\int_r^1 (r\cos\theta + r\sin\theta + z)r\, dz\, dr\, d\theta = \frac{\pi}{2}.$$

17. G is the solid bounded by $z = 4 - x^2$, $y + z = 5$, and the coordinate planes;

$$\iiint_G \text{div}\, \mathbf{F}\, dV = 4\iiint_G x^2 dV = 4\int_{-2}^2\int_0^{4-x^2}\int_0^{5-z} x^2 dy\, dz\, dx = \frac{4608}{35}.$$

19. Refer to Exercise 15 in 18.5 where it is shown that

$$\iint_\sigma \mathbf{F}\cdot\mathbf{n}\, dS = 4\pi a^3;$$

$$\iiint_G \text{div}\, \mathbf{F}\, dV = 3\iiint_G dV = (3)(\text{volume of sphere}) = (3)(\tfrac{4}{3}\pi a^3) = 4\pi a^3.$$

21. (a) Let $\mathbf{F} = f_1\mathbf{i} + g_1\mathbf{j} + h_1\mathbf{k}$ and $\mathbf{G} = f_2\mathbf{i} + g_2\mathbf{j} + h_2\mathbf{k}$ then

$$\mathbf{F} + \mathbf{G} = (f_1 + f_2)\mathbf{i} + (g_1 + g_2)\mathbf{j} + (h_1 + h_2)\mathbf{k} \text{ and}$$

$$\text{div}(\mathbf{F} + \mathbf{G}) = \frac{\partial}{\partial x}(f_1 + f_2) + \frac{\partial}{\partial y}(g_1 + g_2) + \frac{\partial}{\partial z}(h_1 + h_2)$$

$$= \frac{\partial f_1}{\partial x} + \frac{\partial f_2}{\partial x} + \frac{\partial g_1}{\partial y} + \frac{\partial g_2}{\partial y} + \frac{\partial h_1}{\partial z} + \frac{\partial h_2}{\partial z}$$

$$= (\frac{\partial f_1}{\partial x} + \frac{\partial g_1}{\partial y} + \frac{\partial h_1}{\partial z}) + (\frac{\partial f_2}{\partial x} + \frac{\partial g_2}{\partial y} + \frac{\partial h_2}{\partial z}) = \text{div } F + \text{div } G$$

(b) Let $F = f_1 i + g_1 j + h_1 k$ then $fF = ff_1 i + fg_1 j + fh_1 k$

and $\text{div}(fF) = \frac{\partial}{\partial x}(ff_1) + \frac{\partial}{\partial y}(fg_1) + \frac{\partial}{\partial z}(fh_1)$

$$= f \frac{\partial f_1}{\partial x} + \frac{\partial f}{\partial x} f_1 + f \frac{\partial g_1}{\partial y} + \frac{\partial f}{\partial y} g_1 + f \frac{\partial h_1}{\partial z} + \frac{\partial f}{\partial z} h_1$$

$$= f(\frac{\partial f_1}{\partial x} + \frac{\partial g_1}{\partial y} + \frac{\partial h_1}{\partial z}) + (\frac{\partial f}{\partial x} f_1 + \frac{\partial f}{\partial y} g_1 + \frac{\partial f}{\partial z} h_1)$$

$$= f \text{ div } F + (\nabla f) \cdot F.$$

23. Let σ be the surface of the cylindrical solid G bounded by $x^2 + y^2 = a^2$, $z = 0$ and $z = h$. Divide σ into four parts;

σ_1: $z = 0$ with $x^2 + y^2 \le a^2$ and $n = -k$,

σ_2: $z = h$ with $x^2 + y^2 \le a^2$ and $n = k$,

σ_3: $y = \sqrt{a^2 - x^2}$ and σ_4: $y = -\sqrt{a^2 - x^2}$, both with $0 \le z \le h$.

$$\iint_{\sigma_1} F \cdot n \, dS = \iint_{\sigma_1} 0 \, dS = 0, \quad \iint_{\sigma_2} F \cdot n \, dS = h \iint_{\sigma_2} dS = \pi a^2 h,$$

$$\iint_{\sigma_3} F \cdot n \, dS = \iint_{\sigma_4} F \cdot n \, dS = a^2 \iint_{R} \frac{1}{\sqrt{a^2 - x^2}} \, dA$$

$$= \lim_{x_0 \to a^-} a^2 \int_0^h \int_{-x_0}^{x_0} \frac{1}{\sqrt{a^2 - x^2}} \, dx \, dz$$

$$= \lim_{x_0 \to a^-} (2a^2 h \sin^{-1} \frac{x_0}{a}) = \pi a^2 h \text{ so}$$

$$\iint_{\sigma} F \cdot n \, dS = 0 + \pi a^2 h + \pi a^2 h + \pi a^2 h = 3\pi a^2 h$$

and $\text{vol}(G) = \frac{1}{3}(3\pi a^2 h) = \pi a^2 h.$

EXERCISE SET 18.7

1.   curl **F** = **0**   3.   curl **F** = $-xe^{xy}$**k**

5.   curl **F** = $-xye^{xyz}$**i** + $\dfrac{z}{x^2 + z^2}$ **j** + yze^{xyz}**k**

7.   curl **F** = 2**i** + 3**j** + 4**k**;
$$\iint\limits_{\sigma} (\text{curl } \mathbf{F}) \cdot \mathbf{n}\, dS = \iint\limits_{R} (4x + 6y + 4)\, dA$$
$$= \int_0^{2\pi} \int_0^2 (4r\cos\theta + 6r\sin\theta + 4)r\, dr\, d\theta = 16\pi.$$

9.   curl **F** = x**k**, take σ as part of the plane $z = y$ oriented with upward normals, R is the circular reigon in the xy-plane enclosed by
$x^2 + y^2 - y = 0$;
$$\iint\limits_{\sigma} (\text{curl } \mathbf{F}) \cdot \mathbf{n}\, dS = \iint\limits_{R} x\, dA = \int_0^{\pi} \int_0^{\sin\theta} r^2 \cos\theta\, dr\, d\theta = 0.$$

11.   curl **F** = **i** + **j** + **k**, take σ as part of the plane $z = 0$ with $x^2 + y^2 \leq a^2$ and **n** = **k**; $\iint\limits_{\sigma} (\text{curl } \mathbf{F}) \cdot \mathbf{n}\, dS = \iint\limits_{\sigma} dS =$ area of circle $= \pi a^2$.

13.   If σ is oriented with upward normals then C consists of three parts parametrized as
C_1:   **r**(t) = $(1 - t)$**i** + t**j** for $0 \leq t \leq 1$,
C_2:   **r**(t) = $(1 - t)$**j** + t**k** for $0 \leq t \leq 1$,
C_3:   **r**(t) = t**i** + $(1 - t)$**k** for $0 \leq t \leq 1$.

$$\int_{C_1} \mathbf{F} \cdot d\mathbf{r} = \int_{C_2} \mathbf{F} \cdot d\mathbf{r} = \int_{C_3} \mathbf{F} \cdot d\mathbf{r} = \int_0^1 (3t - 1)\, dt = \frac{1}{2} \text{ so}$$

$$\int_C \mathbf{F} \cdot d\mathbf{r} = \frac{1}{2} + \frac{1}{2} + \frac{1}{2} = \frac{3}{2}. \quad \text{curl } \mathbf{F} = \mathbf{i} + \mathbf{j} + \mathbf{k}, \ z = 1 - x - y,$$

R is the triangular region in the xy-plane enclosed by x + y = 1, x = 0, and y = 0;

$$\iint\limits_{\sigma}(\text{curl } F) \cdot n\,dS = 3\iint\limits_{R}dA = (3)(\text{area of R}) = (3)[\tfrac{1}{2}\,(1)(1)] = \tfrac{3}{2}.$$

15. If σ is oriented with upward normals then C can be parametrized as $r(t) = a\cos t\,i + a\sin t\,j$ for $0 \le t \le 2\pi$.

$$\int_{C}F \cdot dr = \int_{0}^{2\pi}0\,dt = 0; \quad \text{curl } F = 0 \text{ so } \iint\limits_{\sigma}(\text{curl } F) \cdot n\,dS = \iint\limits_{\sigma}0\,dS = 0.$$

17. Take σ as part of the plane $z = 0$ for $x^2 + y^2 \le 1$ with $n = k$;
$F = z^2 i + 2xj - y^3 k$, curl $F = -3y^2 i + 2zj + 2k$,

$$\iint\limits_{\sigma}(\text{curl } F) \cdot n\,dS = 2\iint\limits_{\sigma}dS = (2)(\text{area of circle}) = (2)[\pi(1)^2] = 2\pi.$$

19. Let $F = fi + gj + hk$ then

$$\text{curl } F = (\tfrac{\partial h}{\partial y} - \tfrac{\partial g}{\partial z})i + (\tfrac{\partial f}{\partial z} - \tfrac{\partial h}{\partial x})j + (\tfrac{\partial g}{\partial x} - \tfrac{\partial f}{\partial y})k,$$

$$\text{div(curl } F) = \frac{\partial^2 h}{\partial x \partial y} - \frac{\partial^2 g}{\partial x \partial z} + \frac{\partial^2 f}{\partial y \partial z} - \frac{\partial^2 h}{\partial y \partial x} + \frac{\partial^2 g}{\partial z \partial x} - \frac{\partial^2 f}{\partial z \partial y}$$

$$= (\frac{\partial^2 f}{\partial y \partial z} - \frac{\partial^2 f}{\partial z \partial y}) + (\frac{\partial^2 g}{\partial z \partial x} - \frac{\partial^2 g}{\partial x \partial z}) + (\frac{\partial^2 h}{\partial x \partial y} - \frac{\partial^2 h}{\partial y \partial x})$$

but $\dfrac{\partial^2 f}{\partial y \partial z} = \dfrac{\partial^2 f}{\partial z \partial y}$, $\dfrac{\partial^2 g}{\partial z \partial x} = \dfrac{\partial^2 g}{\partial x \partial z}$, and $\dfrac{\partial^2 h}{\partial x \partial y} = \dfrac{\partial^2 h}{\partial y \partial x}$
because of the continuity assumptions so div(curl F) = 0.

21. (a) $\nabla f = \dfrac{\partial f}{\partial x}\,i + \dfrac{\partial f}{\partial y}\,j + \dfrac{\partial f}{\partial z}\,k$,

$$\text{curl}(\nabla f) = (\frac{\partial^2 f}{\partial y \partial z} - \frac{\partial^2 f}{\partial z \partial y})i + (\frac{\partial^2 f}{\partial z \partial x} - \frac{\partial^2 f}{\partial x \partial z})j + (\frac{\partial^2 f}{\partial x \partial y} - \frac{\partial^2 f}{\partial y \partial x})k$$

but the continuity conditions imply equality of mixed second partial derivatives so curl(∇f) = 0.
(b) curl(∇f + curl F) = curl(∇f) + curl(curl F) = curl(curl F) because curl(∇f) = 0 from part (a).

23. Let σ_1 and σ_2 be the upper and lower hemispheres oriented, respectively, by upward and downward normals so

$$\iint_{\sigma}(\text{curl } \mathbf{F}) \cdot \mathbf{n}\,dS = \iint_{\sigma_1}(\text{curl } \mathbf{F}) \cdot \mathbf{n}\,dS + \iint_{\sigma_2}(\text{curl } \mathbf{F}) \cdot \mathbf{n}\,dS.$$

Let C be the boundary shared by σ_1 and σ_2 and suppose that C is positively oriented with respect to σ_1 and hence negatively oriented with respect to σ_2. Then by Stokes' Theorem

$$\iint_{\sigma_1}(\text{curl } \mathbf{F}) \cdot \mathbf{n}\,dS = \int_C \mathbf{F} \cdot d\mathbf{r} \quad\text{and}\quad \iint_{\sigma_2}(\text{curl } \mathbf{F}) \cdot \mathbf{n}\,dS = -\int_C \mathbf{F} \cdot d\mathbf{r} \text{ so}$$

$$\iint_{\sigma}(\text{curl } \mathbf{F}) \cdot \mathbf{n}\,dS = \int_C \mathbf{F} \cdot d\mathbf{r} - \int_C \mathbf{F} \cdot d\mathbf{r} = 0.$$

EXERCISE SET 18.8

1. $$\iint_{\sigma}\mathbf{F} \cdot \mathbf{n}\,dS = 2\iint_{R}(x^2 + y^2 + z)dA = 2\iint_{R}dA = (2)(\text{area of } R) = 2\pi$$

3. $$\iint_{\sigma}\mathbf{F} \cdot \mathbf{n}\,dS = \iint_{R}(x + y + \frac{9 + xy}{\sqrt{9 - x^2 - y^2}})dA$$

$$= \lim_{r_0 \to 3^-}\int_0^{2\pi}\int_0^{r_0}(r\cos\theta + r\sin\theta + \frac{9 + r^2\sin\theta\cos\theta}{\sqrt{9 - r^2}})r\,dr\,d\theta$$

$$= \lim_{r_0 \to 3^-}\int_0^{2\pi}\int_0^{r_0}(r^2\cos\theta + r^2\sin\theta + \frac{9r}{\sqrt{9 - r^2}} + \frac{r^3}{\sqrt{9 - r^2}}\sin\theta\cos\theta)dr\,d\theta$$

$$= \lim_{r_0 \to 3^-}18\pi(3 - \sqrt{9 - r_0^2}) = 54\pi.$$

5. $$\iint_{\sigma}\mathbf{F} \cdot \mathbf{n}\,dS = \iiint_{G}\text{div }\mathbf{F}\,dV = 3\iiint_{G}dV = (3)(\text{volume of sphere}) = 4\pi a^3$$

7. $\displaystyle \iint_{\sigma} F \cdot n \, dS = \iiint_{G} \text{div } F \, dV = 3 \iiint_{G} (x^2 + y^2 + z^2) dV$

$$= 3 \int_{0}^{2\pi} \int_{0}^{2} \int_{0}^{3} (r^2 + z^2) r \, dz \, dr \, d\theta = 180\pi$$

9. div $F = 0$; no sources or sinks.

11. div $F = 3x^2 + 3y^2 + 3z^2$; sources at all points except the origin, no sinks.

13. (a) Take σ as the part of the plane $2x + y + 2z = 2$ in the first octant, oriented with downward normals.
curl $F = -xi + (y - 1)j - k$,
$$\int_{C} F \cdot T \, ds = \iint_{\sigma} (\text{curl } F) \cdot n \, dS$$
$$= \iint_{R} (x - \frac{1}{2} y + \frac{3}{2}) dA = \int_{0}^{1} \int_{0}^{2-2x} (x - \frac{1}{2} y + \frac{3}{2}) dy \, dx = \frac{3}{2}.$$

(b) At the origin curl $F = -j - k$ and with $n = k$,
curl $F(0,0,0) \cdot n = (-j - k) \cdot k = -1$.

(c) The rotation of F has its maximum value at the origin about the unit vector in the same direction as curl $F(0,0,0)$ so $n = - \dfrac{1}{\sqrt{2}} j - \dfrac{1}{\sqrt{2}} k$.

SUPPLEMENTARY EXERCISES CHAPTER 18

1. $x = t, \; y = \sin t, \; 0 \leq t \leq \pi; \; \displaystyle \int_{0}^{\pi} (2 \sin t + 3 \cos t) dt = 4$

3. $x = t, \; y = \ln t, \; 1 \leq t \leq 3; \; \displaystyle \int_{1}^{3} (2t - 1) dt = 6$

5. $x = t, \; y = 2t, \; z = 3t, \; 0 \leq t \leq 1; \; \displaystyle \int_{0}^{1} 4t \, dt = 2$

7. $\partial(y \sin xy)/\partial y = xy \cos xy + \sin xy$
    $\partial(-x \cos xy)/\partial x = xy \sin xy - \cos xy$, not conservative

9. $\partial(3x^2 - y^2/x^2)/\partial y = -2y/x^2 = \partial(2y/x + 4y)/\partial x$, conservative so
    $\partial\phi/\partial x = 3x^2 - y^2/x^2$ and $\partial\phi/\partial y = 2y/x + 4y$, $\phi = x^3 + y^2/x + k(y)$,
    $2y/x + k'(y) = 2y/x + 4y$, $k'(y) = 4y$, $k(y) = 2y^2 + K$,
    $\phi = x^3 + y^2/x + 2y^2 + K$.

11. $\partial(\cos 2y - 3x^2y^2)/\partial y = -2 \sin 2y - 6x^2y$ and
    $\partial(\cos 2y - 2x \sin 2y - 2x^3y)/\partial x = -2 \sin 2y - 6x^2y$ so it is independent
    of path. The line segment from $(1,\pi/4)$ to $(2,\pi/4)$ is $x = 1 + t$,
    $y = \pi/4$, $0 < t < 1$; the line integral along this path is
    $$-\int_0^1 3(\pi/4)^2(1 + t)^3 dt = -7\pi^2/16.$$

13. $\partial(1/y)/\partial y = -1/y^2 = \partial(-x/y^2)/\partial x$, independent of path; $\phi = x/y$,
    $\phi(2,1) - \phi(1,2) = 2 - 1/2 = 3/2$

15. $\int_0^2 \int_0^{2x} (y^2 - 2x)dy\,dx = 0$ 17. $\int_{-2}^2 \int_0^{4-y^2} 3\,dx\,dy = 32$

19. $\iint\limits_R (-3x^2 - 3y^2)dA = -3\int_0^\pi \int_1^2 r^3 dr\,d\theta = -45\pi/4$

21. C: $x = t$, $y = t^2$, $0 \le t \le 1$;
    $W = \int_C \mathbf{F} \cdot d\mathbf{r} = \int_0^1 (2t + 3t^2 + 2t^4)dt = 12/5$

23. C: $x = t$, $y = 2t$, $z = 3t$, $0 \le t \le 1$;
    $W = \int_C \mathbf{F} \cdot d\mathbf{r} = \int_0^1 4t\,dt = 2$

25. By symmetry $\bar{x} = \bar{y} = 0$.

$$\iint_{\sigma} dS = \iint_{R} \sqrt{x^2 + y^2 + 1} \ dA = \int_0^{2\pi} \int_0^{\sqrt{8}} \sqrt{r^2 + 1} \ r \ dr \ d\theta = \frac{52\pi}{3},$$

$$\iint_{\sigma} z \ dS = \iint_{R} z\sqrt{x^2 + y^2 + 1} \ dA = \frac{1}{2} \iint_{R} (x^2 + y^2)\sqrt{x^2 + y^2 + 1} \ dA$$

$$= \frac{1}{2} \int_0^{2\pi} \int_0^{\sqrt{8}} r^3 \sqrt{r^2 + 1} \ dr \ d\theta = \frac{596\pi}{15}$$

so $\bar{z} = \dfrac{596\pi/15}{52\pi/3} = \dfrac{149}{65}$. The centroid is $(\bar{x}, \bar{y}, \bar{z}) = (0, 0, 149/65)$.

27. $\nabla f = \dfrac{x}{x^2 + y^2 + z^2} \mathbf{i} + \dfrac{y}{x^2 + y^2 + z^2} \mathbf{j} + \dfrac{z}{x^2 + y^2 + z^2} \mathbf{k}$;

on σ, $\nabla f = x\mathbf{i} + y\mathbf{j} + z\mathbf{k}$ because $x^2 + y^2 + z^2 = 1$. $D_n f = \nabla f \cdot \mathbf{n}$ so

$$\iint_{\sigma} D_n f \ dS = \iint_{\sigma} \nabla f \cdot \mathbf{n} \ dS = \iint_{R} \frac{1}{\sqrt{1 - x^2 - y^2}} \ dA$$

$$= \lim_{r_0 \to 1^-} \int_0^{\pi/2} \int_0^{r_0} \frac{r}{\sqrt{1 - r^2}} \ dr \ d\theta$$

$$= \lim_{r_0 \to 1^-} \frac{\pi}{2} \left(1 - \sqrt{1 - r_0^2}\right) = \frac{\pi}{2}.$$

29. $D_n \phi = \nabla \phi \cdot \mathbf{n}$ so $\displaystyle\iint_{\sigma} D_n \phi \ dS = \iint_{\sigma} \nabla \phi \cdot \mathbf{n} \ dS = \iiint_{G} \text{div}(\nabla \phi) dV$ by the Divergence

Theorem. $\nabla \phi = \dfrac{\partial \phi}{\partial x} \mathbf{i} + \dfrac{\partial \phi}{\partial y} \mathbf{j} + \dfrac{\partial \phi}{\partial z} \mathbf{k}$ so

$\text{div}(\nabla \phi) = \dfrac{\partial^2 \phi}{\partial x^2} + \dfrac{\partial^2 \phi}{\partial y^2} + \dfrac{\partial^2 \phi}{\partial z^2}$ and $\displaystyle\iint_{\sigma} D_n \phi \ dS = \iiint_{G} \left(\dfrac{\partial^2 \phi}{\partial x^2} + \dfrac{\partial^2 \phi}{\partial y^2} + \dfrac{\partial^2 \phi}{\partial z^2}\right) dV$.

CHAPTER 19

EXERCISE SET 19.1

1. (a) $y = e^{2x}$, $y' = 2e^{2x}$, $y'' = 4e^{2x}$; $y'' - y' - 2y = 0$

   $y = e^{-x}$, $y' = -e^{-x}$, $y'' = e^{-x}$; $y'' - y' - 2y = 0$.

   (b) $y = c_1 e^{2x} + c_2 e^{-x}$, $y' = 2c_1 e^{2x} - c_2 e^{-x}$, $y'' = 4c_1 e^{2x} + c_2 e^{-x}$;

   $y'' - y' - 2y = 0$

3. $m^2 + 3m - 4 = 0$, $(m - 1)(m + 4) = 0$; $m = 1, -4$ so $y = c_1 e^x + c_2 e^{-4x}$.

5. $m^2 - 2m + 1 = 0$, $(m - 1)^2 = 0$; $m = 1$, so $y = c_1 e^x + c_2 x e^x$.

7. $m^2 + 5 = 0$, $m = \pm\sqrt{5}i$ so $y = c_1 \cos \sqrt{5}x + c_2 \sin \sqrt{5}x$.

9. $m^2 - m = 0$, $m(m - 1) = 0$; $m = 0, 1$ so $y = c_1 + c_2 e^x$.

11. $m^2 + 4m + 4 = 0$, $(m + 2)^2 = 0$; $m = -2$ so $y = c_1 e^{-2t} + c_2 t e^{-2t}$.

13. $m^2 - 4m + 13 = 0$, $m = 2 \pm 3i$ so $y = e^{2x}(c_1 \cos 3x + c_2 \sin 3x)$.

15. $8m^2 - 2m - 1 = 0$, $(4m + 1)(2m - 1) = 0$; $m = -1/4, 1/2$ so

    $y = c_1 e^{-x/4} + c_2 e^{x/2}$.

17. $m^2 + 2m - 3 = 0$, $(m + 3)(m - 1) = 0$; $m = -3, 1$ so $y = c_1 e^{-3x} + c_2 e^x$ and

    $y' = -3c_1 e^{-3x} + c_2 e^x$. Solve the system $c_1 + c_2 = 1$, $-3c_1 + c_2 = 5$ to

    get $c_1 = -1$, $c_2 = 2$ so $y = -e^{-3x} + 2e^x$.

19. $m^2 - 6m + 9 = 0$, $(m - 3)^2 = 0$; $m = 3$ so $y = (c_1 + c_2x)e^{3x}$ and

$y' = (3c_1 + c_2 + 3c_2x)e^{3x}$. Solve the system $c_1 = 2$, $3c_1 + c_2 = 1$ to get

$c_1 = 2$, $c_2 = -5$ so $y = (2 - 5x)e^{3x}$.

21. $m^2 + 4m + 5 = 0$, $m = -2 \pm i$ so $y = e^{-2x}(c_1\cos x + c_2\sin x)$,

$y' = e^{-2x}[(c_2 - 2c_1)\cos x - (c_1 + 2c_2)\sin x]$. Solve the system $c_1 = -3$,

$c_2 - 2c_1 = 0$ to get $c_1 = -3$, $c_2 = -6$ so $y = -e^{-2x}(3\cos x + 6\sin x)$.

23. (a) $m = 5, -2$ so $(m - 5)(m + 2) = 0$, $m^2 - 3m - 10 = 0$;
     $y'' - 3y' - 10y = 0$.

 (b) $m = 4, 4$ so $(m - 4)^2 = 0$, $m^2 - 8m + 16 = 0$; $y'' - 8y' + 16y = 0$.

 (c) $m = -1 \pm 4i$ so $(m + 1 - 4i)(m + 1 + 4i) = 0$, $m^2 + 2m + 17 = 0$;
     $y'' + 2y' + 17y = 0$.

25. $m^2 + km + k = 0$, $m = (-k \pm \sqrt{k^2 - 4k})/2$

 (a) $k^2 - 4k > 0$, $k(k - 4) > 0$; $k < 0$ or $k > 4$

 (b) $k^2 - 4k = 0$; $k = 0, 4$

 (c) $k^2 - 4k < 0$, $k(k - 4) < 0$; $0 < k < 4$

27. (a) $\dfrac{d^2y}{dz^2} + 2\dfrac{dy}{dz} + 2y = 0$, $m^2 + 2m + 2 = 0$; $m = -1 \pm i$ so

$y = e^{-z}(c_1\cos z + c_2\sin z) = \dfrac{1}{|x|}[c_1\cos(\ln|x|) + c_2\sin(\ln|x|)]$.

 (b) $\dfrac{d^2y}{dz^2} - 2\dfrac{dy}{dz} - 2y = 0$, $m^2 - 2m - 2 = 0$; $m = 1 \pm \sqrt{3}$ so

$y = c_1e^{(1+\sqrt{3})z} + c_2e^{(1-\sqrt{3})z} = c_1|x|^{1+\sqrt{3}} + c_2|x|^{1-\sqrt{3}}$

29.   (a) $W(x) = \begin{vmatrix} e^{m_1 x} & e^{m_2 x} \\ m_1 e^{m_1 x} & m_2 e^{m_2 x} \end{vmatrix} = m_2 e^{(m_1 + m_2)x} - m_1 e^{(m_1 + m_2)x}$

$$= (m_2 - m_1)e^{(m_1 + m_2)x} \neq 0 \text{ if } m_1 \neq m_2.$$

   (b) $W(x) = \begin{vmatrix} e^{mx} & xe^{mx} \\ me^{mx} & (mx + 1)e^{mx} \end{vmatrix} = e^{2mx} \neq 0.$

31.   (a) The general solution is $c_1 e^{\mu x} + c_2 e^{mx}$;

   let $c_1 = 1/(\mu - m)$, $c_2 = -1/(\mu - m)$.

   (b) $\lim\limits_{\mu \to m} \dfrac{e^{\mu x} - e^{mx}}{\mu - m} = \lim\limits_{\mu \to m} xe^{\mu x} = xe^{mx}.$

EXERCISE SET 19.2

1. $m^2 + 6m + 5 = 0$, $(m + 1)(m + 5) = 0$; $m = -1, -5$ so $y_c = c_1 e^{-x} + c_2 e^{-5x}$.

   Let $y_p = Ae^{3x}$, then $y_p' = 3Ae^{3x}$, $y_p'' = 9Ae^{3x}$,

   $(9A + 18A + 5A)e^{3x} = 32Ae^{3x} = 2e^{3x}$, $A = 1/16$;

   $y = c_1 e^{-x} + c_2 e^{-5x} + \dfrac{1}{16} e^{3x}$.

3. $m^2 - 9m + 20 = 0$, $(m - 4)(m - 5) = 0$; $m = 4, 5$ so $y_c = c_1 e^{4x} + c_2 e^{5x}$.

   Let $y_p = Axe^{5x}$, then $y_p' = (5Ax + A)e^{5x}$, $y_p'' = (25Ax + 10A)e^{5x}$,

   $(25Ax + 10A - 45Ax - 9A + 20Ax)e^{5x} = Ae^{5x} = -3e^{5x}$, $A = -3$;

   $y = c_1 e^{4x} + c_2 e^{5x} - 3xe^{5x}$.

5. $m^2 + 2m + 1 = 0$, $(m + 1)^2 = 0$; $m = -1$ so $y_c = (c_1 + c_2 x)e^{-x}$.

   Let $y_p = Ax^2 e^{-x}$, then $y_p' = (-Ax^2 + 2Ax)e^{-x}$, $y_p'' = (Ax^2 - 4Ax + 2A)e^{-x}$,

$(Ax^2 - 4Ax + 2A - 2Ax^2 + 4Ax + Ax^2)e^{-x} = 2Ae^{-x} = e^{-x}$, $A = 1/2$;
$y = (c_1 + c_2x)e^{-x} + \frac{1}{2} x^2e^{-x}$.

7. $m^2 + m - 12 = 0$, $(m - 3)(m + 4) = 0$; $m = 3, -4$ so $y_c = c_1e^{3x} + c_2e^{-4x}$.

Let $y_p = A_0 + A_1x + A_2x^2$, then $y_p' = A_1 + 2A_2x$, $y_p'' = 2A_2$,

$2A_2 + A_1 + 2A_2x - 12A_0 - 12A_1x - 12A_2x^2$

$= (-12A_0 + A_1 + 2A_2) + 2(-6A_1 + A_2)x - 12A_2x^2 = 4x^2$;

solve the system
$$-12A_0 + A_1 + 2A_2 = 0$$
$$-6A_1 + A_2 = 0$$
$$-12A_2 = 4$$

to get $A_0 = -13/216$, $A_1 = -1/18$, $A_2 = -1/3$ so
$y = c_1e^{3x} + c_2e^{-4x} - \frac{13}{216} - \frac{1}{18} x - \frac{1}{3} x^2$.

9. $m^2 - 6m = 0$; $m = 0, 6$ so $y_c = c_1 + c_2e^{6x}$.

Let $y_p = A_0x + A_1x^2$, then $y_p' = A_0 + 2A_1x$, $y_p'' = 2A_1$,
$2A_1 - 6A_0 - 12A_1x = (-6A_0 + 2A_1) - 12A_1x = x - 1$;
solve the system $-6A_0 + 2A_1 = -1$,
$$-12A_1 = 1$$

to get $A_0 = 5/36$, $A_1 = -1/12$, so $y = c_1 + c_2e^{6x} + \frac{5}{36} x - \frac{1}{12} x^2$.

11. $m^2 = 0$, $m = 0$ so $y_c = c_1 + c_2x$. Let $y_p = A_0x^2 + A_1x^3 + A_2x^4 + A_3x^5$,

then $y_p' = 2A_0x + 3A_1x^2 + 4A_2x^3 + 5A_3x^4$,

$y_p'' = 2A_0 + 6A_1x + 12A_2x^2 + 20A_3x^3$; $A_0 = -1/2$, $A_1 = 0$, $A_2 = 0$, $A_3 = 1/20$

so $y = c_1 + c_2x - x^2/2 + x^5/20$.

13. $m^2 - m - 2 = 0$, $(m + 1)(m - 2) = 0$; $m = -1, 2$ so $y_c = c_1 e^{-x} + c_2 e^{2x}$.

Let $y_p = A_1 \cos x + A_2 \sin x$, then $y_p' = -A_1 \sin x + A_2 \cos x$,

$y_p'' = -A_1 \cos x - A_2 \sin x$,

$-A_1 \cos x - A_2 \sin x + A_1 \sin x - A_2 \cos x - 2A_1 \cos x - 2A_2 \sin x$

$\quad = (-3A_1 - A_2) \cos x + (A_1 - 3A_2) \sin x = 10 \cos x$;

solve the system $-3A_1 - A_2 = 10$,

$$A_1 - 3A_2 = 0$$

to get $A_1 = -3$, $A_2 = -1$ so $y = c_1 e^{-x} + c_2 e^{2x} - 3 \cos x - \sin x$.

15. $m^2 - 4 = 0$; $m = \pm 2$ so $y_c = c_1 e^{-2x} + c_2 e^{2x}$.

Let $y_p = A_1 \cos 2x + A_2 \sin 2x$, then $y_p' = -2A_1 \sin 2x + 2A_2 \cos 2x$,

$y_p'' = -4A_1 \cos 2x - 4A_2 \sin 2x$,

$-8A_1 \cos 2x - 8A_2 \sin 2x = 2 \sin 2x + 3 \cos 2x$, $A_1 = -3/8$, $A_2 = -1/4$;

$y = c_1 e^{-2x} + c_2 e^{2x} - \dfrac{3}{8} \cos 2x - \dfrac{1}{4} \sin 2x$.

17. $m^2 + 1 = 0$; $m = \pm i$ so $y_c = c_1 \cos x + c_2 \sin x$.

Let $y_p = A_1 x \cos x + A_2 x \sin x$,

then $y_p' = (A_1 + A_2 x) \cos x + (A_2 - A_1 x) \sin x$,

$y_p'' = (2A_2 - A_1 x) \cos x - (2A_1 + A_2 x) \sin x$,

$2A_2 \cos x - 2A_1 \sin x = \sin x$, $A_1 = -1/2$, $A_2 = 0$;

$y = c_1 \cos x + c_2 \sin x - \dfrac{1}{2} x \cos x$.

19. $m^2 - 3m + 2 = 0$, $(m - 1)(m - 2) = 0$; $m = 1, 2$ so $y_c = c_1 e^x + c_2 e^{2x}$.

Let $y_p = A_0 + A_1 x$, then $y_p' = A_1$, $y_p'' = 0$; solve the system

$2A_0 - 3A_1 = 0$, $2A_1 = 1$

to get $A_0 = 3/4$, $A_1 = 1/2$; $y = c_1 e^x + c_2 e^{2x} + 3/4 + x/2$.

21. $m^2 + 4m + 9 = 0$; $m = -2 \pm \sqrt{5}i$ so $y_c = e^{-2x}(c_1\cos\sqrt{5}x + c_2\sin\sqrt{5}x)$. Let

$y_p = A_0 + A_1x + A_2x^2$, then $y_p' = A_1 + 2A_2x$, $y_p'' = 2A_2$; solve the system

$$9A_0 + 4A_1 + 2A_2 = 0$$
$$9A_1 + 8A_2 = 3$$
$$9A_2 = 1$$

to get $A_0 = -94/729$, $A_1 = 19/81$, $A_2 = 1/9$ so

$y = e^{-2x}(c_1\cos\sqrt{5}x + c_2\sin\sqrt{5}x) - 94/729 + 19x/81 + x^2/9$.

23. $m^2 + 4 = 0$; $m = \pm 2i$ so $y_c = c_1\cos 2x + c_2\sin 2x$.

$\sin x \cos x = \frac{1}{2}\sin 2x$, let $y_p = A_1x\cos 2x + A_2x\sin 2x$, then

$y_p' = (A_1 + 2A_2x)\cos 2x + (A_2 - 2A_1x)\sin 2x$,

$y_p'' = (4A_2 - 4A_1x)\cos 2x - (4A_1 + 4A_2x)\sin 2x$,

$4A_2\cos 2x - 4A_1\sin 2x = \frac{1}{2}\sin 2x$, $A_1 = -1/8$, $A_2 = 0$;

$y = c_1\cos 2x + c_2\sin 2x - \frac{1}{8}x\cos 2x$.

25. (a) Let $y = y_1 + y_2$, then $y' = y_1' + y_2'$, $y'' = y_1'' + y_2''$ so

$$y'' + p(x)y' + q(x)y = (y_1'' + p(x)y_1' + q(x)y_1) + (y_2'' + p(x)y_2' + q(x)y_2)$$
$$= r_1(x) + r_2(x).$$

(b) $m^2 + 3m - 4 = 0$, $(m - 1)(m + 4) = 0$; $m = 1, -4$ so $y_c = c_1e^x + c_2e^{-4x}$.

For $y'' + 3y' - 4y = x$ let $y_1 = A_0 + A_1x$, then $y_1' = A_1$, $y_1'' = 0$;

$-4A_0 + 3A_1 = 0$ and $-4A_1 = 1$ so $A_0 = -3/16$, $A_1 = -1/4$;

$y_1 = -3/16 - x/4$,

for $y'' + 3y' - 4y = e^x$ let $y_2 = Axe^x$, then $y_2' = (Ax + A)e^x$,

$y_2'' = (Ax + 2A)e^x$, $5A = 1$, $A = 1/5$; $y_2 = \frac{1}{5}xe^x$ thus

$y_1 + y_2 = -\frac{3}{16} - \frac{1}{4}x + \frac{1}{5}xe^x$ is a particular solution.

(c) If $y_i(x)$ is a solution of $y'' + p(x)y' + q(x)y = r_i(x)$ for
   $i = 1, 2, \cdots, n$ then $y_1(x) + y_2(x) + \cdots + y_n(x)$ is a solution of
   $y'' + p(x)y' + q(x)y = r_1(x) + r_2(x) + \cdots + r_n(x)$.

27. $m^2 - 1 = 0$; $m = \pm 1$ so $y_c = c_1 e^{-x} + c_2 e^x$. Let $r_1(x) = 1$ and $r_2(x) = e^x$,
   then $y_1 = A_0$, $y_1' = y_2'' = 0$, $A_0 = -1$; $y_2 = Axe^x$, $y_2' = (Ax + A)e^x$,
   $y_2'' = (Ax + 2A)e^x$, $2A = 1$, $A = 1/2$ so $y = c_1 e^{-x} + c_2 e^x - 1 + \frac{1}{2} xe^x$.

29. $m^2 + 4 = 0$; $m = \pm 2i$ so $y_c = c_1 \cos 2x + c_2 \sin 2x$. Let
   $r_1(x) = 1 + x$ and $r_2(x) = \sin x$, then $y_1 = A_0 + A_1 x$, $y_1' = A_1$
   $y_1'' = 0$, $4A_0 = 1$, and $4A_1 = 1$ so $A_0 = A_1 = 1/4$; $y_2 = B_1 \cos x + B_2 \sin x$,
   $y_2' = -B_1 \sin x + B_2 \cos x$, $y_2'' = -B_1 \cos x - B_2 \sin x$, $3B_1 = 0$ and $3B_2 = 1$ so
   $B_1 = 0$, $B_2 = 1/3$; $y = c_1 \cos 2x + c_2 \sin 2x + \frac{1}{4} + \frac{1}{4} x + \frac{1}{3} \sin x$.

31. $m^2 - 2m + 1 = 0$, $(m - 1)^2 = 0$; $m = 1$ so $y_c = (c_1 + c_2 x)e^x$. Let
   $r_1(x) = \frac{1}{2} e^x$ and $r_2(x) = -\frac{1}{2} e^{-x}$, then $y_1 = Ax^2 e^x$, $y_1' = (Ax^2 + 2Ax)e^x$
   $y_1'' = (Ax^2 + 4Ax + 2A)e^x$, $2A = 1/2$, $A = 1/4$; $y_2 = Be^{-x}$, $y_2' = -Be^{-x}$, $y_2'' =$
   Be^{-x}, $4B = 1/2$, $B = 1/8$; $y = (c_1 + c_2 x)e^x + \frac{1}{4} x^2 e^x + \frac{1}{8} e^{-x}$.

33. $m^2 + 1 = 0$; $m = \pm i$ so $y_c = c_1 \cos x + c_2 \sin x$. Let $r_1(x) = 6$ and
   $r_2(x) = 6 \cos 2x$, then $y_1 = A_0$, $y_1' = y_1'' = 0$, $A_0 = 6$;
   $y_2 = A_1 \cos 2x + A_2 \sin 2x$, $y_2' = -2A_1 \sin 2x + 2A_2 \cos 2x$,
   $y_2'' = -4A_1 \cos 2x - 4A_2 \sin 2x$, $-3A_1 = 6$ and $-3A_2 = 0$ so $A_1 = -2$, $A_2 = 0$;
   $y = c_1 \cos x + c_2 \sin x + 6 - 2 \cos 2x$.

35. (a) $m^2 + \mu^2 = 0$; $m = \pm \mu i$ so $y_c = c_1 \cos \mu x + c_2 \sin \mu x$. Let
   $y_p = A_1 \cos bx + A_2 \sin bx$, then $y_p' = -bA_1 \sin bx + bA_2 \cos bx$,

$$y_p'' = -b^2 A_1 \cos bx - b^2 A_2 \sin bx, \quad A_1 = 0, \quad A_2 = a/(\mu^2 - b^2);$$

$$y = c_1 \cos \mu x + c_2 \sin \mu x + \frac{a}{\mu^2 - b^2} \sin bx.$$

(b) $y = c_1 \cos \mu x + c_2 \sin \mu x + \sum_{k=1}^{n} \frac{a_k}{\mu^2 - k^2 \pi^2} \sin k\pi x.$

37. $m^2 - m = 0$; $m = 0, 1$ so $y_c = c_1 + c_2 e^x$. Let $y_p = A_0 x + A_1 x^2$, then

$y_p' = A_0 + 2A_1 x$, $y_p'' = 2A_1$; $A_0 = 0$, $A_1 = 2$, $y = c_1 + c_2 e^x + 2x^2$. If
$y'(x_0) = y''(x_0) = 0$, then $4 - 4x_0 = 0$, $x_0 = 1$ so $y'(1) = 0$ and
$y''(1) = 0$; $y' = c_2 e^x + 4x$, $y'' = c_2 e^x + 4$, $y'(1) = c_2 e + 4 = y''(1) = 0$ if
$c_2 = -4/e$ so $y = c_1 - 4e^{x-1} + 2x^2$, or simply $y = c - 4e^{x-1} + 2x^2$ where c
is an arbitrary constant.

EXERCISE SET 19.3

1. $m^2 + 1 = 0$; $m = \pm i$ so $y_c = c_1 \cos x + c_2 \sin x$. $u' \cos x + v' \sin x = 0$ and

$-u' \sin x + v' \cos x = x^2$; $u' = -x^2 \sin x$, $v' = x^2 \cos x$ so
$u = x^2 \cos x - 2x \sin x - 2 \cos x$, $v = x^2 \sin x + 2x \cos x - 2 \sin x$,
$y_p = u \cos x + v \sin x = x^2 - 2$, $y = c_1 \cos x + c_2 \sin x + x^2 - 2$.

3. $m^2 + m - 2 = 0$, $(m - 1)(m + 2) = 0$; $m = 1, -2$ so $y_c = c_1 e^x + c_2 e^{-2x}$.

$u' e^x + v' e^{-2x} = 0$ and $u' e^x - 2v' e^{-2x} = 2e^x$; $u' = \frac{2}{3}$, $v' = -\frac{2}{3} e^{3x}$ so
$u = \frac{2}{3} x$, $v = -\frac{2}{9} e^{3x}$, $y_p = u e^x + v e^{-2x} = \frac{2}{3} x e^x - \frac{2}{9} e^x$. But $-\frac{2}{9} e^x$
satisfies the complementary equation so $y = c_1 e^x + c_2 e^{-2x} + \frac{2}{3} x e^x$.

5. $m^2 + 4 = 0$; $m = \pm 2i$ so $y_c = c_1\cos 2x + c_2\sin 2x$. $u'\cos 2x + v'\sin 2x = 0$
 and $-2u'\sin 2x + v'\cos 2x = \sin 2x$; $u' = -\frac{1}{2}\sin^2 2x$,

 $v' = \frac{1}{2}\sin 2x \cos 2x$ so $u = -\frac{1}{4}x + \frac{1}{16}\sin 4x = -\frac{1}{4}x + \frac{1}{8}\sin 2x \cos 2x$,

 $v = -\frac{1}{8}\cos^2 2x$, $y_p = u\cos 2x + v\sin 2x = -\frac{1}{4}x\cos 2x$,

 $y = c_1\cos 2x + c_2\sin 2x - \frac{1}{4}x\cos 2x$.

7. $m^2 + 1 = 0$; $m = \pm i$ so $y_c = c_1\cos x + c_2\sin x$. $u'\cos x + v'\sin x = 0$ and
 $-u'\sin x + v'\cos x = \tan x$; $u' = -\tan x \sin x = \cos x - \sec x$,
 $v' = \tan x \cos x = \sin x$ so $u = \sin x - \ln|\sec x + \tan x|$, $v = -\cos x$,
 $y_p = u\cos x + v\sin x = -\cos x \ln|\sec x + \tan x|$,
 $y = c_1\cos x + c_2\sin x - \cos x \ln|\sec x + \tan x|$.

9. $m^2 - 2m + 1 = 0$, $(m - 1)^2 = 0$; $m = 1$ so $y_c = c_1 e^x + c_2 x e^x$.
 $u'e^x + v'xe^x = 0$ and $u'e^x + v'(x + 1)e^x = e^x/x$; $u' = -1$, $v' = 1/x$ so
 $u = -x$, $v = \ln|x|$, $y_p = ue^x + vxe^x = -xe^x + xe^x \ln|x|$. But $-xe^x$
 satisfies the complementary equation so $y = c_1 e^x + c_2 x e^x + x e^x \ln|x|$.

11. $m^2 + 1 = 0$; $m = \pm i$ so $y_c = c_1\cos x + c_2\sin x$. $u'\cos x + v'\sin x = 0$ and
 $-u'\sin x + v'\cos x = 3\sin^2 x$; $u' = -3\sin^3 x$, $v' = 3\sin^2 x \cos x$ so
 $u = 3\cos x - \cos^3 x$, $v = \sin^3 x$,
 $y_p = u\cos x + v\sin x = 3\cos^2 x - \cos^4 x + \sin^4 x$,
 $y = c_1\cos x + c_2\sin x + 3\cos^2 x - \cos^4 x + \sin^4 x$.

13. $m^2 + 1 = 0$; $m = \pm i$ so $y_c = c_1\cos x + c_2\sin x$. $u'\cos x + v'\sin x = 0$ and
 $-u'\sin x + v'\cos x = \csc x$; $u' = -1$, $v' = \cos x \csc x = \cot x$ so $u = -x$,
 $v = \ln|\sin x|$, $y_p = u\cos x + v\sin x = -x\cos x + \sin x \ln|\sin x|$,
 $y = c_1\cos x + c_2\sin x - x\cos x + \sin x \ln|\sin x|$.

15. $m^2 + 1 = 0$; $m = \pm i$ so $y_c = c_1\cos x + c_2\sin x$. $u'\cos x + v'\sin x = 0$ and

$-u'\sin x + v'\cos x = \sec x \tan x$; $u' = -\tan^2 x$, $v' = \tan x$ so
$u = x - \tan x$, $v = -\ln|\cos x|$,
$y_p = u\cos x + v\sin x = x\cos x - \sin x - \sin x \ln|\cos x|$. But $-\sin x$
satisfies the complementary equation so
$y = c_1\cos x + c_2\sin x + x\cos x - \sin x \ln|\cos x|$.

17. $m^2 + 2m + 1 = 0$, $(m + 1)^2 = 0$; $m = -1$ so $y_c = c_1e^{-x} + c_2xe^{-x}$.

$u'e^{-x} + v'xe^{-x} = 0$ and $-u'e^{-x} + v'(1 - x)e^{-x} = e^{-x}/x^2$; $u' = -1/x$,
$v' = 1/x^2$ so $u = -\ln|x|$, $v = -1/x$, $y_p = ue^{-x} + vxe^{-x} = -e^{-x}\ln|x| - e^{-x}$.

But $-e^{-x}$ satisfies the complementary equation so
$y = c_1e^{-x} + c_2xe^{-x} - e^{-x}\ln|x|$.

19. $m^2 + 4m + 4 = 0$, $(m + 2)^2 = 0$; $m = -2$ so $y_c = c_1e^{-2x} + c_2xe^{-2x}$.

$u'e^{-2x} + v'xe^{-2x} = 0$ and $-2u'e^{-2x} + v'(1 - 2x)e^{-2x} = xe^{-x}$;
$u' = -x^2e^x$, $v' = xe^x$ so $u = (-x^2 + 2x - 2)e^x$, $v = (x - 1)e^x$,
$y_p = ue^{-2x} + vxe^{-2x} = (x - 2)e^{-x}$, $y = c_1e^{-2x} + c_2xe^{-2x} + (x - 2)e^{-x}$.

21. $m^2 + 1 = 0$; $m = \pm i$ so $y_c = c_1\cos x + c_2\sin x$. $u'\cos x + v'\sin x = 0$ and

$-u'\sin x + v'\cos x = \sec^2 x$; $u' = -\sec x \tan x$, $v' = \sec x$ so
$u = -\sec x$, $v = \ln|\sec x + \tan x|$,
$y_p = u\cos x + v\sin x = -1 + \sin x \ln|\sec x + \tan x|$,
$y = c_1\cos x + c_2\sin x - 1 + \sin x \ln|\sec x + \tan x|$.

23. $m^2 - 2m + 1 = 0$, $(m - 1)^2 = 0$; $m = 1$ so $y_c = c_1e^x + c_2xe^x$.

$u'e^x + v'xe^x = 0$ and $u'e^x + v'(x + 1)e^x = e^x/x^2$; $u' = -1/x$,
$v' = 1/x^2$ so $u = -\ln|x|$, $v = -1/x$, $y_p = ue^x + vxe^x = -e^x\ln|x| - e^x$. But

$-e^x$ satisfies the complementary equation so
$y = c_1e^x + c_2xe^x - e^x\ln|x|$.

25. $m^2 - 1 = 0$; $m = \pm 1$ so $y_c = c_1 e^x + c_2 e^{-x}$. $u'e^x + v'e^{-x} = 0$ and

$u'e^x - v'e^{-x} = e^x \cos x$; $u' = \frac{1}{2} \cos x$, $v' = -\frac{1}{2} e^{2x} \cos x$ so

$u = \frac{1}{2} \sin x$, $v = -\frac{1}{10} e^{2x} (2 \cos x + \sin x)$,

$y_p = u e^x + v e^{-x} = \frac{1}{2} e^x \sin x - \frac{1}{10} e^x (2 \cos x + \sin x)$

$\quad = \frac{1}{5} e^x (2 \sin x - \cos x)$, $y = c_1 e^x + c_2 e^{-x} + \frac{1}{5} e^x (2 \sin x - \cos x)$.

27. $m^2 + 2m + 1 = 0$, $(m + 1)^2 = 0$; $m = -1$ so $y_c = c_1 e^{-x} + c_2 x e^{-x}$.

$u'e^{-x} + v'x e^{-x} = 0$ and $-u'e^{-x} + v'(1 - x)e^{-x} = e^{-x} \ln|x|$; $u' = -x \ln|x|$,

$v' = \ln|x|$ so $u = -\frac{1}{2} x^2 \ln|x| + \frac{1}{4} x^2$, $v = x \ln|x| - x$,

$y_p = u e^{-x} + v x e^{-x} = -\frac{1}{2} x^2 e^{-x} \ln|x| + \frac{1}{4} x^2 e^{-x} + x^2 e^{-x} \ln|x| - x^2 e^{-x}$

$\quad = \frac{1}{2} x^2 e^{-x} \ln|x| - \frac{3}{4} x^2 e^{-x}$,

$y = c_1 e^{-x} + c_2 x e^{-x} + \frac{1}{2} x^2 e^{-x} \ln|x| - \frac{3}{4} x^2 e^{-x}$.

29. $m^2 + 1 = 0$; $m = \pm i$ so $y_c = c_1 \cos x + c_2 \sin x$. $u' \cos x + v' \sin x = 0$,

$-u' \sin x + v' \cos x = r(x)$; $u' = -r(x) \sin x$, $v' = r(x) \cos x$ so

$u = -\int r(x) \sin x \, dx$, $v = \int r(x) \cos x \, dx$,

$y = c_1 \cos x + c_2 \sin x - \left[\int r(x) \sin x \, dx \right] \cos x + \left[\int r(x) \cos x \, dx \right] \sin x$.

EXERCISE SET 19.4

1. (a) $M = w/g = 64/32 = 2$, $k/M = 8/2 = 4$; $y'' + 4y = 0$, $y(0) = 1$,
     $y'(0) = 0$.
    (b) $m^2 + 4 = 0$; $m = \pm 2i$ so $y = c_1 \cos 2t + c_2 \sin 2t$,
     $y' = -2c_1 \sin 2t + 2c_2 \cos 2t$, $c_1 = 1$ and $c_2 = 0$, $y = \cos 2t$.

3. (a) $k/M = g/\ell = 980/5 = 196$; $y'' + 196y = 0$, $y(0) = -10$, $y'(0) = 0$.
    (b) $m = \pm 14i$ so $y = c_1\cos 14t + c_2\sin 14t$,

    $y' = -14c_1\sin 14t + 14c_2\cos 14t$, $c_1 = -10$ and $c_2 = 0$,

    $y = -10\cos 14t$.

5. (a) $M = w/g = (1/2)/32 = 1/64$, $k/M = 64$, $y_0 = 2$; $y = 2\cos 8t$.
    (b) $|y_0| = 2$ (c) $T = \pi/4$ (d) $f = 4/\pi$

7. (a) $k/M = g/\ell = 32/(1/2) = 384$, $y_0 = -3/12 = -1/4$; $y = -\dfrac{1}{4}\cos 8\sqrt{6}t$.

    (b) $|y_0| = 1/4$ ft.   (c) $T = \pi/(4\sqrt{6})$ (d) $f = 4\sqrt{6}/\pi$

9. (a) $M = w/g = 32/32 = 1$, $k/M = 8$; $y'' + 4y' + 8y = 0$, $y(0) = -3$,
    $y'(0) = 0$.
    (b) $m^2 + 4m + 8 = 0$; $m = -2 \pm 2i$ so $y = e^{-2t}(c_1\cos 2t + c_2\sin 2t)$,

    $y' = 2e^{-2t}[(c_2 - c_1)\cos 2t - (c_1 + c_2)\sin 2t]$, $c_1 = c_2 = -3$,

    $y = -3e^{-2t}(\cos 2t + \sin 2t)$.
    (d) $\alpha = \beta = 2$, $\omega = \tan^{-1}(\alpha/\beta) = \tan^{-1}1 = \pi/4$; $y = -3\sqrt{2}e^{-2t}\cos(2t - \pi/4)$.
    (e) $T = \pi$ (f) $f = 1/\pi$

11. (a) $\alpha = 1/5$, $\beta = \sqrt{2}/5$; $y = \dfrac{5}{2}\sqrt{6}e^{-t/5}\cos(\sqrt{2}t/5 - \tan^{-1}(1/\sqrt{2}))$.

    (b) $T = 10\pi/\sqrt{2}$ (c) $\sqrt{2}/(10\pi)$

13. (a) $y'' + 64y = 0$, $y(0) = 0$, $y'(0) = -2$.
    $y = c_1\cos 8t + c_2\sin 8t$, $y' = -8c_1\sin 8t + 8c_2\cos 8t$,

    $c_1 = 0$ and $c_2 = -1/4$; $y = -\dfrac{1}{4}\sin 8t$.
    (b) $|y_0| = 1/4$ (c) $T = \pi/4$ (d) $f = 4/\pi$

15. $M = w/g = 1/8$, $k/M = 50$; $y'' + 2y' + 50y = 0$, $y(0) = 1/3$, $y'(0) = -5$.
    $m^2 + 2m + 50 = 0$; $m = -1 \pm 7i$ so $y = e^{-t}(c_1\cos 7t + c_2\sin 7t)$,

    $y' = e^{-t}[(7c_2 - c_1)\cos 7t - (7c_1 + c_2)\sin 7t]$, $c_1 = 1/3$ and $c_2 = -2/3$,

    $y = \dfrac{1}{3}e^{-t}(\cos 7t - 2\sin 7t)$.

17. $T = 2\pi\sqrt{M/k} = 2\pi\sqrt{w/(kg)}$, $T^2 = 4\pi^2 w/(32k)$, $\pi^2 w = 8T^2 k$; for w and w + 4, $\pi^2 w = 72k$ and $\pi^2(w + 4) = 200k$. Solve this system of equations to get

    (a) $k = \pi^2/32$ (b) $w = 9/4$

19. Let ℓ be the depth to which the cylinder is submerged in the water at equilibrium, then $\rho\pi r^2 \ell = \delta\pi r^2 h$ so $\ell = \delta h/\rho$, $M/k = \ell/g = \delta h/(\rho g)$, $T = 2\pi\sqrt{\delta h/(\rho g)}$.

21. (a) $y = y_0\cos(\sqrt{k/M}\,t)$, $y' = -y_0\sqrt{k/M}\sin(\sqrt{k/M}\,t)$; when $\sin(\sqrt{k/M}\,t)$ is 1 or -1, $\cos(\sqrt{k/M}\,t)$ is 0 so $y = 0$ when $|y'| = |y_0|\sqrt{k/M} = 2\pi|y_0|/T$.

    (b) $y'' = -y_0(k/M)\cos(\sqrt{k/M}\,t)$; $|y''|$ is maximum when $\cos(\sqrt{k/M}\,t)$ is 1 or -1, and hence $|y|$ is maximum. The maximum value of $|y''|$ is $|y_0|k/M = 4\pi^2|y_0|/T^2$.

23. Let $\omega = \tan^{-1}(\alpha/\beta)$, then $\cos\omega = \beta/\sqrt{\alpha^2 + \beta^2}$ and $\sin\omega = \alpha/\sqrt{\alpha^2 + \beta^2}$,

$$\beta\cos\beta t + \alpha\sin\beta t = \sqrt{\alpha^2 + \beta^2}\,(\cos\beta t\cos\omega + \sin\beta t\sin\omega)$$
$$= \sqrt{\alpha^2 + \beta^2}\,\cos(\beta t - \omega)$$

so $y(t) = \dfrac{y_0\sqrt{\alpha^2 + \beta^2}}{\beta}\,e^{-\alpha t}\cos(\beta t - \omega)$.

25. (a) $m = -\dfrac{c}{2M} = -\alpha$ so $y = (c_1 + c_2 t)e^{-\alpha t}$, $y' = (c_2 - \alpha c_1 - \alpha c_2 t)e^{-\alpha t}$, $c_1 = y_0$ and $c_2 = \alpha y_0$; $y = y_0(1 + \alpha t)e^{-\alpha t}$.

    (b) $\displaystyle\lim_{t\to+\infty} y_0(1 + \alpha t)e^{-\alpha t} = \lim_{t\to+\infty}\frac{y_0(1 + \alpha t)}{e^{\alpha t}} = \lim_{t\to+\infty}\frac{y_0}{e^{\alpha t}} = 0$.

    (c) $\alpha > 0$ so $y_0(1 + \alpha t)e^{-\alpha t} \neq 0$ for $t > 0$.

APPENDIX 1

EXERCISES, TRIGONOMETRIC FUNCTIONS AND IDENTITIES

1.   (a) $3\pi/2$ (b) $13\pi/6$ (c) $\pi/9$

     (d) $23\pi/30$ (e) $\pi^2/1800$

3.   (a) $12°$ (b) $270°$ (c) $(810/\pi)°$

     (d) $288°$ (e) $900°$

5.   (a)   (b)

     (c)   (d)

     (e)

7.

	θ	sin θ	cos θ	tan θ	csc θ	sec θ	cot θ
(a)	π/6	1/2	√3/2	1/√3	2	2/√3	√3
(b)	−7π/3	−√3/2	1/2	−√3	−2/√3	2	−1/√3
(c)	−5π/4	1/√2	−1/√2	−1	√2	−√2	−1
(d)	−3π	0	−1	0	—	−1	—
(e)	π	0	−1	0	—	−1	—

9. (a) θ = ±nπ, n = 0, 1, 2, ...   (b) θ = π/2 ± nπ, n = 0, 1, 2, ...
    (c) θ = ±nπ, n = 0, 1, 2, ...   (d) θ = ±nπ, n = 0, 1, 2, ...
    (e) θ = π/2 ± nπ, n = 0, 1, 2, ...   (f) θ = ±nπ, n = 0, 1, 2, ...

11. θ = 5π/4 ± 2nπ and θ = 7π/4 ± 2nπ, n = 0, 1, 2, ...

13. θ = π/3 ± 2nπ and θ = 5π/3 ± 2nπ, n = 0, 1, 2, ...

15. θ = π/3 ± nπ, n = 0, 1, 2, ...

17. θ = 4π/3 ± 2nπ and θ = 5π/3 ± 2nπ, n = 0, 1, 2, ...

19. θ = π ± 2nπ, n = 0, 1, 2, ...   21. θ = π/6 ± nπ, n = 0, 1, 2, ...

23. θ = 7π/6 ± 2nπ and θ = 11π/6 ± 2nπ, n = 0, 1, 2, ...

25. θ = π/6 ± 2nπ and θ = 11π/6 ± 2nπ, n = 0, 1, 2, ...

27. (a) from identity (14b) with $\alpha = 45°$ and $\beta = 30°$,

$$\cos 15° = \cos(45° - 30°) = \cos 45° \cos 30° + \sin 45° \sin 30°$$
$$= (1/\sqrt{2})(\sqrt{3}/2) + (1/\sqrt{2})(1/2)$$
$$= (\sqrt{6} + \sqrt{2})/4$$

or from (18a) with $\alpha = 30°$,

$$\cos^2 15° = \cos^2(30°/2) = (1 + \cos 30°)/2 = (2 + \sqrt{3})/4$$

so $\cos 15° = \frac{1}{2}\sqrt{2 + \sqrt{3}}$

(b) from (18b) with $\alpha = 45°$,

$$\sin^2 22.5° = \sin^2(45°/2) = (1 - \cos 45°)/2 = (2 - \sqrt{2})/4$$

so $\sin 22.5° = \dfrac{1}{2}\sqrt{2 - \sqrt{2}}$

(c) from (18a) with $\alpha = 150°$,

$$\cos^2 75° = \cos^2(150°/2) = (1 + \cos 150°)/2 = (2 - \sqrt{3})/4$$

so $\cos 75° = \dfrac{1}{2}\sqrt{2 - \sqrt{3}}$

then from (18b) with $\alpha = 75°$,

$$\sin^2 37.5° = \sin^2(75°/2) = (1 - \cos 75°)/2 = (2 - \sqrt{2 - \sqrt{3}})/4$$

so $\sin 37.5° = \dfrac{1}{2}\sqrt{2 - \sqrt{2 - \sqrt{3}}}$

29.

	$\sin\theta$	$\cos\theta$	$\tan\theta$	$\csc\theta$	$\sec\theta$	$\cot\theta$
(a)	$a/3$	$\sqrt{9 - a^2}/3$	$a/\sqrt{9 - a^2}$	$3/a$	$3/\sqrt{9 - a^2}$	$\sqrt{9 - a^2}/a$
(b)	$a/\sqrt{a^2 + 25}$	$5/\sqrt{a^2 + 25}$	$a/5$	$\sqrt{a^2 + 25}/a$	$\sqrt{a^2 + 25}/5$	$5/a$
(c)	$\sqrt{a^2 - 1}/a$	$1/a$	$\sqrt{a^2 - 1}$	$a/\sqrt{a^2 - 1}$	a	$1/\sqrt{a^2 - 1}$

31. Draw the perpendicular from vertex C
as shown in the figure

$h = 9 \sin 30° = 9/2$

$a = h/\sin 45° = 9\sqrt{2}/2$,

$c_1 = 9 \cos 30° = 9\sqrt{3}/2$,

$c_2 = a \cos 45° = 9/2$,

$c_1 + c_2 = 9(\sqrt{3} + 1)/2$, angle C $= 180° - (30° + 45°) = 105°$

33. From right triangles ADC and BDC,
$$h_1 = b \sin A = a \sin B$$
so $a/\sin A = b/\sin B$.
From right triangles AEB and CEB,
$$h_2 = c \sin A = a \sin C$$
so $a/\sin A = c/\sin C$
thus $a/\sin A = b/\sin B = c/\sin C$.

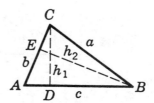

35. $\dfrac{\text{area}}{\text{area of circle}} = \dfrac{\theta}{2\pi}$, area $= \dfrac{\theta}{2\pi}$ (area of circle) $= \dfrac{\theta}{2\pi}(\pi r^2) = \dfrac{1}{2}r^2\theta$

37. $\cos(\pi/2 + \alpha) = \cos(\pi/2 - (-\alpha)) = \sin(-\alpha) = -\sin\alpha$,
$\sin(\pi/2 + \alpha) = \sin(\pi/2 - (-\alpha)) = \cos(-\alpha) = \cos\alpha$.

39. Subtract (15b) from (14b) to get $\cos(\alpha - \beta) - \cos(\alpha + \beta) = 2\sin\alpha\sin\beta$
so $\sin\alpha\sin\beta = \dfrac{1}{2}[\cos(\alpha - \beta) - \cos(\alpha + \beta)]$. To get (19c), add (14b) and (15b).

41. From (14a), $\sin(3\pi/2 - \theta) = \sin(3\pi/2)\cos\theta - \cos(3\pi/2)\sin\theta$
$$= (-1)\cos\theta - (0)\sin\theta = -\cos\theta;$$
similarly, from (15b), $\cos(3\pi/2 + \theta) = \sin\theta$

43. $\tan(\theta/2) = \dfrac{\sin(\theta/2)}{\cos(\theta/2)} = \dfrac{2\sin(\theta/2)\cos(\theta/2)}{2\cos^2(\theta/2)} = \dfrac{\sin\theta}{1 + \cos\theta}$

45. $\tan\theta + \cot\theta = \dfrac{\sin\theta}{\cos\theta} + \dfrac{\cos\theta}{\sin\theta} = \dfrac{\sin^2\theta + \cos^2\theta}{\sin\theta\cos\theta}$
$$= \dfrac{1}{\sin\theta\cos\theta} = \dfrac{2}{2\sin\theta\cos\theta} = \dfrac{2}{\sin 2\theta} = 2\csc 2\theta$$

47. Using (19a), $2\cos 2\theta\sin\theta = 2(1/2)[\sin(-\theta) + \sin 3\theta] = \sin 3\theta - \sin\theta$

49. $\dfrac{\cos\theta\tan\theta + \sin\theta}{\tan\theta} = \dfrac{\cos\theta(\sin\theta/\cos\theta) + \sin\theta}{\sin\theta/\cos\theta} = 2\cos\theta$

51. $\dfrac{\sin\theta + \cos 2\theta - 1}{\cos\theta - \sin 2\theta} = \dfrac{\sin\theta + (1 - 2\sin^2\theta) - 1}{\cos\theta - 2\sin\theta\cos\theta} = \dfrac{\sin\theta(1 - 2\sin\theta)}{\cos\theta(1 - 2\sin\theta)} = \tan\theta$

53. From (19a), $\sin(3\pi/2 - \theta) + \sin(3\pi/2 + \theta) = 2\sin(3\pi/2)\cos\theta$
$$= 2(-1)\cos\theta = -2\cos\theta$$

55. $\sin 3\theta = \sin(2\theta + \theta) = \sin 2\theta \cos \theta + \cos 2\theta \sin \theta$

$\qquad = (2 \sin \theta \cos \theta)\cos \theta + (\cos^2\theta - \sin^2\theta)\sin \theta$

$\qquad = 2 \sin \theta \cos^2\theta + \sin \theta \cos^2\theta - \sin^3\theta = 3 \sin \theta \cos^2\theta - \sin^3\theta;$

similarly, $\cos 3\theta = \cos^3\theta - 3 \sin^2\theta \cos \theta$

57. Consider the triangle having a, b, and d as sides. The angle formed by sides a and b is $\pi - \theta$ so from the law of cosines,

$d^2 = a^2 + b^2 - 2ab \cos(\pi - \theta) = a^2 + b^2 + 2ab \cos \theta,$

$d = \sqrt{a^2 + b^2 + 2ab \cos \theta}.$

59. Let s_1 and s_2 be the length of arcs subtended on circles of radius r_1 and r_2 by a central angle θ, then $s_1 = k_1\theta$ and $s_2 = k_2\theta$ where k_1 and k_2 are constants. If $\theta = 2\pi$ then $s_1 = 2\pi r_1$ and $s_2 = 2\pi r_2$ so $k_1 = r_1$ and $k_2 = r_2$ thus $s_1 = r_1\theta$ and $s_2 = r_2\theta$, $\theta = s_1/r_1 = s_2/r_2$ which proves that s/r depends only on the size of the angle and not on the radius of the circle.

EXERCISES, GRAPHS OF TRIGONOMETRIC FUNCTIONS

1. (a) $2\pi/5$ (b) 6π (c) 8π (d) $\pi/7$
    (e) 2   (f) 10   (g) 1   (h) $2k\pi$

3. (a) 5   (b) 1/3   (c) 1/2   (d) 1

5.

7.

9.

11.

13. (a)

(b)

(c)

15.

17.

19.

21.

23.

25.

27.

29.

31.
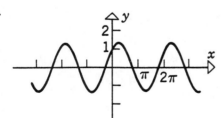

33. (a) odd   (b) even   (c) odd   (d) even
     (e) even   (f) even   (g) odd   (h) even

35. If $|x| \geq \pi/2$ then $|\sin x| \leq |x|$ because $|\sin x| \leq 1 < \pi/2$ for all x. If
     $x = 0$ then $\sin x = 0$ so $|\sin x| \leq |x|$. If $0 < |x| < \pi/2$ then, from

Exercise 10, $\sin|x| \le |x|$. But $\sin|x| = |\sin x|$ if $0 < |x| < \pi/2$ so $|\sin x| \le |x|$.

37. $\tan bx$ repeats when bx changes by π or equivalently when x changes by π/b so the fundamental period is $\pi/|b|$.

EXERCISES, ONE-SIDED AND INFINITE LIMITS

1. If $x > 2$ then $|(x + 1) - 3| = |x - 2| = x - 2 < \epsilon$ when $x < 2 + \epsilon$;
    take $\delta = \epsilon$.

3. If $x > 4$ then $|\sqrt{x - 4} - 0| = \sqrt{x - 4} < \epsilon$ when $x - 4 < \epsilon^2$, $x < 4 + \epsilon^2$;
    take $\delta = \epsilon^2$.

5. If $x > 2$ then $|f(x) - 2| = |x - 2| = x - 2 < \epsilon$ when $x < 2 + \epsilon$;
    take $\delta = \epsilon$.

7. If $x > 0$ then $|1/x^2 - 0| = 1/x^2 < \epsilon$ when $x^2 > 1/\epsilon$, $x > 1/\sqrt{\epsilon}$;
    take $N = 1/\sqrt{\epsilon}$.

9. If $x < -2$ then $|1/(x + 2) - 0| = 1/|x + 2| = -1/(x + 2) < \epsilon$ when
    $x + 2 < -1/\epsilon$, $x < -2 - 1/\epsilon$; take $N = -2 - 1/\epsilon$.

11. If $x > -1$ then $|x/(x + 1) - 1| = |-1/(x + 1)| = 1/(x + 1) < \epsilon$ when
    $x + 1 > 1/\epsilon$, $x > -1 + 1/\epsilon$; take $N = 1/\epsilon$.

13. If $x < -5/2$ then $\left|\dfrac{4x - 1}{2x + 5} - 2\right| = \left|\dfrac{-11}{2x + 5}\right| = \dfrac{11}{\lceil 2x + 5 \rceil} = -\dfrac{11}{2x + 5} < \epsilon$ when
    $2x + 5 < -\dfrac{11}{\epsilon}$, $x < -\dfrac{5}{2} - \dfrac{11}{2\epsilon}$; take $N = -\dfrac{5}{2} - \dfrac{11}{2\epsilon}$.

15. If $x \neq 3$ and $N > 0$ then $1/(x - 3)^2 > N$ when $(x - 3)^2 < 1/N$,
    $|x - 3| < 1/\sqrt{N}$; take $\delta = 1/\sqrt{N}$.

17. If $x \neq 0$ and $N > 0$ then $1/|x| > N$ when $|x| < 1/N$; take $\delta = 1/N$.

19. If $x \neq 0$ and $N < 0$ then $-1/x^4 < N$ when $x^2 < -1/N = 1/|N|$, $x < 1/\sqrt[4]{|N|}$;
    take $\delta = 1/\sqrt[4]{|N|}$.

21. (15): Let f be defined on some open interval extending to the right from a. We will write $\lim\limits_{x \to a^+} f(x) = +\infty(-\infty)$ if given any positive (negative) number N we can find a number $\delta > 0$ such that $f(x)$ satisfies $f(x) > N$ ($f(x) < N$) whenever x satisfies $a < x < a + \delta$.

     (16): Similar to (15) with "right" replaced by "left", a^+ replaced by a^-, and $a < x < a + \delta$ replaced by $a - \delta < x < a$.

23. (a) If $x > 1$ and $N < 0$ then $1/(1 - x) < N$ when $1 - x > 1/N = -1/|N|$, $x < 1 + 1/|N|$; take $\delta = 1/|N|$.

     (b) If $x < 1$ and $N > 0$ then $1/(1 - x) > N$ when $1 - x < 1/N$, $x > 1 - 1/N$; take $\delta = 1/N$.

25. (a) If $M > 0$ then $x + 1 > M$ when $x > M - 1$; take $N = M$.

     (b) If $M < 0$ then $x + 1 < M$ when $x < M - 1$; take $N = M - 1$.

EXERCISES, PROOFS OF LIMIT THEOREMS

1. $|k - k| = |0| = 0 < \epsilon$ when $x > N$ for any $N > 0$.

3. $$|[f(x) + g(x)] - [L_1 + L_2]| = |[f(x) - L_1] + [g(x) - L_2]|$$
    $$\leq |f(x) - L_1| + |g(x) - L_2|.$$

    Given $\epsilon > 0$, there exist negative numbers N_1 and N_2 such that $|f(x) - L_1| < \epsilon/2$ and $|g(x) - L_2| < \epsilon/2$ whenever $x < N_1$ and $x < N_2$, respectively. Let $N = \min(N_1, N_2)$, if $x < N$ then
    $$|f(x) - L_1| + |g(x) - L_2| < \epsilon/2 + \epsilon/2 = \epsilon \text{ so}$$
    $$|[f(x) + g(x)] - [L_1 + L_2]| < \epsilon.$$

5. From Theorem 2, $\lim\limits_{x \to a} (-1) = -1$ so from Theorem 4,

    $\lim\limits_{x \to a} [(-1)g(x)] = \lim\limits_{x \to a} (-1) \lim\limits_{x \to a} g(x) = (-1)L_2 = -L_2$

    thus $\lim\limits_{x \to a} [f(x) - g(x)] = \lim\limits_{x \to a} [f(x) + (-g(x))]$

    $$= \lim\limits_{x \to a} f(x) + \lim\limits_{x \to a} [-g(x)]$$
    $$= L_1 + (-L_2) = L_1 - L_2$$

7. (a) Given $N < 0$, there exist numbers $\delta_1 > 0$ and $\delta_2 > 0$ such that
    $f(x) < N/2$ and $g(x) > -N/2$ whenever $0 < |x - a| < \delta_1$ and
    $0 < |x - a| < \delta_2$, respectively. Let $\delta = \min(\delta_1, \delta_2)$ then $f(x) < N/2$
    and $-g(x) < N/2$ whenever $0 < |x - a| < \delta$ so
    $f(x) - g(x) < N/2 + N/2 = N$.

    (b) No, for example $\lim_{x \to 0} [(-1/x^2) + (1/x^4)] = \lim_{x \to 0} [(1 - x^2)/x^4] = +\infty$

9. If $\lim_{x \to a} f(x) = L$ then given $\epsilon > 0$, there exists a $\delta > 0$ such that
    $|f(x) - L| < \epsilon$ whenever $0 < |x - a| < \delta$. But
    $|f(x) - L| = |[f(x) - L] - 0| < \epsilon$ whenever $0 < |x - a| < \delta$ so
    $\lim_{x \to a} [f(x) - L] = 0$. If $\lim_{x \to a} [f(x) - L] = 0$ then given $\epsilon > 0$, there
    exists a $\delta > 0$ such that $|[f(x) - L] - 0| < \epsilon$ whenever $0 < |x - a| < \epsilon$,
    but $|[f(x) - L] - 0| = |f(x) - L| < \epsilon$ whenever $0 < |x - a| < \delta$ so
    $\lim_{x \to a} f(x) = L$.

11. If $\lim_{x \to a} f(x) = L$ then given $\epsilon > 0$ there exists a $\delta > 0$ such that
    $|f(x) - L| < \epsilon$ whenever $0 < |x - a| < \delta$ or equivalently whenever
    $a < x < a + \delta$ or $a - \delta < x < a$ so $\lim_{x \to a^+} f(x) = \lim_{x \to a^-} f(x) = L$.

EXERCISES, PROOFS USING THE MEAN-VALUE THEOREM

1. In the proof of part (a), reverse the inequality symbols in the
    inequalities $f(x_1) < f(x_2)$, $f'(c) > 0$, and $f(x_2) - f(x_1) > 0$.

3. Suppose $f'(x) > 0$ on both sides of x_0. By the Mean-Value Theorem
    $$\frac{f(x) - f(x_0)}{x - x_0} = f'(c), \quad f(x) - f(x_0) = (x - x_0)f'(c).$$
    If $x < x_0$ then $f(x) - f(x_0) < 0$ so $f(x_0) > f(x)$,

if $x > x_0$ then $f(x) - f(x_0) > 0$ so $f(x_0) < f(x)$,

therefore $f(x_0)$ is not an extremum.

The proof for the case $f'(x) < 0$ is similar; reverse the inequality symbols in $f(x) - f(x_0) < 0$, $f(x_0) > f(x)$, $f(x) - f(x_0) > 0$, and $f(x_0) < f(x)$.

5. (a)

(b)

(c) Let $x_2 > x_1$, where x_1 and x_2 are points in (a,b). f is continuous on $[x_1,x_2]$ and differentiable on (x_1,x_2) so

$$\frac{f(x_2) - f(x_1)}{x_2 - x_1} = f'(c), \ c \text{ in } (x_1,x_2)$$

$f(x_2) - f(x_1) = (x_2 - x_1)f'(c) \geq 0$ because $x_2 - x_1 > 0$

and $f'(x) \geq 0$, thus $f(x_2) \geq f(x_1)$ so f is nondecreasing on (a,b).

(d) Proof similar to that in part (c).

7. (a) Suppose $x_2 > x_1$ where x_1 and x_2 are in (a,b). If both x_1, x_2 are in (a,x_0) or (x_0,b) then $f(x_2) > f(x_1)$ because f is increasing on (a,x_0) and (x_0,b). If $x_2 = x_0$, then

$$\frac{f(x_0) - f(x_1)}{x_0 - x_1} = f'(c), \ c \text{ in } (x_1,x_0)$$

so $f(x_0) - f(x_1) = (x_0 - x_1)f'(c) > 0$, $f(x_0) \geq f(x_1)$.

Similarly, if $x_1 = x_0$ then $f(x_2) > f(x_0)$. The preceding two results yield $f(x_2) > f(x_0) > f(x_1)$, or $f(x_2) > f(x_1)$ if $x_2 > x_0 > x_1$. In all cases then, $f(x_2) > f(x_1)$ whenever $x_2 > x_1$ so f is increasing on (a,b).

(b) Yes, proceed as in part (a) by introducing one point at a time.

(c) $f'(x) = 3x^2 > 0$ except at $x = 0$, so f is increasing on $(-\infty, +\infty)$.

9. (a) From Exercise 8(a) it follows that $f'(x)$ is increasing on some open interval I containing x_0 so f is concave up on I.

   (b) Similar to the proof in part (a).

EXERCISES, CRAMER'S RULE

1. $\begin{vmatrix} 3 & -4 \\ 2 & 1 \end{vmatrix} = 11,$ $\begin{vmatrix} -5 & -4 \\ 4 & 1 \end{vmatrix} = 11,$ $\begin{vmatrix} 3 & -5 \\ 2 & 4 \end{vmatrix} = 22;$ $x = 1,$ $y = 2$

3. $\begin{vmatrix} 2 & -5 \\ 4 & 6 \end{vmatrix} = 32,$ $\begin{vmatrix} -2 & -5 \\ 1 & 6 \end{vmatrix} = -7,$ $\begin{vmatrix} 2 & -2 \\ 4 & 1 \end{vmatrix} = 10;$ $x_1 = -7/32,$ $x_2 = 5/16$

5. $\begin{vmatrix} 1 & 2 & 1 \\ 2 & 1 & -1 \\ 1 & -1 & 1 \end{vmatrix} = -9,$ $\begin{vmatrix} 3 & 2 & 1 \\ 0 & 1 & -1 \\ 6 & -1 & 1 \end{vmatrix} = -18,$ $\begin{vmatrix} 1 & 3 & 1 \\ 2 & 0 & -1 \\ 1 & 6 & 1 \end{vmatrix} = 9,$

   $\begin{vmatrix} 1 & 2 & 3 \\ 2 & 1 & 0 \\ 1 & -1 & 6 \end{vmatrix} = -27;$ $x = 2,$ $y = -1,$ $z = 3$

7. $\begin{vmatrix} 1 & 1 & -2 \\ 2 & -1 & 1 \\ 1 & -2 & -4 \end{vmatrix} = 21,$ $\begin{vmatrix} 1 & 1 & -2 \\ 2 & -1 & 1 \\ -4 & -2 & -4 \end{vmatrix} = 26,$ $\begin{vmatrix} 1 & 1 & -2 \\ 2 & 2 & 1 \\ 1 & -4 & -4 \end{vmatrix} = 25,$

   $\begin{vmatrix} 1 & 1 & 1 \\ 2 & -1 & 2 \\ 1 & -2 & -4 \end{vmatrix} = 15;$ $x_1 = 26/21,$ $x_2 = 25/21,$ $x_3 = 5/7$

9. $\begin{vmatrix} \cos\theta & -\sin\theta \\ \sin\theta & \cos\theta \end{vmatrix} = 1,$ $\begin{vmatrix} x & -\sin\theta \\ y & \cos\theta \end{vmatrix} = x\cos\theta + y\sin\theta,$

$\begin{vmatrix} \cos\theta & x \\ \sin\theta & y \end{vmatrix} = y\cos\theta - x\sin\theta;$ $x' = x\cos\theta + y\sin\theta,$

$y' = -x\sin\theta + y\cos\theta$

NOTES

NOTES

NOTES